The Reception of Darwinian Evolution in Britain, 1859–1909

The Reception of Darwinian Evolution in Britain, 1859–1909

Darwinism's Generations

MARTIN HEWITT

OXFORD
UNIVERSITY PRESS

Great Clarendon Street, Oxford, OX2 6DP
United Kingdom

Oxford University Press is a department of the University of Oxford.
It furthers the University's objective of excellence in research, scholarship,
and education by publishing worldwide. Oxford is a registered trade mark of
Oxford University Press in the UK and in certain other countries

© Martin Hewitt 2024

The moral rights of the author have been asserted

All rights reserved. No part of this publication may be reproduced, stored in
a retrieval system, or transmitted, in any form or by any means, without the
prior permission in writing of Oxford University Press, or as expressly permitted
by law, by licence or under terms agreed with the appropriate reprographics
rights organization. Enquiries concerning reproduction outside the scope of the
above should be sent to the Rights Department, Oxford University Press, at the
address above.

You must not circulate this work in any other form
and you must impose this same condition on any acquirer

Published in the United States of America by Oxford University Press
198 Madison Avenue, New York, NY 10016, United States of America

British Library Cataloguing in Publication Data

Data available

Library of Congress Control Number: 2024942342

ISBN 9780192890993

DOI: 10.1093/9780191982941.001.0001

Printed and bound by
CPI Group (UK) Ltd, Croydon, CR0 4YY

Links to third party websites are provided by Oxford in good faith and
for information only. Oxford disclaims any responsibility for the materials
contained in any third party website referenced in this work.

Acknowledgements

Darwinism's Generations has been a long time in the making. It was conceived in the mid-2010s at the University of Huddersfield, and I am grateful for the support of Bob Cryan, Tim Thornton, and the staff in the School of Music, Humanities and Media, especially Paul Ward, for not looking askance at the research ambitions of a Dean of School. My appointment at Anglia Ruskin University in Cambridge in 2016 brought me within easy access of the riches of the Cambridge University Library, and of the libraries of the Cambridge Colleges, and I am grateful to Iain Martin, Roderick Watkins, Matthew Day, and the staff of the Faculty of Arts, Law and Social Sciences for their support during my time at Anglia Ruskin.

Over the years, this is a project which has suffered greatly from mission creep. It was initially designed as a relatively short article aimed at providing a proof of concept for the Victorian generations schema that I developed in response to a challenge set by my former and much-missed colleague, Rosemary Mitchell. In these early stages, I owe more than can easily be described to Bernie Lightman, who gave unstintingly of his expertise and encouragement in support of an inquiry which he was far more qualified to undertake than I was (or am). As the shorter article became a longer article, and then by incremental stages a 20,000, 30,000, and 40,000 word manuscript, it eventually became clear that only a full-length monograph was going to provide the space needed to make the sort of prosopographical argument that was emerging. Unfortunately, the decision to write a book only served to encourage an open-ended campaign to ferret out archival evidence of responses to Darwin, which has taken me over the past decade to more than fifty different archives and repositories in Britain, North America, and Australia. Without the expertise and the efficiency of the staff of these archives, the breadth of coverage in the following pages would have been impossible. I am grateful to the archivists and archival assistants of all the institutions identified in the Notes, and in particular the staff of the Bodleian Library (Oxford), Cambridge University Library, British Library (London and Boston Spa), National Library of Scotland, University of Edinburgh, Sterling Library (Yale), and Houghton Library (Harvard). It seems invidious to name names, but I am especially appreciative of the

vi ACKNOWLEDGEMENTS

support of Kat Harrington (Royal Botanic Gardens, Kew), Danielle Czerkaszyn, Matthew Barton, and Emily Chen (Oxford Natural History Museum), Maia Sheridan (St Andrews), Jane Harrison (Royal Institution), Anna James (Pusey House Library), Susan Killoran (Harris Manchester College, Oxford), Caroline Lam (Royal Geological Society), Timothy Engels (Brown University, Providence RI), Aisling Lockhart (Trinity College, Dublin), Andrew Isidoro (Burns Library, Boston College), Philip Jeffs (Warrington Library), Leonie Paterson and Graham Hardy (Royal Botanic Gardens, Edinburgh), Alison Diamond (Inveraray Castle), Daniel Lewis (Huntington Library), Rose Pearson (Royal Entomological Society), Sally Jennings (Eton College), Michael C. Weisenburg (Irvin Department of Rare Books and Special Collections), Ernest F. Hollings (Special Collections Library, University of South Carolina Libraries), Harrison Wick of the Charles Darwin collection, and Paul R. Hicks (Memorial Collection, Indiana University of Pennsylvania (IUP) Special Collections and University Archives). Quotations from the Argyll papers are given with the kind permission by His Grace, the Duke of Argyll.

In searching down ownership inscriptions and annotated copies of Darwin texts, I have also tested the patience and generosity of a number of booksellers and auction houses. Many have gone much beyond the call of duty to an academic who confessed to having no interest in buying. Of these I must particularly mention Derek McDonnell at Horden Books (Sydney), Rhianon Knol at Christie's (New York), David Johnson at Pryor-Johnson Rare Books (New York), Delfina Manor at Good Reading Secondhand Books (Benalia, VIC, Australia), Leah Spafford (Spafford Rare Books), and staff at Forum Auctions (London), Gilleasbuig Ferguson Rare Books (Portree), Timkc Books (Hayle), MFR Rare Books, John Michael Lang Fine Books (Seattle), Berkelouw Rare Books (Berrima, NSW, Australia), Bow Windows Books, John Drury Rare Books, and Ernestoic Books.

The project has been supported by several fellowships and conferences which have provided opportunities for writing and archival research which might otherwise have been impossible. The arguments of the book were fully drafted for the first time in the autumn of 2019 at Canberra, as a Visiting Fellow at the Humanities Research Centre at the Australian National University, where I was looked after by Will Christie and the staff and fellows of the Centre. Much of the rewriting and revising was undertaken in Oxford, as a result of the generosity of the Principal and Fellows of St Hugh's College, Oxford, who elected me to the Belcher Visiting

ACKNOWLEDGEMENTS vii

Fellowship in Victorian Studies, and who then in response to the disruptions of COVID-19, allowed me to defer and then extend the Fellowship. Princeton University awarded me a Library Fellowship in 2019, which enabled me to undertake important research in Princeton, Rutgers, and New York, and I am also indebted to the conference committees of Pacific Coast Conference on British Studies at Santa Barbara 2018, and the North American Conference on British Studies at Providence in 2018, for including me on their programmes and in so doing facilitating research in various archival collections in California and New England.

I also owe a great deal to colleagues who have at various (and frequently multiple) stages of the preparation of the manuscript read and commented, providing essential encouragement, advice and criticism which has immeasurably improved the final text. As well as Bernie Lightman, these include R. Steven Turner, Ian Hesketh, Jim Secord, Diarmid Finnegan, Rohan McWilliam, Helen Kingstone, Michael Bentley, David Amigoni, Joanne Shattock, Leslie Howsam, Peter Macdonald, and Patrick Leary. I have also profited from the research assistance of Colin Finlay, Erin Grosjean, Stephan Pigeon, and Rosi Crane.

Writing these acknowledgements within a month of the conclusion of my employment as an academic unavoidably prompts reflections on the academic career which brought me to this point. I hope the Warden and Fellows of Nuffield College, Oxford, who provided an all-but perfect environment in which to undertake doctoral work will appreciate the enduring influence which has ultimately brought a fairly conventional social historian to an interrogation of one of the foundational categories of sociology. I have been fortunate in ways impossible to summarise with the staff and students I have worked with at Trinity and All Saints (Leeds), Manchester Metropolitan University, as well as Huddersfield and Anglia Ruskin, and in collaborations with colleagues in the Leeds Centre for Victorian Studies, the editorial board of the *Journal of Victorian Culture*, and the committee of the British Association for Victorian Studies. In the end, though, neither this book, nor any of the scholarship which preceded it, would have been possible without the support, encouragement, understanding, and occasional scepticism of Peter, Thomas, and Mary-Clare. *Darwinism's Generations* is dedicated to them.

Contents

List of Illustrations	xi
Abbreviations	xiii
Introduction: Victorian Britain and the Problem of Generations	1
1. The Publication of *On the Origin of Species*	40
2. In the Wake of the *Origin*	88
3. The *Descent of Man* and the High Victorians	164
4. The Death of Darwin and After: The 1880s and the Late Victorians	237
5. Darwinian Debates at the *Fin de Siècle*: The Edwardians	325
Conclusion: Continuity, Conversion, and Counter-Example	396
Appendix: Prosopography	443
Select Bibliography	455
Index	477

List of Illustrations

1. 'The Descent of Man', *Punch*, 24 May 1873. — 51
2. 'Piety and Parallel', *Punch*, 30 November 1872. — 90
3. 'Darwin Eclipsed', 'The British Association at Edinburgh—Humours of Science', *The Graphic*, 26 August 1871. — 138
4. 'Baffled Science Slow Retires', *Punch*, 4 January 1873. — 180
5. 'Professor T.H. Huxley, President of the Royal Society', *The Graphic*, 21 July 1883. — 184
6. 'Darwin's Study', *Illustrated London News*, 10 December 1887. — 243
7. 'Punch's Fancy Portraits. Charles Robert Darwin', *Punch*, 22 October 1881. — 245
8. 'The Survival of the Fittest', *Punch*, 3 February 1909. — 397

Abbreviations

APS	American Philosophical Society, Philadelphia
BA	British Association for the Advancement of Science
BJHS	*British Journal for the History of Science*
BL	British Library
BLPES	London School of Economics Library
BMJ	*British Medical Journal*
Bodleian	Bodleian Library, University of Oxford
CLB	Cadbury Library, University of Birmingham
CR	*Contemporary Review*
CUL	Cambridge University Library
DWT	D'Arcy Wentworth Thompson
Edinburgh	University of Edinburgh
FR	*Fortnightly Review*
FRS	Fellow of the Royal Society
ICL	Imperial College London
ILN	*Illustrated London News*
JIC	John Innes Centre, Norwich
JTVI	*Journal of the Transactions of the Victoria Institute*
LMA	London Metropolitan Archives
MJRL	John Rylands Library, University of Manchester
NHM	Natural History Museum, London
NLS	National Library of Scotland
OMNH	Oxford Museum of Natural History
PHL	Pusey House Library, Oxford
PMG	*Pall Mall Gazette*
PUL	Princeton University Library
RBGE	Royal Botanic Gardens, Edinburgh
RBGK	Royal Botanical Gardens, Kew
RI	Royal Institution
SLNSW	State Library of New South Wales, Sydney
SOAS	School of Oriental and African Studies, University of London
TCD	Trinity College, Dublin
TNC	*The Nineteenth Century*
UCL	University College London
UCLA	University of California, Los Angeles

xiv ABBREVIATIONS

UL   University of London
Wellcome   Wellcome Collection Library, London
WTTD   William T. Thiselton-Dyer

A note on birthdates. Where birthdates are rendered as (1830–99) this indicates firm evidence of year of birth. Birthdates rendered (*c.*1830–99) are derived from census entries which rely on ages given and are often inaccurate by a year or two. Birthdates rendered (1830/31–99) are derived from information of age at death only, which only allows birthdate to be narrowed to one of two years.

'Darwinism'
My Gawd, 'ow 'ast Thou made this bloomin' wurld?
I cawn't find out; I've tried until I'm sick;
Is it Thy Will thet Ruin should be 'url'd
On 'arf the race, w'ile 'arf goes prank'd an' curl'd,
Becors the poppylashun is too thick?

Yus, Ive bin readin' Darwin…

('Kennington Cross', *The New Age*, 23 May 1908)

Introduction

Victorian Britain and the Problem of Generations

How old is Roebuck? The question is important on the threshold of a drama of ideas; for under such circumstances everything depends on whether his adolescence belonged to the sixties or to the eighties. He was born, as a matter of fact, in 1839, and was a Unitarian and Free Trader from his boyhood, and an Evolutionist from the publication of the Origin of Species. Consequently he has always classed himself as an advanced thinker and fearlessly outspoken reformer. (George Bernard Shaw, *Man and Superman* (1903), 2.)

Prologue

Riding to Derby on 12 February 1909, Edward Wrench, a surgeon from Baslow, struck up a conversation with a fellow traveller, whom he discovered to be the Rev. John Alexander Clapperton of Nottingham. Perhaps not unsurprisingly—after all it was the very day of the centenary of the birth of Charles Darwin, and only a few months before the fiftieth anniversary of the publication of Darwin's *On the Origin of Species*—their talk turned to evolution. Delighted to hear Clapperton's liberal views on evolution and praise of Darwin and his advocate, Thomas Henry Huxley, Wrench was rather taken aback to discover that he was a Methodist minister.[1] It was an illuminating failure of expectation, for Wrench was clearly, and perhaps not unreasonably, given the well-known hostility of Methodism to Darwinian ideas, assuming that Clapperton's opinions would follow from his denominational affiliation.

Not unreasonably, but still mistakenly. Perhaps because it was not so readily apparent or so easily categorisable, Wrench had not asked the question which George Bernard Shaw observes is critical for an understanding of the drama of ideas—had not ascertained Clapperton's *age*. The tone of Methodism's response to Darwinism was set in the 1860s and 1870s by men

2 DARWINISM'S GENERATIONS

who had grown up in a pre-Darwinian world, for whom the geological and palaeontological discoveries of the first half of the nineteenth century were only just beginning to undermine belief in the literal truth of the biblical account of creation. But Clapperton, born in 1858, was of a different generation; a baby when the *Origin* appeared and not yet in his teens when its sequel *The Descent of Man* followed. He no doubt absorbed a good deal of the predominantly anti-Darwinian tone of his denominational surroundings in the 1870s, but his education at Edinburgh University and Didsbury Training College in the years around Darwin's death in 1882 would have exposed him to the common acceptance of evolutionary ideas amongst his peers, and the newer apologetic literature which was finding ways of reconciling theology and biology. Without the intellectual over-investments of his older brethren, Clapperton was able to square his religious beliefs with the fundamentals of evolutionary science.

Clapperton's reactions to this meeting are not recorded, but had he learned more about Wrench, he too might have been surprised at his travelling companion's evolutionary enthusiasms. For the seventy-six-year-old Wrench, clergyman's son and staunch Anglican, decorated military surgeon, personal physician to the Dukes of Devonshire, and magistrate and pillar of Derbyshire Conservatism, does not at first sight have the typical background of a Victorian Darwinist. And indeed, in the years immediately after the publication of the *Origin*, he had been deeply sceptical of evolutionary ideas. Yet Wrench (1833–1912) was part of a generation born in the years after 1830, still only in their late teens or twenties when the *Origin* was published, who eventually came round to an acceptance of evolution. His meticulously kept diaries for the years from 1860 to his death in 1912 don't record his reading of Darwin's books. Indeed, scattered references suggest that, like many other Victorians, Wrench's exposure to Darwin was largely an indirect and mediated one, and it is impossible to trace the development of his ideas with any precision. But by the 1870s he was offering the audiences of his science lectures a light-hearted evolutionism, and by the later years of the century he was collecting anecdotes of the early advocacy of Thomas Henry Huxley (1825–98), and revelling in the jokes at his opponents' expense which the evolutionary debates had prompted.[2] His was not a vocal commitment, but it was none the less deep-rooted for that.

This fleeting encounter encapsulates the scope and purpose of this study: to use an examination of the responses to Darwin and Darwinian ideas in the fifty years after the publication of *On the Origin of Species* to explore the importance of generational dynamics in the history of nineteenth-century

Britain. It springs from a belief that, in addressing large questions of cultural history, scholars, just like Edward Wrench, have been insufficiently attentive to the potentially powerful ways in which age influenced attitudes and shaped behaviours. And even more that they have failed to recognise both the extent to which these age dynamics were structured generationally, and the implications of this for our understandings of the nature of historical change.

Darwin and his supporters did not share this inattention. At the end of the *Origin*, Darwin confessed that he did not expect to convince experienced naturalists with minds already strongly inclined to a different point of view, and instead 'look[ed] with confidence…to young and rising naturalists, who will be able to view both sides of the question with impartiality.'[3] He constantly reassured his older correspondents that he did not expect to be able to overturn their long-held beliefs. Even after a decade of energetic advocacy, Huxley conceded that while there would be converts amongst 'young men of plastic minds', it would take the passing of a generation for Darwin's ideas to become accepted.[4]

Historians have not entirely ignored generations, and indeed the history of the spread of Darwinian ideas has often been given a generational twist.[5] John Holmes has recently suggested that 'the rapid rise of evolutionism was a victory for Darwin's science, but it was also a consequence of the passing of the generation whose science he had challenged.'[6] Generational analysis offers a useful framework for considering the interrelated impacts of events, of aging and of the common experiences of age-defined cohorts, usually described as 'period effects', 'age effects' and 'cohort effects'. Darwin's and Huxley's predictions drew on all three: the impact of the publication of Darwin's work and the debates it prompted (period effects), the differential openness to new ideas of the young and old (age effects), and the changes which would eventually be wrought by the substitution of one generation by another (cohort effects). But despite a long-standing interest of the social sciences and the proliferation of generational labels in twenty-first-century discourse, generational approaches have never had more than a marginal influence on the scholarship of nineteenth-century Britain. Part of the problem, as Bobby Duffy has argued in his recent survey *Generations. Does When You're Born Shape Who You Are?* (2021), is that generational approaches often walk a narrow line between offering insight into the differences of experience and attitude which different birth cohorts can experience, and constructing reductive stereotypes which are at best simplistic and at worst simply wrong.

4 DARWINISM'S GENERATIONS

Being generous, we might describe previous generational analysis as 'suggestive', but frequently it has been merely crude and perfunctory. Even if historical scholarship has avoided the worst sort of 'snowflake' generalisations of contemporary discourse, all too often generations are invoked when the only discriminations being signalled are an indeterminate opposition of older *versus* younger. And where there is greater specificity, there has been little effort to obtain agreement as to how generations might be defined, never mind their character, or the nature and consequences of their interactions. The sorts of discomforts which ensue are vividly registered in the *faute de mieux* use the literary scholar Franco Moretti makes of generations in his influential *Graphs, Maps, Trees. Abstract Models for a Literary History* (2005).

As a result, before launching into an account of Darwinian reception, we must first tackle the question of how best to establish an effective framework, of both definition and methodology, for exploring the role generational dynamics might have played in the nineteenth century. This involves an investigation of the limits of the theoretical models of generations currently in vogue, and the suggestion of a number of ways in which they might be reworked. Those keen to avoid the abstract theorising which results might simply want to jump straight to the description of the Victorian generational schema which emerges out of this theoretical work (page 12).

A Question of Generations

In truth, the use of generational analysis has been greatly hindered by its conceptual slipperiness, theoretical fragility, and vulnerability to misuse. Definitions often distinguish three usages: family generations (of offspring, parents, grandparents); generations as a cohort of those born at the same time; and generations as a form of collective consciousness: age groups which share a sense of identity, which become in some senses agential. But even these meanings are far from exhaustive. In the case of the Victorians, who used the language of generations incessantly, we need to be especially attentive to syntactical nuances: the play of generation when and generation who. During the nineteenth century 'generation' was used to denote specific historical eras ('the generation of the Napoleonic wars'), or a unit of historical time ('two generations ago'), as well as in a number of conventional formulations, to indicate the here and now ('the present generation'), the young ('the rising generation'), the future ('generations to come'), as well as in loose reference to differences related to age. In the face of these conceptual confusions, it is understandable that many historians dismiss 'generation' as one of those routinely abused terms whose meaning has been all but obliterated by the careless and casual way it is invoked.

Where historians have hesitated, social theorists, sociologists, and anthropologists have not feared to tread.[7] But scholarly thinking on generations remains rooted in the writings of a small number of foundational thinkers, in particular Karl Mannheim, whose relatively brief essay 'The Problem of Generations', originally published in 1928, has exercised a quite extraordinarily enduring influence.[8] Mannheim's essay is endlessly suggestive, but it suffers both from a perhaps inevitable abstraction, and also from a proliferation of positions, a promiscuous use of undefined terms, and a disrupted and at times circular argument. Many of the key concepts and processes presented as part of its historical context remain undeveloped, and Mannheim's preoccupation with the 'lost generation' of 1914–18 has encouraged a series of unhelpful assumptions, including an emphasis on generational agency as particularly associated with the revolt of the youth and the significance of traumatic events, and in the tendency to comprehend generational relations in largely dichotomous ways, not least in the conception of generational succession as the dominant dynamic of historical change, and 'transmission' of cultural values as its central problematic.[9]

Above all, under Mannheim's influence, the equation of the cohort and its self-consciousness has become axiomatic. The tendency has been to assume that, as the historian Samuel Hynes once put it, 'A generation exists when it thinks it does'.[10] More recent deployments of the idea of 'generationality', defined as the 'ensemble of age-specific attributions by means of which people locate themselves within their respective historical periods', have highlighted the conceptual distinction between historical position and consciousness without in practice challenging their conflation.[11] This focus on subjectivities obscures generation as a social and cultural *location*, and the possibility of a generation that has shared social, intellectual or economic characteristics and experiences, but not self-consciousness. While in the analogous case, theorists and historians of class have always retained a productive dialogue between the notions of class 'in itself' and class 'for itself', generational theory has tended to relegate location to the status of 'potentiality', inert and unimportant without self-consciousness.

Emphasis on self-awareness has led, perhaps inevitably, to a focus on intra-familial conflict as a key source of generational formation. From this perspective, the repudiation by young adults of the authority of their

6 DARWINISM'S GENERATIONS

parents encourages them to display attitudes formed in opposition to those of mothers and fathers, who in turn experience a sense of generational eclipse at the loss of influence over their offspring. The facility with which parent–child distinctions offer a clear marker between collective experiences, as for example in the case of some migrant communities, has further reinforced this trend, and encouraged a fundamentally unhelpful conflation of the periodicities of family (genealogical) and cultural (sociological) generations, and the consequent assumption that generations will tend operate on a twenty-five- to thirty-three-year cycle.[12] But the fault lines of family generations cannot simply be writ large onto society as a whole. The coalescence of sociological generations requires structuring forces and fields of attraction which produce discrete collective experiences. Theorists like José Ortega y Gasset and, more recently, Julián Marías have suggested shorter intervals, but except in the context of the generational configurations of the most recent past, these suggestions have rarely been explicitly addressed or adopted.[13]

Perhaps because student-led revolts emerge out of naturally intellectually and institutionally intimate environments—schools, universities, and institutions of professional training—limited attention has been given to the modes by which individual responses are forged into collective positions. Mannheim struggled to provide a model of how the attitudes formed within socially intimate groups (what he calls 'generational units') are adopted by the wider generational cohort, placing a great deal of weight on a *Gestalt* psychology of intellectual growth, and the creation of common views of the world which bound age-groups together. Subsequent theories have focussed on the constitutive significance of shared exposure to events, especially *traumatic* events experienced in early adulthood, that form attitudes and responses which are likely to differ significantly, as Mannheim put it, to those of 'the older generation [who] cling to the re-orientation that had been the drama of *their* youth'.[14] The danger is that this implies an unwarranted fixity of attitude and that it places too much emphasis on generation as a legacy effect, glued together by social memory, rather than something which is constantly being constructed by age-specific experiences and perspectives.

Where approaches have embraced more general cultural dynamics, attention has concentrated on music, student life, or the impacts of war, which tend to foreground youth cultures and 'the revolt of youth' as the crucial motor of generational formation. This was the preoccupation of Mannheim and his contemporaries, further reinforced by the resurgence of

INTRODUCTION 7

interest in generations prompted by the cultural upheavals of the 1960s, rooted in student communities and dominated by a tone of counter-cultural youth rebellion.[15]

Although discussions have often addressed the idea of the 'non-contemporaneity of the contemporaneous' or the 'simultaneity of the non-simultaneous', by severely reducing the potential for multiple coexisting generations the conventional twenty-five- to thirty-three-year-wide generational cohort has also encouraged an unhelpfully polarised or dialectical frame of analysis rooted in conflict more than difference, allowing generational relationships to be reduced to a simple binary process of displacement. Offspring challenge, overthrow, and replace their parents, and then suffer the same fate at the hands of their own offspring; or younger challengers disrupt the status quo and are then dominant until a new set of young bloods come along.[16] This picture of generational concatenation leaves questions of multigenerational interrelationships almost entirely occluded.[17]

Ironically, the dynamics of identity formation are relatively straightforward in comparison to boundary or separation effects, and in particular the explanation of how those born relatively close together might experience the gravitational pull of different generations. Except for accounts of the post-1950 period, generational discussions rarely offer anything other than arbitrary birthdate limits, or any discussion of the way in which generations as age-cohorts are constituted. This despite Alan Spitzer stressing the indispensability of establishing boundaries as far back as 1987, in his important study of *The French Generation of 1820*, quoting Braudel's observation that from this question all others flow: 'To draw a boundary around anything is to define, analyze and reconstruct it.'[18]

Generational approaches are not entirely without traction on the history of Britain in the nineteenth century.[19] Generations have served as structuring devices, from W.L. Burn's *Age of Equipoise*, subtitled 'A Study of the mid-Victorian generation', to Theo Hoppen's *The Mid-Victorian Generation*. Histories of labour, Victorian feminism, even Victorian historiography have at times organised their discussion around generational sequences.[20] For a time, in the 1960s and 1970s, designations like 'the generation of the 1860s', the men who had graduated from the universities during the second half of the 1850s and the first of the following decade (and their contemporaries), were not uncommonly used as an analytical tool for understanding the period.[21] But despite a smattering of studies in the last decade or so, most notably Thomas Otte's *The Foreign Office Mind: The Making of British*

8   DARWINISM'S GENERATIONS

Foreign Policy, 1865–1914 (2011), Heather Ellis, *Generational Conflict and University Reform. Oxford in the Age of Revolution* (2012), and Richard Fulton's *Warrior Generation, 1865–1885. Militarism and British Working Class Boys* (2020), unlike the twentieth century, or even the Romantic era, there is no stable or regularly deployed understanding of what Victorian generational formations might look like.

One of the paradoxes of the Victorians is that their profound investment in generational languages was combined with a lack of interest in theorising about generations, and perhaps for this reason generational usages were profuse but inconsistent. Victorian autobiographers in particular repeatedly defined their relationship with their parents in generational and historical terms.[22] Of course such dramatisations, as developed for example by Edmund Gosse's *Father and Son* (1907), Samuel Butler's *The Way of All Flesh* (1903), and subsequently by Virginia Woolf and the Bloomsbury set, were fundamental to early twentieth-century repudiations of 'Victorianism'. But they reflected a particular consciousness, the distances of age, of the sort articulated in Arnold Bennett's reflections on his father in 1897. Bennett, born in 1867, confessed to noticing 'the approaches of middle-age upon him. I felt acutely that he and I were of different generations; that parent and child, be they never so willing, can never come intellectually together, simply because one time of life differs crudely and harshly from another.'[23] Such sensibilities rarely produced explicit generational identities or schemas. What they did do, as the essays of the contemporary historian Walter Bagehot show, was to encourage a diffuse recognition of generational dynamics. Bagehot's essays are littered with generational languages, but are also underpinned by a model of historical causation and change which is explicitly generational. He affirmed the power of parent–sibling division and the critical influences of student years to create generational formations.[24] He accepted the imperative of transmission, and the responsibility of each generation 'to translate into the language of the living, the truths first discovered by the dead'. He interpreted historical figures as embodiments of their generation, and historical change, for example in 1830 and in 1865, as occurring through generational succession.[25]

For all this, we must be careful not to overstate the congruences of the sensibilities of Victorians and twentieth-century commentators. Applications of generational designations to contemporary conditions were often rooted in dynamics which bore little relation to the ideas of Mannheim and his disciples. For example, it was frequently not the traumatic which underpinned Victorian generational identities but the elegiac remembrance of the golden friendships of early manhood. Gosse, in 1897, mused on the way the protracted lives of men of genius could form weirs in literary history, and generational progression was as often construed as a matter of necrology, of the passing of the old as of the emergence of the young.[26] Above all, despite the intermittently powerful generational identities visible in nineteenth-century Europe, before W.B. Yeats' 'tragic generation' and the Bloomsbury group, Victorian Britain displays neither widespread generational self-consciousness nor stably imputed generational identities. And for this reason, we cannot base an effective generational approach to the Victorians on their own generational consciousness any more than on the standard sociological conventions.

Rethinking the Generational Model

Drawing in part on the unduly neglected insights of Marías and Ortega y Gasset, this study adopts five key adjustments to conventional generational approaches.[27] First, the recognition that generation is primarily a shared *position* rooted in the structuring force of contemporaneity, a similarity of age-location within historical time. Second, that its significance is a matter of *effect* and not merely of consciousness or agency. Third, that the dynamics of generational formation involve the production of *commonalities* rather than identities: shared experiences, but also shared opportunities, languages, reference points and perspectives. Fourth, that the size and particular intervals of generational cohorts are determined not primarily by biological but by *cultural* periodicities, and that in the nineteenth century although they vary in breadth, generations tend to resolve into cohorts with roughly fifteen-year intervals. Fifth, that in consequence, intergenerational relationships are not sequential and bilateral but *stratigraphical* and *multivalent*, and that rather than merely providing a line of transmission, generations constitute a complex field of debate and interaction.

Fundamentally, a generation is a non-arbitrary age cohort, the outcome of the 'stucturating force' (in Antony Giddens' term) of a common location in history, a similar age of encounter with key events or new ideas, a shared experience of life-cycle specific institutions (for example education), a similar pattern of memory as against mediated history, and the shared perspectives of a common age. In this respect it is equivalent to categories of class, gender or race, and worthy of attention alongside them. And like all social categories, it is an objective and not merely subjective phenomenon: that is,

10 DARWINISM'S GENERATIONS

although the extent to which generations develop a sense of their own identity is important in determining their historical visibility and agency, and although part of the distinctive characteristic of a generation will rest in its consciousness, each generation is produced first and foremost through material circumstances which operate at least in part unnoticed.

The overriding measure of generational significance is not self-conscious identity but *character*: even where this character is not apparent to members of the generation themselves. To the extent that such character is manifested in some sort of a generational worldview, this is composed less of specific beliefs and values and more by what we might think of as intellectual 'repertoires' (akin to Bourdieu's 'habitus', or Raymond Williams' 'ways of feeling'): modes of reading, vocabularies and idioms, habitual actions and behaviours, emotional responses or strength of intellectual and emotional investment. This worldview includes the intertwined propositions, evidences, authorities and reference points, characteristic questions and modes of argument which shape intellectual engagements: who and what people read, what texts were particularly influential or foundational, what ideas were retained and redeployed. We can think of these as component parts of the sorts of composite photographs that the Victorians themselves produced to display physiognomic types, or as reflective of the observation of Oxford don Mark Pattison (1813–84) that an age is characterised not by 'its peculiar opinions, but the complex elements of moral feeling and character in which, as in their congenial soil, opinions grow'.[28]

Generational configurations were contingent on the irregular operation of the forces of attraction and distinction which created the necessary degrees of coalescence and separation, and are ultimately brought into being not by borders but by their central pole(s) of attraction. The key dynamic was not the centrifugal force of genealogical relations, which have no means within themselves to extend out to general cultural formations (although children could react against parents in ways which reinforced generational distinctions). Rather, it was the centripetal forces of particular events and experiences, and the shared perspectives on them, that similar age brought which were crucial, perhaps especially but certainly not exclusively those encountered in late adolescence and early adulthood, when recipients were generally most malleable and receptive, including social and economic conditions, intellectual and technological novelties, political and military conflicts. The sort of dynamics acknowledged by the Cambridge philosopher Henry Sidgwick (1838–1900) in discussing the influence of Comte and Mill over his mind in the early sixties: 'I say "in my mind," but you will understand that it was largely derived from intercourse with others of my generation, and that at the time it seemed to me the only possible ideal for all adequately enlightened minds'.[29] The suggestion is not that such beliefs ossified, but that thereafter they become more and more viscous, solidifying values, setting a trajectory and narrowing the bounds of future development.

The historical happenstance of these processes means that generations appear irregularly in chronological time. And the difficulty of all but the most all-encompassing of events creating a consolidating experience over a range of ages wider than between fifteen and twenty years produces a periodicity which is shorter than the conventional biologically derived one. The evolutionist and town planner Patrick Geddes (1854–1932) sometimes wrote of this as the 'half-generation—waves every 15, 16, or 17 years'.[30] This narrowing is easier to accept if we recognise that we cannot treat generations as compartments or boxes.[31] Their boundaries are not absolute; the edges are blurred, and generational experiences and attributes interleave. Robert Wohl likened generational pull to a magnetic field, at the centre of which lay an experience or set of experiences, accepting that the force of attraction, even its chronological location, could shift over time.[32] A figure born in or around the boundaries of generational birth-ranges may most usefully be understood as belonging to the adjacent generation, not purely on the basis of attitudes or behaviour, but as a reflection of inevitable variabilities in experience, such as age of university education, the weakened force of generational identities in the face of strong countervailing identities, and the fact that generational differences are always partial, coexisting with degrees of continuity across generations. Here again, we can quote Bagehot: 'The work of nature in making generations is a patchwork—part resemblance, part contrast. In certain respects each born generation is not like the last born; and in certain other respects it is like the last'.[33]

Accepting generations whose normal breadth is significantly narrower than the conventional twenty-five- to thirty-three-year span requires a fundamental reshaping of understandings of the pattern and progression of history. It both accelerates and thickens historical time. Once life expectancy rises, periods become, in Wilhelm Pinder's term, generationally 'polyphonous'. Relationships between generations do not involve a linear sequence of challenge and supersession, but rather a complex cycle of coexistence and interaction. Even allowing for the relatively restricted life expectancy of Victorian Britain, the population included significant numbers living into their seventies and eighties, so that at most times, four if not

12 DARWINISM'S GENERATIONS

five generations of adults coincided. And, although both Victorians and later commentators have tended to see the transformations in the position of Darwinism in the thirty years from 1859 as occurring within a single generation, with the implication that changes occurred primarily as a result of changes in opinion not replacement of the opinion-bearers, if the generational cycle actually operates with a roughly fifteen-year periodicity, this expands the potential role of generational displacement in the explanation of historical change, even in relatively short time frames.

Victorian Generations: A Tentative Schema

This study uses a detailed empirical investigation of the response to Darwin and Darwinian evolution in Britain to show how far the application of a framework along these lines might uncover a stable and broadly applicable generational pattern in the Victorian period. It builds on preliminary work which sought to establish a tentative generational schema for Victorian Britain, using several approaches to identify birthdates which might mark the boundaries between generations. These included an exploration of the dividing lines of particular groupings (for example, poets or historians) proposed by contemporaries, the age structure of Victorian coteries and groups of collaborators where similarity of age appears to have been a significant influence, and the age profile of the authorship of volumes of collected essays published in the period. For example, the age characteristics of a series of artistic groups with relatively clearly defined memberships, from the Pre-Raphaelite Brotherhood to the Decadents and the Camden Town Group, were mapped to see how far the distinctive concentration and range of their memberships offered any potential lines of generational division. From this process a tentative best-fit set of dividing lines was identified, around the years 1812/3, 1829/30, 1844/5, 1860/1 and 1875/6.

The resulting cohorts were then used to organise a dataset of contributions to the Victorian periodical press in each year from 1824 to 1901, drawn from the Wellesley Index to Victorian Periodicals. This analysis demonstrated that the tentative generational cohorts identified produced a much more defined and regular pattern of cohort emergence, pre-eminence and decline (as indicated by the relative proportions of contributions of each generation to the total number of articles published each year in the sample periodicals) than alternative fifteen-year cohorts. It also, and perhaps more significantly, suggested that the points of transition between the periods, when one generation provided the largest number of contributions to the period of preponderance of the next, aligned remarkably tightly to the conventional boundaries of the sub-periods of Victorian Britain: the early 1830s, 1851, the later 1860s, and the mid-1880s. So the cohort born between 1814 and 1829 provided by far the largest share of articles of any generation for the years of the 1850s and first half of the 1860s, overtaking those born between 1796 and 1813, who had been the most numerous contributors in the later 1830s and 1840s, and in turn being superseded from the later 1860s by those born between 1830 and 1844. It seems logical to label the cohorts by their period of greatest influence, and so to name those most prominent before 1851 (that is those born between c.1795 and c.1813) early Victorian, and the succeeding groups mid Victorian, high Victorian, late Victorian and Edwardian. This alignment of generational nomenclature and particular historical periods is not without its drawbacks, and it is worth emphasising here that, in the approach followed in the rest of the book, the description 'mid-Victorian' or 'Edwardian' does not imply the productions or characteristics of a particular period, but of a specific cohort, which gives them much a much more enduring presence than the period with which they are especially associated.

What is suggested, therefore, is a series of generations stretching across the nineteenth century, beginning with a Georgian (in literary contexts perhaps 'late-Romantic') generation, including in their number Lord John Russell (1792–1878) and Sir Robert Peel (1788–1850), the essayist Leigh Hunt (1784–1859), the poet Percy Bysshe Shelley (1792–1822), John Keble (1792–1866), and Thomas Carlyle (1795–1881), indelibly marked by the consequences of the French Revolution and the wars against Napoleon. An early Victorian generation whose members include Charles Darwin (1809–82) and Charles Lyell (1797–1875), W.E. Gladstone (1809–98), and Benjamin Disraeli (1804–81), and also John Stuart Mill (1806–73), Alfred Lord Tennyson (1809–92), Elizabeth Gaskell (1810–65), and Charles Dickens (1812–70), whose formative experiences came in the post-Napoleonic War years, the years of repression and radicalism, of the last gasp of the English *ancien regime*, and who experienced the social problems of the 1830s and 1840s as paradigmatic of their maturity. To be set against a generation of mid Victorians, including Queen Victoria (1819–1901) herself, George Eliot (1819–80), John Ruskin (1819–1900), Matthew Arnold (1822–88), Wilkie Collins (1824–89), Walter Bagehot (1826–77), and Thomas Henry Huxley (1825–95), shaped by the experience of the reform agitations of 1830–32 and their legacies, by the visible emergence of an urban-industrial society and

14 DARWINISM'S GENERATIONS

its social and economic challenges, its infamous complacency rooted in the perspective from which it viewed the real if modest economic gains of the years after 1848. The high Victorians, for example Robert Gascoyne-Cecil, Marquess of Salisbury (1830–1903), Leslie Stephen (1832–1904), Lewis Carroll (1832–98), William Morris (1834–96), Thomas Hardy (1840–1928), and Walter Pater (1839–94), with their limited memories of the defeat of Chartism and of the hungry forties, could afford to be less satisfied with the material gains symbolised by the Great Exhibition, and more nervous about the social and cultural implications of the rise of democracy. While the late Victorians, think Arthur J. Balfour (1848–1930), Herbert Asquith (1852–1928), Robert Louis Stevenson (1850–94), Oscar Wilde (1854–1900), George Bernard Shaw (1856–1950), Beatrice Webb (1858–1943), and Cecil Rhodes (1853–1902), maturing in the shadow of the rise of Germany and the decline of British economic hegemony from the later 1860s, unsurprisingly addressed themselves more to the extent of government responsibility and the attractions of empire. In turn the Edwardians, David Lloyd George (1863–1945), Walter Sickert (1860–1942), H.G. Wells (1865–1946), John Galsworthy (1867–1933), Beatrix Potter (1866–1943), Ford Madox Ford (1873–1939), and G.K. Chesterton (1874–1936), confronted with the weakening of the traditional elites and more direct and visible challenges to Victorian cultural norms, embraced a new materialism and a new scepticism.

Darwinism as Case Study

The reception of Darwinian thought in Britain provides a particularly promising opportunity for testing the appropriateness of these generational cohorts and the extent of their significance. Darwin's thought—and we will be attending particularly to Darwin's writings and ideas rather than the broader corpus of evolutionary theory—had an extraordinarily broad cultural reach. Admittedly, throughout the years covered by this study, there was a strand of educated culture, often trained in the Classics, or devoted to literature, which displayed a combination of indifference and contempt for 'stinks', or anything scientific; but we shouldn't underestimate the enthusiasm natural history inspired in Victorian Britain, from beetle-collecting children to parson-naturalists. The fascinations of a few Victorian literary figures: Tennyson, Ruskin, Charles Kingsley, Hardy, Samuel Butler, or Grant Allen are well known, but the net drew in from far wider afield. The publisher William Longman (1813–77), Edward Lear (1812–88), illustrator turned nonsense writer, the philologist and first editor of the Oxford English Dictionary, Sir James Murray (1837–1915), and Henry Fox Talbot (1800–77), the pioneer of photography, all professed a passion for natural history. And it is easy to lose sight of the number of eminent Victorians who passed through a phase of significant work or research in the post-*Origin* natural sciences, including the town planner Patrick Geddes, Beatrix Potter (1866–1943), the designer Christopher Dresser (1834–1904), and reproductive rights campaigner Marie Stopes (1880–1958), whose first career was as a palaeobotanist; not to mention the much larger number whose intellectual formation included time spent in lecture halls or laboratories, notoriously H.G. Wells (1866–1946), but also the poet Constance Naden (1858–89), the theosophist Annie Besant (1847–1933), the philosopher and politician R.B. Haldane (1850–1928), and the sociologist L.T. Hobhouse (1864–1929).

Nor was the reception of Darwin in Victorian Britain merely a matter of a few highbrow periodicals or the technical treatises of a narrow university-educated elite. Interest in Darwin knew few social or spatial bounds. Direct evidence of working class responses survives only in snatches, and indirect evidence is infected by a tendency to poke fun at the supposed misapprehensions of the less educated, but there is no doubt that Darwinian ideas did percolate into working class culture, and direct exposure to Darwin's own writings was increasingly common.[34] In the immediate aftermath of the *Origin*, Cornish coastguards scouted anxiously for updates to debates, and in the later 1860s the future prime minister Arthur Balfour (1848–1930) reported receiving a Darwinian diatribe from his barber, 'the doctrine of evolution, Darwin and Huxley and the lot of them—hashed up somehow with the good time coming and the universal brotherhood, and I don't know what else.'[35] One cooper in Brentford used the public library in the early 1890s to read two of Darwin's books, one of Henry Drummond's Christian recastings of evolution, Paul Du Chaillu's sensationalised account of his 'discovery' of the gorilla, leavened by a solitary novel.[36] Discussion was not confined to the rarefied columns of the intellectual periodicals. One intervention which went through three editions between 1871 and 1873 was apparently issued with a cover illustration of a hideous monkey on a donkey, blowing a trumpet from which emerged the words "Homo Sum", underneath which was a quote of Darwin's belief in man's descent from a hairy quadruped, and it was reported in 1912 that the convicted poisoner Frederick Seddon (1872–1912) was using his final days in Brixton prison to read *Descent* and discuss evolution at great length with the prison chaplain

16 DARWINISM'S GENERATIONS

and doctor.[37] And if the intensity of scrutiny waned, the penetration of Darwinian texts waxed. J.M. Barrie (1860–1937)'s tale *Auld Licht Idylls* (1888) described a highland bothy, complete with a weekly newspaper and several books, including one of Darwin's, and Robert Falcon Scott (1868–1912) recorded that the sailors on his Antarctic expeditions in the early 1900s read a good deal, and that some of them were 'very fond of Darwin's "Origin of Species"'.[38]

Nor, despite the gendered nature of these examples, should we think of the engagement as exclusively masculine, or even adult. Complementing the well-documented evidence of an involvement in nature study and writing on the part of Victorian women, there is rich evidence of shared, domestic, and indeed female cultures of consumption and debate which existed as the obverse of this masculine record: the reading aloud of Darwin's works in the family of William Thomson, Lord Kelvin (1824–1907);[39] the intimate correspondence which bound Julia Wedgwood (1833–1917), Mary Everest Boole (1832–1916), and Victoria, Lady Welby (1837–1912); and the discussions about Darwin and evolution between the future anthropologist A.C. Haddon (1854–1940) and his mother and sisters, even the educationalist Catherine Winkworth's dreams of Huxley lecturing the palaeontologist Richard Owen.[40] Ethel Chamberlain, who having worked her way through A.R. Wallace's survey *Darwinism* (1889), wrote to her brother, the future prime minister Neville, that she was longing to talk it over with him and to try experiments in the cross-fertilisation of plants and pigeons, was unusual but certainly not unique.[41] Darwin was read by young as well as old, and indeed over time seems to have become a staple of the reading of young adults of the educated and self-educating classes. By the 1880s, the teenaged future publisher Grant Richards (1872–1948) was being issued with Darwin's *Voyage of the Beagle* as part of the library distribution at his school.[42] Things had almost certainly been less relaxed twenty years earlier, but that didn't stop some schoolboys, like one 15-year-old pupil at Bootham School in York in 1865, who was reported to have got hold of a copy of the *Origin* and been converted.[43] Most of this history, of course, remained unrecorded; but the patchiness of the survivals cannot disguise Darwin's remarkable ubiquity.

The cultural history of Darwinism was also long, not merely a legacy of ideas but a constant process of encounter and response, of reading Darwin and of reading about Darwin. No reception history can ever be contained in the immediate readings and responses to a text, but this is particularly true of Darwin. The *Origin* and *Descent* were books not just read, but read and reread. For serious readers, new editions necessitated new encounters. The evidence of the steady reprinting of Darwin's works, of their distribution as school and university prizes, their gradual spread into the public libraries of late-Victorian Britain, and ultimately onto the bookshelves of serious readers of all classes, bears witness to the uninterrupted augmentation and renewal of his readership. New issues meant new audiences, but also nostalgic rereadings; the banker and amateur entomologist John Lubbock (1834–1913), it was said, frequently went back to Darwin's writings as 'a wonderful cordial' with which to brush away the cobwebs.[44] Even in the midst of his political life in the 1930s, Neville Chamberlain returned repeatedly to Darwin's *Life and Letters*, reverencing Darwin's modesty, and the 'never ceasing speculation' which left everything allocated its due weight.[45] Such encounters constantly refuelled argument and debate, conversations in common rooms and around dinner tables, in church halls and public house snugs, omnibuses and railway carriages, street corners and, as we have seen, highways.

This longevity registered the fertility of Darwin's ideas for all areas of Victorian life and thought: the way, as the liberal journalist and politician John Morley (1838–1923) put it, Darwin's ideas circulated under a hundred disguises.[46] But it was also a reflection of the depth of challenge that Darwin posed to the world views of the Victorians, producing not merely a moment of heightened intellectual crisis, but also a set of enduring problematics. Darwin did not merely modify the prevailing orthodoxy, he offered a fundamentally different world view. Contemporaries were forced to re-examine their existing understandings, and to come to a decision about the extent to which they needed to adjust previous beliefs to accommodate the new Darwinian ones, an adjustment which in many cases took decades to effect.

Making Sense of Darwinism

The choice of the Darwinian debates as a case study has its challenges. There was no fixed set of 'Darwinian' ideas, even for Darwin himself. The eponymous labelling of the ideas developed in the *Origin* and its successors implied a relationship to a stable and internally coherent authorial position, which of course was illusory, both in terms of Darwin's own writings and the meanings imputed to them. Nor was there any stability in the referent of 'Darwinism': for some, and especially in the immediate aftermath of the publication of the *Origin*, Darwinism might be synonymous with a belief in evolution, although Darwin was not the only nor the first to articulate an evolutionary past; for others, and increasingly, it was more likely to indicate the primacy given to a particular theory of the cause of evolution; for others again it would come to stand for a particular form of metaphysics, which denied the supernatural. In the same way, it could be represented by its allies or supporters, as coterminous with or analogous to the positions of Huxley or of the German Darwinist Ernst Haeckel. Responses to Darwin were never purely a question of belief, acceptance or rejection of a set of linked propositions. They were a matter of emotion, immediacy, strategies of accommodation, forms of utilisation. As a matter of the history of ideas, a binary approach which seeks to apply some fixed qualification for the label 'Darwinian' will inevitably mask as much as it reveals. Nevertheless, Darwin's thought was often experienced in almost tribal terms, and alignment in the debates was often primarily a matter not of substance but of affiliation: did the individuals involved feel themselves to be generally for or against the broad Darwinian position? Many who enrolled under the Darwinian banner on close inspection accepted without reservation very little of what intellectual historians would now understand the term to mean, while many who vehemently rejected the identity would if pushed have accepted much of what it stood for. But for all these confusions, the issues at stake were sufficiently pointed and urgent to demand the making of important choices, choices sufficiently scouted to enable the reconstruction of individual stances and the charting of underlying age and generational patternings.

Participants were not necessarily conscious of some of the finer distinctions of the evolutionary argument, and even where they were, the mediations of second-hand summary often leave us only blurred representations of their views. In fact, and especially in the years after his death in 1882, as James Moore has demonstrated, it was not just 'Darwin' and 'Darwinism' which operated as signifiers whose meanings were deliberately made and remade, but 'neo-Darwinism', 'ultra-Darwinism' and 'Lamarckism'. The conflict over Darwinism was fought out semantically and emotionally as much as propositionally, as a struggle to control or define the language and the tenor of the debate.[47] Take the notion of 'design', a key marker of the division between religious and non-religious view of evolution, and hence of Darwin, but whose meaning ranged broadly, from belief that evidence of the adaptation of plants and animals to their environment demonstrated a divine creator to interpretations of the shifting forms of natural evolution as proof of purposeful change and an overarching plan. And nice distinctions were beyond many of the participants in the debates, who instead took broad brush approaches, as in the remark of one contemporary that he was inclined to 'sail under a Spencerian cum Huxley cum Darwin flag'.[48] Hardly surprising then, that there is a temptation, most recently explored in Ian Hesketh's collection, *Imagining the Darwinian Revolution* (2021), to abandon this sort of personalisation entirely. As the philologist Max Müller (1823–1900) reflected in 1878, considering the difficulty of pinning down exactly what was at issue, 'The Darwinians are much worse than Darwin himself, and I think the word "Darwinism" ought either to be sharply defined or should be replaced'.[49] And yet contemporary debate was couched in precisely these terms and they cannot simply be jettisoned. What follows seeks to explore the response to Darwin and the positions which came to be propagated in his name as part of a wider set of responses to evolutionary thought, remaining sensitive to shifting contemporary usages in both the substantive and labelling senses without abandoning the attempt to ascertain individual beliefs.

Darwin developed his ideas in a number of works, including *The Descent of Man* (1871), *The Variation of Animals and Plants Under Domestication* (1868), and *The Expression of the Emotions in Man and Animals* (1872), and this and his almost obsessive revisions of his key works (the *Origin* alone went through six editions between 1859 and his death in 1882) ensured that there was no single stable Darwinian evolutionary position. In parallel, Darwin became the figurehead of a coterie of like-minded evolutionary champions, especially Huxley and the physicist John Tyndall (1820–93), who along with Darwin's later followers offered personal interpretations or added further glosses. In particular, it is impossible for the first three decades after the *Origin* to isolate the impact of Darwin from the influence of Huxley. Huxley mediated. He remediated. His rhetorical act of subsuming his personality into Darwin's created a further lamination. In the popular mind Huxley stood for Darwin; but in practice both his science and his sociology were distinct. What for Darwin was a scientific proposition was for Huxley a philosophical assertion, and in his hands Darwinism acquired if not a cosmogeny then certainly an anti-theological edge quite beyond Darwin's own. Darwinism as a set of biological propositions became bound up with Darwinism the philosophy of scientific naturalism, if not materialism. In response, there was always a tendency for (religious) opponents to seek to inflate the claims Darwinism made in order to oppose them.

20 DARWINISM'S GENERATIONS

Yet many Victorians probably continued to encounter Darwin primarily as a naturalist-traveller, or as a specialised botanical or zoological observer.[50] The print runs of his various works give an indication of the circulation of these various personas. By 1889 there were over 36,000 *Origins* and over 20,000 copies of *The Descent of Man* in circulation in the editions of John Murray alone, but also at least 20,000 copies of Darwin's account of his voyages in H.M.S. Beagle, and already more than 11,000 copies of his book on *The Action of Worms*, although less than a decade old, a figure which dwarfed the sales of most of Darwin's other volumes, which had generally sold in the low thousands. Darwin's death in 1882 and the publication of the three volume *Life and Letters of Charles Darwin* (1887) opened him up to a further series of essentially biographical engagements, as the humble man of science, the patient and devoted student of nature, a paragon for who he was and how he worked, rather than for what he said. And once the broad principles of evolution began to acquire the appearance of scientific orthodoxy, and extended into questions of the age of the earth, natural history, evolutionary mechanisms, historical anthropology, philology, ethics, and astronomy, Darwinism fragmented.

All this means that codifying what Darwin said is a futile exercise, albeit one which is still too often at the centre of discussions of Darwinism. But nor is it enough to simply argue that as long as someone describes a position as Darwinian, or claims to be a 'Darwinian', then we should accept it as such; or is it entirely satisfactory to focus on struggles to define Darwinism in particular ways at the expense of investigation of what individuals and groups were actually prepared to commit to as their own beliefs. We can answer the question as to whether there was a Darwinian revolution after 1859 by debating in what ways and to what extent the revolution can be described as 'Darwinian'; we can also ask in what ways and to what extent there was a 'revolution' in evolutionary thinking. And any satisfactory answer to either question cannot simply sidestep the question of individual belief.

The discussion that follows will attend primarily to Darwin's evolutionary work and to the particular evolutionary propositions he advanced, or were advanced under his aegis. At the heart of this were the crucial distinctions between the fact, the nature, and the cause or causes of evolution: that is, first, the veracity or otherwise of the claim that evolution had taken place; second, the specific nature of that historical process; and third, the mechanisms or dynamics of this process. In the contemporary debates these distinctions were routinely confused or conflated, but they remain vital for an understanding of the nature of the reception, interpretation and development of Darwin's ideas.

The most common, and in some respects the most troublesome, of these conflations is the tendency to equate 'Darwinism' and the simple fact of evolution, the existence of a long natural history in which the world's flora and fauna had developed through a series of stages to the complex diversity visible in the nineteenth century. This was especially a feature of debate in the decades immediately after the publication of the *Origin*, but it never entirely disappeared and indeed continues to be a problematic feature of recent treatments. The confusion is understandable. After all, for Darwin himself, belief in the fact of evolution was the 'great thing', and it remained the primary focus of his scientific investigations and rhetorical effort. But Darwin's wasn't the only evolutionary theory in play in these debates. Evolutionary concepts could be traced back to the classical thought of Lucretius, to the eighteenth-century philosophies of Kant and Hegel, to the astronomy of Laplace and the biological theories of Jean-Baptiste Lamarck (1744–1829), and the broad evolutionary philosophy of Herbert Spencer (1820–1903).

All these writers offered different versions of the nature and mechanics of the evolution they were claiming. Crudely, Kant and Hegel envisaged it as a fundamentally spiritual process, the progressive development of human consciousness, and had little truck with ideas of biological evolution. Laplace and Spencer's theories of evolution were cosmic in scope, aspiring to a general evolutionary account of the universe. Darwin's was a theory of the natural history of the world, with, at least for our purposes, three central propositions: (i) gradualism, that this process had occurred across the long stretches of the world's history through the slow transmutation of one form or species into another; (ii) common descent—the fact that the contemporary flora and fauna had descended through entirely natural processes in a single continuous process from earlier organisms, and ultimately from one or at best a very small number of primordial life forms; and (iii), in the face of religious accounts of a specific act of divine creation, that the world was the product of entirely natural, and indeed material, processes, and that humankind had emerged from these processes in exactly the same way as all other life. Here again, there was nothing fundamentally novel about these positions. Darwin's gradualism drew heavily on the work of 'uniformitarian' geologists of the early and mid-nineteenth century, who had rejected previous 'catastrophist' theories based on violent upheavals, and argued that the evolution of the earth had taken place entirely through the normal

22 DARWINISM'S GENERATIONS

natural processes still visibly operating in the world. Ideas of common descent, including for humankind, were shared by other biological theories which had argued for the 'transmutation' of one species into another, not least those of Lamarck and Spencer and their followers, although Spencer's version of evolution, while subsuming Darwin's, was much broader and more schematic; a general movement from simple to complex, or from homogeneity to heterogeneity, operating across biology, psychology, and sociology in ways which created fundamental ambiguities of meaning in the debates which ensued. In particular, and although this was one of the areas in which Darwin's precise position was itself neither fixed nor unambiguous, Spencer's view of evolution was always much more explicitly directional and progressive than Darwin's branching and implicitly random process. Yet Darwin never decisively articulated these differences, and to insist on them as a requisite marker of Darwinian belief would be to exclude many, perhaps most, of those who would have identified themselves in years after 1859 as Darwinists.

The crucial line of distinction focused on differences of *mechanism*, and it was in this respect that Darwin's theoretical originality primarily lay. His achievement was to construct an account of the natural causes of evolutionary history which seemed amenable to demonstration in ways that Lamarck's and Spencer's did not. The explosive novelty of *On the Origin of Species* was the theory of natural selection—that the cause (or at least the primary cause) of evolution was the process whereby the constant struggle for survival between organisms led, via the greater success of those better adapted to their environment, to their 'selection' at the level of species, populations, and ultimately, as it operated from generation to generation over immense tracts of time, to the emergence of distinct species.[51] As we will see, the precise load that natural selection was required to bear in accounting for evolution remained a matter of debate, not just between Darwin and his followers and those like Spencer, who saw it operating alongside other processes, especially the inheritance of characteristics developed by organisms in the course of adapting to conditions (for example, most famously, the giraffe's longer neck), but also for Darwin and the Darwinians themselves. Even within Darwin's own writings, working alongside natural selection were discussions of the mechanisms of adaptation, variation, and transmutation, in which Darwin conceded some ground to scientific opponents who questioned the capacity of natural selection to serve single-handedly as an explanation of evolution, including ideas of 'sexual selection' as a supplement to natural selection. As we will

INTRODUCTION 23

see, by the 1880s these had created lines of division within Darwinism almost as fundamental, and certainly as fraught, as the differences between the Darwinians and those who refused the label.

In simple terms, opposition could and variously did involve the rejection of any or all of these propositions or a challenge to their status as knowledge: the extent to which Darwin or his supporters were able to provide sufficient evidence to 'prove' the theory scientifically, or whether it remained a 'hypothesis', or even worse, merely a 'speculation'. In the early stages, for many refusal to accept Darwinism was rooted in rejection of the fact of evolution, sometimes associated with commitment to the conventional biblical computation of the age of the earth and usually to the Genesis account of creation. The extent to which this sort of complete denial of evolution persisted is one crucial question for any assessment of the impact of Darwin. But it was only part of the wider conspectus of anti-Darwinian positions.

Even where individuals acknowledged some form of evolutionary history, it was possible for them to subscribe to positions more or less antagonistic to Darwin's account of the nature of evolution. The more hostile countered with belief in 'successive creations', the acceptance of the palaeontological record of a series of periods with distinct flora and fauna, but understood not as a single sequence, but as a set of entirely discrete worlds, produced by successive wholesale extinctions and then the divine creation of a fresh natural order (a version of deep time which allowed various metaphoric reconciliations of the Genesis creation account and natural history). In its less antagonistic version, this position could be combined with the acceptance of a limited degree of evolution within constraints for each species, which allowed for the development of species over time, but not the transmutation of one species into another, and/or, as a third distinction, saltationism, the belief that evolution occurred not slowly and gradually but through sudden jumps and ruptures.

As the evidence for the evolutionary sequence filled out, and Darwin's ideas became the scientific orthodoxy, repudiation, even strident repudiation, of Darwinism was increasingly likely to signal not the rejection of the fact of a single gradual evolutionary succession, but rather refusal to accept the primacy of natural selection as the cause of this, and was likely to involve (to extend the alternatives outlined above) either (iv) a renewal of Lamarckian ideas of the inheritance of characteristics acquired as a result of the efforts of organisms to adapt to their environment; (v) belief that this adaptation was evidence that the natural world was directly designed by divine intervention; or (vi), what we might call divine naturalism, the sense

24 DARWINISM'S GENERATIONS

that God worked indirectly through natural laws (including natural selection), but that evolution was the working out of some progressive divine purpose. A particularly powerful version of the role of direct divine intervention was (vii) 'directed selection',[52] that is to say natural selection except for the more important evolutionary steps, including the creation of humankind, or versions thereof, the most important version of which was the belief that humanity had not evolved (or not simply evolved) from precursor animals, and that at some point divine action had infused a soul or the capacity for reason into humanity and created something qualitatively distinct from the rest of creation. At the margins, antagonism might indicate simply hostility to the materialism of the scientific naturalism of many Darwinians and its anti-religious implications, of the sort that led some sceptics to take refuge behind 'final causes' (a retreat to the argument that natural science could not touch the question of the initial creation of life or the universe).

Was There a Darwinian Revolution?

In recent years 'Darwin Studies' has almost become a field in itself, and in the History of Science the focus of some of the best work of the last thirty years has been on the development of evolutionary science. As a result, we know a great deal about Darwin's own intellectual genealogies, and about the scientific ideas and practices of the group of evolutionary scientists clustered around him, most notably Wallace, Spencer, Joseph Hooker (1817–1911), and of course Huxley.[53] We have a rich body of work which explores the rippling out of Darwinian influences into fields such as philology, anthropology, physiology, and psychology.[54] We also have a detailed picture of the popularisation and dissemination of evolutionary ideas across Victorian culture, and the ways in which they were implicated in what has come to be described as 'social Darwinism'.[55] The more narrowly intellectual focus of James Moore's foundational *The Post-Darwinian Controversies* (1979) has been extended by the recent work of Bernard Lightman, Thomas Dixon, Gowan Dawson, Rob Boddice, Piers J. Hale, and Jonathan Conlin.[56] And the contributions of David Livingstone and others have deepened our understandings of the ways place and local culture could inflect the shape of responses.[57]

In parallel, under the influence of works such as Gillian Beer's *Darwin's Plots* (1983) and George Levine's *Darwin and the Novelists* (1988), an extensive literature has accumulated which examines with extraordinary skill and not a little ingenuity the broad, often indirect, and occasionally entirely unconscious influence of not just Darwinian and evolutionary ideas, but also of Darwin's literary techniques in Victorian culture.[58] This work has done much to illuminate the mutually constitutive exchanges of literary imagination and scientific knowledge. Two recent books by John Holmes, *Darwin's Bards. British and American Poetry in the Age of Evolution* (2009) and *The Preraphaelites and Science* (2018), exemplify the continued richness and ambition of this literature.

There are several peculiar consequences to these approaches. One is the unsatisfactory position in which it leaves one of the central questions raised by Darwin's career, namely the extent to which the years after the publication of the *Origin* in 1859 saw a 'Darwinian revolution'. As Hesketh's *Imagining the Darwinian Revolution* has again demonstrated, this is a question which raises crucial, and still inadequately answered, questions not just for Darwin scholars, but for everyone interested in the cultural history of Victorian Britain. The idea that *On the Origin of Species* effected an intellectual revolution has an impressive pedigree. There is no doubt that contemporaries were quickly convinced of the fact. In 1879 the *Saturday Review* remarked that 'broadly speaking, ten years before the *Origin of Species* was published nobody believed in evolution, and ten years after everybody did'.[59] Three years later the *Times* observed that the previous twenty years had turned a brilliant speculation into an established and unquestionable truth.[60] Twentieth-century studies largely followed suit. Titles such as Gertrude Himmelfarb's *Darwin and the Darwinian Revolution* (1959) and Michael Ruse's *The Darwinian Revolution* (1979) indicate the approach, and studies of Victorian culture abound with passing judgements which imply the cataclysmic impact of Darwinism, its 'demolition' of previous conceptions, its transformation of the intellectual landscape.[61] There was often an overt whiggism at work here, constructing a conflict between Darwinian truth-seekers and a motley group of obstructionists clinging to theories they largely knew themselves to be exploded, an uneven battle in which the forces of knowledge regularly 'trounced' their ignorant opponents.[62]

Another consequence is a tendency to narrow the focus to Darwin and the circle around him, to adopt the perspective of those on the 'right' side of the debate, to suggest that the reception of the *Origin* and *Descent* was an encounter between science and superstition. A third is to encourage a set of essential inferential studies of figures and cultural phenomena for whom the evidence of direct engagement with Darwin's work is patchy at best, which have been able to demonstrate awareness but only infer influence or conviction.[63] It may well be, as Beer has suggested, that Darwin's influence became more significant as evolutionary ideas became embedded assumptions rather than nodes of debate, and it may also be that the advantage of this approach is its ability to read for the unacknowledged, and to identify analogies and figures of speech which would have resonated, even if only subconsciously, with the evolutionary preoccupations of readers. But the search for influence, especially where only implicit, is fraught with conjecture, not least because of the extent to which Darwin himself built on a set of existing ideas, most famously those of Malthus. Too often the search for intellectual genealogies succumbs to the danger of forcing arguments, of conflating analogy and influence, or of failing to recognise the extent to which the beliefs of many contemporaries were shadowed rather than shaped by Darwin. As the strictures of the Victorian critic Andrew Lang (1844–1912) in his study of Tennyson in 1901 show, this is a long-established temptation. But take the discussion of George Meredith's *Modern Love* in Holmes' *Darwin's Bards*, which pushes to and perhaps beyond the limits of what might usefully be described as Darwinian influence, reading it into any sense of the cruelty or purposelessness of nature, or anything biological rather than spiritual.[64] As Levine's work has so effectively demonstrated, characteristics identified as Darwinian, in the sense that they are congruent with a Darwinian view, might just as easily derive from intellectual or figurative resources available before or beyond Darwin.

Inevitably there was eventually a reaction, complicating the simple picture of Darwin's originality, pointing out that opposition to evolutionary science was often formidable and persistent, and that shifts in opinion operated across a broad front rather than a narrow line.[65] But this work never entirely dispelled 'revolutionary' frames, and it is notable that the scholarship of the last decade has shown signs of a return to belief in a rapid and comprehensive Darwinian revolution, not merely on the basis of the far-reaching significance and enduring legacy of Darwin's ideas, but as a profound and relatively abrupt caesura in intellectual attitudes. We are told that Darwinism in Britain triumphed rapidly; in some senses as early as the British Association meeting in 1868, and more broadly by the 1870s Darwinism is often presented as the established orthodoxy.[66] This is especially true of literary and cultural historians. John Holmes in particular has sought to move the balance away from the reaction, arguing that opposition to the principle of evolution was short-lived among Victorian scientists, and that the poets who engaged with evolution in the 1860s 'had in mind a set of ideas that were Darwinian by Darwin's own standards'.[67] The *Oxford Dictionary of Biology* continues to assert that the *Origin* created a furore, but that 'the wealth of evidence presented by Darwin gradually convinced most people', an account which is problematic both in its claim that 'most people' were 'gradually convinced', and that the root of their conversion was 'the wealth of evidence presented by Darwin' (even if we allow this argument to encompass the work of those who 'popularised' and 'disseminated' Darwin's ideas).[68]

The extent to which these approaches have become dominant seems strangely wilful given the extraordinary range of explicit comment that Darwin and evolutionary theory generated, and the extent to which the examination of the ways Darwin's works were read and his ideas encountered, confronted, absorbed, and employed is far from exhausted. There is a danger (it is a crude rendition, but one that crystallises an essential truth) that we are left with a sleight of hand which emphasises the crucial significance of Darwin's identification of a credible natural mechanism of evolutionary change, because it was this that enabled the transformation of contemporary belief in evolution, while simultaneously suggesting that the effort to uncover the precise extent of the acceptance of Darwinian ideas is largely irrelevant (and in some cases even acknowledging that few contemporaries (and even relatively few professed 'Darwinians') actually accepted his theory).[69] In the absence of any evidenced sense that Darwin's analysis of the *causes* of evolution was convincing, it is difficult to demonstrate the central role of Darwinian ideas in effecting a change which had been in process, albeit in the face of strong resistance, for decades before the *Origin*, and gathered pace after the *Origin* in the context of a revolution in knowledge, not just in biology, but across anthropology, archaeology, palaeontology, and philology.

Although in recent years our understanding of the reception of Darwin has been strengthened by several studies of individual responses, not least in the work of Jonathan Smith, Robert W. Smith, and Piers Hale,[70] and by a steady stream of collections of essays on specific texts and individuals, it is notable that overwhelmingly this scholarship has concentrated on the Darwinian side of the debate. Although work continues in exploring the diffusion of Darwinian ideas in Europe and America, for Britain the foundational studies of Ellegard, Hull and of Thomas Glick remain largely unsuperseded.[71] There has been no general effort to take advantage of the opportunities provided by the extensive programmes of digitisation of the

28 DARWINISM'S GENERATIONS

last twenty years to revisit questions of the broad reception of Darwinism, or to mirror Jim Secord's study of the first book to bring evolutionary ideas to a broad readership, Robert Chambers' *Vestiges of the Natural History of Creation* (1844), and try to assess the wider spread of Darwinian beliefs. Historians of science have taken the view that a focus on the question of the conversion of individuals, and the nature and extent of their commitment to Darwinism, oversimplifies the range of positions adopted by supporters and opponents, and falls into the trap of forcing this reception history into crude conflict models. Instead, they have embraced what we might describe as the 'functionalist' approach of much history of science, which sees beliefs (and the purchase of beliefs) in purely purposive terms—in terms of the ways in which specific positions will serve individuals, for example in terms of their professional or institutional development, rather than as having a dynamic of their own. In his recent study of Hooker, Jim Endersby went as far as to argue explicitly that the question of conversion, the if, when, and to what extent Victorians accepted Darwin's ideas, is 'the wrong question'. Rather, Endersby suggested, we should be asking what made Darwinism and its component arguments useful and how it was used.[72]

The discussion which follows proceeds from a belief that the older approaches are not without continued potential, especially in respect of what they might reveal about generational dynamics. We cannot ignore, however arid and ultimately tedious much of the debate came to be, just how far the Victorians' engagements with Darwin involved the calibration of their commitment. Of course, contemporary alignments were complicated, and of course, once Darwinian evolution became hegemonic within scientific circles, then opposition is increasingly just a matter of noises off. But for most, and especially beyond the narrow reaches of scientific practice, what mattered first and foremost was whether Darwin was right, how far his arguments were scientifically credible, to what extent he was able to prove his propositions, the extent to which it was legitimate to apply his insights. On none of these questions was there ever anything approaching a general consensus, and for much of the period the powerful emotional freight they carried made it inevitable that discussions were conducted within a broadly conflictual frame, even if ultimately the front line was not between those who agreed with Darwin and those who disagreed, but between those who judged their disagreements as insignificant in comparison with their agreements, and those who felt the need to emphasise differences. Debates over Darwin did not always take place within a rhetorical battlefield, but they were rarely entirely distinct from the campaigns of

INTRODUCTION   29

hostile factions. Just as the history of the first half-century begins with what one Darwinian described as the 'battle of truth against the torrent of ignorance and abuse', so it concludes in the context of the almost equally venomous 'war between the tribe of the Mendelians and the tribe of the Darwinians' which marked the Edwardian years.[73] To embrace the almost inevitable pun, the Darwinian debates were as much guerrilla as gorilla warfare.

Accepting conflict modes does not of itself involve denying the existence of various middle grounds, and even if this is only part of a much richer story, it is the one which brings generational differences into the sharpest focus.[74] Ultimately, such an approach is justified by the enduring centrality of judgements about the emotional and intellectual response to Darwin's works, the pace and extent of the acceptance of his ideas, and the degree to which this involved the *conversion* of contemporaries to some sort of Darwinian position, or alternatively, the eclipse of opponents and their replacement by supporters.

Holmes' careful language and several of the essays in Hesketh's collection alert us to the treacherous semantics of these debates. It may well be true, as Thomas Dixon has argued, that the Darwinian theory of evolution 'spread rapidly through Victorian culture'.[75] But this is not the same as suggesting a rapid subscription to Darwinian science, although it certainly leaves open (and perhaps even encourages) the drawing of that conclusion by a hasty reading, and it fuels an impression that in the years after 1859 there was a rapid process of conversion of individuals to belief in Darwinian evolution, in Piers Hale's assertion that 'with only a very few exceptions those who embraced evolution in Britain in the years after Darwin's *Origin of Species* was published, identified themselves as "Darwinians"', or that as David Oldroyd put it, that 'better-educated readers...came to accept the theory fairly quickly', or as Virginia Richter has recently suggested, we can think in terms of 'Darwin's cataclysm' because readers 'were convinced or simply overwhelmed by the sheer mass of facts'.[76]

The problem is that we just don't know: the fundamental question set up by this historiographical debate—just how quickly, broadly, and in their key components, Darwinian ideas were *accepted* in Britain in the years after the publication of *On the Origin of Species*—is still an open question. As it stands, the argument presented, for example by John Holmes, in which figures like E.R. Lankester (1847–1929), Raphael Meldola (1849–1915), and Edward Bagnall Poulton (1856–1943) are paraded to provide evidence of a Darwinian 'solid majority' in late-Victorian Britain, gives nowhere near

30 DARWINISM'S GENERATIONS

enough attention to the position within Darwinian debates that these figures occupied, to their typicality, never mind to the question of their age.
The irony is that Holmes is commendably sensitive to the generational
dynamics of the debates, and in many ways the arguments presented in the
rest of this book build on his work. But the judgements he seeks to make are
amenable to demonstration only by the sort of prosopographical work
which scholars have hitherto been unable or unwilling to undertake.

Methodology

In seeking to explore responses to Darwin through the frame of generations, this study unapologetically takes up this prosopographical challenge. Without attempting to reduce individual responses to tabulated
categories, it seeks to interrogate how and when individuals from across
the culture encountered, engaged with, accepted and employed Darwinian
ideas. In particular it asks how far these processes were shaped by considerations of age, and thereby of generation. It seeks to take up a number of
approaches already well-established in book history and the histories of
reading and reception, not least attention to the materiality of the book and
the ways in which the bound volume operated physically as well metonym-
ically as symbolic of the words inside, the specificities of reading acts and
their autobiographical, textual, and institutional contexts, but also to those
who not only know about, but debate Darwin on various forms of scanty
or even entirely indirect evidence.[77] I have very deliberately avoided the
conventional approach of selecting a small number of exemplary cases.
The leading figures are prominent often because of their idiosyncrasy not
their typicality; no account of generational effects can be wrought from a
Huxley or a Hardy, although in interesting ways, as we will see, they too
exhibited aspects of their generational identities. Exceptions abound: they
neither prove nor disprove the rule. The intention here has been to start
from the other end, to explore those individuals for whom evidence of their
explicit engagements survive, but at a scale which allows for age and generational patterns to be made visible. This might initially be seen as 'distant
reading', but because of its intent to treat each individual as a specific case
as well as representative of a larger collective, it is perhaps more appropriately
described as 'extensive reading'.

Interrogation of age effects imposes a coarse filter on the materials which
can be included. Unattributed interventions, or those attributable to

INTRODUCTION 31

obscure figures whose birthdates cannot be ascertained with a reasonably high degree of confidence, have largely been excluded because they cannot illuminate generationally.[78] Even so, use of a range of printed and online biographical databases, as well as the digital archives, has enabled the identification of over 1,500 different individuals whose views on Darwinian evolution in the fifty years between 1859 and the 1909 commemorations of *Origin* and Darwin's birth can be ascertained (in excess of 200 individuals per generation, ten to twelve individuals per year; see Appendix 1). Although this is a 'haphazard' rather than random sample, a largely successful effort has been made to keep a rough evenness of spread across the period, and to ensure some diversity of social background and gender. Inevitably, the sample is skewed towards scientists and the professionally religious, and university-educated males overwhelmingly preponderate, but there is a significant representation of women and of the lower middle and working classes who were not university educated (more so as the period progresses). It is worth emphasising that the bias that remains is less an indication of the balance of interest in evolution and more a reflection of the overwhelming dominance of these voices in the surviving record and differential barriers to obtaining birthdates.

No attempt at statistical aggregation has been made. This would efface the gradations of response to which the discussion needs to attend, and would suggest a fixity of position which is rarely borne out by the evidence. Instead (accepting that they are rarely given the attention they deserve as individuals, and are interrogated primarily as spokesmen and women for their wider generation), a deliberate attempt has been made to ensure each individual is identifiable. Where possible, this involves reference in the main text, though the priority has been to maintain its focus on the argument supported by specific illustrative examples. The endnotes significantly expand the evidentiary base, especially the cumulation of instances which underpin the argument.

The material on which the argument is founded has been gathered gradually. A deliberate attempt has been made to go beyond the Darwinian inner circle and their notorious opponents. To the victors, the sources: or at least the most accessible printed texts and the most-used archival collections,[79] and this has encouraged a history of Darwinism refracted primarily through the prism of the winners.[80] The correspondents of the leading Darwinists were not entirely intimidated, but these collections offer dissent moderated by deference, not the frank forthrightness of opposition shared. The markedly different tone of the botanist George Dickie (1812–82), in

32 DARWINISM'S GENERATIONS

writing to Hooker and to the much more sceptical John Hutton Balfour (1808–84), Professor of Botany at Edinburgh, the contrast between William Boyd Dawkins' deferential (even fawning) tone in letters to Darwin and the robustness of the critique of Darwin's works in his anonymously published reviews, reminds us that while private letters offer opinion less trammelled by publicity, they were not without their own constraints.[81] And a full account of the history of Darwinism also requires attention to the waverers and the bystanders, to the appropriators and the subverters: figures like the Oxford anatomist George Rolleston (1829–81), so useful for the conventional accounts of the marginalisation of Richard Owen (1804–92) in the early 1860s, but thereafter allowed to fall from sight, just as his position becomes more complex and interesting, or indeed like Patrick Geddes, left to the historians of town planning, while his place in late nineteenth-century transitions in evolutionary thinking remains largely unexplored. In consequence, in addition to significant use of the searchable corpus of digitised nineteenth-century newspapers and periodicals, and libraries of contemporary publications provided by the Internet Archive, the Hathi Trust, and the Biodiversity Heritage Library, the discussion is supported by an extensive programme of archival research in the papers of nineteenth-century scientists, intellectuals and men and women of letters, not just the substantial collections of the Darwinists, but also the papers of sceptics and opponents, as well as by sampling the popular science magazines, the proceedings of local natural history societies, and the provincial press. Although this is not in any specialised sense a digital humanities project, it has been made feasible by the transformation of the digital landscape over the past decade, and so contributes in a small way to the extension of approaches in the field.

One final caveat. The focus here is exclusively on British opinion. This is not to underestimate the increasingly transnational nature of the cultures of science, both popular and professional, in the second half of the nineteenth century, and the very significant role that German and then American writers in particular had on the debate in Britain, nor to preclude the possibility of transnational generational parallels. But as historians of science recognise, even within narrow disciplinary arenas, histories of ideas and knowledge are powerfully shaped by personal and institutional forces which tend to be nationally distinct. All the more so, the broader generational dynamics whose presumed operation underpins the cohorts being explored here. In this light it should not be surprising if the debates outside Britain do not work in entirely synchronous ways, and that the additional complexity that a transnational discussion would entail remains unexplored.[82]

Conclusion

The following six chapters can be conceived as offering a series of increasingly 'hard' versions of my argument. At the least demanding level, they reinforce the views of those who have argued that there is no 'Darwinian revolution' in Britain in the decades after 1859, suggesting that the progress of acceptance of evolutionary theories was slower and more sectional than is often argued and even more frequently implied. They enforce the importance of age effects in understanding the nature of responses to the *Origin* and the evolutionary debates which followed, and indeed argue that these age effects were at least as significant as ideology, religion, geography, or class in shaping Darwinian cultures. But the argument goes further, vindicating the significance of not just age but of generational alignments, and the necessity of conceptualising historical change, at least in broad intellectual and cultural terms, generationally. And in doing so it provides support for the specific generational schema outlined above. First in that it suggests that the birth years around 1830 composed a powerful watershed in the readiness of individuals to adopt Darwinian beliefs and identities: the overwhelming majority of those born after this date adopting, adapting or accommodating some version of Darwinian evolution and thinking of themselves as in alignment with Darwinian ideas, whereas the vast bulk of those born before 1830 resisted Darwinism, perhaps reluctantly accommodating some of its elements, but generally considering themselves as dissenting from the Darwinian point of view. Moreover, the evidence indicates that the 1830 dividing line is just one of a series; that the age effects visible align not universally, but with remarkably strong correlations, with a multigenerational schema of roughly fifteen-year-wide cohorts, which structured a set of intergenerational exchanges that contoured the public debate. As a result, although there was a fundamental shift in the balance of attitudes in the thirty or so years after the appearance of the *Origin*, this was the result of conversion only in the sense that those born after 1830 often came to support Darwinism by a process of abandoning their previous ideas relatively early in adult life, and those born in the decade and a half before 1830 generally come to acknowledge some sort of evolutionary past, though not usually a 'Darwinian' one. For the rest, contemporaries might have adjusted,

34 DARWINISM'S GENERATIONS

turned away, temporised, but they did not throw off their initial views: the emotional and intellectual effort required was too great; rather they aged and died and were succeeded by others with different opinions. To the extent that there was a Darwinian cultural revolution, it was achieved not by individual conversion but by generational substitution.

Notes

1. Diary of Edward Wrench, 12 February 1909, Wr./D/54, University of Nottingham.
2. For an account of Wrench's career, see *Derby Daily Telegraph*, 13 March 1912; for his shifting attitudes to evolution, see scattered entries and insertions in his diaries, for example 29 January, 19 February 1861, 1 September 1863, 30 October 1872, pasted in matter in the diary for 1901, and notes on Huxley, Wrench Papers, Wr./D/46/13 and/87, University of Nottingham.
3. Charles Darwin, *On the Origin of Species* (1859), 482; Darwin to Hooker, 3 March [1860], DCP-LETT-2719, Darwin Correspondence.
4. Huxley to Busk, 17 July 1860, ff.14–15, Avebury Papers, Add. MS 49639, BL. For Huxley in 1880, 'in another twenty years, the new generation, educated under the influences of the present day, will be in danger of accepting the main doctrines of the *"Origin of Species"* with as little reflection, and it may be with as little justification, as so many of our contemporaries twenty years ago rejected them', Thomas H. Huxley, 'The Coming of Age of *The Origin of Species*', (1880), Huxley, *Darwiniana* (1896), 229.
5. Most systematically in James Moore, '1859 and All That: Remaking the Story of Evolution and Religion', in R.G. Chapman and C.T. Duval, eds., *Charles Darwin, 1809–1882: A Centennial Commemorative* (1982), 167–94; see also Melanie Baldwin, *Making Nature: The History of a Scientific Journal* (2015). This said, there has been no systematic application of generational approaches of the sort provided by Henrika Kuklick's *The Savage Within: The Social History of British Anthropology, 1885–1945* (1991).
6. John Holmes, *The Preraphaelites and Science* (2018), 239. Holmes, *Darwin's Bards: British and American Poetry in the Age of Evolution* (2009) distinguishes between the generation of Darwin and 'the next generation, born from the late 1820s to the 1840s', and so offers a discussion which in some respects aligns with the description of what I describe as the 'high Victorians' here, see 22; Marsha L. Richmond, 'The 1909 Darwin Celebration: Re-Examining Evolution in the Light of Mendel, Mutation, and Meiosis', *Isis* 97 (2006), 447–84, especially 472–3.
7. For example, Lucien Febvre, 'Generations', in *Bulletin du centre international de synthese. Section de synthese historique, no.7, Revue de synthese historique* 47 (1929), Herbert Butterfield, *The Discontinuities between the Generations in History: Their Effect on the Transmission of Political Experience* (1971), P. Nora, 'Generations', in P. Nora, *Realms of Memory: Rethinking the French Past* (1996), 499–531. Reinhart Koselleck, *Futures Past: On the Semantics of Historical Time*, trans. Keith Tribe (1985), and Koselleck, *Zeitschichten* (2003), Norbert Elias, *Studies on the Germans* (2013, or. 1989).
8. Karl Mannheim, 'The Problem of Generations' (1928), in Paul Kecskemeti, ed., *Karl Mannheim: Essays* (1972, or. 1952), 276–322; for arguments for its continued valency see Judith Burnett, *Generations: The Time Machine in Theory and Practice* (2010).
9. Robert Wohl, *The Generation of 1914* (1979). Note how Wohl's argument is taken up and used as an operating assumption in, for example, Jason Scott Smith, 'The Strange History of the Decade: Modernity, Nostalgia, and the Perils of Periodization', *Journal of Social History* 32.2 (1998), 263–85, as one example among very many.
10. Samuel Hynes, *A War Imagined* (1991), 331.

INTRODUCTION 35

11. Ute Daniel, *Kompendium Kulturgeschichte: Theorien, Praxis, Schlüsselwörter* (2001), 331, translation and citation, A. Erll, 'Generation in Literary History: Three Constellations of Generationality, Genealogy and Memory', *New Literary History* 45.3 (2014), 387.

12. This is equally true of works such as Norbert Elias' *The Germans: Power Struggles and the Development of Habitus in the 19th and 20th Centuries* (1996), or Randall Collins' *The Sociology of Philosophies: A Global Theory of Intellectual Change* (1998).

13. See Julián Marías, *Generations: A Historical Method* (1970, or. 1967).

14. Mannheim, 'Problem of Generations', 301.

15. Lewis Feuer, *The Conflict of Generations: The Character and Significance of Student Movements* (1969). For a later example see D. Wyatt, *Out of the Sixties: Storytelling and the Vietnam Generation* (1993).

16. S.N. Eisenstadt, *From Generation to Generation: Age Groups and Social Structure* (1956).

17. See the discussion in Marianne Hirsch's *The Generation of Postmemory* (2012).

18. F. Braudel, *The Mediterranean and the Mediterranean World in the Age of Philip II*, trans. S. Reynolds (1972), I, 18. Spitzer rightly pointed out that despite a long tradition of hostility to the construction of boundaries in what in purely biological terms is a single flow of births and deaths, 'the problem of establishing generational boundaries should inhibit us no more than does the problem of marking off the categories in any continuum—such as class, ideology, or political movement—where there is a shading or ambiguity at the boundaries', Alan Spitzer, *The French Generation of 1820* (1987), 7. A recent article, Oksana S. Karashchuk et al., 'The Method for Determining Time-Generation Range', SAGE Open, Oct. 2020, doi:10.1177/2158244020968082, simply reads off generational cohorts from phases in birth rate and GDP data.

19. For some recent approaches, see the 'Born in 1819' Roundtable, *Journal of Victorian Culture* 24.4 (2109), 415–58.

20. See Royden Harrison, *Before the Socialists* (1965), Michael Bentley, *Modernizing England's Past: English Historiography in the Age of Modernism* (2005).

21. Sheldon Rothblatt, *The Revolution of the Dons: Cambridge and Society in Victorian England* (1968), Christopher Harvie, *The Lights of Liberalism: University Liberals and the Challenge of Democracy, 1860–1886* (1976); the approach is used frequently in Collini et al., *That Noble Science of Politics: A Study in Nineteenth Century Intellectual History* (1983).

22. D. Epstein Nord, 'Victorian Autobiography: Fathers and Sons', in Maria DiBattista and Emily Whitman, eds., *The Cambridge Companion to Autobiography* (2014), 87–101, 91; widely noted but not really developed, as in Julie-Marie Strange, *Fatherhood and the British Working Class, 1865–1914* (2014), Trev L. Broughton, *Men of Letters, Writing Lives: Masculinity and Literary Auto/biography in the Late Victorian Period* (1999).

23. *The Journal of Arnold Bennett* (1932), 53, quoted by Ruth Robbins, *Pater to Forster, 1873–1924* (2003), 67; or Sir William Rothenstein's 'Portrait of My Father and Mother', exhibited at a portrait exhibition in London in 1900/1, *The Outlook* 6.154 (12 January 1901), 764.

24. This is visible throughout both his more coherent volumes, *The English Constitution* (1867) and *Physics and Politics* (1872), but also in the miscellaneous collections of *Literary Studies* (1879), *Biographical Studies* (1881), and *Economic Studies* (1880). Bagehot was no doubt drawing on Mill's discussions of generations in his *System of Logic* (1843). See L. Dowling, *Charles Eliot Norton: The Art of Reform in Nineteenth Century America* (2007), 85–6, Walter Bagehot, 'Malthus', *Economic Studies*, 149.

25. Walter Bagehot, 'Bishop Butler', *Literary Studies*, III, 129; Bagehot, 'Lord Althorp and the Reform Act of 1832', *FR* ns 20 (November 1876), 574–600.

26. E. Gosse, 'Ten Years of English Literature', *North American Review* 165 (August 1897), 139–48, 140. Gosse is another interesting case of Victorian generational sensibilities; see also his remark to Vernon Lee, 'Passion Sunday', 1906, remarking that despite the difference of their ages, 'we belong to the same generation. We belong to a generation which—to be blunt—has passed away', quoted in V. Colby, *Vernon Lee: A Literary Biography* (2003), 270.

36 DARWINISM'S GENERATIONS

27. Marías, *Generations*; for Ortega y Gasset see the discussion of his ideas on generations in the *Stanford Encyclopedia of Philosophy*, https://plato.stanford.edu/entries/gasset/#Conc GeneTempHistReasCritPhilHist, Andrew Dobson, *An Introduction to the Politics and Philosophy of José Ortega y Gasset* (2009).
28. Quoted in James Mavor, *My Windows on the Street of the World* (1923), 6–7.
29. Quoted in A.C. Benson, *Life of Edward White Benson* (1900), I, 249–50.
30. See P. Geddes to Mavor, 6 January 1903, MS119, Mavor Papers, Thomas Fisher Library, University of Toronto; see also the discussion in Marías, *Generations*, 28–33.
31. In this respect, it might make sense to represent generations by their mid-point or centre of gravity, as Marías does, Marías, *Generations*, 186–7. His dates (rendered as middle dates, 1811, 26, 41, 56, 71, 86, 1901) align surprisingly well with the schema for the Victorian period outlined below.
32. Wohl, *Generation of 1914*, 210.
33. Walter Bagehot, 'The Use of Conflict', *Physics and Politics* (1872), 54.
34. Take the case of one W.J. Barber, who wrote to George G. Stokes in the months after Darwin's death in 1882 to seek his assistance in understanding Darwinism and the ways it could be reconciled with his religious beliefs; although clearly Barber had struggled to get access to Darwin's books, and his knowledge relied to a considerable extent on second-hand accounts and debates, he told Stokes, 'as the atheistic party are today using the theory of Darwin as a great lever to endeavour to uproot divine revelation it behoves every man however humble his position to endeavour to understand the subject for themselves…', W.J. Barber (1841–?) to G.G. Stokes, 13, 16 October 1882, MS Add. 7656/1B/113–114, Stokes Papers, CUL.
35. William Pengelly to Lyell, 27 March 1862, Lyell Papers, COLL 203/1/4692–24, Edinburgh; W.E. Houghton, *The Victorian Frame of Mind* (1957), 38.
36. *Huddersfield Daily Chronicle*, 6 September 1892.
37. William Penman Lyon (1812–77)'s *Homo versus Darwin: A Judicial Examination of Statements Recently Published by Mr Darwin Regarding 'The Descent of Man'*, described in *The Freethinker*, 23 November 1890, and see his letters in *Eastern Daily Press*, 23 May, 18 November 1872; *Reynolds Newspaper*, 14 April 1912.
38. *Otago Witness*, 18 January 1905, an extract of an interview by Blathwayt of Scott.
39. Sylvanus Thompson, *The Life of William Thompson, Baron Kelvin of Largs* (1910), II, 1091–2.
40. Margaret Shaen, *Memorials of Two Sisters: Susanna and Catherine Winkworth* (1908).
41. Ethel Chamberlain to Neville Chamberlain, 27 September 1891, NC1/14/33, Neville Chamberlain Papers, CLB.
42. G. Richards, *Memories of a Misspent Youth* (1932), 47.
43. *History of Bootham School York* (1926); cited in T.C. Kennedy, *British Quakerism, 1860–1920* (2001), 93; Oscar Browning's anonymised survey of boys published in the *Chemical News*, 22 May 1868, included reference to a 14-year-old, very much interested in fossils and the works of Lyell, who 'read Darwin's Cruise of the "Beagle" twice, and the Origin of Species with avidity and intelligence'; see Oscar Browning, *Memories of Sixty Years at Eton, Cambridge and Elsewhere* (1910), 127.
44. Mrs Adrian Duff Grant, *The Life and Work of Lord Avebury* (1924), 26. See also H.E. Roscoe to J.D. Hooker, 21 September 1906, JDH/1/2/18/161, Hooker Papers, RBGK.
45. Keith Feiling, *Life of Neville Chamberlain* (1946), 234.
46. Adrian Desmond, *Huxley: Evolution's High Priest* (1997), 139.
47. James Moore, 'Deconstructing Darwinism: The Politics of Evolution in the 1860s', *Journal of the History of Biology* 24.3 (1991), 353–408.
48. H. Salmon to Geddes, 30 September 1884, ff.199–199a, MS10523, Geddes Papers, NLS.
49. Müller to Prof. Noire, 8 February 1878, G.A.G. Müller, *Life and Letters of Friedrich Max Müller* (1902), II, 42.
50. A. Conan Doyle, *Through the Magic Door* (1907), 242. Similarly, when Richard Garnett (1835–1906), Keeper of Printed Books at the British Museum, was asked in the mid-1890s which two or three books they would like the public to know more about, he named Darwin's *Voyage of the Beagle*, *Westminster Budget*, 1 March 1895.

51. For discussions of the variations of Darwinian theory see E. Mayr, 'Darwin's Five Theories of Evolution', in D. Kohn, *The Darwin Heritage*, 755–72, and D. Hull, 'Darwinism as a Historical Entity', ibid., 773–812.

52. The phrase is A. Ellegard's, *Darwin and the General Reader* (1958), 32.

53. Frank M. Turner, *Between Science and Religion: The Reaction to Scientific Naturalism in Late Victorian England* (1974), Angelique Richardson, *After Darwin: Animals, Emotions and the Mind* (2013), J. David Pleins, *In Praise of Darwin: George Romanes and the Evolution of a Darwinian Believer* (2014), Matthew Stanley, *Huxley's Church and Maxwell's Demon: From Theistic Science to Naturalistic Science* (2015). The potential narrowness of this work is exemplified by Jean Gayon's *Darwinism's Struggle for Survival* (1998), which while offering a conspectus of the development of evolutionary science across the late nineteenth and twentieth century, does so within a compass so narrow that the leading late nineteenth-century British Darwinian Edward B. Poulton gets only one substantive mention, and even Huxley merits only a single five-line discussion.

54. G. Radick, *The Simian Tongue: The Long Debate about Animal Language* (2007), R.J. Richards, *Darwin and the Emergence of Evolutionary Theories of the Mind* (2014).

55. Bernard Lightman, *Victorian Popularizers of Science: Designing Nature for New Audiences* (2007), Bernard Lightman and Bennett Zon, eds., *Evolution and Victorian Culture* (2014), and also Jane Goodall, *Performance and Evolution in the Age of Darwin: Out of the Natural Order* (2002), Barbara Larson and Fae Brauer, eds., *The Art of Evolution: Darwin, Darwinisms, and Visual Culture* (2009). For social Darwinism see Greta Jones, *Social Darwinism and English Thought* (1980), David Crook, *Darwin's Coat-tails: Essays on Social Darwinism* (2007), Crook, *Darwinism, War and History* (1994), Mike Hawkins, *Social Darwinism in European and American Thought, 1860–1945* (1997).

56. James R. Moore, *The Post-Darwinian Controversies: A Study of the Protestant Struggle to Come to Terms with Darwin in Great Britain and America, 1870–1900* (1979), M. Fichman, *Evolutionary Theory and Victorian Culture* (2002), Gowan Dawson, *Darwinism, Literature and Victorian Respectability* (2007), Thomas Dixon, *The Invention of Altruism: Making Moral Meanings in Victorian Britain* (2008), Bob Boddice, *The Science of Sympathy: Morality, Evolution, and Victorian Civilization* (2016), Jonathan Conlin, *Evolution and the Victorians: Science, Culture and Politics in Darwin's Britain* (2014), P.J. Hale, *Political Descent: Malthus, Mutualism and the Politics of Evolution in Victorian England* (2014).

57. David N. Livingstone, *Dealing with Darwin: Place, Politics and Rhetoric in Religious Engagements with Evolution* (2014), Ronald L. Numbers and John Stenhouse, *Disseminating Darwinism: The Role of Place, Race, Religion and Gender* (1999).

58. Gillian Beer, *Darwin's Plots: Evolutionary Narrative in Darwin, George Eliot and Nineteenth Century Fiction* (1983, 3rd ed., 2009), Beer, *Open Fields: Science in Cultural Encounter* (1996), George Levine, *Darwin and the Novelists* (1988), Levine, *Darwin the Writer* (2011). For their influence see amongst many David Amigoni, *Colonies, Cults and Evolution: Literature, Science and Culture in Nineteenth Century Writing* (2007), Joseph Carroll, *Evolution and Literary Theory* (1995), John Glendening, *The Evolutionary Imagination in Late Victorian Novels: An Entangled Bank* (2016), Goodall, *Performance and Evolution*, Sally Shuttleworth, *The Mind of the Child: Child Development in Literature, Science and Medicine, 1840–1910* (2010), Jessica Straley, *Evolution and Imagination in Victorian Children's Literature* (2016).

59. [Frederick Pollock], *Saturday Review*, 31 May 1879, quoted in Ian Hesketh, 'Imagining the Darwinian Revolution in the Nineteenth Century', in Hesketh, *Imagining the Darwinian Revolution: Historical Narratives of Evolution from the Nineteenth Century to the Present* (2022), 21–36.

60. *The Times*, 21 April 1882.

61. T.S. Kuhn, *Structure of Scientific Revolutions: Fiftieth Anniversary Edition* (2012), 151; for discussions of the Darwinian revolution from a Kuhnian point of view, see E. Mayr, 'The Nature of the Darwinian Revolution', in Mayr, *Evolution and the Diversity of Life* (1997), 277–96, and John C. Greene, 'The Kuhnian Paradigm and the Darwinian Revolution in Natural History', in his *Science, Ideology and World View: Essays in the History of*

38 DARWINISM'S GENERATIONS

Evolutionary Ideas (1981), 30–59; M. Ruse, 'The Darwinian Revolution: Rethinking its Meaning and Significance', *Proceedings of the National Academy of Sciences* 106 (2009), 10040–7; also D.R. Oldroyd, *Darwinian Impacts: An Introduction to the Darwinian Revolution* (1980). See also the conclusions of those like Jonathan Hodge who argue that 'The book's two big ideas, the tree of life and natural selection, were both radically novel and together constituted a fundamental and disturbing challenge to most readers within and beyond the scientific community', M.J.S. Hodge, 'Darwin's Book: On the Origin of Species', *Science and Education* (2012), 2279.

62. Hence the doubly problematic judgement of R.A. Slotten, *The Heretic in Darwin's Court: The Life of Alfred Russel Wallace* (2006), 260, that Wallace's 'Creation by Law' article in *Quarterly Journal of Science* (April 1867) 'trounced the outmoded views' of Argyll's *Reign of Law*.

63. For example, Jude Nixon's study of Hopkins, *Gerard Manley Hopkins and His Contemporaries: Liddon, Newman, Darwin and Pater* (1994).

64. Holmes, *Darwin's Bards*, 202–10.

65. Dawson, *Darwinism, Literature and Victorian Respectability*, 8; see M.J.S. Hodge, 'Against "Revolution" and "Evolution"', *Journal of the History of Biology* 38 (2005), 101–24, Ruse, 'The Darwinian Revolution'; Peter Bowler, in particular, especially his *The Non-Darwinian Revolution: Reinterpreting a Historical Myth* (1992), and *Reconciling Science and Religion: The Debate in Early-Twentieth Century Britain* (1996); Nicholas Jardine, 'Writing off the Scientific Revolution', *Journal of the History of Astronomy* 22 (1991), 311–18.

66. Ian Hesketh, *Of Apes and Ancestors: Evolution, Christianity and the Oxford Debate* (2009); Richard Bellon, 'Inspiration in the Harness of Daily Labour: Darwin, Botany and the Triumph of Evolution, 1859–68', *Isis* 102 (2011), 393–420, 417.

67. Holmes, *Darwin's Bards*, 8–9; Holmes, *Pre-Raphaelites*, 164. Joseph Carroll has relatively recently suggested that within a decade the *Origin* had 'almost completely changed the general view of evolution in the minds of the educated public'. To give one other recent example, see the suggestion of Oliver Hochadel, 'Darwin in the Monkey Cage: The Zoological Garden as a Medium for Evolutionary Theory', in Dorothee Brantz, ed., *Beastly Natures: Animals, Humans and the Study of History* (2010), 82, that after the *Origin*, 'the perception of apes changed abruptly and fundamentally'.

68. 'Evolution', *Oxford Dictionary of Biology* (2019 edition).

69. As Michael Ruse has observed, 'people became evolutionists in droves, but that number of "selectionists"' [i.e. Darwinians] was 'very few'. Ruse, 'Darwinian Revolution', 10042.

70. Jonathan Smith, 'Alfred Newton: The Scientific Naturalist Who Wasn't', in B. Lightman and M. Reidy, eds., *The Age of Scientific Naturalism* (2015), 137–56; see R.W. Smith, 'The "Great Plan of the Visible Universe": William Huggins, Evolutionary Naturalism and the Nature of the Nebulae', ibid., 113–35, Piers J. Hale, 'Darwin's Other Bulldog: Charles Kingsley and the Popularisation of Evolution in Victorian England', *Science and Education* 21 (2012), 977–1013.

71. Ellegard, *Reception*; also D. Hull, *The Reception of Darwin's Theory of Evolution by the Scientific Community* (1973), T.F. Glick, ed., *The Comparative Reception of Darwinism* (1972), Kohn, *The Darwinian Heritage*. The exceptions are generally fragmented essay collections, including Eve-Marie Engels and Thomas F. Glick, eds., *The Reception of Charles Darwin in Europe* (2008), and Thomas Glick and Elinor Shaffer, *The Literary and Cultural Reception of Charles Darwin in Europe* (2014), Eckart Voigts, Barbara Schaff, and Monika Pietrzak-Franger, eds., *Reflecting on Darwin* (2014).

72. J. Endersby, *Imperial Nature: Joseph Hooker and the Practices of Victorian Science* (2009), 320–6.

73. The quotes are from John Lubbock, *Essays and Addresses, 1900–1903* (1904), 8, and D'Arcy Wentworth Thompson, from his review of Poulton's *Essays on Evolution* (1908) in *Mind* 17 (1908), 571.

74. For one recent example of this productivity, see the essays in B. Lightman, ed., *Rethinking History, Science and Religion* (2019).

75. Dixon, *Altruism*, 151.

INTRODUCTION   39

76. Hale, 'The Politics of the Darwinian Revolution', 105, Oldroyd, *Darwinian Impacts*, 198. Virginia Richter, *Literature after Darwin: Human Beasts in Western Fiction, 1859–1939* (2011), 18, 31. Richter is a good example of the sorts of misleading impression of an almost automatic transformation that careless condensation of style can produce, as in her suggestion that 'After Darwin [she explicitly notes this means after the *Origin*], the human being was just an animal like any other', *Literature after Darwin*, 3.

77. See Mark Towsey's *Reading History in Britain and America, c.1750–c.1840* (2019), and the idea of life-cycle reading proposed by Leslie Howsam, *Past into Print: The Publishing of History in Britain, 1850–1950* (2009).

78. Studies like Angelique Richardson's '"The Book of the Season": The Conception and Reception of Darwin's *Expression*', in Richardson, *After Darwin: Emotions and the Mind* (2013), 51–88, or Thomas Yorty, 'The English Methodist Response to Darwinism Reconsidered', *Methodist History* 32.2 (January 1994), 116–25, use virtually entirely unsigned newspaper or magazine reviews, and so are completely blind to age or generational effects.

79. Ruth Barton is one who has noted that the history of evolutionary thought in Britain is dominated by the victors' versions. For one example of a richly researched study which nevertheless relies on a highly skewed archival research base, see Hale's 'Darwin's Other Bulldog'.

80. There is some sign of greater attention to the non-Darwinians, as in Mark McCartney, Andrew Whitaker, and Alastair Wood, eds., *George Gabriel Stokes: Life, Science and Faith* (2019).

81. For Dawkins, see Henry-James Meiring, 'Scientific Patronage in the Age of Darwin: The Curious Case of William Boyd Dawkins', *Studies in the History and Philosophy of Science* 89 (2021), 267–82.

82. See Bowler, *Non-Darwinian Revolution*, 124–5 and *passim*. See also G.L. Geison, *Michael Foster and the Cambridge School of Physiology: The Scientific Enterprise in Late Victorian Society* (1978), especially 335–7.

The Reception of Darwinian Evolution in Britain, 1859–1909: Darwinism's Generations. Martin Hewitt, Oxford University Press. © Martin Hewitt 2024. DOI: 10.1093/9780191982941.003.0001

1

The Publication of *On the Origin of Species*

Although I am fully convinced of the truth of the views given in this volume…, I by no means expect to convince experienced naturalists whose minds are stocked with a multitude of facts all viewed during a long course of years from a point of view directly opposed to mine…[B]ut I look with confidence to the future—to young and rising naturalists, who will be able to view both sides of the question with impartiality. (Charles Darwin, *On the Origin of Species* (1859), 482.)

On the Origin of Species by Means of Natural Selection, or the Preservation of Favoured Races in the Struggle for Life was published in a single volume by the publishing house of John Murray on 24 November 1859. In appearance it was an unremarkable book, with no illustrations, no tricks of typography, no flashy gilt lettering. It was not cheap, costing 14 shillings, and if initial sales were encouraging, they were dwarfed by the 7,000 copies requested of the simultaneously published *Narrative of the Discovery of the Fate of Sir John Franklin*. Despite its 502 pages, Darwin intended the *Origin* merely as a preliminary statement of an argument about the development of the natural world, which he expected to extend to three large volumes. By the end of the year, the book had already been superseded by the publication of a second revised edition. Yet this prefatory and provisional manifesto was destined to become probably the most influential book of the Victorian era.

As Gillian Beer has demonstrated, Darwin, saturated in the narrative techniques of Victorian literature, had laboured to craft a book that would persuade. *On the Origin of Species* presented 'one long argument' which sought to overthrow the deeply rooted scientific and religious orthodoxy of the permanency of species and their origin in divine 'special creation', on the basis of indirect evidence that was at best fragmentary, a task which required a complicated rhetorical effort of logic-based demonstration and persuasion, challenging prevailing belief in special creation and the fixity of

THE PUBLICATION OF *ON THE ORIGIN OF SPECIES* 41

species with an account of the transformation or 'transmutation' of species which required no supernatural intervention, but could be explained by 'natural selection'. Darwin demonstrated the heterogeneity of all life forms and their tendency to vary from parent to offspring. He explored the incessant Malthusian struggle for survival which operated throughout nature. He pointed out the capacity of 'artificial selection' by breeders to modify the forms of existing species, and its implications for the greater survival rates and reproductive capacity of variations better adapted to their environment. Proceeding through geology, palaeontology, comparative anatomy, embryology and the geographical distribution of species, he pointed out the observable phenomena that seemed random and inexplicable until explained by 'descent by modification', which was the phrase he used throughout, rather than 'evolution'. He was frank about the problems of the evidence, not least the imperfection of the geological record which meant that there was (as yet) no material evidence of much of the long sequences of development his argument was suggesting. He was deliberately not frank about the place of humanity or the role of God in the evolutionary process, although the implications that humankind was equally a product of evolution rather than of divine creation were all too clear.

Immediate Readings

The *Origin* had been extensively trailed in scientific circles, not least in consequence of Darwin and Wallace's combined presentation to the Linnean Society in October 1858, and was eagerly anticipated. The first edition was sold out by the publication day amidst a scramble for copies. Ninety of Darwin's scientific friends and acquaintances were sent a copy direct from the publisher. The rest percolated to the reading public via personal purchases and subscription libraries. The *Origin* was one of the 'books of the season', and remained so for several years.[1] From the very day of publication the educated public was agog. Fifty years later the independent minister J.A. Picton (1832–1910) could still remember vividly a discussion of the *Origin* he had while walking through Manchester a few days after publication.[2] It was not easy reading, even for Darwin's closest scientific confidantes. Joseph Hooker described it as 'the very hardest book to read, to full profit, that I ever tried'; other readers judged it 'a hard book', 'heavy reading', 'sadly wanting in illustrative facts'.[3] Conceptually, it was challenging. The notion of natural selection as it operated at the level of a population rather than the

42 DARWINISM'S GENERATIONS

individual organism was particularly difficult to grasp.[4] Thirty years later, reading the *Origin* slowly again 'for the nth time', with the view of picking out the essentials of the argument, T.H. Huxley himself commented wryly that 'Nothing entertains me more than to hear people call it easy reading'.[5] Responses repeatedly compared it unfavourably with the earlier evolutionary treatise, *Vestiges of the History of Natural Creation*, which had been much more popular in appeal.[6] Robert Wight (1796–1872), an expert on Indian botany who had retired to England in 1853,[7] confessed that it was only through reading the reviews of the American botanist Asa Gray that he came to understand.[8] It might have made 'a good many converts', observed the sympathetic Oxford anatomist George Rolleston (1829–81), had it been better arranged.[9]

In many respects, given these challenges, the early responses were surprisingly positive. Indeed, the geologist and palaeontologist Hugh Falconer (1808–65) remarked in April 1860 that '[n]ot the least remarkable circumstance is the toleration with which it has been received by the great mass of the thinking public'.[10] Readers were dazzled by the sweep and ambition of Darwin's argument, and the sheer weight of observation he brought to bear.[11] 'What a work it is!' enthused the Unitarian writer Harriet Martineau (1802–76), 'overthrowing (if true) revealed religion on the one hand, and natural (as far as Final Causes and Design are concerned) on the other. The range and mass of knowledge take away one's breath'.[12] Even those without obvious predisposition were often bowled over. In July 1860 the Scottish scholar John Stuart Blackie (1809–95) was wandering through the grounds of the hydropathy establishment at Sudbrooke Park in Surrey (where Darwin was also occasionally a visitor) absorbing the *Origin* and meditating on the inadequacy of orthodox views of a once-and-for-all creation.[13] The banker Samuel Jones Loyd (1796–1877) was full of admiration, attracted to the simplicity of evolutionary dynamics against belief in innumerable acts of individual creation, although reassuring his sceptical friend G.W. Norman that 'Neither you or I are lineal descendants of Apes or Asses'.[14]

Unsurprisingly, Wallace praised it as the 'Principia' of natural history, and Robert Chambers (1802–71), the still anonymous author of the *Vestiges of the Natural History of Creation*, welcomed it, albeit as a vindication of the evolutionary argument he had first advanced in 1844, rather than on its own terms.[15] For those like Huxley's colleague at South Kensington, Andrew Crombie Ramsay (1814–91), and the Oxford mathematician Baden Powell (1796–1860), who had already come to doubt special creation, the *Origin* bolstered their existing beliefs, and in natural selection offered a convincing

THE PUBLICATION OF *ON THE ORIGIN OF SPECIES*   43

mechanism in place of 'the vague gropings towards the light' in which they had previously indulged.[16] It 'makes an epoch', wrote George Eliot (1819–90), despite her criticism of Darwin's style, while her close friend Sara Hennell (1812–99) commented that it 'writes out my own presentiments' as the 'due process both the working out of a new species & religion'.[17]

Not everyone could immediately plunge into the intricacies of Darwin's writing. For many, cost or distance meant even a loan from Mudie's subscription library was out of reach (it was said that Mudie himself loaned his copy of the *Origin* to many a poor clergyman unable otherwise to access it).[18] The *Origin* remained a divisive book, still not always considered a safe purchase for the libraries of mechanics' institutes and literary associations.[19] At his home in Moffat, the avid amateur botanist, John G. Macvicar (1800–84) had still not managed to get sight of a copy of the *Origin* twelve months after publication.[20] Not that this did much to constrain the conversations. Familiarity with the book's contents was readily attained from the reviews, editorials and commentaries. Many pitched in without any first-hand knowledge at all. Robert Dick (1811–66), the baker-naturalist of Thurso, seems also to have been unable to obtain a copy of the *Origin* by the summer of 1860, but was quite prepared to engage in vigorous rebuttal in his private correspondence.[21] The Anglican clergyman and hymn writer Godfrey Thring (1823–1903) later recalled a 'hammer and tongs' argument about the *Origin* in which his disputant only belatedly realised that Thring had not read the book, and was sustaining his argument from general knowledge and the material supplied by his opponent.[22]

Anticipating the storm ahead, the inner circle of Darwin's friends and supporters had swung immediately into action, Huxley telling Darwin soon after reading the *Origin* that he was 'sharpening up my claws and beak'.[23] Huxley seized the chance to publish a positive review in the *Times*, followed up in April by a longer plug in the *Westminster Review*. Within a few weeks of its publication, he was endorsing the *Origin* on the platform of the Royal Institution.[24] As the newspaper reviews appeared in December, and the longer commentaries in the intellectual periodicals followed in the first half of 1860, the history of the natural world became the topic of the day. Darwin, before the publication of the *Origin* a largely unknown geologist and traveller, was catapulted into celebrity.[25] The aristocrat-naturalist Sir Charles Bunbury found himself surrounded by Darwinian discussion, marvelling 'how much it has been talked about by un-scientific people'; his friend Charles Lyell could speak of little else, and on a visit in December the Christian socialist clergyman and novelist Charles Kingsley was full of talk

44 DARWINISM'S GENERATIONS

of the book.[26] What struck observers was the breadth of the *Origin*'s appeal. In the hothouse intimacy of undergraduate circles the excitement was especially intense. It was reported from Oxford in 1860 that everybody, even the unscientific undergraduates, was discussing it 'with the greatest ardor', and it was the subject of animated debate in university societies and discussion groups.[27] Public associations and private spaces were equally absorbed. As one observer put it, the *Origin* achieved a notoriety no other scientific work had achieved, 'Royal Societies discuss it,…it is talked over at clubs. It is received with smiles in drawing rooms and frowned upon in churches.'[28] The young banker Edward Clodd (1840–1930) recalled how the *Origin* 'roused all his faculties to a state of joyous activity'. He went to Huxley's lectures, and he discussed the wonders of evolution and the destiny of man's soul endlessly.[29]

On closer inspection, the picture was less positive. Despite exceptions, the predominant tone of the initial reviews was hostile. The Oxford botanist Charles Daubeny (1795–1867), although fiercely supportive of Darwin's right to investigate such questions scientifically, rejected not just the lineage of man and ape, but also the evolution of the consciousness of the vertebrates from 'the dull vegetative faculties' of the invertebrates; and assisted the Bishop of Oxford, Samuel Wilberforce (1805–73), in his wholesale attack on the *Origin* in the *Quarterly Review*.[30] Inevitably, despite Darwin's efforts to conciliate him, Richard Owen (1804–92), without doubt the leading palaeontologist in the country, also quickly expressed his opposition, albeit behind the anonymity provided by the *Edinburgh Review*. And despite the many accounts published subsequently which claimed retrospectively a rapid conversion, the archival record, notwithstanding the enormous amount of extant material, is remarkably thin in unambiguous endorsement. Even the scientists of Darwin's circle, including the Oxford geologist John Phillips (1800–74), Leonard Jenyns (1800–93) (an old college friend, and son in law of John S. Henslow (1796–1861), Darwin's mentor), and the conchologist John Gwyn Jeffreys (1809–85), while often writing encouragingly to Darwin, could not bring themselves to concur.[31] Meetings at the Royal Society in Edinburgh and the Cambridge Philosophical Society in June 1860 heard forceful repudiations.[32] John Phillips' May 1860 Rede Lecture, *Life on the Earth. Its Origin and Succession* (1860), while largely skirting round the *Origin*, concluded with a clear rejection.[33] There were plenty of clerical lecturers and pamphleteers alert to the dangers that a book like the *Origin* might pose.[34]

THE PUBLICATION OF *ON THE ORIGIN OF SPECIES*   45

Darwin himself was notoriously disheartened by the early response to the *Origin*, and he was repeatedly left disappointed by the lack of unqualified support. Even Huxley's public endorsements left Darwin unhappy because their enthusiasm was less for his evolutionary arguments than for the right of science to trench on matters where it had previously deferred to religion. And indeed Huxley confessed in private in July 1860 that he was 'not a Darwinian' but simply concerned that justice be done to Darwin's argument.[35] In the immediate aftermath of publication, Darwin was prepared to identify only one 'believer', his longstanding confidante Hooker, and even his belief was only 'partial'.[36] By the following March, the list had expanded to eleven, including the group emerging as the inner core of the Darwinian party: Hooker, Lyell, Huxley, Ramsay, Lubbock, the University College London physiologist, W.B. Carpenter (1813–85), along with several other correspondents, the geologist Joseph Beete Jukes (1811–69), the American Henry Darwin Rogers (1808–66), Regius Professor of Natural History and Geology at Glasgow University, the phrenologist Hewett C. Watson (1804–81), and George Henry Kendrick Thwaites (1812–82), superintendent of the Peradeniya botanical gardens in Ceylon.[37] And of course Darwin, who was worrying at what he saw as a 'storm of hostile reviews' and the 'excessively slow' progress of opinion, could have added Wallace, and perhaps Herbert Spencer (1820–1903), who acknowledged to Darwin in February 1860 that his evolutionary ideas had undergone a 'considerable modification' as a result of reading the *Origin* (although Spencer's ideas were never terribly closely aligned to Darwin's, as Darwin was quick to recognise).[38] Many of these, like Lyell, Hooker, and Lubbock, had been familiar with Darwin's ideas before 1858, and yet even then, as Ruth Barton has shown, they offered at best a highly attenuated agreement.[39] Most of the others on the list had done no more than write encouragingly to Darwin. The case of Jukes illustrates just how difficult it is to pin down precise positions; although in letters to Darwin in the first half of the 1860s he expressed a general belief that the geological record demonstrated a sequence of species, and elsewhere he suggested he was coming round to the idea that species form was determined by environmental pressures, his endorsement, even in this private context, was at best oblique.[40] In the same way Thwaites, although he talked to Hooker in April 1860 about the prospect of becoming 'a perfect convert to Darwin's doctrines bye and bye', was hardly a clear disciple. He later told Hooker that he thought that the more Darwin's views were left 'floating in the mind' the more worthy they seemed of acceptance, but that much yet required clearing up from observation of facts, especially as regards more minute organisms.[41]

Beyond this group, individual responses were often fiercely hostile. Many established scientists were of course ordained clergymen and members of a narrow and exclusive group of Oxford and Cambridge fellows, confirmed in their anti-evolutionary views by a sustained campaign against the message of the *Vestiges*. The geologist Adam Sedgwick (1785–1873) was appalled by what he read. Passages in the book, he told Darwin in November 1859, 'greatly shocked my moral taste.'[42] The liberal philosopher Alexander Bain (1818–1903) recalled meeting with Sedgwick in late 1859, fresh from reading the *Origin*, along with William Whewell (1794–1866), Master of Trinity College Cambridge (the leading British man of science of the early Victorian period), and being treated to 'a vehement diatribe against Darwin—in which Whewell concurred—for setting aside the Creator.'[43] Jane Carlyle (1801–66) dismissed the book contemptuously, despite the ecstasy the scientific world was in, confessing that the argument that humanity was descended from 'the great original Oyster' didn't provoke the slightest curiosity, or offer any illumination for her practical life.[44]

The sense of conflicted enthusiasm is very obvious in the response of George Douglas Campbell, 8th Duke of Argyll (1823–1900). Argyll, who was to become one of Darwin's most persistent opponents, thought the *Origin* 'a delightful [book], suggesting endless subjects for discussion and enquiry', a 'perfect storehouse of knowledge.'[45] So enthusiastic indeed was he in praising its suggestiveness that the *Times* at one point labelled him 'the Darwinian Duke.'[46] But Argyll's doubts were visible from the outset. He felt the *Origin* failed, both in its discussion of the process by which new species were created, and in its ingenious attempts to explain the deficiencies of the geological record. This was, of course, especially the case for the possible evolution of humans; Argyll was not deflected by Darwin's reticence: he recognised the implications, but told Lyell that all the existing facts were against such a theory for man.[47] As Argyll clearly demonstrates, positivity cannot, especially at this early stage, be taken for persuasion.

Context

At times after 1859, Darwin's supporters seemed determined to exaggerate the extent of the break engendered by the *Origin*. But claims like Huxley's that before 1859 there was only one biologist who offered any support for

THE PUBLICATION OF *ON THE ORIGIN OF SPECIES* 47

evolution are misleading. The publication of the *Origin* was only one point in a long intellectual history. It helped to crystallise debates over evolution which stretched back to Darwin's grandfather Erasmus Darwin in the eighteenth century, and had been building ever since the geological and palaeontological discoveries of the early nineteenth century, with their implications of 'deep time', and of a great history of now extinct life, discoveries widely publicised by Charles Lyell's *Principles of Geology* (1830–33). Darwin had himself famously first conceived of the idea of natural selection as early as the mid-1830s. Those who had been undergraduates at mid-century could recall the agitation created by the geological challenge to the Genesis account, and the idea of evolution 'floating about in men's minds for some time'.[48] The historian W.E.H. Lecky, at Trinity College, Dublin between 1856 and 1859, remembered an intellectual atmosphere 'much agitated' by discoveries of geology and challenges to the Mosaic cosmogeny, and Huxley's own reflection in 1860 was that his rejection of special creation, but inability to believe in Lamarck, 'two absurdities', had left him well before the *Origin* with only 'uneasy scepticism'.[49] Already by the 1850s, geological discoveries were forcing Christians to rework their orthodoxies; pressures which led Kingsley in 1855, in echo of Matthew Arnold's famous poem of doubt, 'Dover Beach', to describe the feeling of standing 'on a cliff which is crumbling beneath one, and falling piecemeal into the dark sea'.[50]

It is now recognised that scientific naturalism did not emerge in *de novo* in 1859, just as the traditions of natural theology had not survived unscathed into the second half of the century.[51] It is true that in the 1830s the publication of the Bridgewater Treatises helped cement a version of natural theology in which the adaptation of plants and animals to their environment was celebrated as incontrovertible evidence of the existence and the goodness of God. But in the 1840s the *Vestiges of the Natural History of Creation* had offered an accessible melange of evolutionary ideas which challenged this static frame.[52] Vigorously condemned by the scientific establishment, it had nonetheless been read widely, selling more than 20,000 copies by 1860. On the whole, the scientific elite had remained firmly opposed to evolutionary theories, but developmental frames had been visibly extending their influence. Adrian Desmond long ago demonstrated quite how firmly Lamarckian ideas had taken a hold in the radical circles of London's medical schools during the 1840s, producing what Piers Hale has called 'swaths' of radical evolutionists.[53] But the influence of the evolutionary hypothesis spread much more widely. A decade before the *Origin*, Hugh Miller was regretting that it was scarcely possible to travel by railway or encounter any

48   DARWINISM'S GENERATIONS

group of intelligent working men 'without finding decided trace of its ravages'.[54] With extreme caution, the palaeontologist Richard Owen was developing and articulating a derivatist theory which accepted the origin of species in an initial primordial state, and their gradual development into their current forms through the long ages of geologic time. Even Owen could not accept the common ancestry of man and animals, but its possibility had prompted study of the comparative anatomy of humans and primates from the later 1840s onwards.[55]

Probably the most important incubator of evolutionary ideas was the radical salon gathered around the publisher John Chapman (1821–94) at 152 The Strand in the early 1850s.[56] Here figures including Herbert Spencer, George Eliot, George Henry Lewes (1817–78), and also Huxley gathered to discuss new ideas and provide the intellectual milieu for the radical thought of the *Westminster Review*. The ideas of this circle really only came to the attention of the general public after 1859, but by the later 1850s it is not hard to find enthusiastic progressivists, including the mathematicians William Hopkins (1793–1866) and Baden Powell (1796–1860), whose *Philosophy of Creation* (1855) defended the possibility of transmutation, and even Richard Owen himself.[57] Evolutionary perspectives spread widely. In astronomy, nebular theories traced the origin and development of the universe over long ages. In philosophy, the French Positivist Auguste Comte (1798–1857) had developed a comprehensive system based on general laws of evolution. In Britain, the *System of Logic* (1843) of John Stuart Mill (1806–73), an almost 'sacred book' for undergraduates of the later 1840s and 1850s, offered an implicitly evolutionary approach.

In scientific circles, there was an increasing recognition of the variation and fluidity of species.[58] Joseph Hooker's study of the flora of New Zealand, published in 1853, was quite explicit about the increasing diversity of beliefs about the development of species, and reviewers were quick to see it as a conclusive argument for transmutation.[59] In philology, the belief that languages had descended from a common stock offered a sufficiently powerful Darwinian analogy for the leading philologist of the period, Max Müller, to go as far as to describe himself as 'a Darwinian before Darwin' in this broader evolutionary sense.[60] The momentum that evolutionary ideas were gathering was perhaps most forcibly registered in the number of rebuttals being published, not just Hugh Miller's *Footprints of the Creator* (1850) and *The Testimony of the Rocks* (1857), with their attempts to reconcile revelation and geology by arguing for a series of successive creations, but also the more hysterical *Ophthalmos* (1857) of P.H. Gosse (1810–88), which argued

THE PUBLICATION OF *ON THE ORIGIN OF SPECIES* 49

that the fossil evidence was a sort of grand *tromp l'oeil*, created by God to test humanity's faith.[61]

Before 1859 these evolutionary tendencies remained marginalised, constrained by powerful forces of conservatism evident in the widespread resistance to the accumulating evidence of the age of man, and the clinging of many older Victorians even to the computations of Bishop Ussher that the creation account of Genesis could be dated precisely to 4004 BCE. Both British archaeology and conventional geology continued to discount the evidence of prehistoric, or pre-Adamite, man through much of the 1850s, despite a number of potentially crucial discoveries. Many, including Charles Lyell himself, felt safe in recurring to the position of the palaeontologist and theologian William Buckland (1784–1856) that any association of human and prehistoric animal remains was merely accidental;[62] but the exploration from 1858 of Brixham cave in Dorset, revealing numerous human artefacts *in situ* with extinct animals, with no credible grounds for rejecting this as evidence that they were contemporaneous, demanded fresh thinking.[63] Earlier finds were re-evaluated, and previous positions retracted, as Lyell did at the British Association in 1859. Relatively suddenly, a time frame opened up which made the evolution of man, as well as the rest of the animal kingdom, much more likely.

The power of the implicit evolutionism gathering pace in the 1850s helps to explain Wallace's development, almost entirely independently of Darwin, of the theory of natural selection. Wallace and Darwin had corresponded, and Wallace's published work in these years, not least essays such as his 'On the Law which has regulated the Introduction of New Species' (1855), was clearly developing components of a general theory of evolution.[64] As the decade progressed, Darwin's friends had pressed him with growing urgency to establish his primacy in developing such ideas, and their premonitions were borne out in June 1858, when Darwin received from Wallace, who was exploring in Indonesia and Malaya, the brief outline of his theory of natural selection. Faced with the sudden loss of his claim to authorship of the theory, Darwin's friends quickly concocted a plan to allow Darwin to claim credit for his long-held ideas, while still dealing fairly with Wallace. A presentation of Wallace's paper and of Darwin's theories from his notebooks, at the Linnean Society on 1st July 1858 meant Darwin was not obliged to give up entirely his priority in the discovery.

For the Cambridge zoologist Alfred Newton (1829–1907), it was the publication of these papers in the Linnean Society's journal rather than the appearance of the *Origin* which marked the start of the Darwinian era:

50 DARWINISM'S GENERATIONS

'for I and others accepted it from the moment when the no. of that year's journal reached us,...—I shall never forget how I felt reading that paper the night it arrived, & went to bed happy!'[65] Even before the *Origin* appeared, lines of controversy were being drawn. Naturalists who had previously toyed with evolutionary ideas rushed to publicise their positions and establish their own claims to priority. Richard Owen hurried out the publication of his 1859 Rede lecture, to promulgate his 'latest convictions on the question of the extinction, succession and alleged transmutation of species.'[66] Others received the new ideas with puzzlement and perplexity. The entomologist Thomas V. Wollaston (1822–78) was perturbed at the ecological and developmental assumptions of Wallace's and Darwin's papers, and found it impossible to accept several of their key principles: indefinite variation, the tendency for diversification and disappearance of the original type species, even that the comparative abundance of species depended on their capacity to obtain food; 'and as for the "selecting power of Nature" (if such indeed is more than a fancy)', he told Lyell, 'I suspect that if some one organ of a species were to be (from peculiar circumstances) unusually developed, the animal would suffer seriously from the compensating detraction of something else.'[67]

Public Controversies

The *Origin* was in this sense an intervention in a long-standing debate, in which the lines of division were well established and the attitudes of contemporaries already firmly embedded. And yet, coming as it did from a respected gentleman scientist, and offering as it did a carefully constructed argument for evolution, the *Origin* clearly brought the evolutionary debate into the cultural mainstream. At the very least, outright dismissal had become more difficult to sustain, and as the months passed, perhaps evidence did start to accumulate that a fundamental change in attitude was occurring. In this regard, much has been made of the proceedings of the annual meeting of the British Association in Oxford in the summer of 1860, which was the occasion of the exchange between Huxley and Samuel Wilberforce, the Bishop of Oxford, in which Wilberforce's ill-judged quip that he wondered if Huxley were descended from the ape on his grandfather's or grandmother's side was supposed to have been turned back on him with decisive effect by Huxley's rejoinder that he would rather have an ape for an ancestor than a man who prostituted his learning in the defence of error.

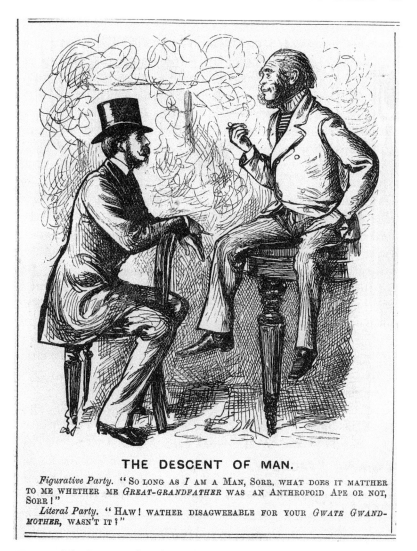

Figure 1 'The Descent of Man', *Punch*, 24 May 1873.

In subsequent years the encounter acquired a mythic status as a watershed in the debates over the *Origin*, the presage of, if not the moment at which, the chains of clerical science were thrown off once and for all, and a regular reference in evolutionary squibs (see Figure 1).

Wilberforce has gone down in history as 'Huxley's hapless opponent', but he was no neophyte: his scientific grounding was sufficient for him to act as

52 DARWINISM'S GENERATIONS

a vice-president of the Zoological Society in the early 1860s, and his criticisms homed in on some of the *Origin's* weakest links.[68] And he could draw on considerable support, including Owen, Daubeny (who despite adopting a very different tone, had made his own opposition clear in a paper delivered the day before), the physiologist Sir Benjamin Brodie (1783–1862), Richard Greswell (1800–81), fellow at Worcester College,[69] and of course Adam Sedgwick, manoeuvring as chair of the Geological section to 'keep the Darwinian theories out of the room'.[70] An impression certainly spread that the bishop had been publicly humiliated, and a tradition developed of Huxley's smashing of 'Soapy Sam' which fed into the enthusiasms of younger Darwinians in the later 1860s and 1870s; but this was not the way it had looked at the time to the physicist Balfour Stewart (1828–87), who reported that 'the bishop had the best of it'.[71] Notwithstanding the fun, Stewart was not alone amongst older onlookers in dismissing the whole proceedings of the meeting as comparatively unimportant.[72] Tellingly, the most positive attributable verdict, itself noticeably circular in its assertion, that the meeting closed 'leaving the impression that those most capable of estimating the arguments of Darwin in detail saw their way to accept his conclusions', came from one of the high Victorians present, William Henry Fremantle (1831–1916), later Dean of Ripon.[73] Even the high Victorian Anglican philologist F.W. Farrar (1831–1903) saw the outcome as 'a victory of manners, not of substance'.[74]

Certainly, beyond the theatre of this one encounter, there were few signs of a seismic shift in attitudes. The winter sessions for 1860–61 brought further evidence of hostility. At the Free Church College, Glasgow, William Keddie (1809–77) opened his series on 'Natural Science in Relation to Theology' with a rebuttal.[75] The Sheffield Literary and Philosophical Society was told that, with a few valuable exceptions, the *Origin* would leave the question of the origin of species exactly where Lamarck had left it. The Tyneside Naturalists' Field Club, and the Liverpool Literary and Philosophical Society were treated to detailed accounts of *Origin's* weaknesses.[76] Although Argyll at the Royal Society of Edinburgh was cautious, accepting that Darwin's discussion of distribution and the struggle for existence had cast much light on the question, a debate towards the end of 1860 at the Cotteswold Naturalists' Field Club apparently produced unanimous rejection.[77] Spring 1861 brought more of the same. The mineralogist Thomas Rowney (1817–93) told the Hull Literary and Philosophical Society in January that species were created by God, while the audience at the Church of England Literary Institute in Derry were urged to wait for the scientific disproof to come, and

THE PUBLICATION OF *ON THE ORIGIN OF SPECIES* 53

trust in revelation.[78] And so it went on. In March 1862 the York Philosophical Society was warned against the false philosophies of the day, including Darwin, Lamarck and *Vestiges*.[79] Later in that year the Glasgow naturalist John Scouler (1804–71) pointed out the inadequacy of Darwin's evidence to the Paisley Philosophical Society.[80]

Huxley continued to proclaim evolution both in London and farther afield, and his lectures to working men in the Queen Street Hall in Edinburgh in January 1862, in which he pulled no punches in his account of the common descent of man and apes, generated just the sort of frenzied horror he was hoping for.[81] But at the same time, stalwarts of the lecture platform, including Benjamin Waterhouse Hawkins (1807–94), David Page (1814–79), Daniel Mackintosh (1815–91), and the former Chartist Thomas Cooper (1805–92) continued to serve up varieties of anti-Darwinism to provincial audiences. Hawkins took advantage of lectures like 'The Use of Natural History' to controvert evolutionary theory in general, and Darwinism in particular, using the blackboard sketches for which he was celebrated to demonstrate the lack of affinity of the spines of men and apes.[82] Cooper's lectures, especially popular with Dissenting audiences, peddled reworkings of Paleyite natural theology.[83] Page and Mackintosh offered less religious apologia, and more popular science, and tended to take a less directly hostile stance, but the implications were made clear enough. Page told his audiences not to be afraid of the *Origin*, but to read it alongside a slew of uncompromisingly anti-Darwinian interventions, and to think about the importance of inductive reading, and then make up their own minds.[84] Of course, audience reactions varied: the response of Henry Jeffs (1819–88), accountant and freemason of Gloucester, to a lecture in December 1860 given by Hawkins was to acquire two monkeys, 'dabble in comparative anatomy', and as he later recalled, 'almost [lose] myself in a maze of philosophy and metaphysics'.[85]

The nature of human relationships to the higher mammals became the primary question on which the early skirmishes over Darwinism were fought. The arrival in London in early 1861 of Paul Du Chaillu, a French-American explorer-cum-adventurer, with dead specimens of gorillas, previously an almost fabulous creature, along with dramatic accounts of his explorations, inevitably fanned the debate.[86] Du Chaillu was the sensation of the season, and the suggestion of the affinity of man and ape was widely discussed, intruding the Darwinian debate even into the pages of *Punch*. But Du Chaillu was also drawn into the bitter and often highly technical debate between Owen and Huxley about how far it was possible to demonstrate

54 DARWINISM'S GENERATIONS

vital anatomical differences between the two, the great hippocampus debate, which Charles Kingsley satirised in *The Water Babies* (1862). Owen argued that human anatomy indicated fundamental differences of structure of the cranium, both in the size of the brain and in the existence of the 'hippocampus major'. Huxley disputed this, and set the younger Darwinians to work dissecting simian brains to prove it. In the spring of 1861, while Huxley was lecturing on the topic to working men at the Royal School of Mines, the two gave rival courses to the Royal Institution, and crossed swords in the pages of the *Athenaeum*.[87] Max Müller and Rolleston offered differing slants on the question for the Royal Institution in the following winter, Müller arguing that language could not have evolved from the sounds made by animals, Rolleston urging the affinities of the brains of men and animals.

Largely because of the absence of Huxley, the controversy did not make much of an impact at the 1861 British Association meeting at Manchester, although the gathering did hear an important intervention from William Thomson arguing that the age of the earth during which it could have supported life could only be a fraction of the time Darwin required for evolution by natural selection.[88] (A year later Thomson was still expressing his disdain for the 'pigeon fanciers and beetle collectors'.[89]) Combat was renewed, however, at Cambridge in 1862. Looking back, Alfred Newton identified the Cambridge meeting as marking the last determined resistance of the anti-Darwinians and their ultimate defeat; 'the "crucial struggle" when "the victory of the new doctrine was then declared in a way that none could doubt"'.[90] Owen gave two anti-Darwinian papers. His contribution to the Biology section, chaired by Huxley, on the aye-aye, a Madagascan lemur, which he used to argue for the clear distinction between the higher mammals and humans, not just on the basis of the hippocampus and brain sizes, but also contrasts in the structure of their feet,[91] brought him a battering from the younger Darwinists, W.H. Flower (1831–99) and Rolleston (Rolleston with a vehemence for which he later apologised). Leaving the session, Kingsley assured a somewhat doubtful companion that Owen deserved the thrashing he got.[92] But Owen was unrepentant, and he continued to enjoy the authority as Britain's greatest living naturalist and a considerable constituency for his views. He made sure he was visible through the following months, delivering prizes in Liverpool, opening the new Philosophical Hall in Leeds (where he gave four lectures on the Mammalia), and lecturing in Norwich. Into the autumn, he was pressing his position in the opening lecture to the winter course of YMCA lectures at Exeter Hall. And he continued to be a standard citation for those marshalling arguments

THE PUBLICATION OF *ON THE ORIGIN OF SPECIES*   55

and authorities against Darwin in lectures, pamphlets and newspaper interventions.[93] Around the country, there were perhaps incipient signs of pushback against wholesale theological dismissal of Darwin, but little in the way of general endorsement of his views.

Late Georgian Antipathies

The extent of this immediate hostility to the *Origin* has been long recognised. What has been less apparent is the degree to which these responses were generationally patterned. Darwin's most vocal and implacable opponents were predominantly born before the mid-1810s ('late Georgians' and 'early Victorians' in the schema outlined above). For the late Georgians in particular, it is only very occasionally possible to locate responses which indicate that Darwin was not immediately rejected, and then generally all we have is the absence of condemnation, rather than direct statements of support.[94] With a few visible exceptions, the late Georgians distanced themselves fundamentally from Darwin. Some were unconfrontational but firm. John Bird Sumner (1780–1862), Archbishop of Canterbury since 1848, was already comfortable with the possibility of a pre-Adamite world, which might have included 'creations having the qualities of man'; but he was happy, until this was proved, to continue to adhere to Paley, 'and not suffer what I do know, to be distracted by what I do not know'.[95] Often opinion was more outspoken. Thomas Carlyle (1795–1881), whose antipathy towards Darwinism became notorious, commented that the *Origin* was 'Wonderful to me, as indicating the capricious stupidity of mankind; never could read a page of it, or waste the least thought upon it'.[96] Richard Whately (1787–1863), philosopher and Archbishop of Dublin, reading the book in 1861 was almost as caustic, wondering to a friend on a visit to the Geological Museum at Cambridge, how long the huge Dinornis fossil had taken to develop from a mushroom, while Charles Hippuff Bingham (1805/6–75), belatedly 'wading through' the *Origin* in 1863, dismissed it peremptorily as 'unmitigated bosh'.[97]

The older members of the scientific community were scarcely less blunt. Thomas Bell (1792–1880), President of the Linnean Society, who had downplayed the significance of the presentation of the Darwin-Wallace papers in 1858, strongly repudiated the *Origin* on its appearance.[98] Benjamin Brodie, despite his sympathy for evolutionary ideas, found the *Origin* unconvincing, while the scientific polymath Sir David Brewster (1781–1868) apparently

56 DARWINISM'S GENERATIONS

went as far as suggesting that the *Origin* should be burnt as 'unscriptural', Darwin's hypothesis having been 'long ago refuted by the most distinguished of our naturalists'.[99] William Hopkins, the astronomer John F.W. Herschel (1792–1871), the geologist Roderick Murchison (1792–1871), who had done so much to map out the fossil record, even John Henslow, Darwin's old tutor, all moved quickly to make clear their dissent.[100] Murchison's carefully annotated copy of the *Origin* was strewn with exclamations and protestations, especially severe on Darwin's efforts to explain away the imperfections of the fossil record.[101]

Henslow is a good example of the difficulties of situating contemporaries. He had defended Darwin at the hostile meeting of the Cambridge Philosophical Society in May 1860, arguing that he was motivated simply by the search for truth and was exalting rather than debasing the position of the Creator, and perhaps on this basis he continued to be cited as a supporter late into the century. But his correspondence speaks of deep scepticism. While conceding that the *Origin* was 'a marvellous assemblage of facts & observations—& no doubt contains much legitimate inference', he told his brother-in-law, 'it pushes hypothesis (for it is not a real theory) too far'.[102] Henslow was to die in 1861, but in February 1861 he was still affirming his confidence in the biblical chronology of man, and the ultimate reconciliation of scientific discovery to it.[103]

Various types of intellectual constraint were at work here. In some quarters, the *Origin* was interpreted as just a fresh gloss on Lamarckism, 'the system of the author of the *Vestiges*, stripped of his ignorant absurdities', as Adam Sedgwick described it.[104] John G. Macvicar, even before reading the *Origin*, coupled it with Lamarck's 'most daring system of atheism'.[105] Even Lyell, still known by many for his anti-Lamarckism, read the *Origin* as a development of the older work.[106] Otherwise, the late Georgians of the Anglican scientific establishment interpreted the *Origin* as contributing to a tradition of subversive anti-religious texts against which they had long been struggling, as in the case of Marion Bell (née Shaw, 1787–1876), who confessed in the mid-1860s to being guided by her sense of what her late husband, Sir Charles Bell (1774–1842), would have thought, believing that from his treatise on hands 'he understood "types"—but not the new Theory of Development or of Man's creeping up'.[107] Even the theological liberals, who might have accepted Darwinism, generally didn't. Whewell dismissed the *Origin* as 'unphilosophical', and famously refused even to allow a copy into the Trinity College library.[108] Opposition was bolstered by an underlying conservative horror at Darwin's apparent repudiation of each species'

THE PUBLICATION OF *ON THE ORIGIN OF SPECIES*   57

rightful place in nature. This comes through in the sarcasm of someone like Sedgwick, who apparently presented the evolutionary theory to his Cambridge students in the early 1860s in terms of 'Fishes, dissatisfied no doubt with their condition, got flexable [sic] noses and were changed into the merry creature called a monkey. And the monkey, having rubbed his tail off—no doubt with sitting on wooden chairs—was changed into a man.'[109] Sedgwick vehemently attacked the *Origin*'s marginalisation of divine power, denying the ability of science to address questions not amenable to empirical verification, and sought to rally anti-Darwinian sentiment by circulating accounts of his lecture to his circle.[110] One, John Kelly (1791–1869), Assistant Geologist in the Geological Survey of Ireland, welcomed his interventions as having 'broken down Darwin'.[111]

The Cornish naturalist and antiquarian Jonathan Couch (1789–1870) left a series of manuscript essays revealing his engagements with Darwinian evolution in the aftermath of the *Origin*. Couch was an archetype of the older provincial naturalist, a trained physician whose eclectic but intensely local interests ran from the life sciences to geology, from Cornish folk beliefs to local history. Although intrigued by its parallels with philology, he was clear that Darwin's theory was not only 'doubtful & gratuitous', but rejected by most naturalists.[112] Even the partial acceptance of evolutionary ideas of a figure like the Duke of Argyll was a step too far for Couch.[113] He remained convinced of the fixity of species: 'one species can no more be turned into another than the ear be made to act as an organ of sight'.

Early-Victorian Rebuttals

The initial hostility of the slightly younger early-Victorian generation, born roughly between the mid-1790s and the defeat of Napoleon, Darwin's own generation, was equally uncompromising. The journalist Lionel Tollemache (1838–1919) recalled his father, John Tollemache (1805–90), observing that a friend to whom he had promised to lend a copy of the *Origin* should first borrow and then burn it.[114] The *Origin* outraged the children's botanical writer, Margaret Gatty (1809–73); despite the arrayed authority of Hooker, Huxley and Lyell, she was convinced that its 'madness' would be '*found out*…and exposed as a great man's *blunder*'.[115] Amongst the early Victorians there was more genuine admiration for the scope and ambition of Darwin's work, and a recognition that it would set the agenda for debate, but still little by way of endorsement. In addition to those already mentioned,

58 DARWINISM'S GENERATIONS

early objectors included the Scottish botanist Andrew Murray (1812–78) and the entomologist John O. Westwood (1805–93).[116] Although many early Victorians had already conceded that the Mosaic cosmogeny could not be sustained, the further transformation of fundamental beliefs called for by the *Origin* continued to prove too much. Confessing his scepticism to Darwin in January 1860, Charles Bunbury softened the blow by praising Darwin's astonishing labour and patience, range of reading, and extraordinary power of observation and condensation; but the bottom line was still the bald 'I am not convinced.'[117] Likewise the Scottish naturalist Sir William Jardine (1800–74), who told John Hutton Balfour that he considered natural selection 'an <u>assumed</u> fact and dont think it exists and his case is not proved.'[118] By way of contrast, the Anglo-Australian farmer-politician, Charles James Griffith (1808–63), made it clear in the observations he bound into a first edition of the *Origin* which he seems to have bought on an extended visit to England between 1858 and 1860, that although he was prepared to accept some sort of 'natural selection' operating analogously to the artificial selection of breeding, he rejected entirely the suggestion that it could have the sort of cumulative impact necessary to create new species.[119]

Those early Victorians embedded in institutional religion, whether non-scientist or scientist, often found it impossible to separate the credibility of Darwin's science from its potential consequences. The Tractarian divine E.B. Pusey (1800–82) saw evolutionary thought as a threat which would 'wreck the faith of many', fearing that in the struggle to reconcile evolution and Genesis, 'with most, Genesis would have to give way.'[120] Pusey's objections were fundamentally theological, but he should not for that reason be dismissed. He attempted to keep abreast of Darwinian developments, not only in England but also in Germany, where he saw an entirely materialist and anti-Christian doctrine, and he was quite prepared to bandy references to de Quatrefages and Cuvier.[121] But he sought to intervene as a theologian not a scientist, and ultimately it was the clear incompatibility of Darwinian thought with his Christianity which was the foundation of his rejection. Robert Dick, even though sympathetic to the Bishop of Natal, John Colenso, accused of heretical views of biblical inspiration, and although prepared to concede the largeness of Darwin's conception, nevertheless remained loyal to the approach of Hugh Miller, telling his friend C.W. Peach (1800–86) that while Peach might be inclined to go along with Darwin, and was 'welcome to [his] own opinions…I'll not part with one of mine.'[122] (Peach, as it happens, later claimed that he 'never fell in with Darwin's views and conclusions altogether': although Darwin's 'facts are stubborn and not easily got over.'[123])

THE PUBLICATION OF *ON THE ORIGIN OF SPECIES* 59

For the pious provincial bryologist William Wilson (1799–1871), Darwinism was always as much a philosophy as a biological theory, and even at the end of the 1860s, he could confess to pausing on the threshold of a book like Wallace's *Contributions to the Theory of Natural Selection* (1870) 'with fear and trembling'.[124] Early Victorians like Alexander Thomson of Banchory (1798–1868) championed natural science in Scottish universities as a vital defence of natural theology; and where they recognised the corrosive implications of the spread of Darwinian ideas, they searched for ways to broaden natural theology's base.[125] Works like Ebenezer Brewer's *Theology in Science* (1860) attempted to extend natural theology from simple Paleyite adaptation to consideration of broader evidences of beneficence provided by the physical world: the volcano as 'safety valve', the balance of sea and land, the way erosion and decomposition created fertile soils.[126] Even progressive divines like James Martineau (1805–1900) and devout scientists like Lyell were reluctant to give up all the core principles of natural theology.[127] Lyell remained committed to a 'preconceived plan' which demonstrated much greater 'power, wisdom, design or forethought' than a series of 'separate, special, and miraculous acts of creation'.[128] In like manner, Westwood vigorously resisted efforts to argue that mutual adaptation undermined theories of design.[129] As he told his students, 'The marks of design are too strong to be gotten over—design must have had a designer. That designer must have been a person—That person must have been God'.[130]

The Edinburgh botanist J.H. Balfour was also unconvinced, though his sympathy with Asa Gray's cautious welcome made him one of the most plausible early-Victorian candidates to be numbered as one of Darwin's advocates. For his part, even accepting that there might be no inherent incompatibility with natural theology, Hutton saw 'many difficulties' in carrying the theory to its logical conclusions, refusing any central explanatory force for natural selection, affirming the strict limitation of variation, dismissing the ability of the geological record to sustain any of Darwin's key propositions, and falling back on the core watchmaker positions: 'The eye…is a decided stumbling block. I believe that it was formed complete as regards its functions'.[131] Many in his circle agreed. 'I am told you made a stand up fight against Darwin's infidel book', commented his equivalent at the University of Glasgow, George Walker Arnott (1799–1868), 'for the honour of Scotland, he hath not one Botanical defender that I know of'.[132]

Yet, while the Georgians were stalled in their shock at the trespassing of science into matters of revelation, early Victorians were more likely to engage directly with Darwin's argument, not least its definitions and logic.[133]

60   DARWINISM'S GENERATIONS

They approached the *Origin* through the lens of the debates over scientific method, and the rival claims of inductive and deductive approaches which had simmered in the years after the publication of Whewell's *The Philosophy of the Inductive Sciences* (1840), whose third edition had appeared in 1857. Quite a few early-Victorian responses to Darwin reflected Whewell's uneasy acceptance of both conceptual deduction and evidential induction. Swayed by the range and weight of Darwin's examples, many conceded the plausibility of his arguments, but remained troubled by his suggestion of the malleability, even the non-existence, of fixed species, which appeared to strike at the heart of beliefs in the necessity of such divinely ordained concepts and phenomena for the production of knowledge, which Whewell reiterated in his *On the Philosophy of Discovery* (1860).[134] This was true also of J.F.W. Herschel, although he was one of the late Georgian cohort; Herschel's annotations to his copy of the *Origin* make it clear that he engaged with the text primarily as an exercise in logic, searching for the internal inconsistencies in Darwin's argument, and rejecting natural selection as an adequate explanation because of Darwin's inability to explain variability.[135] And although the language the early Victorians used showed little consistency in the way 'induction' or 'deduction' was deployed, the essence of their criticisms was generally consistent: the conclusions that Darwin attempted to draw from his examples crossed the line separating legitimate reasoning from illegitimate speculation; his arguments were either insufficiently inductive (in that as the botanist William Henry Harvey (1811–66) put it, they attempted to force all the evidence to converge on the single fact of evolution), or they were too narrowly deductive, theories which one single contradictory fact would upset.[136] Even a sympathetic figure like Darwin's cousin Hensleigh Wedgwood (1803–91) questioned not merely the evidence for evolution but the possibility of making any such argument in the first place.[137] Complaining of Darwin's failure to define 'species', and his inconsistency in the use of 'natural selection', the Whig politician George Cornewall Lewis (1806–63) lamented that 'His mind is of the German type, speculative, laborious, and unsound.'[138]

Logical deficiency became a staple of early-Victorian anti-Darwinism. While Darwin and his supporters discounted the arguments of revealed religion because they were not scientific, his 'science' was equally called to account: for straining homologies into identities, for 'heaping supposition upon supposition'.[139] Margaret Gatty poked fun at Darwin's attempts to substitute the credibility of his ideas for any actual proof.[140] The clerical naturalist Francis Orpen Morris (1810–93), the moral philosopher Thomas

THE PUBLICATION OF *ON THE ORIGIN OF SPECIES*   61

Rawson Birks (1810–83), and the naturalist and physician Charles Robert Bree (1811–86), who were to remain three of Darwin's most voluble critics, all focused on what they saw as failings of inductive demonstration.[141] They also wrote with a vehemence which few later commentators were prepared to hazard: of the *Origin* Morris remarked that 'a more inconclusive, illogical book…I had never read' (although *Variation* trumped it for 'absolute unmitigated inconclusiveness').[142]

It was a position, if not an emotional pitch, which was widely shared. William Sharpey (1802–80), liberal physician and friend of Darwin, baulked at the extrapolations required by Darwin's homologies: 'I confess while I have great reliance to put on homologies as applied to form and construction,—I have not the same faith in homology as a guide to elementary origin and histogeny,—Adam was formed out of the dust of the earth but Eve was a metamorphosed rib'.[143] Owen, Wilberforce, Martineau, and the radical journalist Thornton Hunt (1810–73) all expressed doubts about the soundness of Darwin's argumentation.[144] They were convinced neither by Darwin's evidence nor his premisses. J.E. Gray (1800–75), Keeper of Zoology at the British Museum, who tellingly validated his opposition to species mutability, as opinion based on his forty years of experience, dismissed the *Origin* as full of 'fallacious reasonings'. In April 1860 he suggested that the *Origin* made little progress amongst scientific men, and was 'Lamarck's doctrine and nothing else'. This was the most damning inconsistency, he thought, given that Darwin had been attacking Lamarck's position for twenty years. Gray rejected the reliability of the fossil record, and was apparently bold enough to tell Darwin that, having read the *Origin*, he had found nothing original in it.[145] John Stuart Mill (1806–73) observed that Darwin did seem to have proved 'that it *may* be true', but this was a narrowly rhetorical judgement arising from contemporary ideas of *vera causa* (a cause that has its existence demonstrated by evidence independent from the phenomenon it is supposed to cause); in fact, despite sharing much of the materialist positions of younger Darwinists like Huxley and Alexander Bain, Mill was never able to reconcile his thinking to Darwinism.[146]

Despite Darwin's deliberate obfuscation, the implications of his argument that man too had evolved was a further stumbling block for many. For J.H. Balfour this was 'the great difficulty'.[147] Darwin's close friend Leonard Jenyns, while less horrified than most early Victorians at the idea of some evolution from earlier fossil forms to contemporary species, baulked at the implications for man's evolution from animals, refusing to

62 DARWINISM'S GENERATIONS

accept that reasoning faculties and moral sense 'c[oul]d ever have been obtained from irrational progenitors.'[148]

Many early Victorians continued to conceive of the challenge of 'science' to religion as primarily geological rather than biological.[149] This was significant because, in the early 1860s, it was the naturalists (especially those like Wallace and H.W. Bates (1825–93), with experience of the tropics) who were by far the likeliest to support Darwinian positions. The early Victorians, with their preoccupation with geologic time and the palaeontological record, were much less favourably placed to accept evolutionary thought, despite the fossil evidence of the sequences of extinct life. The geologists of the pre-Victorian period, like Sedgwick and William Buckland, had been happy to abandon biblical literalism with its genocidal flood, and savage Hebrew blood-letting, while holding onto the benign watchmaker; and they were unwilling to give up the comfort of design to return to the destructive waste of natural selection, and so continued to try to reconcile design and Darwinism. For someone like J.B. Mozley (1813–78), later Regius Professor of Divinity at Oxford, the reliance on design allowed contemplation of the possibility of transmutation, and even the mechanism of natural selection, but only by recognising the 'enormous chasm' left in Darwin's theory, especially in the creation of successions of progressive variations that were needed to create an organ like the eye, and the perversity of not accepting that this gap was filled by divine action.[150] This strain of religious resistance remained a powerful force well into the 1870s and beyond.[151]

It would be easy to categorise this opposition as based on religious dogma and political conservatism. But it seems also to have been true even for advanced radicals. The progressive phrenologist John Epps (1805–69) does not seem to have been able to apply his developmental philosophy to support for the *Origin*. Surprisingly, there is no reference to Darwinism, in his *Life*, including in the extensive diary extracts presented there. Likewise John Elliotson (1791–1868), mesmerist and materialist, offers no evidence of any evolutionary commitments either before or after the *Origin*.[152]

Deeply rooted intellectual and emotional investments in non-evolutionary ideas were a formidable obstacle. The Aberdeen anatomist, John Struthers (1823–99), reflected in 1874 on the way the jury trying Darwin's ideas was prejudiced by its absorption in early life of the belief in the original separation of species.[153] Owen's suggestion to Sedgwick that he was too old now to form any definite ideas, was no doubt self-serving, but it was a position he came increasingly to rely on.[154] Even where there was considerable sympathy and an inclination to follow Darwin, as Edmund Gosse

THE PUBLICATION OF *ON THE ORIGIN OF SPECIES*   63

(1849–1928) suggested, at the brink, mid-career figures like his father Philip, born in 1810, simply lacked the plasticity of mind to abandon fundamental beliefs and embrace new frameworks.[155] Acceptance no doubt looked a great deal easier for the young son. 'I can scarcely think that due allowance is made', complained the botanist George Bentham (1800–84), writing in 1863, 'for those who like myself [after] a long course of study of the phenomena of organic life...have now felt their theories rudely shaken by the new light opened on the field by Mr Darwin, but who cannot surrender at discretion so long as many important outworks remain contestable'.[156]

Mid-Victorian Dilemmas

For the mid-Victorian generation, at the height of its influence when the *Origin* was published, there was less intellectual overcommitment, but the rethinking required to accept Darwin's ideas remained profound. Although in some instances mature convictions were demolished, more common was simply greater caution: a more balanced recognition of the dangers of a hasty or intemperate rejection of the new ideas simply because they appeared to threaten established religious positions, coupled with a slow and strictly limited accommodation. The publisher Alexander Macmillan (1818–96) was one of those who felt that opposition or cries of alarm were fruitless of any real good; in his circle, the *Origin* produced lively discussion without tension, 'Darwin and conundrums and general jollity pleasantly intermixed', as he described one evening in April 1860.[157]

The apparently personal animus of some of the early Victorians' attacks brought Darwin considerable sympathy from many mid Victorians. They frequently drew the lesson from the history of responses to *Vestiges* that hysterical criticism only brought attention and greater importance, and urged the need for space to debate the theory, without fear of the consequences.[158] 'I wish someone could bring out the other side', Macmillan told Tennyson, 'But surely the scientific men ought on no account to be hindered from saying what they find are facts'.[159]

Sir Henry W. Acland (1815–1900), the Regius Professor of Medicine at Oxford, offers a case in point. On evolution, Acland temporised, accepting (at least to Owen) that there was as yet 'no-one...rash enough' to argue that even if some species are demonstrated to have passed one into the other 'this is the general law'.[160] But he was dismayed at the bitterness of Owen's opposition to Darwin in the years after 1859, recognising that any attempt

64 DARWINISM'S GENERATIONS

to shore up religious truth by rejecting scientific arguments because of their religious implications simply placed religion at the mercy of scientific advances.[161] This point of view pushed Acland into intervening in the controversy around Owen's debates with Huxley in the early 1860s. Anxious at the feelings of perplexity and regret prompted by Owen's address on 'Characters of Man and the Higher Monkeys' at the Cambridge meeting in 1862, Acland got up a circular letter urging him to clarify his position on evolution, and avoid giving the impression of resistance on grounds which he knew were not defensible.[162] Acland was not alone: the clerical geologist Samuel W. King (1821–68) was another who felt that Owen should have acknowledged his error gracefully and backed down.[163] Owen resisted, maintaining that his anatomical argument was scientifically sound, and that as it had been his position since at least 1857, it could not be dismissed as anti-Darwinian special pleading. He would not, he told Acland, bow to 'young "claqueries"' and their 'inarticulate noises [nor] to each sophism of my adversary in the <u>Section</u> at Cambridge'.[164] These tensions were reprised after Owen lectured to the YMCA in 1863 on the power of God as evidenced by animal creation.[165] But Owen's transparent dissembling outraged even firmly anti-Darwinian mid Victorians like Whitwell Elwin (1816–1900), the Anglican clergyman and editor of the *Quarterly Review*, who as one of Murray's readers had initially opposed the publication of the *Origin*, and Acland was deeply troubled.[166] 'Even you', he told Owen, 'cannot afford to be indifferent to the goodwill and regard of the generation that follows after you'.[167]

The mid Victorians seem to have invested in much more careful and conscientious reading of the *Origin*. Wollaston confessed to going through the *Origin* with 'the greatest possible care', with a view to ascertaining the precise point in Darwin's argument where he felt he must disagree. The Calcutta-based naturalist, Thomas Thomson, reserved his opinion after a first reading, waiting until a second before indicating he was unconvinced.[168] Similarly, the palaeontologist Thomas Davidson (1817–85) 'read over and over' the copy of the *Origin* that Darwin presented to him. The slightly older Giuseppe Gagliardi (1812–81), recently installed as Professor of Natural History at Cardiff, confessed that going through the *Origin* for the second time in 1867 he was 'rather bewildered', still finding Darwin's ideas 'a mystery, the more profound the more I am plodding in his work'.[169]

But evidence that Darwin had persuaded his readers is in remarkably short supply. Reflecting no doubt his own progress through the text, the Anglican R.W. Church (1815–90) suggested that 'if the reader pursues

THE PUBLICATION OF *ON THE ORIGIN OF SPECIES* 65

the work to the end, he will rise from the perusal,…certainly supplied with many new thoughts, and strengthened and expanded by the sensation of contact with a powerful, philosophical, and richly stored mind', but 'not, perhaps, convinced by Mr. Darwin's arguments'.[170] We might say that the group closest to Darwin, Huxley, Hooker, Tyndall, Lubbock, and other members of the circle which coalesced into the X-Club, who became almost synonymous in the public mind with Darwinism, supported Darwin the scientist far more than the arguments of the *Origin*.[171] Lubbock sanctified Darwin without any attempt to commit to his theory. Tyndall avoided any sort of biological pronouncement until 1870; Spottiswoode even longer. Even for Huxley, Darwin's 'Bulldog', the *Origin* was always primarily artillery for the war of science against religion (a 'Whitworth gun in the armoury of liberalism').[172] He rejoiced especially in Darwin's challenge to conventional ideas of design and its tendency to extend the authority of science over new regions of thought, and he used the accusation that the opposition to the *Origin* was an unwarranted inhibition of science by theology to take the fight to Darwin's opponents.[173] He defended Darwin's inductions from the attacks of early-Victorian logicians, and he was persuaded that Darwin's morphology demonstrated beyond doubt the derivation of more recent fauna from older fossil forms. But in respect of Darwin's actual theory of natural selection he was much more ambivalent. His review of the *Origin* in the *Westminster Review* was full of doubts.[174] One 'stumbling block', as he described it, was that Darwin could not show that mutating infinite varieties could be produced by selection from a common stock; so that selection was not a *vera causa* for the physiological character of species.[175] Another was his unwillingness to abandon a place for sudden jumps as part of the evolutionary process.[176] Huxley remained a 'heretic' on this point to the end 'much to Mr Darwin's disgust', he confessed in 1894.[177]

The efforts of Darwin's supporters to take advantage of the new spaces for debate being opened up by the new shilling monthlies to generate controversy speaks to their interest at least as much in securing freedom from theological constraint as in endorsing Darwin's views.[178] Without persecution, Lubbock argued, there was little chance of progress, but 'if they will only argue and persecute we shall soon have it all our own way'.[179] In 1860 and 1861 Huxley and Lubbock went touting for attacks on Darwin from known opponents, no doubt for precisely this reason.[180] In the same way, Huxley's lectures were deliberately provocative. The regret expressed by the 'Declaration of the Students of Natural Sciences' in 1864 at the 'perversion' of science into occasions for casting doubt on revealed religion was a

66 DARWINISM'S GENERATIONS

godsend to the Darwinians in this respect, provoking just the sort of back-lash against the claims of religious orthodoxy that they had been working so hard to induce.[181]

Once we tease support for the pursuit of scientific truth untrammelled by theological proscriptions apart from commitment to Darwin's specific arguments, it is difficult to find anything more than scattered evidence of acceptance amongst the mid Victorians. Some, like the geologist David Page, who were no doubt already operating within an implicitly evolution-ary framework on palaeontological grounds and who were liable to be iden-tified as Darwinian, were actually careful to position themselves as rejecting Darwin's theories.[182] In retrospect, W.B. Carpenter, George Rolleston, and the Cambridge zoologist Alfred Newton might serve as other examples of fully fledged Darwinians. All three seem to have been immediately con-vinced that Darwin had not only provided a suggestive hypothesis, but a 'vera causa'. Ultimately, Newton's ideas most closely aligned with Darwin's, although as Jonathan Smith has demonstrated, his journey to this position was a great deal slower and more tortuous than some of his later pro-nouncements suggested, and although he came to accept natural selection, he was never fully subscribed to the scientific naturalism of the X-Clubbers. In the same way, although Carpenter was happy to give unqualified endorse-ment to mutability and the efficacy of natural selection, he was much more cautious in his support of the larger questions of descent across all geo-logical time, suggesting only that it had at least as valid a foundation in a broad basis of phenomena as the theory of successive creations.[183] Nor did he accept that natural selection had destroyed traditional theistic arguments of design in creation, and when it came in 1864, he turned down an invita-tion to join the X-Club.[184] In like manner, what chiefly motivated the Broad Church Anglican Rolleston in the years between the publication of the *Origin* and Huxley's *Man's Place in Nature* (1863) seems to have been out-rage at the self-serving sophistries of some of Darwin's scientific rivals, combined with repugnance of the ignorant hostility of many of the clergy,[185] and yet all the signs are that he himself had always been uncomfortable with the vehemence of Huxley's anticlerical rhetoric.[186] Rolleston developed a finely tuned facility for tortured equivocation. He confessed in January 1860 that 'Without being a Darwinite to the entire length he goes, I cannot avoid being one as far as man goes', and even at the end of the decade he can be found writing that 'where verification is, *ex hypothesi*, impossible, such a theory cannot be held to be advanced out of the region of probability',

THE PUBLICATION OF *ON THE ORIGIN OF SPECIES* 67

leaving acceptance or rejection to 'the particular constitution of each individual mind to which it is presented'[187] Although in the early 1860s he was regarded by Huxley as a key member of the Darwinian group, and can be found happily writing in support of Darwin,[188] there are signs that by September 1861 his enthusiasm for the latest revised edition of the *Origin* was waning.[189] From the outset, Darwin was unsure of his commitment, describing one of his essays of 1861 as written in 'fear and trembling of...God, man, Owen, and monkeys'[190]

The skittish amateur naturalist Charles Kingsley, frequently cited as an example of early Darwinian enthusiasm, shared Rolleston's vacillation.[191] Long interested in natural history, Kingsley recognised the power of the *Origin*, and although it 'startled many preconceived judgements', as early as the summer of 1860 he was describing himself as 'to a great extent a Darwinite' (although his friend Charles Bunbury noted that Kingsley was favourable to Darwin's 'speculations, without plunging into them with the headlong zeal of Huxley and Lubbock').[192] His version of evolution was always more progressive than Darwin's, a vision of infinite and perpetual upward development, and inevitably more divine.[193] And his commitment was always ambivalent: within a year he was not so sure, 'differing now and now agreeing'[194] Like Rolleston and many of his contemporaries, Kingsley was less troubled by technical difficulties than by the belligerence of Darwin's supporters. Purely negative attacks alienated him. He was (as he told Lubbock in 1861) trying to find something to assert, 'to show men what there is in the Bible, instead of what there is not'[195]

Whether convinced or not, mid Victorians generally acknowledged and frequently praised the *Origin* as a piece of compelling scientific writing.[196] Echoing the Duke of Argyll, the chemist John Hall Gladstone (1827–1902) read the *Origin* on its appearance with 'interest and pleasure'. Previously sceptical about evolution, Gladstone found that Darwin's argument, especially the notion of natural selection, led him to 'entertain a different idea of the probabilities of the case'[197] But in both instances difficulties soon mounted up. Argyll was troubled by the lack of geological evidence of transitional species, and unconvinced by Darwin's ingenious explanations. He dwelt on the apparent tendency of artificial breeds to revert to type, and the lack of evidence of species change over the 3,000 years of recorded history. He could not reconcile single centres of creation with Darwin's account of transmutation, and was not convinced Darwin had explained the process of transmutation from one species to another.[198] Gladstone likewise found that

68   DARWINISM'S GENERATIONS

as the early controversies developed, doubts solidified, notably along the lines of the objections of Argyll and another mid Victorian, the Catholic biologist St George Jackson Mivart (1827–1900), and like many mid Victorians in the years immediately after 1859, he did not progress beyond a broad acceptance of the significance of the survival of the fittest and limited development through the working of some creative law, combined with a wait and see attitude to a theory 'in the early stage of crude guesses'.[199] Even the more sympathetic entomologist Edwin Brown (1818/9–76), who conceded in 1863 that Darwin's view of nature explained more difficulties than any other writer on evolution, and accepted the force of his remarks on the imperfection of the geological record, was still concerned that the gaps in the series remained 'astoundingly great'.[200]

Immediate mid-Victorian responses were more often inclined to resistance. Even a contributor to *Essays and Reviews* like Rowland Williams (1817–70) read the *Origin* 'with doubt'.[201] Initial responses among the amateurs and enthusiasts, although frequently tempered by academic diffidence, remained generally unconvinced. One amateur ornithologist, George Dawson Rowley (1822–78) confessed that the *Origin* did not convince him, although Darwin appeared to be 'a most patient observer'.[202] The Episcopalian minister Gilbert Rorison (1821–69) rehearsed what would become a staple of mid-Victorian objections—the gulf in brain power of man and animal—in lectures in 1860 and 1861, subsequently published as *The Three Barriers: notes on Darwin's Origin of Species* (1861). Already in 1861, William Thomson had determined that the geological evidence did not allow of the time required for natural selection.[203] Thomson's notebooks of 1861 show that his strategy was 'to undermine the indefinitely long timescale maintained by geologists such as Lyell rather than to attack Darwin directly'.[204]

Despite his hopes, Darwin was unable to convince Thomas Wollaston. Wollaston was prepared to accept a degree of variation but only within the *a priori* limits of each type, despite the fact that his research into the insects of Madeira was entirely consistent with separate evolution from a common stock, arising out of the separation of Madeira from Europe.[205] Similarly, Thomas Davidson could only be drawn to observe that although he was prepared to go some way with Darwin, 'I do not see that he has in any way proved his theory to rest upon a solid foundation', or that it was supported 'with those proofs which science absolutely demands, and which the present state of science is not able to afford'.[206]

THE PUBLICATION OF *ON THE ORIGIN OF SPECIES* 69

High-Victorian Enthusiasm

Where there was sustained enthusiasm in the immediate aftermath of publication beyond Darwin's immediate circle, it came from high Victorians confronting the *Origin* as young adults, free from entrenched beliefs, and open to the allure of the intellectual upheaval and the inspiration which the *Origin* brought. The high Victorians were less visible in the early debates over the *Origin*, notwithstanding a smattering of notices, including two by Darwin's cousin Julia Wedgwood (1833–1917) in *Macmillan's Magazine* in June 1860 and July 1861, an enthusiastic review from the economist Henry Fawcett (1833–84) in the same magazine in December 1860, and other interventions by the 25-year-old Sandhurst cadet Frederick Wollaston Hutton (1836–1905), and the young secularist John Watts (1834–66), who led the discussion in the *National Reformer* newspaper.[207] High Victorians with books in press occasionally took the opportunity to make passing acknowledgement of the new ideas. Rev. Hugh Macmillan (1833–1903), who was later to demonstrate a firm, albeit religiously inflected Darwinian commitment, offered a temporising note in his *Footnotes from the Page of Nature* (1861). Where we find pro-Darwinian interventions in the provinces, closer investigation suggests that these were usually instances of high Victorians with access to print or platform. At Nottingham, one high Victorian cautiously endorsed the credibility of Darwinian evolution to the newly formed Nottingham Naturalists' Society, and reiterated his position in the newspaper correspondence which followed.[208] But these voices were initially only scattered and occasional.

In contrast, in private conversations and discussions there was considerable excitement, particularly if we can believe later autobiographical accounts. In the words of Leslie Stephen (1832–1902), '"Evolution"…was revealing itself as a demon horned and hoofed.'[209] The Cambridge circle of the mathematician W.K. Clifford (1845–79) 'were carried away by a wave of Darwinian enthusiasm', as one later recalled.[210] At Oxford, where Walter Pater (1839–94) was similarly moved, the enthusiasm was equally intense.[211] The clerical geologist T.G. Bonney (1833–1924) 'speedily read' the *Origin* on its publication and 'saw at once' that Darwin had made as great an advance in natural history as Copernicus had in astronomy.[212] At Trinity College Dublin, Sir Robert Ball (1840–1913) claimed he was an 'instantaneous convert', later recalling 'the intense delight with which I read it.'[213] Students railed impatiently at their teachers for failing to recognise what the geologist

70 DARWINISM'S GENERATIONS

Archibald Geikie (1835–1924) described as the 'new revelation'.[214] At Glasgow, the young ministerial student George Matheson (1842–1906) was very quickly using the *Origin* to ruminate from the pulpit on the implications of spiritual evolution.[215] Nor was it just undergraduates. The young Frankfurt-based diplomat Wilfrid Scawen Blunt (1840–1922), the expatriate Opporto merchant and ornithologist, William Chester Tait (1844–1928), and Samuel Butler (1835–1902), on a remote sheep station in New Zealand, three day's ride from the nearest bookshop, all recorded the transformation of their ideas produced by reading the *Origin*.[216] In March 1860, the 24-year-old J. Reay Greene, off to take up an academic post in Dublin, was preparing to follow the advice of Joseph Hooker as to how to approach the staunchly hostile botanist William Harvey (1811–66), vowing to be 'careful and patient in discussing the Darwinian views', but hoping to talk him round.[217]

The appearance of the *Origin* made the naturalist Thomas Belt (1832–78) an 'ardent evolutionist', and led him to the enthusiastic search for facts that would back up the theory.[218] And along the same lines, a reading of the *Origin* prompted Frederick DuCane Godman (1834–1919) and Osbert Salvin (1835–98) to launch an expedition to study the fauna and flora of Central America to investigate the distribution of species and its bearing on evolution, travels which ultimately formed the basis for the multi-volume *Biologia Centrali Americana* (1879–1915).[219] Hooker's warning to Lubbock in 1861, that the group of supporters he was gathering was too narrow, ('they represent the young progressionists in Science', he counselled, and 'their opinions are of no weight in religious matters'; what was needed was men of 'older standing and opposite tendencies'), is suggestive of the Darwinians' initial reliance on the high Victorians.[220]

Conclusion

In some accounts, the transformation to be wrought by the *Origin* was rapidly completed, and the triumph of Darwinian principles assured almost at once. By 1863 Kingsley was celebrating that Darwinism was 'conquering everywhere, and rushing in like a flood, by the mere force of truth and fact'.[221] Except for the responses of the high Victorians, it is difficult to find any evidence to support his optimism. Unthinking hostility garnered sympathy for Darwin, but this rarely had any bearing on underlying beliefs.[222] In metropolitan science circles progress was being made. The publication of Huxley's School of Mines *Lectures to Working Men* at the very start of the

THE PUBLICATION OF *ON THE ORIGIN OF SPECIES* 71

year was met with general admiration and considerable enthusiasm, repeated when his *Man's Place in Nature* appeared a couple of months later. Biblical literalism, both as regards the history of man and of creation, was becoming increasingly difficult to sustain.

But even at a national level, the advance of Darwinian sentiment was limited and uncertain. Outside scientific circles, Darwin's new-found status is easily exaggerated. When in 1861 Antonio Panizzi (1797–1879), Principal Librarian at the British Museum, was asked to identify living men distinguished in science and literature, his list, even after additions from Lord John Russell, contained Whewell, Owen, Lyell, Murchison, Max Müller, Sir William Hooker, and Charles Babbage, but not either Darwin or Huxley.[223] Despite the tribulations of Owen, and Huxley's success in mobilising resentment at clerical 'interference', it is far from clear that in these years the British Association manifested a decisive shift towards evolutionary views. Younger attendees remembered fierce battles in which 'whatever Darwin advanced was viewed with hatred and suspicion',[224] and Sir William Armstrong pulled few punches in describing Darwinism in his presidential address at the 1863 meeting at Newcastle as a hypothesis 'pushed beyond the limits of reasonable conjecture', and according to one account at least, his remarks were well received.[225] Huxley enjoyed it: just fighting enough, he thought, to keep him interested; but his response did not suggest any sense of a fundamental shift in support.[226] For all the mid Victorians' anxiety, Owen remained a figure of authority. Even for those unsympathetic to his wholesale opposition to Darwinism, his expertise could not be ignored, and despite his acceptance of some form of evolution, he continued to serve as beacon of design arguments for older figures, like the geologist George William Featherstonhaugh (1780–1866), and the Cornish gentleman scientist Sir Richard Rawlinson Vyvyan, 8th Baronet (1800–79).[227] It is perhaps easy to overestimate the penetration the *Origin* achieved in its first few years. By 1863 around 7,000 copies of the *Origin* were in circulation; but Murray's sales were plateauing, and only a further 3,000 were issued over the rest of the decade, despite the publication of a 4th edition in 1866, and a fifth in 1869.

Outside London, not just to YMCA audiences or congregational societies, but also for the larger provincial literary and philosophical societies, interest in the relationship between man and ape, and the controversies over the antiquity of man stoked by Lyell and Huxley's works, were still being played out in an overwhelmingly anti-Darwinian register. Just as for initial responses to the *Origin*, in part this reflected the predominance of

72 DARWINISM'S GENERATIONS

religious voices, but it extended also to the scientific voices present. Hawkins and Mackintosh were only the most visible of an apparently inexhaustible supply of lecturers willing to confront the Darwinian challenge to natural theology in all its various guises.[228] The most that can be said is that there was a fuller acknowledgement of the great sequence of fossil life, and a greater willingness to acknowledge the need to treat Darwin's writings as worthy of careful consideration. At Worcester in September 1863, Edwin Lankester (1814–74) urged humility, while seeking to show how erroneous Darwin's ideas were; while in a number of appearances the palaeontologist J.W. Salter (1820–69) steered a narrow course between denying Darwin's ideas of a connected evolutionary process, and conceding only the possibility of some 'slight changes' being brought about through the processes he outlined.[229] Robert Wight recognised the value of the *Origin* in informing procedures for identifying species, and he was prepared to accept that species were perhaps 'not quite <u>so</u> immutable as we short-sighted naturalists have /hitherto\ supposed'.[230] Yet, despite his desire to find a position from which to judge favourably, eighteen months of anxious rumination brought little movement. In December 1862, while conceding the basis of Darwin's theory in the fact of variation, Wight was even more categorical that the *Origin* was 'lost labour for it leads to nothing, does not advance our knowledge of the origin of vitality and only claims for it powers which all our experience goes to disprove'. He remained convinced that only limited modification was possible: breeders might obtain many varieties of pigeon, 'but can't change a pigeon into a hawk'.

Where the initial responses of *Origin's* older readers had been favourable, this did not always survive a period of reflection. The clerical ornithologist, H.B. Tristram (1822–1906), often cited in evidence of the rapid spread of Darwinian ideas, offered early enthusiasm, but after the Oxford meeting he fell in behind the doubts he heard being expressed.[231] Hugh Falconer, who in April 1860 had been ardently recommending the *Origin* as a work of vast research, supported by an unprecedented array of facts and arguments, was by 1863 expressing doubts and distancing himself from any suggestion of endorsement.[232] As J.H. Balfour put it in 1863, 'We may indulge in conjectures as to the mode in which species have been formed, but we have not yet been able to fathom the depths of God's doings in the creation and perpetuation of species'.[233] Fear of the theological implications dragged many clerical scientists back from the cusp of commitment. The first instinct of the future biblical scholar Fenton Hort (1828–92), who had a long-standing interest in the natural sciences, was 'that the theory is unanswerable'; but his enthusiasm quickly ebbed to scepticism (confessing that the 'scientific

THE PUBLICATION OF *ON THE ORIGIN OF SPECIES* 73

question is a very complicated one—far more complicated than Darwin seems to have any idea'). Eventually, he decided against the review of the *Origin* he had considered, and thereafter seems to have studiously avoided any direct endorsement of Darwinian positions.[234]

The force of convention was still very strong. In the summer of 1862, Kingsley had described himself as 'a Darwinite solus' in Chester, and a few months later Darwin was being told of the atmosphere of casual ridicule of the *Origin* in Dublin, and so we should not be surprised that, dining with fellows at Trinity College Dublin in May 1862, Thomas Archer Hirst found ignorance and hostility; one fellow, John Hewitt Jellett (*c.*1817–88), celebrated Wilberforce's triumph at Oxford, complaining that the inordinate interest being excited in the controversy was merely the result of the heterodox tendency of Darwin's views. Hirst's protest was met with silence.[235] A couple of months earlier, a similar scene had been played out at a dinner in Hereford, where the Rev. W.S. Symonds found himself listening to an excited tirade against modern science from the Dean, Richard Dawes (1793–1867), one of Whewell's Cambridge circle.[236]

By May 1863 R.H. Hutton (1826–97) and the editorial staff of the liberal and Broad Church *National Review* had decided that the Darwinian controversy had been sufficiently ventilated.[237] They were premature. The long road to acceptance, not just for natural selection but for an evolution of common descent, was only just beginning.

Notes

1. Still being advertised as such in *Nottingham Journal,* 9 March 1861.
2. Picton to Clodd, 23 July 1908, In-letters, Box 6, Edward Clodd Papers, BC MS 19c Clodd, Brotherton Library.
3. Quoted in Huxley's obituary of Darwin, *Proceedings of the Royal Society of London* 44 (1888), 287; Frances Joanna Bunbury, *Life of Sir Charles J.F. Bunbury, Bart* (1906), II, 154; Rolleston, *Scientific Papers and Addresses*, I, xxii; Eliot to Barbara Bodichon, J.W. Cross, *George Eliot's Life* (1885), II, 108.
4. See for example the construction of natural selection as a 'power...inherent in all vital organisms', 'Notes on the *Origin of Species*', f.130, MS2217, Leonard Horner Papers, NLS. (The same document suggests that Horner understood breeding in Lamarckian, rather than Darwinian terms, that breeding was a matter of 'plac[ing] the animals in circumstances which would 'produce the modification they wished for'.)
5. Huxley to Michael Foster, 14 February 1888, *Life and Letters of T.H. Huxley*, II, 190–1.
6. Carpenter, 'Charles Darwin', *Modern Review* 3 (July 1882), 512; see also Carpenter, 'Darwin on the Origin of Species', *National Review* 10 (1860), 188–214.
7. See H.J. Noltie, *The Life and Work of Robert Wight* (2007).
8. Robert Wight to Asa Gray, 31 March/1 April 1861, Asa Gray Correspondence, Wi–Wy, Harvard; the bryologist William Wilson (1799–1871) likewise described the *Origin* as 'soporific', Wilson to Richard Spruce, June [1870], ff.185–8, Wallace Papers, Add. MS 46435, BL.
9. Rolleston, *Scientific Papers and Addresses*, I, xxii.

74 DARWINISM'S GENERATIONS

10. Falconer to Mary Somerville, 12 April 1860, MSF–1, Somerville Papers, dep c.370, Bodleian.
11. Alexander Macmillan (1818–96) to F.J.A. Hort, 28 January 1860, in *Letters of Alexander Macmillan* (1908), 36–7; he later told A.A. Vansittart that the *Origin* 'has excited some noise here, but on the whole has been tolerantly received', ibid., 43; for Karl Marx it was 'absolutely splendid…Never before has so grandiose an attempt been made to demonstrate historical evolution in Nature', Geoffrey M. Hodgson, *Economics in the Shadow of Darwin and Marx* (2006), 12–13.
12. Martineau to G.J. Holyoake, nd [1859?], Martineau Papers, Add. MS 42726, BL.
13. *The Letters of John Stuart Blackie to His Wife* (1909), 144.
14. Jones Loyd to R.W. Norman, 29 January [1860], in D.P. O'Brien, ed., *Correspondence of Lord Overstone* (1971), II, 904–5.
15. Wallace to George Silk, 1 September 1860, Wallace Online; Robert Chambers, *Vestiges of the Natural History of Creation* (11th ed., 1860), 'Proofs, Illustration and Authorities', lxii–lxiv.
16. A.C. Ramsay to Darwin, 21 February 1860, DCP-LETT-2706A, Darwin Correspondence; F. Burkhardt, 'England and Scotland: The Learned Societies', in Thomas F. Glick, *The Comparative Reception of Darwinism* (2nd ed., 1994, or. 1974), 32–74, 43. For Powell, Darwin's 'masterly volume' 'must soon bring about an entire revolution in opinion in favour of the great principle of the self-evolving powers of nature', Powell, 'The Study of the Evidences of Christianity', *Essays and Reviews* (1860), 139.
17. Eliot to Barbara Bodichon, J.A. Cross, *George Eliot's Life* (1885), II, 108; Sara Hennell to Martineau, 13 April 1860, HM/428, Martineau Papers, CLB; see also Hennell to Martineau, 24 June 1860, HM/429. Devin Griffith argues that Eliot's 'is no compliment', noting various of Eliot's characters who have been 'crippled' by just such an adhesion, D. Griffith, *The Age of Analogy: Science and Literature between the Darwins* (2106), 206. Hennell was concerned to find a natural understanding which could be reconciled with her religious beliefs, 'to gain a scheme of nature that shall truly answer to the instincts of religion that certainly are native to the mind of women especially' as she told Frances Power Cobbe; 'When religion has been effectively naturalized and systematized, and only then, I believe it will be free to grapple with the hard results to wh[ich] uncontrolled Science must inevitably come', Hennell to Cobbe, nd [poss 1870?], Frances Power Cobbe Papers, Box 3 (1870–73), Huntington Library, San Marino CA.
18. *The Sphere*, 1 January 1910.
19. In July 1863 the committee of the Elgin Mechanics' Institute had to justify refusing to purchase the *Origin*, despite requests from a body of members, on the basis that they considered the majority would be opposed, *Elgin and Morayshire Courier*, 17 July 1863.
20. John G. Macvicar to J.H. Balfour, 3 November 1860, M87, John Hutton Balfour Papers, RBGE.
21. See the letters from Dick to C.W. Peach, Robert Dick Correspondence, Gen863/1, Edinburgh.
22. Letter in *The Guardian*, 13 August 1890.
23. Huxley to Hooker, 31 December 1859, quoted in Browne, *Darwin*, II, 105.
24. The audience included Charles Babbage, JT/2/32c, Thomas Hirst Journal, tps, RI.
25. G.O. Trevelyan, *Life and Letters of Lord Macaulay* (1876), II, 403–4.
26. Bunbury, *Life of Sir Charles J.F. Bunbury* II, 151, 154.
27. James Bryce, 'Personal Reminiscences of Charles Darwin and of the Reception of the "Origin of Species"', *Proceedings of the American Philosophical Society* 48 (193) (September 1909), iii–xiv; George Rolleston, *Scientific Papers and Addresses* (1884), I, xxii. See also the discussion in Hesketh, *Of Apes and Ancestors*, 28–9.
28. John Duns (1820–1909), review in *North British Review* 32 (1860), 455–86, 455.
29. Quote from Amy Cruse, *Books and the Victorians* (1935), 96. In July James Brooke (1803–68), Rajah of Sarawak, was said to be enjoying the *Origin* exceedingly, Woolner to Lady Tennyson, 25 July 1860, Amy Woolner, *Life and Letters of Thomas Woolner* (1917), 197; though Brooke remained unconvinced, at least of the evolution of man, see Spenser Buckingham St John (1825–1910), *Life of Sir James Brooke* (1879), 374.

THE PUBLICATION OF *ON THE ORIGIN OF SPECIES* 75

30. See Daubeny, *Remarks on the Final Causes of Sexuality in Plants, with Particular Reference to Mr Darwin's Work on The Origin of Species* (1860).

31. Jenyns was noted in the *BMJ*, 16 March 1861 as having given his 'emphatic condemnation', 27; his obituaries noted a figure who 'was cautious in accepting the theories of brilliant but oft-times erratic theorists', *Journal of Microscopy and Natural Science* 12 (1893), 418; John Gwyn Jeffreys wrote to challenge some of Darwin's geological details; see Darwin's response, 29 December [1859]; Jenyns to Darwin, 4 January 1860, DCP-LETT-2637A, Darwin Correspondence; W.J. Sollas, *The Age of the Earth and Other Geological Studies* (1906), 251–4.

32. Discussion in Hesketh, *Oxford Debate*, 28–9. For Clark see Browne, *Darwin*, II, 117, citing letter of Hooker.

33. Phillips to Sedgwick, 1 May 1860, MS Add. 7652/II/O/43, Sedgwick Papers, CUL.

34. Examples include Rev. T. Lawson to the Bacup Mechanics' Institute, *Bury Times*, 28 April 1860, and lectures that furnish published interventions, including Gilbert Rorison at Peterhead, *Peterhead Sentinel*, 11 May 1860.

35. Alexander Macmillan to Daniel Wilson, 25 July 1860, *Letters of Alexander Macmillan*, 58; Matthew Stanley, *Huxley's Church and Maxwell's Demon: From Theistic Science to Naturalistic Science* (2015), 57–9.

36. See Darwin to W.B. Carpenter, 19 November 1859, DCP-LETT-2536, Darwin Correspondence.

37. For Watson, see Frank N. Egerton, *Hewett Cottrell Watson: Victorian Plant Ecologist and Evolutionist* (2003).

38. Darwin to Hooker, 30 May 1860, DCP-LETT-2818, Darwin Correspondence. For the relations of Spencer and Darwin see B. Lightman, 'The "Greatest Living Philosopher" and the Useful Biologist', in Hesketh, ed., *Imagining the Darwinian Revolution*, 37–57.

39. Barton, 'The Darwinism of the X Club', in Hesketh, ed., *Imagining the Darwinian Revolution*, 58–79. For further discussion of this point, see above, p. 65.

40. Jukes to Darwin, 27 February 1860, DCP-LETT-2716A, Darwin Correspondence.

41. G.H.K. Thwaites to J.D. Hooker, 13 February, 24 April 1860, 28 September 1863, DC/162/126, 134,204, Directors' Correspondence, RBGK; Thwaites to Darwin [14 February 1860], DCP-LETT-2697, 24 September 1863, DCP-LETT-4303, Darwin Correspondence.

42. Sedgwick to Darwin, 24 November 1859, DCP-LETT-2548, Darwin Correspondence.

43. A. Bain, *Autobiography* (1904), 257–8.

44. Jane Carlyle to Anna Thornhill [28 January 1860], K.J. Fielding and D.R. Sorensen, eds., *Jane Carlyle: Newly Selected Letters* (1988), 247–8, and to Mary Dods, 21 February 1863, ibid., 289.

45. I. Campbell, ed., *George Douglas, Eighth Duke of Argyll, K.G., K.T. (1823–1900): Autobiography and Memoirs* (1906), II, 482.

46. *The Times*, 8 August 1863.

47. Argyll to Lyell, 29 February 1860, COLL 203/1/63–68, Lyell Papers, Edinburgh.

48. Archbishop Temple's letter to his son William Temple, 3 June 1898, E.G. Sandford, *Memoirs of Archbishop Temple by Seven Friends* (2 vols, 1906), II, 686–8.

49. See W.E.H. Lecky, 'Formative Influences', *The Forum* 9 (1890), 380; 'Darwin on the Origin of Species', *Westminster Review* 17 (April 1860), 541–70.

50. Quoted in Hesketh, *Oxford Debate*, 9.

51. John Brooke, 'The Natural Theology of the Geologists: Some Theological Strata', in L.J. Jordanova and Roy Porter, *Images of the Earth: Essays in the History of Environmental Sciences* (1979), 39–64; J.R. Topham, *Reading the Book of Nature: How Eight Best Sellers Reconnected Christianity and the Sciences on the Eve of the Victorian Age* (2022).

52. George W. Stocking Jnr, *Victorian Anthropology* (1987), 38; James Secord, *Victorian Sensation: The Extraordinary Publication, Reception, and Secret Authorship of Vestiges of the Natural History of Creation* (2003).

53. Adrian Desmond, *The Politics of Evolution: Morphology, Medicine and Reform in Radical London* (1989); Piers J. Hale, 'The Politics of the Darwinian Revolution', in Hesketh, ed., *Imagining the Darwinian Revolution*, 108.

54. Hugh Miller, *Footsteps of the Creator* (1861, or. 1849), 17–18.

76 DARWINISM'S GENERATIONS

55. See Nicholas A. Rupke, *Richard Owen: Biology without Darwin* (2nd ed., 2009), 183–208.
56. See Rosemary Ashton, *142 Strand: A Radical Address in Victorian London* (2011).
57. For Baden Powell, see his 'On the Law Which Has Regulated the Introduction of New Species', *Annals and Magazine of Natural History* ns 16 (1855), 184–96, and the discussion in Pietro Corsi, *Science and Religion: Baden Powell and the Anglican Debate, 1800–1860* (1988), and Martin Fichman, *An Elusive Victorian: The Evolution of Alfred Russel Wallace* (2004), 66–138; Rupke, *Richard Owen*, 86, 159.
58. See Alfred Newton, 'The Early Days of Darwinism', *Macmillan's Magazine* 57 (February 1888), 241–9; reprinted in A.F.R. Wollaston, *Life of Alfred Newton* (1921). Amongst the extensive secondary literature, see Crosbie Smith, *The Science of Energy: A Cultural History of Energy Physics in Victorian Britain* (1998).
59. Endersby, *Imperial Nature*, 319.
60. Quoted in Nixon, *Gerard Manley Hopkins*, 145.
61. Perhaps more as a rebuttal of ideas of the great age of the earth than of evolution per se; Kingsley apparently told Gosse that in publishing the book 'he had done more for the cause of infidelity than anyone in the last 20 years', Kingsley to Dear Sir [from context likely W.S. Symonds], 10 August 1860, Box 23, Folder 3, Charles Kingsley Papers, M.L. Parrish Collection, PUL.
62. Edward Vivian (1808–93) to Lyell, 14 January 1860, COLL 203/1/5857–5862, Lyell Papers, Edinburgh. In 1867 Vivian indicated his refusal to accept Darwinian accounts, or successive creations, committing instead to the notion of the 'days' of Genesis as eras; see *Exeter and Plymouth Gazette*, 11 January 1867.
63. Stocking, *Victorian Anthropology*, 71–4; A. Bowdoin van Riper, *Men among the Mammoths: Victorian Science and the Discovery of Human Prehistory* (1993).
64. See P. Kitcher, *The Advancement of Science* (1995), 17.
65. Newton to Sydney Vines, 18 May 1901, MS Add. 8580/60, Sydney Vines Papers, CUL.
66. Owen to John Robert Parker Jnr, 17 June 1859, MS5786/158, Richard Owen Papers, Wellcome. Hesketh observes that although Owen's archetype conception had clear links with opposition to transmutation, it was not necessarily anti-mutationist; Owen believed evolution resulted from mutations aligned to an overarching divine plan.
67. Thomas Wollaston to Lyell, 31 July [1859], COLL 203/1/6180–82, Lyell Papers, Edinburgh. 'I admire your pluck in proportion as I detest your theory' Wollaston told Darwin, [16 September 1860], DCP-LETT-2919, Darwin Correspondence.
68. R.A. Ackerman, *The Myth and Ritual School: J.G. Frazer and the Cambridge Ritualists* (1991), 75; Hesketh, *Oxford Debate*, 30–46; Rupke points out he had studied under Buckland, and also attended Owen's lectures, Rupke, *Richard Owen*, 162, and was already in February 1860, according to Lyell, 'seiz[ing] on the most vulnerable points with no small skill', Lyell to Darwin [13–14 February 1860], DCP-LETT-2694, Darwin Correspondence.
69. Hesketh, *Oxford Debate*, 80.
70. Segdwick to Alexander Thomson, 30 July 1860, THO/72, Thomson Papers, New College, Edinburgh.
71. Joseph Crompton (1823/4–78), from Norwich, pressing Hooker to come and lecture in Norwich, 'after the neat "backhander" you gave a certain episcopal dignitary at the Huxley and Darwin discussion there', J. Crompton to Hooker, 29 August 1861, JDH/1/2/15/121, Hooker Papers, RBGK. See T. Okey, *A Basketful of Memories* (1930), 42. Balfour Stewart to J.D. Forbes, 5 July 1860, msdep7, Incoming letters 1860, no.133, Forbes Papers, St Andrews. Huxley's response to Wilberforce's jibe was, thought Stewart, 'most impudent'; in turn Forbes reported to Whewell that he had 'heard that Huxley's attack on the B[isho]p of Oxford in re Darwin was most indecent', James D. Forbes to Whewell, 15 July 1860 c204/133, Trinity College, Cambridge. Tuckwell presented it as 'the younger men...on the side of Darwin, the older men against him', *Reminiscences of Oxford* (2nd ed., 1908), 56.
72. John G. Maciver to J.H. Balfour, 11 August 1860, M82, John Hutton Balfour Papers, RBGE. See also the diary account of the Methodist divine Benjamin Gregory (1820–1900), J. Robinson Gregory, *Benjamin Gregory, D.D.: Autobiographical Recollections* (1903), 414.
73. In a recollection provided to Frank Darwin for his *Charles Darwin: His Life Told in an Autobiographical Chapter, and in a Selected Series of his Published Letters* (1892), 238. The

THE PUBLICATION OF *ON THE ORIGIN OF SPECIES*   77

obituaries of the Conservative MP Percy Wyndham (1835–1911) reported that he was present at the debate and was 'an early believer in Darwin's theories'; see Max Egremont, *The Cousins* (1977), 21.

74. F.W. Farrar to Leonard Huxley, 12 July 1899, ff.13–19, Box 16, Huxley Papers, ICL. For one exception see Rev. S.W. Earnshaw of Sheffield, who returned to report that some of the attenders had spoken very favourably of Darwin, and only a few condemned him, *Sheffield Independent*, 14 July 1860.

75. *British Standard*, 23 November 1860.

76. *Transactions of the Tyneside Naturalists' Field Club. V. 1860–62* (1862), 19, *Liverpool Mercury*, 11 December 1860.

77. Duke of Argyll, 'Presidential Address', *Proceedings of the Royal Society of Edinburgh 1858–64*, 350–76; *Proceedings of the Cotteswold Naturalists' Field Club III* (1865), 27–8.

78. *Cork Constitution*, 16 January 1860.

79. Samuel W. North (1826–94), nevertheless, asking for careful consideration, *York Herald*, 28 November 1863.

80. *Paisley Herald*, 11 October 1862.

81. D. Livingstone, 'Public Spectacle and Scientific Theory: William Robertson Smith and the Reading of Evolution in Victorian Scotland', *Studies in the History and Philosophy of Biological and Biomedical Sciences* 35.1 (March 2004), 1–29, 6–7.

82. *Gloucester Journal*, 22 December 1860. Hawkins' lectures included 'The Gorilla' which he was giving in 1862, and 'On the unity of plan and varied adaptation of the animal frame...viewed as evidence of Design and Fore-Knowledge of the Omniscient Creator', which he was giving at YMCAs and elsewhere in 1863; see *Norfolk News*, 12 December 1863.

83. As in 'The Being of God Demonstrated from Natural History', *Clerkenwell News*, 21 November 1860.

84. *Arbroath Guide*, 15 December 1860.

85. Letter of Jeffs, *Land and Water*, 18 February 1871.

86. For details see Rupke, *Richard Owen*, 270–86; Desmond, *Huxley*, 281–311, Jim Secord, 'Shop Talk', 42–3; gorillas were only a scientific rumour before du Chaillu turned up in 1861 with his stuffed examples. For the register of Du Chaillu's gorilla book as one of the sensations of the season, see *Punch*, 1 June 1861, 226.

87. Macmillan to Geikie, 19 June 1861, Sir Archibald Geikie Papers, Gen 524/10, Edinburgh: 'the Gorilla book has been obscuring everything, and "all the world has been wondering after the beast"'.

88. Thomson, 'Physical Considerations Regarding the Possible Age of the Sun's Heat'. See also his questioning of Darwin's account in *Macmillan's Magazine*, Smith, *Science of Energy*, 172–3.

89. Quoted in David B. Wilson, 'A Physicist's Alternative to Materialism: The Religious Thought of George Gabriel Stokes', *Victorian Studies* 28 (1984), 94.

90. J. Smith, 'Alfred Newton: The Scientific Naturalist Who Wasn't', in Lightman and Reidy, eds., *The Age of Scientific Naturalism* (2014), 139.

91. Owen, 'On the Characters of the Aye-aye as a Test of the Lamarckian and Darwinian Hypothesis of the Transmutation and Origin of Species', *Report of the British Association for the Advancement of Science Held at Cambridge in October 1862* (1862), 114–16.

92. Lyell to Huxley, 11 October 1862, ff.76–7, Box 6, Huxley Papers, ICL.

93. Rev G.T. Perks, *London City Press*, 7 February 1863, George Field, *Bradford Review*, 5 February 1863; editorial, *Newcastle Chronicle*, 29 August 1863, *Newcastle Guardian*, 22 August 1863 ('the greatest comparative anatomist of his age').

94. Hence Sir Francis Palgrave (1788–1861), in the immediate aftermath of the publication of the *Origin*, noted to a friend that 'Darwin's Unity [sic] of Species is very interesting and has sold out', Palgrave to Dawson Turner, [1859], MC 2847/N2/4/124–139, Norfolk Record Office; while Mary Somerville cagily remarked that *Origin* was 'certainly...the most profound investigation of an extremely difficult subject, and will no doubt give rise to much discussion', Mary Somerville to [Hugh] Falconer, 27 April 1860, printed in *Lady Grace Prestwich, Essays Descriptive and Biographical* (1901), 132.

78 DARWINISM'S GENERATIONS

95. Sumner to Horner, 1 April 1861, *Memoir of Leonard Horner*, II, 303. The entomologist W.S. Macleay (1792–1865) simply concluded that there was nothing to be gained in terms of the argument of the *Origin* in denying the constant intervention of a creative power, *Life of Lord Sherbrooke*, II, 205–6.

96. Froude, *Reminiscences by Thomas Carlyle* (1881 ed.), 434. Carlyle had much more time for Owen, who he (apparently) described as 'one of the few men who was neither a fool nor a humbug', *Life of Richard Owen*, I, 198.

97. E.J. Whately, *Life and Correspondence of Richard Whately, D.D., Late Archbishop of Dublin* (1866), II, 438; Bingham to Richard Owen, 20 January 1863, ff.153–4, Gen Corr V, Owen Papers, NHM.

98. *Address of Thomas Bell, Esq., F.R.S. etc., the President...Read at the Anniversary Meeting of the Linnean Society, on Monday May 24th, 1859* (1859); see reference in Lyell to Darwin, 13–14 February 1860, DCP-LETT-2694, Darwin Correspondence.

99. See Benjamin Brodie, *Autobiography* (1865), 15, Timothy Holmes, *Sir Benjamin Collins Brodie* (1898), 205–6; Brodie was one of the speakers at the 1860 Oxford British Association, opining baldly that he could not subscribe to the Darwinian thesis, Hesketh, *Oxford Debate*, 80. For Brewster's dismissal of Darwin as no more scientific than table turning, see his opening address to Edinburgh University in 1861, *Inverness Courier*, 7 November 1861.

100. Alexander Macmillan to Daniel Wilson, 25 July 1860, *Letters of Alexander Macmillan*, 58; *Macmillan's Magazine* 3 (1860–1), 336; J. Herschel, *Physical Geography* (1861), 11–12, where Herschel stands by his successive creations view. See also Murchison to George Gordon, 1 December 1859, quoted in Michael Collie, *Murchison in Moray: A Geologist on Home Ground* (1995), 184. Murchison sided firmly with Kelvin and those who questioned the magnitude of geological time required by Darwinian theory, as in his *Siluria* (4th ed., 1867), 499–500 (C. Smith and M.N. Wise, *Energy and Empire: A Biographical Study of Lord Kelvin* (1998), 590). For Hopkins see 'Physical Theories of the Phenomena of Life', *Fraser's Magazine* 61 (June 1860), 739–52, and 62 (July 1860), 74–90; discussed in M.J.S. Hodge, 'England', in Thomas F. Glick, ed., *The Comparative Reception of Darwinism* (2nd ed., 1994), 23–34.

101. Murchison's copy is in private hands, but see description in auction listing, Bonhams, 25 May 2004, https://www.barnebys.co.uk/realised-prices/lot/darwin-charles-on-the-origin-of-species-by-means-of-natural-selection-YSHSoxCaD [accessed 27 January 2021].

102. Henslow to L. Jenyns, 26 January 1860, quoted in Armstrong, *English Parson-Naturalist*, 64, 69; he was even less positive to Adam Sedgwick, telling him that [Asa Gray's] review in the *Annals of Natural History* 'squares perfectly with my ideas,...I wrote to Darwin lately expressing my conviction that he had quite overrated the bearing of his facts', Henslow to Sedgwick, 7 February 1860, MS Add. 7652/II/O/25, Sedgwick Papers, CUL. Darwin confessed to Lyell in February 1860 that Henslow, like Bunbury, would go 'only a very little way with us', 15–16 [February 1860], DCP-LETT-2700, Darwin Correspondence.

103. Henslow to [Daniel Nihill], 14 February 1861, John S. Henslow Papers, H382, APS.

104. Letter to Miss Gerard, 2 January 1860, in J.W. Clark and T.M. Hughes, *Life and Letters of the Rev Adam Sedgwick* (1890), II, 359 ('a dish of rank materialism, cleverly cooked and served up'). See his paper to the Cambridge Philosophical Society in 1860, discussed in Burkhardt, 'England and Scotland', 70–1, and Sedgwick, 'Objections to Mr Darwin's Theory of the Origins of Species', *Spectator*, 24 March 1860, 285–6, 7 April 1860, 334–5.

105. Macvicar to J.H. Balfour, 8 February 1860, M80, John Hutton Balfour Papers, RBGE. This was true also of Charles Lyell; see M. Bartholomew, 'Lyell and Evolution: An Account of Lyell's Response to the Prospect of an Evolutionary Ancestry for Man', *BJHS* 6.3 (1973), 261–303, 275.

106. See Huxley to Lyell, 17 August 1862, f.31, Box 30, Huxley Papers, ICL, which commented that 'Natural selection is a great step in making Lamarck's "variation" hypothesis intelligible.' Lyell also praised passages in which Huxley accepts 'the Darwinian hypothesis, with a reserve', Box 6, Huxley Papers, ICL.

THE PUBLICATION OF *ON THE ORIGIN OF SPECIES* 79

107. Marion Bell (née Shaw) (1787–1876) to Kingsley 6 April [1866–72?,] ff.186–7, Kingsley Papers, Add. MS 41299, BL.

108. E.B. Poulton, *Charles Darwin and the Origin of Species* (1909), 15; John Willis Clark, *Old Friends at Cambridge* (1910), 75; William Whewell to J.D. Forbes, 24 July 1860, Incoming letters 1860, no. 145 (a,b), Forbes Papers, St Andrews; and his 'Comte and Positivism', *Macmillan's Magazine* 13 (1866), 388–93. For the complicated relationship of Whewell's thought and Darwin's, see Levine, *Darwin among the Novelists*, 24–55. Whewell's engagement was largely on the general grounds of the relations of science and religion, and largely a repetition of his earlier response to *Vestiges*; see Whewell to John Evelyn Denison, 15 November 1863, Os/1303, Denison Collection, University of Nottingham.

109. See lecture notes of H.G. Seeley [n.d. but probably *c*.1862], MS Add. 7652/II/GG/9, Sedgwick Papers, CUL.

110. Sedgwick subsequently set his students the examination task of repudiating Darwin's ideas, Browne, *Darwin*, II, 108. One reviewer of his *Life* remarked that 'Darwinism he loathed and attacked in a review so bitter, that as Mr Darwin remarked, "no one else could have used such abusive terms"', *London Evening Standard*, 4 September 1890; for his drumming up of opposition, see the various acknowledgements and responses in MS Add. 7652/II/II, Segdwick Papers, CUL.

111. John Kelly to Sedgwick, 13 June 1860, MS Add. 7652/II/II/7, Segdwick Papers, CUL; Kelly to Phillips, 18 November 1863, JP/C/1863/20a, John Phillips Papers, OMNH.

112. 'Language', Jonathan Couch Papers, APS.

113. 'Recent Speculations on Primeval Man by the Duke of Argyll—March 1868', Couch Papers, APS. Crouch was willing to draw from Argyll various arguments against the evolution of man; see his notes to 'On Primeval Man, by the Duke of Argyll. No.1 Origin of Man'.

114. Lionel Tollemache, *Benjamin Jowett: Master of Balliol* (1904), 67.

115. As quoted in Lightman, *Victorian Popularizers*, 156–8, which also outlines her subsequent expressions of hostility to William Harvey, including her short story, 'Inferior Animals' (1861).

116. Andrew Murray, 'On Mr Darwin's Theory of the Origin of Species', *Proceedings of the Royal Philosophical Society of Edinburgh* 4 (1860), 274–91. See Clark, *Bugs*, 113; Duncan M. Porter and Peter W. Graham, *Darwin's Sciences* (2016), 76; for the persistence of these views, see Westwood's lecture in Nottingham, *Nottinghamshire Guardian*, 24 March 1871. For another see remarks of William Keddie (1809–77), *Glasgow Herald*, 20 November 1860.

117. C.J.F. Bunbury to Darwin, 30 January 1860, DCP-LETT-2669, and Darwin's response, Darwin to Bunbury, 9 February [1860] DCP-LETT-2690, Darwin Correspondence. In the weeks before the *Origin* had appeared, Bunbury had confided in his diary that it was a book sure to 'cause no little combustion', but also wrote that 'however mortifying it may be to think that our remote ancestors were jelly fishes, it will not make much difference practically to naturalists who deal with recent plants and animals', Bunbury, *Life of Sir Charles J.F. Bunbury* II, 160.

118. Jardine to J.H. Balfour, 7 January 1860, J90, John Hutton Balfour Papers, RBGE. See also Jardine to Darwin, 20 December 1859, DCP-LETT-2590, and Darwin to Lyell, 25 February 1860, DCP-LETT-2714, Darwin Correspondence.

119. Griffith's copy of the *Origin*, 1st edition, 5th thousand, courtesy of Horden Books, Sydney, inventory number 5000782. Griffith, who was born in Kildare and was a devout Anglican, had initially emigrated as a 22-year-old in 1840; see C. Woods, 'Griffith, Charles James (1808–63)', *Australian Dictionary of National Biography*.

120. Elder, *Chronic Vigour*, 35, quoting H.P. Liddon, *Life of Edward Bouverie Pusey* (1897), IV, 336. See also the response of William Ullathorne (1806–89), primarily concerned with the implications for religious argument of conceding the truth of 'the ape theory', Ullathorne to Brown, 28 May 1863, quoted in Judith F. Champ, *William Bernard Ullathorne, 1806–1889: A Different Kind of Monk* (2006), 273–4.

80 DARWINISM'S GENERATIONS

121. Pusey to H.N. Ridley, 20 November [1868], HNR/2/1/5/113–14, Ridley Papers, RBGK.
122. Dick to C.W. Peach, 15 April 1860, Robert Dick Correspondence, Gen863/1/89, Edinburgh.
123. Peach's later annotation of Dick's letter of 15 April 1860, Robert Dick Correspondence, Gen863/1/89, Edinburgh.
124. Wilson to R. Spruce, June [1870], ff.185–8, Wallace Papers, Add. MS 46435, BL. (Wilson was clearly fundamentally opposed to Darwinism on one level; but wrote of 'Darwin's theory being <u>one truth</u> in conjunction with <u>another</u> (and perhaps higher) truth'.) Wilson's letter is an interesting example of the kind of correspondence almost certainly under-represented in the archive, surviving only because Spruce as recipient, whose papers do not survive, forwarded the letter on to Wallace.
125. See correspondence in THO 20, Alexander Thomson Papers, New College, Edinburgh; and the work of Ebenezer Cobham Brewer (1810–97), including his *Theology in Science*, discussed in Lightman, *Popularizers*, 64–71.
126. Lightman, *Popularizers*, 64–71.
127. See Martineau to Knight, 11 July 1871, ff.13–14, Carpenter MS 6, Harris Manchester, Oxford.
128. Lyell journals, cited in Bartholomew, 'Lyell and Evolution'.
129. J.O. Westwood to Lubbock, 28 December 1874, ff.75–6, Avebury Papers, Add. MS 49644, BL. What is notable here is not just the opinion, but also fact that Westwood is deter-mined to present it so forcefully in a thank-you letter.
130. Notes for 1864 lectures, Box 48, Westwood Papers, OMNH. Blackie's sardonic agreement is visible in the short ditty reprinted in W.D. Adams, *Quips and Quiddities* (1881).
131. John Hutton Balfour to Gray, 12 December 1860, Asa Gray Correspondence, B–Ba, Harvard. By April 1861, Balfour was prepared to concede that Darwin was not inevitably opposed to natural theology, but 'if you carry out Darwin's hypothesis to its limits you will find many difficulties'.
132. Arnott to Balfour, 11 May 1860, A234, Hutton Balfour Papers, RBGE; see a later dispara-ging aside about 'the Darwinian theory', Arnott to William Wilson, 18 July 1865, MS 52, William Wilson Papers, Warrington Library.
133. As in the intervention of Charles Perry, Bishop of Melbourne (1807–91), at the annual meeting of the Royal Society of Victoria, *The Age*, 9 April 1861, *The Argus*, 12 July 1861. For this and Australian responses generally, see Barry W. Butcher, 'Darwin Down Under: Science, Religion, and Evolution in Australia', in R.L. Numbers and J. Stenhouse, eds., *Disseminating Darwin* (1999), 39–59, 43. Writing to Sir George Stokes in 1880, Perry told Stokes that his remarks on the Darwinian theory of evolution 'are just such as I should have expected from you, and from every intelligent man, who has under-stood the principles of inductive philosophy', 8 December 1880, MS Add. 7656/1P/236, Stokes Papers, CUL.
134. See P.M. Duncan to Lyell, 11 January 1870, COLL 203/1/1142–43, Lyell Papers, Edinburgh.
135. Charles H. Pence, 'Sir John F.W. Herschel and Charles Darwin: Nineteenth Century Science and Its Methodology', *HOPOS: The Journal of the International Society for the History of Philosophy of Science* 8.1 (2018), 108–40.
136. William Henry Harvey to Asa Gray, 3 November 1860, quoted in Glick, *What about Darwin*, 176; for the extent to which Harvey engaged constructively with Darwin's ideas, see P.J. Bowler, 'In Retrospect: Charles Darwin and His Dublin Critics: Samuel Haughton and William Henry Harvey', *Proceedings of the Royal Irish Academy* 190C (2009), 409–20; but according to the recollection of Sir Robert Ball, although Harvey later regretted the extent of his initial hostility, 'he never became a real Darwinian. Indeed, I don't think he ever truly grasped what Darwinism meant', *Reminiscences of Sir Robert Ball* (1915), 44; Bellon, 'Inspiration', 403–4. William Thomas Denison (1804–71) commented that 'When [Darwin] passes this true deductive inference, and proceeds to build further inductions on it, and to force all things to converge on one point, then I draw back…A good deal of Darwin reads to me like an ingenious dream', Denison to Tait, 11 June 1861, TAIT/79/253–7, Lambeth Palace Library. This was true also of Sedgwick, see for example his letter to Livingstone, *Letters of Adam Sedgwick*, 411–12; see Richard Bellon, 'The

THE PUBLICATION OF *ON THE ORIGIN OF SPECIES* 81

Moral Dignity of the Inductive Method and the Reconciliation of Science and Faith in Adam Sedgwick's *Discourse*, *Science and Education* 21 (2012), 937–58.

137. Wedgwood to Darwin [January? 1860], DCP-LETT-2389, Darwin Correspondence. Interestingly his daughter Katherine Euphemia Wedgwood (1839–1931) was by the 1870s still comfortable reconciling evolution with the account of creation in Genesis, K.E. Farrer to unidentified correspondent, 17 February [?1878], 9609/4/1/4, Farrer Family Papers, Surrey History Centre.

138. Lewis to Mrs Robert Lowe, 27 February 1860, in M.A. Pratchett, *Life and Letters of the Rt Honourable Robert Lowe*, (1893), II, 203, quoted in Glick, *What about Darwin*. See also for example Dr Thomas Gough (1804–80), GP and amateur naturalist, who commented to Adam Sedgwick that the *Origin* was 'very interesting, well written, tho' not logical[,] not inductive, I think', Gough to Sedgwick, 30 June 1860, MS Add. 7652/II/II/8, Sedgwick Papers, CUL.

139. Francis O. Morris, *All the Articles of the Darwinian Faith* (1875), 30. 'It is always the same', wrote one critic, '*facts* on the one side, *theory* on the other', J.E. Howard, *Creation and Providence* (1878), 3; Birks, *Scripture Doctrine of Creation* (1872), 223–42. For a sustained discussion of these sorts of attacks on the implications of Darwin's thinking, see Dawson, *Darwinism, Literature and Victorian Respectability*. E.D. Girdlestone (1805–84) was still engaging Huxley in debates around induction in the late 1880s, see Girdlestone to Huxley, 10 December 1886, and Huxley to Girdlestone, 19 September 1890, Huxley Papers, APS.

140. M. Gatty to W. Harvey, 14 February 1861, HAS48/149, Sheffield City Archives.

141. For Morris, see Lightman, *Popularizers*, 43–8; C. Bree, *Species not Transmutable, nor the Result of Secondary Causes, Being a Critical Examination of Mr Darwin's Work Entitled 'Origin and Variation of Species'* (1860). Bree conceded that this pamphlet 'was written too hastily and contains several mistakes', and that Asa Gray's more sympathetic response 'makes out a plausible case'; but maintained that Gray could 'never prove the truth of Darwin's wild and unscientific hypothesis', Bree to J.H. Balfour, 27 March 1861, B374, John Hutton Balfour Papers, RBGE. See the similar vehemence of Bree to Newton, 19 May 1860, MS Add. 9839/1B/969, Newton Papers, CUL.

142. M.C.F. Morris, *Francis Orpen Morris: A Memoir* (1897), 221–2. He was described by his nephew as 'a naturalist of an early Victorian type, and being sixty years of age when the *Origin of Species* was published, it was natural that he should retain to the end of his life his biblical belief in special creation', F.E. Bower, *Sixty Years of Botany in Britain* (1938), 2.

143. William Sharpey to Lubbock, 25 January 1862, ff.84–7, Avebury Papers, Add. MS 49639, BL.

144. *Quarterly Review* (1860), see Wilberforce, *Essays*, I; [Richard Owen], 'Darwin on the Origin of Species', *Edinburgh Review* 111 (1860), 487–532. James Martineau, 'Science, Nescience and Faith', *National Review* 15 (October 1862), 395. Hunt observed that although he read the *Origin* with 'great interest...I was disappointed in it...especially at the immense lacunae in its facts', Hunt to Adam Sedgwick, 2 April 1860, MS Add. 7652/II/O/32, Sedgwick Papers, CUL. Likewise, John Bullar, London barrister (1807–67), told Owen that 'I am too easy about my own os [coccysis?] and those of my ancestors to care whether or not Mr Darwin's fidget about his own "whistlebone"...has any foundation in fact: but I do care for a piece of close reasoning', Bullar to Owen, 2 November 1860, ff.174–7, Owen Collection, Gen Corr VI, NHM.

145. According to Lyell to Dr [William?] Sharpey, 25 November 1865, Sir Charles Lyell Papers B L981, APS; J.E. Gray to Asa Gray, 12 April 1860, Gray Corr, La–Le, Harvard, and correspondence in Albert Gunther, *A Century of Zoology at the British Museum* (1975).

146. Mill to Alexander Bain, 11 April 1860, in *Letters of John Stuart Mill* (1910), I, 236, extracted in Glick, *What about Darwin*. For his approval of the method, but caution as to proof, see likewise Mill to Hewett C. Watson, 30 January 1869, *Letters of John Stuart Mill*, II, 181. Donald Winch points out that the incompatibility of Mill's thought with the social implications of Darwin's biology led younger political economists such as W.S. Jevons to dismiss his 'ignorance of the principles of evolution', see Winch, 'Darwin Fallen among the Politicians', *Proceedings of the American Philosophical Society* 145.4 (2001), 415–37.

147. J.H. Balfour to Darwin, 14 January 1862, DCP-LETT-3387, Darwin Correspondence. Whitwell Elwin's criticism of Huxley's abandonment of his normal lucidity when

82 DARWINISM'S GENERATIONS

referring to opinions he only 'darkly hints at', suggests that Elwin took a similar view, Elwin to Lyell, 4 March 1863, COLL 203/1/1169–70, Lyell Papers, Edinburgh.

148. Leonard Jenyns to Darwin, 4 January 1860, DCP-LETT-2637A, Darwin Correspondence; Jenyns to Newton, 7 April 1885, MS Add. 9839/1B/733, Newton Papers, CUL.

149. For example, Daniel Moore (1809–99); see his *The Age and the Gospel* (1864), and 'The Unsearchableness of God' in his *Sermons on Special Occasions* (1880), 396–415. Likewise the concerns of Thomas Story Spedding (1800–70); see Spedding to Lyell, 4 October 1864, COLL 203/1/5418–9, Lyell Papers, Edinburgh. (This is at times apparent also for some mid Victorians. So for example the formative idea for Westcott seems to have been the geological record and antiquity of man, rather than natural selection; see W. Lant Carpenter to Lady Welby, 2 March 1889, 1970–010/003(04), Lady Victoria Welby Papers, York University, Canada.)

150. Mozley, 'The Argument of Design' (July 1869), in his *Essays, Historical and Theological* (1878), 363–413.

151. Mavor, *My Windows*, 72. Harold Browne (1811–91), Bishop of Winchester, was still affirming the efficacy of the argument from design in Cambridge in 1882, *The University Pulpit: Supplement to the Cambridge Review*, 10 May 1882.

152. Desmond, *Politics of Evolution*, 288.

153. Address to Aberdeen University Medical Students Society, *Aberdeen Daily Press*, 24 February 1874. Struthers was opposed by Rev. Alexander Anderson (1806/7–84), in his *Science—Theology—Religion, with Notices of the Teaching of Professor Struthers and Others* (1874), which strenuously defended the argument from design.

154. Owen to Adam Sedgwick, 31 March 1860, MS Add. 7652/II/O/23b, Sedgwick Papers, CUL. The appeal to age as an excuse for not shifting to more evolutionary positions became a common one for Owen; see also R. Owen to Charles Moore, 28 September 1870, Owen/5, Charles Moore Papers, Geological Society, London.

155. E. Gosse, *Father and Son* (1907), 119–20.

156. J. Reynolds Green, *A History of Botany in the United Kingdom from the Earliest Times to the End of the 19th Century* (1909), 498. Compare to the obstacles of 'training and associations' noted by T. Spencer Cobbold (1828–86) to the BA in 1870, *Liverpool Courier*, 20 September 1870.

157. Macmillan to A.A. Vansittart, 14 April 1860, C.L. Graves, ed., *Letters of Alexander Macmillan* (1910), 43, 160.

158. See Edwin Lankester to the Coventry Institute, *Coventry Herald*, 23 March 1861, discussed in R. England, 'Censoring Huxley and Wilberforce: A New Source for the Meeting that the *Athenaeum* "wisely softened down"', *Notes and Records* 71 (2017), 371–84.

159. Graves, *Letters of Alexander Macmillan*, 140. R.H. Hutton (1826–97) was another, warning the Church Congress in 1869 that the theological opposition to Darwin had 'done far more to give a false impression of the weakness of theology' and to encourage the very atheism they seek to prevent. For Hutton, even if Darwin had not proved his general hypothesis, he had demonstrated the importance of the struggle for existence in the modification of species, (although this operates 'to perfect gradually the organisation of each tribe of animals'); see *Liverpool Daily Post*, 7 October 1869, and his essay 'The Materialists' Stronghold', *Spectator* 47 (1874), 1169.

160. Acland to Owen, 2 October 1862, ff.18–19 MSS Acland d.99, Bodleian.

161. J.B. Atlay, *Sir Henry Wentworth Acland…A Memoir* (1903), 305. See also R.W. Church (1815–90), *Life and Letters of Dean Church* (1894), 184, 307.

162. Acland to Owen, 4 October 1862, printed letter, in MSS Acland d.200, Bodleian. For a more detailed account of his episode, see Rupke, *Richard Owen*, 218–20.

163. S.W. King to Charles Lyell, 6 October [1862], 16 October 1862, and later 9 February 1863, Lyell Papers, COLL 203/1/3208–9, 3212–3, 3218–9, Edinburgh.

164. Owen to Acland, 13 October 1862, MSS Acland d.200, Bodleian.

165. Charles Carter Blake (1840–97), to Richard Owen, 22 December 1863, ff.202–3, Owen Papers, Gen Corr IV, NHM. Owen claimed that the YMCA lecture was 'supressed as being heretical, very many copies being bought up and burnt', Owen to Dear Madam, 7 May 1864, Owen Collection/200, Wellcome Library.

THE PUBLICATION OF *ON THE ORIGIN OF SPECIES*   83

166. See Elwin to Lyell, 4 March 1863, COLL 203/1/1169–70, Lyell Papers, Edinburgh. Elwin initially opposed the publication of the *Origin* as one of Murray's readers, and was even less complimentary about *Descent*, describing it as 'little better than drivel'. Browne, *Darwin*, II, 347.

167. Acland to Owen, 4 October 1862, MSS Acland d.200, Bodleian, quoted in Rupke, *Richard Owen*, 296.

168. Thomson's hesitations are recorded in Thwaites' letters to Hooker, 12 April 1860, 26 June 1860, DC/162/131,139, Directors' Correspondence, RBGK.

169. As reported by Francis Walker to A.H. Haliday, 30 March 1868, #35, Box 4, Haliday Papers, Royal Entomological Society.

170. [R.W. Church], *The Guardian*, 8 February 1860, see http://darwin-online.org.uk/content/frameset?itemID=A512&viewtype=text&pageseq=1 [accessed 3 February 2022].

171. For a fuller articulation of the argument here and in the following sentences, see Barton, 'The Darwinism of the X-Club', 58–79, and Barton, 'The Scientific Reputation(s) of John Lubbock, Darwinian Gentleman', *Studies in History and Philosophy of Science* 95 (2022), 185–203.

172. Huxley to Hooker, 31 December 1859, quoted in Browne, *Darwin*, II, 105.

173. Huxley, 'Darwin on the Origin of Species', 541.

174. An early opportunity for unequivocal vindication at a lecture at the Royal Institution in February 1860 fell flat, and Darwin, who had travelled especially to attend, left wishing that Huxley hadn't quite so convincingly dwelled on the weak points of natural selection; see Darwin to J.D. Hooker, 14 February [1860], Darwin Papers, DCP-LETT-2696, Darwin Correspondence, Browne, *Darwin*, II, 105.

175. See Huxley to William Sharpey, 4 December [1862], https://www.bonhams.com/auctions/21764/lot/117 [accessed 15 March 2023].

176. For example, Huxley, *Lay Sermons*, 342; see discussion in M. Di Gregorio, *T.H. Huxley's Place in Natural Science* (1984), 65–6; F. Galton, *Hereditary Genius* (1869), and *Natural Inheritance* (1889), 18–34.

177. Quoted in A. Cock and D.R. Forsdyke, *Treasure Your Expectations: The Science and Life of William Bateson* (2008), 131, see also comments of E.R. Lankester in *The Darwin Wallace Celebration* (1908), and also of one former student recalling that warnings about adopting natural selection 'as a creed' were one of his central messages in the 1870s, Charles Herbert Hurst to Huxley, 7 September 1894, ff.201–3, Box 30, Huxley Papers, ICL.

178. See the discussion in B. Lightman, 'Creating a New Space for Debate: The Monthlies, Science and Religion', in Lightman, ed., *Rethinking History, Science and Religion* (2019), 85–109.

179. Lubbock to Lyell, 21 February 1863, COLL 203/1/3636–37, Lyell Papers, Edinburgh. In this sense Bowler, *Non-Darwinian Revolution*, 70–1, is quite wrong in suggesting that the X-Clubbers proceeded by deliberately avoiding controversy.

180. Huxley encouraged John Phillips, see letter of 20 November 1860, quoted in Desmond, *Huxley*, 4188; T.V. Wollaston to Lubbock, 2 December [1861], ff.71–2, Avebury Papers, Add. MS 49639, BL.

181. Barton, *The X-Club*. Interestingly, Herbert McLeod (1841–1923), the initiator of the Declaration, did not intend it as part of the opposition to the *Origin*, which he had read in 1860, largely with agreement; see Hannah Gay, '"The Declaration of Students of the Natural and Physical Sciences", Revisited: Youth, Science and Religion in Mid-Victorian Britain', in William Sweet and William Feist, eds., *Religion and the Challenges of Science* (2017), 19–37, 21; and decisions to sign seem to have owed more to attitudes to the autonomy of science from clerical intervention as to any substantial views about either natural science or biblical theology.

182. For Page, see his *Past and Present Life of the Globe* (1861), and the account of his lecture, *Montrose Review*, 14 December 1860; obituaries spoke of his suspicion of the idea of separate creations, but also his consciousness of the lack of evidence of transmutation, *Dundee Advertiser*, 11 March 1879. Discussed in Lightman, *Popularizers*, 223–38.

183. Carpenter, 'Darwin on the Origin of Species', 213. Carpenter saw his review in the *British and Foreign Medical and Chirurgical Review* as in large part a 'vindication of Lamarck',

84 DARWINISM'S GENERATIONS

Carpenter to Hooker, 9 March 1860, JDH/1/2/4/18, Hooker Papers, RBGK. What Desmond calls Carpenter's 'almost an afterthought' of unhappiness with full scale ideas of descent, in this respect was absolutely consistent with mid-Victorian positions; see Desmond, *Politics of Evolution*, 195. The position was restated in 1872; see 'On Mind and Will in Nature', *Contemporary Review* 20 (1872), 738–62, which offers a fuller endorsement, including evolution of humanity, and the moral consciousness of animals.

184. See Barton, *The X-Club*, section 3.2; writing to Frank Darwin in 1882, Carpenter acknowledged his 'general acceptance' of Darwin's views, Carpenter to F. Darwin, 13 June 1882, MSS. DAR.198/35, Darwin Papers, CUL.

185. This was a constant refrain of his correspondence with Huxley in these years; see George Rolleston, 13 April 1860, f.142, and [December 1860], ff.150–2, Box 25, Huxley Papers, ICL.

186. See Huxley to Busk, 17 July 1860, ff.14–15, Huxley to Lubbock, [nd, 1863?] ff.53–4, Avebury Papers, Add. MS 49640, BL; and Rolleston to Lubbock, 8 June 1861, ff.25–6, Avebury Papers, Add. MS 49639, BL. Edward Tylor remarked that although it 'altered the current of its force', the *Origin* 'was destined not indeed to carry Rolleston's mind altogether in its stream', Tylor, 'Life of Dr Rolleston', in Rolleston, *Scientific Papers and Addresses* (1884), I, xxxii.

187. See letter 19 January 1860, in Rolleston, *Scientific Papers*, xxxiv–xxxv; also his letter in the *Medical Times and Gazette*, 18 October 1862, and Rolleston, *Forms of Animal Life* (1870). For a similar sense of cautious 'probable but not proven' see W.H. Dallinger (1839–1909), 'Atheism, Evolution and Theology', *London Quarterly Review* 49 (1878), 322–57.

188. Rolleston to Lockyer, 25 September 1863, Norman Lockyer Papers, EUL110, Exeter University.

189. Confessing in September 1861 that he had not quite finished reading 'all through' the third edition Darwin had sent him in the spring, Rolleston to Darwin, 1 September 1861, DCP-LETT-3241, Darwin Correspondence.

190. Darwin to Hooker, 23 [April 1861], DCP-LETT-3098, Darwin Correspondence.

191. Cited by Lightman, *Popularizers*, 71–81, as offering sense of complexity and continued commitment to natural theology and evolution.

192. Kingsley to Huxley, 7 December 1859, ff.160–1, Box 19, Huxley Papers, ICL; Kingsley to Bates, 13 April 1863; Bunbury, *Life of Sir Charles J.F. Bunbury* II, 195.

193. Kingsley to Huxley, 21 September 1860, ff.161–8, Box 19, Huxley Papers, ICL. See also F.E. Kingsley, *Charles Kingsley: His Letters and Memories of His Life* (1877), II, 156, where it is suggested that the *Origin* 'opened a new world to him, and made all that he saw around him, if possible, even more full of divine significance than before'.

194. Kingsley to Darwin, 31 January 1862, DCP-LETT-3426, Darwin Correspondence.

195. Kingsley to Lubbock, 6 March 1861, ff.43–50, Avebury Papers, Add. MS 49639, BL; Kingsley to T.G. Bonney, 13 June 1868, Box 45, Folder 8, M.L. Parrish Collection, PUL.

196. John Braxton Hicks (1823–97), 'On the Diamorphosis of Lyngbya, Schizogonium and Prasiola, and Their Connection with the So-Called Palmellaceae', *Quarterly Journal of the Microscopical Society* (1861), 157.

197. John H. Gladstone, 'Points of the Supposed Collision between the Scriptures and Natural Science', in *Faith and Freethought: A Second Course of Lectures Delivered at the Request of the Christian Evidence Society* (1872), 155.

198. See Argyll's correspondence with Charles Lyell, not least Argyll to Lyell, 2 March 1861, COLL 203/1/69–70, Lyell Papers, Edinburgh; see also Argyll address in *Proceedings of the Royal Society of Edinburgh* 5 (1866), 264–92.

199. Gladstone, 'Points', 159.

200. Brown to Lubbock, 19 December [1863], ff.104–5, Avebury Papers, Add. MS 49640, BL.

201. *Life of Rowland Williams, D.D.* (2 vols, 1874), II, 104. His subsequent verdict on Darwin's later volumes of the 1860s was that they were 'instructive, and not uninteresting; but…painfully materialistic and theologically negative in tendency', *Life*, II, 277–8.

202. Rowley to Alfred Newton, 26 April 1860, MS Add. 9839/1R/308, Newton Papers, CUL; see also Thomas Southwell (1831–1909) to Newton, 7 May 1860, MS Add. 9839/1S/1769, Newton Papers, CUL.

THE PUBLICATION OF *ON THE ORIGIN OF SPECIES* 85

203. William Thomson to J.D. Forbes, 19 June 1861, incoming letters 1861, no. 83, Forbes Papers, St Andrews.

204. See also William Thomson to Phillips, 7 June 1861, JP/C/1861/10, John Phillips Papers, OMNH.

205. Wollaston to Lyell, 13 April [1860], COLL 203/1/6183–84, Lyell Papers, Edinburgh; Wollaston, *Coleoptera Atlantidum* (1865), sustained in *Coleoptera Hesperidum* (1868), where Wollaston contrasted the possibilities of artificial breeding and the 'feral world' beyond.

206. Davidson to Lyell, 21 November 1862, and 26 November 1862, COLL 203/1/712–13, 722–23, Lyell Papers, Edinburgh.

207. Julia Wedgwood, 'The Boundaries of Science', *Macmillan's Magazine* 2 (1860), 134–8, 4 (1861), 237–47; H. Fawcett, 'A Popular Exposition of Mr Darwin on the Origin of Species', *Macmillan's Magazine* 3 (1860), 81–92. F.W. Hutton, 'Some Remarks on Mr Darwin's Theory', *The Geologist* (1861), 132–6, 183–9, and letter ibid., 286–9. For Watts see *National Reformer*, 28 December 1861, 4 January 1862.

208. The young solicitor Hugh Browne (1834–1914), *Nottingham Daily Express*, 9, 15 April 1861.

209. Leslie Stephen, 'Some Early Impressions [2]', *National Review* 42 (1903–1904), 208.

210. Frederick Pollock (1845–1937) in Glick, *What about Darwin*, xxiv, 'Natural Selection was to be the master-key of the universe; we expected it to solve all riddles and reconcile all contradictions'; see also Dawson, *Darwin, Literature and Victorian Respectability*, 166–8.

211. Thomas Wright, *Life of Walter Pater* (1907), I, 203. For Pater, Darwinian evolution was 'old Heracliteanism awake once more in a new world and grown full of proportions'. Glick has an interesting extract of a letter from J.R. Green to Boyd Dawkins from 1860, which speaks of the 'utter weakness' of Darwin's 'dead stops' theory, Glick, *What about Darwin*, 158, quoting L. Stephen, ed., *Letters of J.R. Green* (1902), 43. For the role of classical traditions of evolutionary thought in providing a context through which Darwinian ideas could be approached see Turner's 'Ancient Materialism and Modern Science: Lucretius and the Victorians', in his *Contesting Cultural Authority* (1993), 262–83.

212. T.G. Bonney, *Memories of a Long Life* (1921), 37.

213. '...and I have felt their influence so much during all my subsequent life...', Robert Ball, 'The Relation of Darwinism to other Branches of Science', *Longman's Magazine* 3 (1883), 76–7; this was a feature of Ball's construction of himself; see the account of dinner conversation in 1890 in 'The Greenwich of Ireland: A Visit to Sir Robert Ball's Observatory', *Globe*, 8 October 1890. This contrasts in interesting ways with the gradual process by which Ball described his loss of religious faith; see Lightman, *Popularizers*, 400–1.

214. A. Geikie, *A Long Life's Work* (1924), 72, and also John Morley's comment that the words evolution, adaptation, survival, and natural selection appeared as 'so many patent pass-keys that were to open every chamber', Morley, *Critical Miscellanies: Volume III* (1886), 126.

215. D. Macmillan, *Life of George Matheson* (1907), 50–1. Recollections of Matheson described him as having a deep interest in scientific study, and a frequent engagement with the works of Huxley, Tyndall, Kelvin, and Müller.

216. See E. Longford, *A Pilgrimage of Passion* (2007), 26; Tait to Darwin, 26 January 1869, DCP-LETT-6577, Darwin Correspondence; S. Butler, *Unconscious Memory* (1880), 17. For another high-Victorian response, see Henry Gyles Turner (1831–1920), who later recalled being converted by the *Origin* in 1862, not least because of the overwhelming logic of the book, Butcher, 'Darwin Down Under', 47.

217. J. Reay Greene (1836–1903) to Hooker, 12 March 1860, JDH/1/2/9/290, Hooker Papers, RBGK.

218. See introduction of Anthony Belt, in Thomas Belt, *The Naturalist in Nicaragua* (1911 ed.), xv.

219. 'Biographical Notices' in *Ibis: Jubilee Supplement*, 9th ser, 2 (1908), 84.

220. See Barton, *The X-Club*, citing Hooker to Lubbock, 29 February and 4 March 1861, JDH/2/3/10/240–41, 242–44, Hooker Papers, RBGK.

86 DARWINISM'S GENERATIONS

221. Kingsley to F.D. Maurice [nd 1863?], *Kingsley: His Letters and Memories*, 171. See Hesketh, *Oxford Debate*, 99–100, which suggests that evolution, if not Darwinism, was an established orthodoxy by the 1863 BA.

222. Daubeny to John Phillips, JP/C/1864/13, John Phillips Papers, OMNH; and his conciliatory paper on 'The Decay of Species' at the 1864 BA, for example, *Gardeners' Chronicle*, 8, 15 October 1864.

223. A. Panizzi to A. Layard, 4 February [1861?], ff.59–60, Layard Papers, Add. MS 38987, BL.

224. Marmaduke Alexander Lawson (1840–96), *Morning Post*, 26 August 1882.

225. Kingsley to Huxley, 18 July 1862, f.205, Box 19, Huxley Papers, ICL; Jukes to Darwin, 3 November 1862, DCP-LETT-3794, Darwin Correspondence; *Report of the Meeting of the British Association for the Advancement of Science, Held at Newcastle in August and September 1863* (1864), lxiii, and account in *Dundee Advertiser*, 29 August 1863. John Crawfurd's attack on Darwinism in his paper on Lyell's *Antiquity of Man* also elicited some support in the discussion which followed; see account in *Dundee Courier*, 31 August 1863.

226. Huxley to A. Geikie, 13 October 1862, Sir Archibald Geikie Papers, Gen 525/5, Edinburgh.

227. See Featherstonhaugh to Owen, 24 April 1865, ff.192–3, Owen Collection, Vol. 12, NHM; Sir Richard Rawlinson Vyvyan, 8th Baronet (1800–79) to Sir R. Owen, 29 December 1863, f.427, Richard Owen Papers, Add. MS 39954 (part 2), BL.

228. A combination Mackintosh could create even when lecturing on volcanoes; see *Shrewsbury Chronicle*, 18 December 1863.

229. *Worcestershire Chronicle*, 30 September 1863. For Salter see *Western Times*, 21 August 1863, (Edinburgh) *Daily Review*, 2 December 1863. For Salter's rather confused response see Salter to Sedgwick, 8 June 1860, MS Add. 7652/II/II/5, Sedgwick Papers, CUL. By 1867 Salter was confessing to Darwin that even though more and more of his observations were consistent with evolutionary theory, he still found it difficult to accept the breaks in the fossil record, Salter to Darwin, 4 January [1867], DCP-LETT-4969, Darwin Correspondence.

230. T.E.T. Bond, 'Robert Wight (1796–1872), Dr Freke and the "Origin of Species"', *Nature*, 4 November 1944, 566–9. See also William Bence Jones (1812–82) to Francis H. Dickinson, 3 March 1860, DD/DN/4/4/146/225–26, Francis and Caroline Dickinson Correspondence, Somerset Heritage Centre.

231. See his memoir on birds of North Africa, quoted in Gayon, *Darwinism's Struggle for Survival*, 185–6, and his paper in *Ibis* 1 (1859), 429–32. But later, 'The more I look into this renovation of Lamarck, the more I see it is one blind plunge into the gulph of atheism and the coarsest materialism'; see *Life of Alfred Newton*, 119–20. It was not that Tristram was not still drawn to the theory; see his description of it as 'beautiful, ingenious and self-consistent', in his 'Recent Geographical and Historical Progress in Zoology', *Contemporary Review* 2 (1866), 124, quoted in D. Livingstone, *Darwin's Forgotten Defenders: The Encounter between Evangelical Theology and Evolutionary Thought* (1984), 133. But his later career gives the lie to Gayon's observation that his comments 'show how quickly the selectionist explanation and all its characteristic ingredients...took root in naturalist circles'. Likewise the suggestion (Hesketh, *Oxford Debate*, 85–6) that it was the Oxford debate which prompted Tristram to abandon Darwinism is not born out by other evidence. Although Tristram might briefly have wavered after the BA meeting at Oxford, the sceptical stance of his 'President's Address' to the Tyneside Naturalists' Field Club in March 1860, *Transactions of the Tyneside Field Naturalists' Club*, 4 (1858–60), 219–28, is quickly visible again in his comments of the mid-1860s, a position which seems thereafter to have been maintained for the rest of his life, see the discussion below, pp. 136–37. See reference to Tristram's 'unfavourable criticism', Newton to H.B. Tristram, 9 April 1860 [copy], MS Add. 9839/1T/208, Newton Papers, CUL. For another convinced that 'every new fact I get hold of' told 'dead against' the *Origin*, see T.V. Wollaston to Lubbock, 2 December [1861], ff.71–2, Avebury Papers, Add. MS 49639, BL.

232. Falconer to Mary Somerville, 12 April 1860, MSF-1, Somerville Papers, dep c.370, Bodleian. His nomination of Darwin for the Copley Medal balanced appreciation of the *Origin of Species* with recognition that 'I am far from thinking that Charles Darwin has

THE PUBLICATION OF *ON THE ORIGIN OF SPECIES*   87

made out all his case', Hugh Falconer to William Sharpey, 25 October 1864, DCP-LETT-4644, Darwin Correspondence; his January 1863 article in the *Natural History Review* raised questions about the adequacy of natural selection as an explanation of species mutation.

233. J.H. Balfour, *A Manual of Botany* (1863), 196.

234. See Hort to John Ellerton, 3 April 1860, in A.F. Hort, ed., *Life and Letters of Fenton John Anthony Hort* (1896), I, 416, and material in the following letters, ibid., 430–4, 445; G. Patrick, *F. J. A. Hort: Eminent Victorian* (2015). There is a tendency to present Hort as largely accepting Darwin, having merely 'reservations on points of detail' (G. Neville, 'Science and Tradition: F.J.A. Hort and His Critics', *Journal of Theological Studies* 50.2 (1999), 560–82, 564), but what is notable about Hort's subsequent career is his almost complete avoidance of any engagement in evolutionary discussions despite his deep interest in science.

235. 4 May 1862, JT/2/32c, Thomas Hirst Journal, tps, RI; the typescript actually renders this 'inadequate' interest, but the context makes it clear that this is almost certainly a transcription error.

236. Symonds to Lyell, 5 March 1863, COLL 203/1/5648–51, Lyell Papers, Edinburgh.

237. Charles H. Pearson (1830–94) to Newton, 12 May [1863], MS Add. 9839/1P/117, Newton Papers, CUL.

The Reception of Darwinian Evolution in Britain, 1859–1909: Darwinism's Generations. Martin Hewitt, Oxford University Press. © Martin Hewitt 2024. DOI: 10.1093/9780191982941.003.0002

2

In the Wake of the *Origin*

The impulse of the new spirit introduced by Darwin did not stimulate Balfour as it might have done a younger man. His religious beliefs always in evidence were showing then the influence of his early environment, and whilst Darwin's work was incorporated in his teaching, the acceptance of Darwin's theory appeared too near the negation of faith. On Balfour indeed, as on others with like views, the immediate effect of the *Origin* was the opposite of vivifying. It gave a shock. (I. Bayley Balfour, 'The Edinburgh Professors', in F.S. Oliver, *Makers of British Botany* (1913), 298)

For those who have not already consigned resistance to Darwinism to the dustbin of lost causes by 1863, the decade of the 1860s is often portrayed as seeing the rapid and dramatic triumph of Darwinism.[1] The award in 1864 of the Royal Society's Copley Medal to Darwin, despite determined opposition from figures like Edward Sabine (1788–1883), the Society's president, has been seen as a significant milestone towards general acceptance.[2] Joseph Hooker confidently told the British Association in 1868 that natural selection had become 'an accepted doctrine with every philosophical naturalist', and Huxley frequently used his lecture appearances to proclaim the rapid growth of support for Darwinian ideas.[3] If he had been alone in Chester in 1862, by 1867 Kingsley could report both Oxford and Cambridge swinging rapidly behind Darwinism, even allowing for generational differences: not just, he told Darwin, 'the younger MAs…greedy to hear what you have to say', but even if mid Victorians like John Couch Adams (1819–92) and Arthur Cayley (1821–95) were 'fighting desperately', 'the elder (who have, of course, more old notions to overcome) are facing the whole question in a quite different tone from what they did three years ago.'[4] A few months later he was exulting that 'I have actually found a Darwinian Marchioness!!!!'[5] William Tuckwell, who had provided one of the most vivid of the accounts of Huxley's confrontation with Wilberforce at the British Association, drew a sharp contrast with the general response ten years later at Exeter in 1869, when, as he later recalled, the audience, so generally hostile in 1860, cheered on Huxley's dismissal of a clerical anti-Darwinian.[6] Huxley's selection as President of the British Association for its 1870 meeting at Liverpool, and the award of an honorary doctorate to Darwin by Oxford in 1870, apparently confirmed the transformation.

As the narrative of Darwin's reception was constructed in subsequent years it was frequently presented as a sudden, revolutionary transformation. Accounts were couched in terms of conversion (even if, in Kingsley's description of Cambridge, 'an honest, but "funky" stage of conversion'); the acceptance of evolutionary thought conceived of as the rapid abandonment of previous ways of thinking.[7] Darwinism 'made headway among the educated classes', was 'adopted', contemporaries 'came to agree'. Wallace reassured Darwin in 1871 that he had 'a noble army of *converts*', while Alfred Newton recalled how one by one he found most of his naturalist friends gradually coming to accept Darwinism as 'a true creed'.[8]

Darwinism in the 1860s

This picture of transformation is supported by the appearance during the 1860s of a series of studies which promoted evolutionary understandings. Darwin's own publications, especially *The Fertilisation of Orchids* (1862) and *The Variation of Animals and Plants Under Domestication* (1868), provided further articulation of his ideas about evolution, supplementing his theory of natural selection with the subordinate process of sexual selection. Herbert Spencer's *First Principles* (1860–2) and *Principles of Biology* (1863–4) developed a synthetic evolutionary philosophy. Familiarity was increased by H.W. Bates' *The Naturalist on the River Amazons* (1863) and Wallace's *Malay Archipelago* (1869), which offered the extensive readership of this sort of naturalists' travelogue enthusiastic applications of natural selection.[9] A string of publications addressed, in the title of Huxley's 1863 intervention, *Man's Place in Nature*. John Lubbock's *Prehistoric Times* (1865), and in particular Charles Lyell's *Antiquity of Man* (1863), publicised the archaeological discoveries of the previous decade and their demonstration of the great age of humanity. Lyell's *Antiquity* also offered a full summary and general endorsement of Darwin's arguments.[10]

Opponents looked on aghast as Darwinian ideas in physical form circulated widely and Darwinian language and phraseology seeped into public discourse (Figure 2). No matter that mostly this was a sort of evolutionary

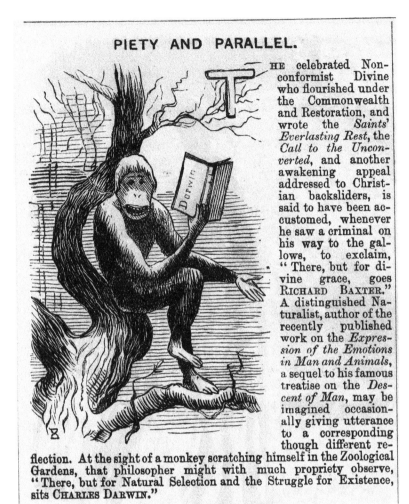

Figure 2 'Piety and Parallel', *Punch*, 30 November 1872.

irony, which simultaneously smiled with but also at Darwin. Nor that in some respects books like *The Antiquity of Man* were less a symptom of a shift to Darwinism than a cause of hardening anti-Darwinian attitudes. The notion of development was increasingly naturalised, even if in ways which fell far short of any Darwinian conception, and the idea of the survival of the fittest (not of course itself a formulation of Darwin's) proved especially seductive for its ready application to all forms of competition, as in the essay which G.H. Lewes contributed to the first edition of the *Fortnightly Review*

IN THE WAKE OF THE *ORIGIN* 91

in 1865, which conceived the literary marketplace as a biological ecosystem, with texts as organisms, and survival being the only proof of worth.[11] Contemporaries drew on the figurative power of the warfare of nature Darwin had explored. 'The London world is a very Darwinian sort of scramble', the essayist Walter Bagehot (1826–77) told a friend in December 1870.[12] The fiction of the 1860s, and in particular the sensation novels of Wilkie Collins (1824–89) and Mary Elizabeth Braddon (1835–1915) persistently invoked Darwinian themes, and relatively favourable passing references can be found even in fiction serialised in the religious magazines.[13] Poets were also wrestling with evolutionary themes. The high Victorian Thomas Hardy (1840–1928), who described himself in his autobiography as 'among the earliest acclaimers of *The Origin of Species*', was exploring Darwinian perspectives, in for example, 'Hap', written in 1866, with its denial of divine purpose in a universe marked merely by 'Crass Casualty' and 'dicing Time', and only just after the end of the decade, A.C. Swinburne (1837–1909) published his powerfully evolutionary *Songs Before Sunrise* (1871).

Accepting all this, care is still needed: the creep of Darwinian similes and evolutionary metaphors into everyday language registered fascination or preoccupation rather than necessarily acceptance; the ebbing of the initial furore reflected the inevitable decline in emotional charge brought by the loss of novelty. For many—Dickens is a good example—engagement with Darwinian ideas remained indirect and generally prompted by doubts.[14] Collins' invocations need to be read alongside the fact that none of Darwin's works could be found in his library, and the passages in his novels which suggest a scepticism at the claims of evolutionary science.[15] 'Hap' remained unpublished until 1898. Darwinian responses playing out in the literature of George Meredith (1828–1909) were resigned and unsatisfied. As he later put it in in his prelude to *The Egoist* (1879), science provided no antidote to contemporary anxieties, it merely offered 'the extension of a tail.... We were the same, and animals into the bargain. That is all we got from Science'.[16] In the same way, although recent criticism would see a poem like Tennyson's 'Lucretius' (1868), with its concern with the 'monkey-spite' below the surface of human rationality, as expressing a specifically Darwinian philosophy, it probably does no more than reflect a much more general evolutionary sensibility.[17] (Despite Tennyson's status as the evolutionary poet of the period, his intellectual position remained complicated and conflicted; in later life he 'spoke of Darwin, and of the great truth in Evolution' but 'only [as] one side of a truth that had two sides', and in 'Locksley Hall Sixty Years After' (1886) he offered a sustained protest at the implications of evolutionary thought.)[18] Even Swinburne, it can be argued, presents a largely pseudo-Darwinian view of evolution as merely the 'principle of growth'.[19]

Unless one is willing to accept uncritically the self-serving triumphalism of Huxley and his acolytes, or the over-anxious worrying of some of Darwin's opponents, the record suggests that shifts of opinion during the decade were slow, cautious, and heavily circumscribed. Individual Darwinists remain in short supply; genuine converts even rarer. It is notable that the argument of Richard Bellon, that there was a relatively sudden Darwinian triumph in the wake of the publication of Darwin's volume on *The Fertilisation of Orchids* in May 1862, relies almost entirely for substantiation on a few contemporary comments about the changing atmosphere of the British Association, and is really an argument about the standing of Darwin's science, rather than demonstrating changing convictions as to evolution.[20] In fact, even the reception of *Orchids* suggests a more complicated history. Kingsley described it as 'a most valuable addition to natural theology', and many readers interpreted the book as continually pointing out proofs of design.[21] Although there is some evidence of plant breeders and field naturalists latching on to *Orchids*, a more convincing case in this regard might be made for *The Variation of Animals and Plants Under Domestication*, for which the reviews were generally more sympathetic than they had been for the *Origin*, and which did prompt new enthusiasms for the implications of Darwinism for improved animal stocks and new plant hybrids.[22] By 1871, seduced by the promise of almost limitless improvement by careful breeding, even Sir Willoughby Jones (1820–84), Norfolk landowner and Tory politician, was proclaiming that he was 'a great believer in Darwinism'.[23]

We cannot take such comments at face value, nor succumb to the temptation of reading into public discourse the existence of an educated class almost all of whom had read their Darwin in the first few years after publication. It seems hard to believe, but looking back from the 1890s, the staunch Darwinist Thomas Lauder Brunton (1844–1916) recalled that in the mid-1860s, even though he had completed a science degree and qualified in Medicine, he had not heard of the *Origin* until he encountered it while studying in Vienna.[24] Despite being taught by Alexander Bain, students at Aberdeen in the later 1860s barely registered evolutionary developments.[25] The physicist Oliver Lodge (1851–1940) recalled that it was not until he was sixteen or seventeen, in the later 1860s, that he was introduced to Darwin and Huxley, although he had a long-standing interest in science and had attended lectures at King's College.[26] This was true even of sympathetic younger naturalists who could still reach their later twenties in the early 1870s without having read the *Origin*.[27] The autodidact bookbinder Frederick Rogers (1846–1915) claimed that until he heard J.A. Picton preaching on the relations of science and religion in 1870, he had not the slightest idea what the doctrine of evolution meant.[28] (Rogers also recalled that subsequently it took him four months to read *Descent of Man* every day in his lunch break, a regime from which his digestive system apparently never recovered.)[29] There is plenty of scattered evidence which suggests that even keen naturalists came to the actual reading of the *Origin* surprisingly late. The influential mid-Victorian intellectual Benjamin Jowett seems only to have embarked on a serious effort to understand Darwinism in the final years of the decade.[30]

Indeed, frequently Darwinian thought was initially treated as just a minor coda to the challenge of *Essays and Reviews*, merely one of the heterodox texts cited by the essayists. (Significantly it was the recollection of at least one high-Victorian Congregationalist minister looking back from 1896 that the *Origin* had appeared after *Essays and Reviews*.)[31] The radical Thomas Hodgskin (1787–1869) failed to mention Darwin or the *Origin* in the science column he wrote throughout the 1860s.[32] Reading the correspondence of the decade of the Presbyterian principal John Cairns (1818–92) opens out a world of intellectual crisis in which the questions of the 1840s seem like 'a faded controversy, a geological debate when the earth and the works thereof have been burnt up', but where the new challenges are almost entirely theological, driven by Strauss and German criticism.[33] Even as the *Essays and Reviews* crisis blew over, Darwinism did not assume centre stage. For many early-Victorian divines it was John Seeley's *Ecce Homo* (which significantly outsold the *Origin* in the 1860s) which was the most dangerous book of the times, while for Seeley himself, as for James Martineau, it was the 'narrow negations of the physicists', or Comtism, not Darwinism, which provided the most threatening intellectual dissolvent.[34]

Scholars have often seized on the *Fortnightly Review* essays of Walter Bagehot, collected in *Physics and Politics* (1873), as evidence of how widely Darwinian ideas had saturated contemporary thought by the end of the decade; but although Bagehot was certainly intrigued by the implications of the *Origin*, and unthreatened by its biological fundamentals, his applications were self-confessedly analogic, and it is clear from the otherwise only scattered references in his writings that he saw social development as a process governed by its own laws rather than natural selection, and was not

94 DARWINISM'S GENERATIONS

enough of a Darwinian even to subscribe to the transmutation of species.[35] To find a wholehearted application of natural selection to social questions we need to look to high-Victorian Darwinists, as in 'The Natural History of Morals', in the *North British Review* in December 1867, which was the work of the journalist James Macdonell (1841–79).[36] A few months before, the *Gloucestershire Chronicle* published a long letter from a tea grower in Upper Assam, who can be tentatively identified as another high Victorian, Samuel Edward Peal (1834–97), who had gone out to India in 1862, and who confessed that it was the reading of the *Origin* before he left England that enabled him to make sense of life in the Indian forest: 'The book is the soul of the coming ages'.[37] Another Darwinian exile was the Edinburgh botanist John Scott (1836–80), who had been foreman of the Edinburgh Botanical Garden, and went out to India in 1864 with Darwin's active involvement.[38]

As the decade progressed, Darwin's correspondence features several younger men enthusiastic about his ideas who were beginning to publish work within an evolutionary framework; medics like James Crichton-Browne (1840–1938) and academics like William Boyd Dawkins (1837–1929), 'unspeakably grateful' for the *Origin*.[39] Dawkins told Darwin that all the active researchers he knew endorsed his ideas, adding that he couldn't see how any naturalist could work on any other hypothesis, but his only explicit references were to high Victorians like the geologist Henry Woodward (1832–1921).[40]

Otherwise, the dominant tenor of press and periodicals, and the context of discussion in public sphere and private spaces, remained highly sceptical, if not downright hostile. It was hardly surprising that, in public, the note struck was often one of caution. Despite the close affinity of his ideas, the historian W.E.H. Lecky (1838–1903) was anxious to stress in his *History of the Rise and Influence of the Spirit of Rationalism in Europe* (1866) that his progressivism was not derived from biological models based on the mutability of species.[41]

This is not to say that there were no older Darwinians in evidence in the 1860s. These mid- and occasionally early-Victorian Darwinists fall primarily into three groups. Those who had largely accepted evolution before the publication of the *Origin*, often under the influence of Lamarckian or Spencerian ideas, but also on the basis of the more generalised ideas of the sort embodied in *Vestiges*, and who largely persisted in these pre-Darwinian beliefs, perhaps grafting on some role for natural selection. Those whose primary concern was the struggle for intellectual freedom from the control of church establishments, including secularists and freethinkers, but also university radicals whose engagement with the details of Darwinian thought often appears superficial, but who lined up vigorously against Darwinism's opponents. And finally, the relatively small number who, while they might not have accepted Darwin's ideas *in toto*, nevertheless firmly identified as Darwinists and accepted all the central components of his theory as having at least significant credibility.

The core of this latter group, the Darwinians 'proper', coalesced around T.H. Huxley and the primarily mid-Victorian coterie organised formally in 1864 as the X-Club, an intimate and enormously influential dining society of pro-evolutionists.[42] As well as Hooker, Tyndall, and Lubbock, the X-Club membership comprised Herbert Spencer, George Busk (1807–86), zoologist and palaeontologist, the chemist Edward Frankland (1825–99), the printer and mathematician William Spottiswoode (1825–83), and Tyndall's friend and protégé, the mathematician Thomas Archer Hirst (1830–92). Several of the 'Club' members show little evidence of close engagement with the biological intricacies of Darwin's thought: their scientific interests lay elsewhere, and their ties to the Club were as much personal and political as intellectual. Over the decade others were drawn into the X-Club network: figures like the scientific journalist W.S. Dallas (1824–90), the paper manufacturer and archaeologist Sir John Evans (1823–1908), and the zoologist Philip Lutley Sclater (1829–1913), who, while disclaiming any particular Darwinian expertise, affirmed himself a 'derivatist'.[43]

Otherwise, the evidence suggests that most of those born before the end of the 1820s who expressed approval of Darwin's ideas (in public or in private) during the 1860s had already accepted pre-Darwinian versions of evolution, including John Henry Newman (1801–90), who was able to accommodate the *Origin* with equanimity into his existing understanding of 'development', and the early-Victorian palaeontologist Searles Valentine Wood (1798–1880), who had been convinced of organic evolution well before 1859.[44] Hooker was a loyal supporter; but then, as he acknowledged, he had known Darwin's views for fourteen years before he had adopted them, and even then, as we shall see, his endorsement was less wholehearted than his general protestations might suggest.[45] By and large, these early supporters seized on the *Origin* as clarifying or coalescing existing ideas; as Lewes put it, the *Origin* gave 'articulate expression to the thought which had been inarticulate in many minds'.[46] The same was true of Bates, and of Kingsley who, for all his cautions, read the *Origin* in the context of a long-held doubt of the doctrine of the permanence of species.[47] W.B. Carpenter later reflected that several influences had predisposed him

96 DARWINISM'S GENERATIONS

to Darwin's ideas, including his own work on the single-cell marine foraminifera which had led him to a conception of the range of variation in that group beyond anything previously suspected, so that when he saw that Darwin 'had got hold of a definite scientific doctrine, [he] was fully prepared to give it [a] favourable consideration.'[48] Likewise, it was later said of Newton that the way Darwin's ideas solved the difficulties he had long identified with the theory of special creations meant that he 'simply adopted the new philosophy, not being in need of conversion.'[49]

A number of anthropologists, for instance Augustus Henry Lane-Fox Pitt Rivers (1827–1900), who had already seized on evidence of cultural progression before 1859, saw in gradual variation and natural selection further scientific endorsement of their ideas, but their application of Darwin was diffuse and contradictory.[50] It is clear that *Vestiges* had provided a conduit to evolutionary beliefs for many, despite the fierce opposition of the scientific establishment. The historian J.A. Froude (1818–94), famous for his autobiographical loss-of-belief account, *Nemesis of Faith* (1849), later remarked that that he had been in no doubt of evolution 'since I read the *Vestiges*.'[51] Similarly, F.W. Newman (1805–97) 'was already ready to believe in "Evolution"', he recalled, 'while our scientists were spitting contempt at it.'[52] (Newman made no attempt to incorporate Darwin into his thorough-going evolutionism, and remained dismissive both of scientists in general and of Darwin in particular.) Lamarck and Spencer offered other often intertwined routes to evolutionary beliefs, as was the case for Robert E. Grant (1793–1874), Professor of Comparative Anatomy at University College, and one of his protégés, Thomas Laycock (1812–76), physician and Professor of Medicine at Edinburgh University, both of whom remained alienated from the Huxley group despite their long-standing evolutionary commitments.[53]

Indeed, many of these pre-*Origin* evolutionists had a tendency to treat Darwin largely as a version of their existing Lamarckism. The early response of the Scottish progressivists around Robert Chambers and the Edinburgh Philosophical Society mixed exasperation at the sudden conversion of the scientific establishment with a tendency to read the *Origin* as confirming Lamarck.[54] Richard Vyvyan, who had been widely suspected of writing *Vestiges*, accepted that Darwin had gone far beyond Lamarck in assigning causes, but nevertheless concluded that Darwin remained inferior.[55] For Lewes, Darwin's theory was a larger and more philosophic view of Lamarck's law of adaptation.[56] Benjamin Jowett's belated attention to Darwin and Wallace in 1871 prompted him to confess to Florence Nightingale that 'I should suspect there is quite as much in Lamarck's theory as in theirs,' although 'neither unlocked the mystery of animal life'.[57] Lyell's persistent failure to acknowledge the distinction between Darwinism and Lamarckism was a source of irritation for Darwin for much of the decade.[58]

In much the same way, the small band of mid-Victorian Spencerians read Darwin through Spencer, even if at this point Spencer had little of the wider public readership of Darwin.[59] Alexander Bain was explicit in 1863 that, for him, Spencer was more important than Darwin, and looking back the philosopher Alexander Campbell Fraser (1819–1914) recognised that his work over the decade had explored responses primarily to Comtean positivism and the evolutionism of Spencer, and only to a much lesser extent Darwin's biology.[60] The unorthodox Baptist minister James Hinton (1822–75) had read the *Origin* enthusiastically in 1860, delighted that the struggle for existence offered an explanation of evolutionary change hitherto lacking; but the development of his thought in the following years owed much more to Spencer, whose *Principles of Biology* he was even more enthusiastic about.[61]

Much of this evolutionary commitment had at best a diffuse connection to Darwin's ideas. Jowett is a striking example of the way the mid Victorians felt themselves falling under the influence of Darwin without ever being willing to formally accept much of his position. Long influenced by Hegelian ideas, Jowett was apparently prepared to accept the natural origin of all species, and Darwin as one of the three men who had moulded and formed his mind, describing the *Origin* as 'one of the greatest and more far-reaching books' of the century, a book which had opened a new conception of ethics to him.[62] And yet, younger thinkers remained conscious of Jowett as approaching Darwinism through a framework of already settled convictions, and although he warmed to the implications of the *Origin* and *Descent* for his views on biblical inspiration and orthodox theology, he also confessed that in regards to Darwinism he was 'rebellious', and found much of it 'deeply distasteful', 'speculative and inexact'.[63] Looking back from 1912, the jurist A.V. Dicey (1835–1922) reflected that, ignorant as he was himself about the physical sciences, he had 'observed that Jowett was less impressed than have been most even of the ignorant men of my age with the force of the general Darwinian argument'.[64]

For many others, the conflicts over Darwin were just an incidental feature of the broader conflict between liberal thought and biblical literalism.[65] For example, the reforming Oxford don, Mark Pattison, was said to have accepted evolution 'probably' because 'he regarded Evolution as a convenient stone to throw at his orthodox antagonists and would-be persecutors'.[66]

98 DARWINISM'S GENERATIONS

W.R. Greg, a former classmate of Darwin's, and strongly connected to the intellectual set around John Chapman, was probably drawn to arguments about human evolution by Wallace's writings in the mid-1860s, which Greg quoted extensively to bolster his Manchester School laissez-fairism and hostility to protective social legislation.[67] Certainly there is only enigmatic Darwinian reference in Greg's writings, even in his frequently noticed essay 'On the Failure of Natural Selection in the Case of Man' (1868).

The Narrow Bounds of Conversion

Despite a grudging recognition of the suggestiveness of Darwin's ideas, there is very little evidence of a decisive move from denial of evolution to its acceptance amongst Darwin's older opponents. This appearance of pertinacity may perhaps be an effect of the pattern of surviving evidence. Most early Victorians were already beyond their mid-50s when the *Origin* appeared, and as their participation in the public debates dwindled, so it becomes harder to trace shifts in their views. Even so, such evidence as survives offers no indication of substantial movement. Some simply detached themselves from contemporary debate, like the Anglican cleric Archibald Boyd (1803–83), of whom it was said that his reading stopped with the *Essays and Reviews*, so that he knew nothing of Darwin, Huxley, or Spencer,[68] or the Christian Socialist F.D. Maurice (1805–72), who was said to have understood 'Darwinism...as little as a traveller newly arrived in some distant land understands the purport of its most idiomatic and hurried conversation.'[69] James Martineau, who did have a solid grounding in the natural sciences and retained an active engagement with the progress of thought right up to his death in 1900, demonstrated only the most limited acknowledgement of some form of evolutionary process, without any acceptance of the science of evolutionary mechanism.[70] Robert Browning (1812–89), like Coventry Patmore (1823–96), kept up with scientific advances primarily as a source of metaphor. Despite the widespread tendency to read an evolutionary sensibility into his *Dramatis Personae* poems, especially 'Caliban upon Setebos', Browning hardly progressed beyond a sort of vague spiritual developmentalism; as Leslie Stephen commented, 'He let Strauss or Darwin go their own way, and apparently they did not in the least trouble his mind.'[71]

Much of the 'evidence' of rapid conversion is retrospective, based on biographical or autobiographical accounts which claim early adhesion, often in the face of all contemporary indications. The account of Francis

Galton (1822–1911) that the *Origin* 'demolish[ed] a multitude of barriers by a single stroke' tallies with his article 'Hereditary Talent and Character', published in *Macmillan's Magazine* in 1865, and the tribute he offered to Darwin himself in 1869, but is otherwise unsubstantiated by contemporary materials, despite his attempts to align himself to the Darwinians.[72] In the same way, Alfred Newton's narrative of his embracing of Darwinism as he read the Linnean Society paper of 1858 is almost certainly a retrospective exaggeration.[73] Newton's extensive private papers are noticeably, even suspiciously, thin on correspondence from the period immediately after the publication of the *Origin*, but there is a surviving letter to H.B. Tristram, from July 1860, in which Newton noted that he might have wavered for a considerable time as a 'Darwinoid', had not Wilberforce's speech at the British Association that year pushed him on 'into pure and unmitigated Darwinism', and his later commitments were seen within Darwin's circle as cautious if not pusillanimous.[74] Those whose own scientific knowledge was limited often found it difficult to avoid being bowled over by the force of the arguments of Darwinists. And yet, once the initial impact had passed, they often reverted back to their previous certainties.[75]

So where are the converts? The symbolic case has often been taken to be Charles Lyell's conversion, embodied in the 1868 edition of his *Principles of Geology*. And this avowal of his conversion was the subject of much private self-congratulation within Darwinian circles in 1868.[76] As a potential convert, Lyell had a lot going for him. The earlier editions of the *Principles* had already served as a foundational text in the development of evolutionary thinking, and he had been exposed to Darwin's ideas well in advance of the publication of the *Origin*. Those who knew him suggest that, into old age, he retained a notable openness to new arguments.[77] As early as November 1859, it was reported that Lyell was 'very full of Darwin's book and quite a convert to his theory'.[78] However, while conceding in 1861 that there was no rival hypothesis in the field, Lyell's private correspondence through the 1860s registers an extended and at times fevered uncertainty, as he lurched from one stumbling block to another, and his tortured progress is registered in Darwin's despairing initial response to the caution of the *Antiquity of Man*, which he described as 'an elaborate, obscure and protracted exercise in beating about the bush'.[79] In the early 1860s Lyell bombarded Darwin with questions. A chance conversation with Owen about recent fossil discoveries, consultations with the botanists William Carruthers (1830–1922) and Bunbury, or reading the arguments of Argyll and John Gould (1804–81) about the lack of evidence of transitional species, all brought doubts

100   DARWINISM'S GENERATIONS

bubbling to the surface.[80] In 1866 he was mired in the question of the impact of the ice age on South American fauna in which he felt Darwin's own ideas transgressed Darwinian principles.[81] In 1867 it was the question of barriers to species distribution and the difficulties this presented to monogenetic arguments which prompted the suggestion that confirmation of the transmutation of species might require 'several hundred thousand years'.[82] By 1868 Lyell had accepted evolution, but a narrowly providential version which continued to question the weight Darwin put on natural selection, and always denied human evolution.[83] It is only in much later testimony that we get a picture of Lyell as fully within the Darwinian fold; and this was a matter of society and sympathy as much as science.[84]

The botanist George Bentham (1800–84) is a more credible model for conversion. His retrospective account was of being 'throughout one of [Darwin's] most sincere admirers' who 'fully adopted his theories and conclusions notwithstanding the severe pain and disappointment they at first occasioned in me'.[85] Less intimate with Darwin, Bentham had been fully aware of the gathering challenges to fixity of species in the 1850s, but not with the specific development of Darwin's ideas. Present at the Linnean Society in July 1858, he had immediately realised his long-standing commitment to the fixity of species must be abandoned. His 1863 presidential address to the Society and his report to the British Association in 1874 were later cited as acknowledging his 'full adoption of Mr Darwin's views'.[86] But if in time Bentham did come round to a fully Darwinian view, his record during the 1860s suggests again a much more hesitant and tortuous process than his telescoped reminiscences might suggest. In the months before the *Origin* appeared he had described himself as 'too old and too lazy' to adapt to the new ideas,[87] and in his private correspondence with the Harvard botanist Asa Gray it is noticeable how cautious, indeed coy, he remained. In the immediate aftermath of the *Origin*, the furthest he would go was to say that 'some of my views have become partially modified' by Darwin's ideas.[88] In the following years, although he explored Darwinian themes in his Linnean Society presidential addresses, Bentham carefully avoided any categorical commitment to them, and was clearly disturbed at the sort of speculative developments being essayed by Darwin's German disciple Ernst Haeckel.[89] In 1874 while acknowledging that the intervening fifteen years had sufficed to 'establish' the theory, he did so in the context of his own sense of 'belong[ing] to a past generation of botanists', and unwillingness to engage in wider theoretical controversies; and he could not resist adding the proviso, that the proof stood 'at least for botany'.[90] Perhaps scarred by his

IN THE WAKE OF THE *ORIGIN* 101

treatment at the Linnean Society (for which see below, pp. 110–11), Bentham's correspondence in the 1870s and 1880s showed little active commitment to Darwinian developments.[91]

The crabbed progress of Lyell and Bentham, despite all the circumstances which should have eased conversion, bears witness to the considerable barriers of intellectual investment faced by the older generations in embracing evolution. Unsurprisingly, it is difficult to identify other examples of wholehearted conversion amongst the early Victorians. One was perhaps the Scottish anatomist Allen Thomson (1809–84), whose presidential address to the British Association in 1877 spoke of the triumph of Darwinian principles (Thomson was quoted as remarking that 'I consider it impossible for anyone to be a faithful student of embryology, in the present state of science, without at the same time becoming an evolutionist').[92] The Irish botanist and zoologist, G.J. Allman (1812–98), enthusiastic about Huxley's confrontation with Wilberforce in 1860, was publicly committed to evolution by the early 1870s.[93] Otherwise, early Victorian endorsements tended to be deliberately vague and/or hedged about with reservations. In 1867, Isaac Anderson Henry (1800–84) told the members of the Edinburgh Botanic Society that he believed 'much of the Darwinian theory', but his clarifications limited this to the development of an already fully created natural world.[94] William Pengelly (1812–94), the explorer of Kent's Cavern, was described in reminiscences by an old friend as 'an early convert to the theory of creation by evolution', but it is not easy to see exactly how far he went,[95] while although James McCosh (1811–94), who moved from his Professorship of Moral Philosophy at Belfast to Princeton in 1868, recounted how he had come to accept evolution while crossing the Atlantic, this was clearly a fairly natural extension of the principles that he had articulated for the previous twenty years, and was itself laced with severe limits, including design, a moral end, a view of causation which involved a divine power not just natural law, considerable doubts as to the primacy of natural selection, and direct rejection of evolution of man from animals without at least a point at which divine intervention creates the soul.[96]

There was greater scope for conversion amongst the succeeding generation, where the lines of division were less absolute, the convictions not so ingrained, and the circles of intimacy less discrete. Few mid Victorians felt the need to offer the categorical dismissals characteristic of the early Victorians. They were aware of the force of Darwin's arguments, and as a result, the challenge of the *Origin* frequently produced a sustained quandary, a state of turmoil typified by Kingsley's confession to F.D. Maurice in

102 DARWINISM'S GENERATIONS

1863 that he was busy working out the implications of Darwinism for natural theology, but 'that he did not think this would bear fruit for seven years'.[97] Yet, for the mid Victorians too, where acceptance of evolution is visible, it was drawn out, conflicted, and rarely unambiguous. Hooker, who had been made familiar with Darwin's ideas in the mid-1840s while still in his twenties, recalled in 1860 that they were 'gradually forced on an unwilling convert', and only in the face of his own strenuous resistance and unremitting study.[98] And a recent detailed study of his position has characterised it as 'deliberately equivocating'.[99] Other scientific converts who might be named include the Aberdeen anatomist John Struthers, and the parasitologist T. Spencer Cobbold (1828–86). During the 1860s, Struthers' lectures embraced the Darwinian image of the tree of life and branching descent, and helped give Aberdeen the reputation for being more Darwinian than any other British university.[100] On Darwin's death he professed his appreciation of the 'complete overturn' Darwin had achieved in the interpretation of animal structures in which he had been educated.[101] But Struthers did not go far enough for orthodox Darwinians, and his observations were often strangely obscure and non-committal.[102] Cobbold's initial opposition dissipated and he clearly accepted the evidence of evolution; but despite corresponding with Darwin seems in his public statements to have deliberately steered clear of any direct identification with 'Darwinism'.[103]

In the case of the astronomer William Huggins (1824–1910), dismay at the obscurantist resistance of the clerical establishment seems to have ultimately won the day. By the 1890s Huggins was a convinced evolutionist. His pioneering work employing spectrum analysis demonstrated the consistent chemical basis of the universe, and its own evolutionary dynamics, and closer to home he had accepted the fact of progress by means of evolution as having been established by Darwin, 'and his prophet Huxley'.[104] This had not always been the case. During the 1860s, Huggins was 'disgusted' by scientists' tendencies to dismiss the idea of a supreme intelligence, and it was only later that the intransigence of theological resistance to evolutionary thought, particularly the evidences of entrenched anti-evolutionary prejudices in the Church of England, became his overriding concern.[105] By this stage, as his strenuous efforts to oppose the anti-scientific lectures of Samuel Kinns (discussed below, pp. 250–51) demonstrated, he was clearly a fully subscribed evolutionist, though it is not clear exactly how Darwinian.[106]

The way in which the coin fell frequently depended on accidents of religious conviction. Part of Lyell's difficulties was that he shared none of

Huxley's anti-religious temperament. His strong faith produced a powerful instinct to resist the full implications of Darwin's thought.[107] Frances Power Cobbe recalled asking Lyell how attacks on the argument from design in nature touched him religiously, to which he replied, 'Not at all', delighting in the incapacity of even sceptical scientists to describe evolution without using language implying divine oversight.[108] He was appalled at the combative tone of W.B. Carpenter's attack on the clergy in a Sunday lecture in 1866, and never quite made it into the X-Club inner circle.[109]

Similarly, in the case of the anatomist William Kitchen Parker (1823–90), whose work did much to advance the embryological evidences of links across widely separated orders, although he seems eventually to have overcome powerful religious scruples to calling himself a 'Darwinian', the testimony of his son was that he retained his 'adherence to the theory of creation'.[110] Clarifying Parker's precise stance is difficult. His writing was obscure: literally—his faint and spidery handwriting defied deciphering—and because of its playful self-deprecation. A self-taught anatomist, fiercely opposed to Owen, he continued to resist those he called the 'Agnostic rogues', agreeing with Sir George Gabriel Stokes (1819–1903) that any scientific theory was 'a small thing compared with the truth of the "Everlasting Gospel of our *Lord & Saviour*"'.[111] So, while the even more sceptical Stokes believed Parker was an unequivocal evolutionist, albeit viewing it as God's mode of working, the high-Victorian Darwinian Arabella Buckley (1840–1929) seems to have seen him rather as hesitating on the fringes of evolutionary commitment, reluctant to take the final step.[112]

Similar constraints were at work for Charles Kingsley, often identified as one of Darwin's earliest converts. For the X-Clubbers, Kingsley was the epitome of the 'enlightened clergyman'. More than anyone else before the interventions of the greatly more scientifically accomplished Anglican priest Aubrey Moore (1848–90) in the 1880s, Kingsley became a personal troth of the compatibility of Christianity and Darwinism.[113] As we have already seen, his correspondence of the 1860s is full of bursts of enthusiasm for evolutionary thought and of breathless accounts of the advance of Darwinian principles. But even Kingsley's most indulgent friends conceded that he could be incoherent, and this shows in his attempts to work through his Darwinism in these years.[114] As with Lyell, the surviving letters are enough to show Kingsley making sustained attempts through a network of correspondents within Darwinian circles, including Huxley, Wallace, Bates, and others, as well as Owen and Argyll, to try to navigate through various anxieties, including the scope and efficacy of natural selection, and the wider question of the role that might be ascribed to God in the absence of arguments of design and a single creation.[115] Before the publication of *Descent*, Kingsley was able to reconcile these challenges with his 'Darwinite' identity. Dogmatic atheism was blamed on 'Comte, Herbert Spencer, & the rest',[116] and Huxley's candid confession of his own doubts about natural selection helped. But it is clear that Kingsley's views diverged from Darwin's in several key respects. He continued to speak of a providential God, and refused to accept the existence of purely physical 'natural laws': '[n]othing exists but <u>Will</u>; all physical laws and phenomena are but the manifestations of that will, one orderly, utterly wise, utterly benevolent'.[117] And inevitably, he skirted round the question of the commonality of man and animal, grasping with relief Wallace's renunciation of the application of natural selection to man in 1869–70.[118] In this context the publication of *Descent* only put further strains on Kingsley's allegiances. Thereafter he remained 'ready to follow Darwin and the "evolution" doctrines, as far as we see good', but found it increasingly difficult to express unambiguous enthusiasm.[119]

Beyond these well-documented cases the search, even among the mid Victorians, for examples of other 'converts' is surprisingly unfruitful, and the cases of Lyell, Bentham, and Kingsley should alert us to the challenges of treating the often cursory testimony we have entirely at face value. Others can be identified, although the precise nature of their evolutionary beliefs is often unclear, and they comprise only a tiny proportion of the early and mid Victorians whose position on Darwin can be ascertained. Many of the possible additions come from within Darwin's own network of associates and correspondents, pushed forward by sustained private persuasion and personal regard, whose precise commitments are often obscured rather than illuminated thereby. This is certainly the case for the ornithologist William Bernhardt Tegetmeier (1816–1912), a regular correspondent of Darwin's in the later 1850s and 1860s, who by 1871 was singing the praises of the 'singular genius' of Darwin, for stripping the veil of familiarity from the world and showing the unseen beauties of the spirit of its forms.[120] Tegetmeier almost certainly accepted evolution, but his precise position is far from clear, despite a long career as journalist and correspondent; for his biographer, 'his life was too fully occupied with facts, with living interests— with men and women and other animals, to worry about abstruse metaphysical ideas'.[121] Much the same could be said of Sir James Paget (1814–99), Darwin's physician, and on friendly terms with the Darwinian inner circle. Like Acland, Paget praised Darwin's method and condemned those who from positions of limited knowledge contested Darwin's arguments, but his writings noticeably avoid explicit avowal of belief in Darwin's doctrines, and he frequently took refuge behind the premise that questions of science are never 'settled'.[122] W.B. Carpenter was also able to shift a great deal further than most, by 1872 rejecting special creations, accepting the common ancestry of the animal kingdom.[123] At the same time, he had a conflicted relationship with the Darwinian label: he angrily denied the suggestion of Owen in a review in *Athenaeum* in March 1863 that he was a Darwinian (and indeed he observed that his position might be seen as 'anti-Darwinian').[124]

The Vagaries of Institutional Darwinism

The tortuous struggles of Lyell, Bentham, and Kingsley help to explain the challenges Darwin's supporters faced in the 1860s. The institutional history of the 1860s offers no stronger support for a decisive Darwinian shift in these years. There was no natural or easy extension of influence, no transformation of institutional cultures. There was progress, but it was slow, partial, and achieved only by a deliberate and sustained campaign, orchestrated by Huxley and his allies in the X-Club, to manoeuvre themselves into positions of influence and authority.[125] Huxley was the driving force. For the thirty years after the appearance of the *Origin* it seemed that he was everywhere; travelling the country to give provocative lectures, sermonising from the pages of the intellectual magazines, holding the administrative reins of scientific societies or installing his friends, pressing the claims of his younger followers. The allies he gathered around himself worked assiduously during the 1860s to secure their influence, promote their protegés, and secure official sanction for Darwinian science.[126] This was a double-edged sword. Huxley's robust anticlericalism and determination to exclude all standing for theology on matters scientific compounded the anxieties of the religious-minded, and despite the care with which he (and his allies) steered clear of anything that could be seen as aligning them with the formal organisations of secularism, this associated 'Darwin' indelibly with atheism. There were dark mutterings about the tyranny of the Darwinians as early as 1860, and within a few years, an older pre-Darwinian naturalist like the distiller James Scott Bowerbank (1797–1877) could already feel that he was rubbing up against the 'Busk–Huxley clique'.[127] X-Club members leant on potential opponents, warning them that any challenge to Darwinian ideas would be taken as a personal attack.[128] At the Royal

106   DARWINISM'S GENERATIONS

Institution, the most prestigious of the London lecturing institutions, the sympathy of the secretary, Henry Bence Jones (1812–82), and Tyndall's insider status as Professor of Natural Philosophy, meant Darwinians were always well represented in the lecture lists. Lubbock appeared almost every other season through the 1860s, and he was joined by others, such as W.K. Clifford, by 1868 giving his own evolutionary lectures (albeit an unruly concoction of Darwinian, Lamarckian, and predominantly Spencerian notions).[129] Energetic efforts were also made to spread the word across the provinces, where Huxley in particular was always in demand. After 1865 the emergence of the Sunday Lecture movement brought together sympathetic audiences of liberals and progressives.

Even so, the forces of resistance remained formidable, if often obscured by Huxley's pyrotechnics. Royal Institution audiences also heard bluntly non-Darwinian lectures from, for example, J.O. Westwood in 1867.[130] As H.W. Bates found, English scientific cultures remained dominated by the old empirical approaches, and despite the vast amount of material he had collected during his eleven years abroad which spoke to evolutionary developments, he was largely ignored by the scientific establishment; he wrote of the staff of the British Museum responding like 'a nest of hornets' to his South American discoveries.[131] The careers of Bates and Wallace illustrate the sluggishness of Darwinian advance outside the inner circle of active X-Club promotion. Wallace never achieved a recognised scientific appointment, and after several years looking, Bates could only obtain the fairly menial position of assistant secretary of the Royal Geographical Society. This gave him some time for research, but thereafter he wrote little on Darwinism or geographic distribution.[132] In both cases Darwinian credentials were unable to overcome social exclusions.

Even in the universities, constantly replenished with undergraduates supposedly devouring the *Origin*, change was slow. At Oxford, the High Church early Victorian E.B. Pusey led a powerful resistance. For Pusey, natural science 'manufacture[d] Atheists', and although he failed to block Darwin's honorary doctorate, he succeeded in quashing awards proposed at the same time for Tyndall and Froude.[133] Even a culturally conservative undergraduate like Andrew Lang (at Balliol from 1864) found Oxford 'a lumber room of ruinous philosophies, decrepit religions, forlorn beliefs', and Alfred Newton remained concerned about how 'heretical' even a modest endorsement of Darwinism would appear to a Cambridge audience.[134] The established academic staffing changed at a snail's pace, and the weight of older appointments, and even fresh appointments of early Victorians, meant that

IN THE WAKE OF THE *ORIGIN* 107

movement in the curriculum was almost imperceptible. Exam papers gave little attention to evolution, and even Huxley's lectures at the School of Mines almost entirely ignored the topic.[135] Meanwhile, training at the country's theological colleges studiously avoided Darwin and Spencer, amidst efforts to strengthen the teaching of natural theology in the face of the 'brooding' danger of the *Origin* and *Essays and Reviews*.[136]

The X-Club's most striking success was the control it came to exercise within the Royal Society, and the influence that it could exert as a result at the very apex of British scientific culture. For much of the 1860s the Royal Society was decidedly cold in its response to Darwin. In 1868 Sabine as president and virtually the entire slate of officers chose not to attend the British Association meeting, apparently in no small part as a result of their anticipations of a Darwinian presidential address from Hooker.[137] Things began to change when Spottiswoode was manoeuvred into the treasurership in 1870, replacing the chemist William Allen Miller (1817–70), firmly aligned with the North British Physicists, who had felt compelled to attend the British Association meeting at Edinburgh a few months before his death 'to combat the Heresy of Tyndall and Huxley'.[138] Thereafter, several X-Clubbers were able each year to gain election to the Council. As Huxley was installed as Biological Secretary in 1872, the physicist George Gabriel Stokes looked on askance, perturbed at objectionable passages in the first paper subsequently proposed for publication in the *Transactions*.[139] Not that this can be interpreted primarily as a Darwinian takeover. Here, as elsewhere, a significant part of Huxley's success was his ability to present himself as the champion of unfettered science against the intrusions of religion. The successful nomination of Darwin for the Copley medal owed as much to the way in which it could be presented as (in Hugh Falconer's words) 'a determined protest' against 'the profession of religious <u>against</u> scientific faith'.[140] Although Henry Tristram was anxious by the later 1860s that even his sort of lukewarm Darwinism might be seen as so limited as to scupper his chances of being elected a fellow, the Royal Society's *Philosophical Transactions* managed to avoid any mention of evolution for the whole of the decade.[141]

A similar picture emerges from the British Association, because of its broad constituency, its peripatetic links with provincial science, and its widely publicised annual congress, perhaps an even more significant barometer of scientific culture. Hence the importance of the clashes which occurred at Oxford in 1860 and Cambridge in 1862. The BA was 'an oligarchy, presiding over a severely limited democracy', in which the Council

108 DARWINISM'S GENERATIONS

and the presidents of the sections it nominated exerted considerable power to shape the agenda, but could not always guide debate.[142] By the 1870 Liverpool meeting, the section heads included several Darwinian sympathisers, including Rolleston, the Cambridge physiologist Michael Foster (1836–1907), and Sir John Evans. The presidential addresses of the 1860s displayed some Darwinian sympathies. At the 1866 meeting at Nottingham where the early-Victorian physicist Sir William Robert Grove (1811–96) took a generally favourable stance, the Anglican weekly, *The Guardian*, remarked that 'the Darwinian theory...was everywhere in the ascendant'.[143] Here again the X-Clubbers made skilful use of the rhetoric of scientific freedom. At Bath in 1864, T.A. Hirst was encouraged by 'the applause with which every protest against fettering science by religious dogmas was received'.[144] Yet in many ways the Bath meeting was all about the fossil record, with Darwinism on the defensive as regards gaps in the evolutionary sequence. As the decade progressed, the contributions of the more febrile opponents of Darwinism, like F.O. Morris, were given increasingly short shrift, but anti-Darwinian presenters and papers continued to feature on the annual programmes, and there was plenty of scope for opposition from the floor; as was illustrated by John Crawfurd (1773–1868)'s blunt dismissal of Darwinism at Nottingham in 1866.[145] Observers continued to see the BA as 'the arena of the fight about the Development theory', not its pulpit, and the recollections of some, even if favourably disposed, was that the dominant response to Darwinian views was hatred and suspicion.[146] The 1867 meeting in Dundee (though described as 'capital' by Lubbock) showed little sign of a decisive shift towards Darwinian thought, while in 1868 although press coverage was dominated by Hooker's stridently pro-Darwinian opening address, there was also significant pushback.[147] For Rolleston, who had intervened to challenge F.O. Morris' paper, the meeting marked a deterioration in the courtesies of discussion; recognising the greater assertiveness with which Darwinism was advanced, he feared that this was only further alienating opponents.[148] In 1869, despite Tuckwell's satisfied verdict on the improved tone, newspaper reports presented the anti-Darwinian papers less as the lunatic fringe and more as the event of the meeting, and even the committed anti-Darwinite John Eliot Howard thought there was too much crude opposition.[149] The meetings at Exeter, Liverpool, and Edinburgh saw natural selection on the defensive, losing ground as geologists drew back, alarmed by the implications of William Thomson's limit of 100 million years for the age of the earth. Tyndall dodged and there was little Darwinian resistance.[150]

The sceptical flavour of the BA meetings was perhaps disproportionately a function of the unruliness of the anthropological section. The history of the fierce competition of the Ethnological Society and its breakaway, the Anthropological Society, illustrates the chaotic landscape of evolutionary debate in these years, but also the limits of the institutionalisation of Darwinism. As several studies have shown, there was actually very little difference of scientific position or generational composition between the two societies.[151] Huxley clashed with both, describing the Ethnological as an 'organised stupidity' and the Anthropological as a 'nest of imposters'.[152] The Ethnological was the established organisation, and boasted a number of members from Darwin's circle, but the active membership also included vocal opponents; while the Anthropological Society, formed in 1863 as a breakaway group, although its council included figures like the Spencerian psychologist John Hughlings Jackson (1835–1911), provided a platform for anti-Darwinians like James Hunt (1833–69), who owned and edited the *Anthropological Review*, and Joseph Barnard Davis (1801–81), whose hostility to Lubbock and 'the rest of the Darwinians' comes through clearly in his private correspondence.[153] The amalgamation of the two associations into the Anthropological Institute in 1871 (after lengthy labour by Huxley and his allies) did eventually allow evolutionist approaches to predominate, but only after more difficulties in the early 1870s, when the obstructionism of a rump of the former Anthropologicals drove Lubbock to resign the presidency.[154]

The situation was no more promising in entomology. Admittedly, bug-hunting was notorious as the 'last bastion of amateur science'.[155] Yet as the locus for the development in the 1860s of ideas of protective mimicry, one of the most accessible arguments in favour of natural selection, the Entomological Society should have offered fertile ground for Darwinism. Wallace, Bates, and Lubbock all served as president between 1860 and 1871, and some younger entomologists[156] and the *Entomologist's Monthly Magazine* which they established in 1864, attempted to promote a more theoretical and evolutionary approach. Yet they made little progress. In the 1860s and 1870s the Entomological was riven with infighting and disputes between older and younger members, and not a place where a convinced evolutionist felt himself in the congenial society of those who agreed with him in principle, recalled one younger member.[157] As late as 1873 J.O. Westwood, probably the most senior academic entomologist in England, was pressing for the appointment of a Professor or Reader in Natural Theology in Oxford to 'counteract the atheistical demoralization

110 DARWINISM'S GENERATIONS

resulting from the unlimited teaching of Darwinism'.[158] Popular entomology continued to be shaped by the emphasis on discovery and classification typified by Edward Newman (1801–76), long-time editor of the *Zoologist* and fiercely hostile to Darwinian ideas.[159] Bates was so disillusioned with the conservatism of the entomologists that he gave up entomological work for several years, and in 1871 one high Victorian vented his frustration that the field was retrogressing.[160]

It is the Linnean Society, however, which provides the most telling case. Not least because, in the later 1860s and early 1870s, George Bentham's presidency has been seen as leading the Society decisively in a Darwinian direction. Yet the Linnean was a recurring headache for the Darwinians for much of the 1870s and 1880s. Apparently several members resigned after the publication of the *Origin* because the Society refused to eject Darwin, but old-time naturalists like Bowerbank continued to treasure it as a refuge from the X-Club-dominated Royal Society.[161] Recognising the conservatism of the ordinary members, Bentham explicitly discouraged the discussion of 'theoretical' questions.[162] Some sense of how unsympathetic the Society was is offered by the history of Richard Spruce's paper on modifications in plant structure caused by the agency of ants read in April 1869, subsequently set in type but never published, apparently because of anxieties about its Darwinian frame, and Spruce's refusal to strike out his views of the operation of the process over 'thousands of ages'.[163]

Ultimately, Bentham's leadership prompted a revolt partly driven by a feeling amongst the general membership that the Society's shift towards Darwinism had gone too far.[164] The precise history is only dimly visible, and was a reflection of various tensions, including the institutional rivalries of the Royal Botanic Gardens at Kew and the Natural History Museum, but unhappiness at Bentham's 1873 presidential address seems to have prompted him to look to step down; and his departure was then hastened by an ugly row at a special meeting in January 1874, where William Carruthers, Keeper of Botany at the Natural History Museum, organised an opposition to proposals to transfer some powers from the membership to the Council.[165] Both age and intellectual difference were at play here. Carruthers' old-school evangelicalism and suspicion of evolutionary thought cannot have helped, but there were also tensions between the old 'species-makers' and the new, more theoretical and scientific naturalists.[166] In the aftermath, Hooker remarked that it was now for the 'younger men' to push the society forwards, but nominations of younger Darwinists like E.R. Lankester continued to face opposition from Carruthers, and Hooker despaired at the state of affairs. In 1878 there was talk of a small band of conspirators, operating by 'secret canvas', trying to mobilise Huxley and Darwin to get the 'better class of fellows' to move. The weakness of the Darwinians was confirmed by their failure to get William Thiselton-Dyer (1843–1928), a young botanist closely aligned to Darwin and Hooker (whose daughter he married and whom he succeeded as Director of the Royal Botanic Gardens at Kew), elected to the secretaryship in 1870, or to successfully oppose Hooker or Thiselton-Dyer to Carruthers for the presidency in 1886, which brought further private lamentations at the unwillingness of the 'new school of botanists' to rally around the Society.[167] It was a group of late Victorians, increasingly frustrated with the old-fashioned science of the Linnean Society's *Proceedings* who launched the *Annals of Botany* soon thereafter.[168] Into the twentieth century younger biologists were still complaining about 'Linnean approaches', with their pre-Darwinian views of species.[169]

There is little evidence that circumstances were more favourable elsewhere. Certainly not at the Royal Geographical Society, where the convinced anti-Darwinian Roderick Murchison was president through the 1860s. Nor in the Geological Society, where Peter Martin Duncan (1821–91) seems to have been the only positive voice, despite Darwin's initial hopes for support from the geologists.[170] The replacement of the *Geologist*, edited by the anti-Darwinian catastrophist Samuel Joseph Mackie (1823–1902) by the *Geological Magazine* under the editorship of Henry Woodward (1832–1921) suggested movement, and by 1868 Murchison was aware that 'the younger Geologists' were likely to be 'much more of a Darwinian than I can <u>possibly be!</u>', but his account of the acclamation which greeted the anti-Darwinian tone of a paper from the Duke of Argyll in February of that year strongly suggests that the sympathies of the bulk of the ordinary membership were with him.[171] Early-Victorian geologists seemed keen to line up to proclaim their opposition.[172] Younger geologists recalled 'the heart-searchings and misgivings with which the new doctrine was received, and the long time that elapsed before many of us undertook to investigate and apply the new doctrine', experiencing geology as a science where the older authorities were 'permanently warped by their clinging to obsolete theories'.[173] It was the same in ornithology. Frederick DuCane Godman recalled of the British Ornithological Union that initially there were only a handful of adherents to Darwin's ideas and 'the new faith...grew, very slowly at first'.[174] Even in 1910 it could be suggested that alarmingly little evolutionary work had been done within ornithology, in contrast to the continued investment in identifying new species.[175]

112   DARWINISM'S GENERATIONS

The London press remained largely unsympathetic throughout the 1860s. The leading journal of general culture, the *Athenaeum*, accused Bates' *Amazons* of twisting its facts to Darwin's theory, and sneered at *Animals and Plants under Domestication*.[176] The tone of the provincial press also remained predominantly hostile. Even a publishing house like Chambers Brothers, notwithstanding Robert Chambers' authorship of *Vestiges*, deliberately catered to the 'dull and vapid' general reader, rather than, as Chambers himself confessed, 'addressing ourselves to Darwin's [sic], Huxleys, Thompsons, etc etc'.[177] Attempts by the X-Clubbers to secure a reliable publishing outlet of their own foundered, first in the difficulties of the *Natural History Review*, which had given an opportunity for Norman Lockyer (1836–1920) and his contemporaries to obtain a platform, and then in the failure of the *Reader*, which was briefly backed as a vehicle for articulating the claims of Darwinian science in the mid-1860s. Neither could attract enough of a readership to thrive, never mind the broad public that had been hoped for.[178] At the start of the 1860s editorial chairs were largely occupied by early Victorians. Over the decade the generational composition of the press shifted, partly as a result of the death or resignation of older editors and the appointment of younger ones, and partly through the launch of new titles. The emergence of the mid Victorians is especially visible in the popular science journals: Henry J. Slack (1818–96) at the *Intellectual Observer* (1862–8), James Samuelson (1829–1918) as editor of the *Popular Science Review* (1862–) and the *Quarterly Journal of Science* (1864–), Mordecai C. Cooke (1825–1914) for *Hardwicke's Science Gossip* (1865–), and Philip Lutley Sclater (1829–1913), editor of *The Ibis*. Often personally ambivalent towards Darwinian ideas, these mid Victorians were at least, as Bernie Lightman has argued, determined to open up spaces for unfettered discussion of evolution.[179] But their reach and influence paled into insignificance in the face of the enormous circulation of titles like the *Sunday Magazine*, launched by Thomas Guthrie (1803–73) and William Garden Blaikie (1820–99), which from the outset sought to sustain natural theology approaches.[180]

At times generational dynamics emerge in stark relief. The short-lived *Journal of Travel and Natural History*, edited by the early Victorian Andrew Murray, very much as a vehicle for his anti-Darwinian views, was effectively ostracised by the younger physiologists and anatomists who aligned themselves instead with the more Darwinian *Journal of Anatomy and Physiology*, edited by G.M. Humphry (1820–96) and William Turner (1832–1916).[181] The history of the naturalists' press is equally instructive.

IN THE WAKE OF THE *ORIGIN* 113

The well-established *Gardener's Chronicle*, edited until 1865 by John Lindley (1799–1865), was prepared to give some space to Darwinists, although Lindley himself remained unconvinced, and under his successor, Maxwell T. Masters (1833–1907), the paper offered a platform for younger Darwinian naturalists.[182] But it was balanced by *The Field*, where Edward Newman was natural history editor, and from 1866 by *Land and Water*, edited by Frank Buckland (1826–81), son of William Buckland, which offered a consistently conservative Paleyite tone, celebrating practical studies and anathematising 'philosophy'.[183] *Land and Water's* regular correspondents included a number of Darwin's prominent early-Victorian opponents, including F.O. Morris and Charles Bree, and was welcomed by Bowerbank as 'really writing Natural History', in contrast to 'the transcendental rubbish that I often read in *Nature* and other serials'.[184] By 1871 the late Victorian Raphael Meldola, then twenty-two, was just beginning to take the Darwinian fight to the paper's unsympathetic readers.[185]

All the evidence suggests that non-metropolitan amateur natural history cultures remained predominantly hostile: older members obstructing the influx of Darwinian ideas along with their wider resistance to atheism. Even the relatively innocuous topic of the antiquity of man in the hands of as cautious an evolutionist as W.B. Carpenter could unnerve provincial lecture managers.[186] In 1867 the Philosophical Institution in Edinburgh even came under pressure because of its willingness to give a platform to the almost ubiquitous Huxley.[187] Members could find themselves urged to withdraw proposed contributions on Darwin, for fear that any sort of advocacy would be fatal to the good order of the society.[188] The hesitancies of the Warrington Field Naturalists' Club in the 1860s are telling of a desire to avoid controversies that might split the membership. So readings from Darwin's *Orchids* were balanced by extracts from Hugh Miller's *Footprints of the Creator*. In 1865 the Warrington club did hear a paper on insect mimicry, but one which explicitly avoided going into theoretical explanations.[189] When in 1867 the Alloa GP John Duncanson (1826–?) addressed the local Natural History and Archaeological Society on Darwinism, although he was clearly disposed to accept, the thrust of his paper was primarily simply to claim legitimacy for it as a theory, to be tested purely by scientific arguments.[190]

This was at least in part a result of generational forces. In localities where scientific associations were well established in the 1860s, the dominant tone was set by the Georgians and early Victorians who had taken the lead in their formation. In Gloucestershire, for example, the Cotteswold Naturalists' Field Club, formed 1846, was still dominated through the 1860s by some of

114 DARWINISM'S GENERATIONS

the original early-Victorian founders who resisted evolutionary ideas and continued to deploy the natural theology that had been central to their initial foundation.[191] Not until 1879 did the Cotteswold hear a wholeheartedly evolutionary paper. At Belfast, where a Field Club was established in 1863 on the initiative of a number of younger naturalists looking for greater freedom than was possible in the long-established Natural History and Philosophical Society, the early years were marked by furious discussions on Darwin, as older figures led by the Church of Ireland minister Rev. William MacIlwaine (1807–85) moved quickly to impose orthodoxy.[192] In contrast, the Woolhope Field Naturalists' Club (Hereford), established in 1851, where W.S. Symonds (1818–87) was influential, avoided immediate rejection and at least heard a number of more non-committal comments in the later 1860s.

It was in the more generationally advanced young men's societies, even within theologically conservative denominations, where Darwinian ideas were more freely discussed. John G. McKendrick (1841–1926), Professor of Physiology at Glasgow, recalled being introduced in the early 1860s to Darwin by a paper given by another high Victorian, John Brebner (1833–1902), at the Young Men's Association associated with the Presbyterian Free West Church in Aberdeen.[193] And where advocates of Darwinism are visible on the provincial lecture platform, they tended to be high Victorians, challenging the orthodoxy of their elders. At Huddersfield, the young local naturalist Charles C.P. Hobkirk (1837–1902), who had been rehearsing the evidence of the antiquity of man as early as the mid-1850s, gave a paper in favour of the *Origin* to the Naturalists' Society, to a hostile reception.[194]

The establishment of the Victoria Institute in 1865 was a further expression of this resistance.[195] The Institute, which in its first years also went under the title of the 'Philosophical Society of Great Britain', aped the operations of the other London scientific associations, holding monthly meetings to hear papers on scientific topics. Although it attracted members only slowly at first, there were 300 by 1871. Its generational identity was not cut and dried. The driving force for its formation, and its first secretary, James Reddie (1819–71) was an older mid Victorian, who published several anti-Darwinian tracts; and after Reddie's death, when Capt. F.W.H. Petrie (1831–?) became secretary, the Institute set out deliberately to appeal to a broader range of potential members. But for at least the first half dozen years of its existence in the later 1860s, its appeal was very much to the early Victorians. Its president, the Earl of Shaftesbury, and most of its other leading figures were early Victorians, and although Reddie's concerns had been at the

IN THE WAKE OF THE *ORIGIN* 115

encroachments of science *tout court*, the Institute came to be seen as a primarily anti-Darwinian body, aiming at the subordination of science to theology.[196] As Walter Raleigh Browne (1842–84), civil engineer and Christian controversialist, observed in 1883, 'instead of regarding [Darwinism] as a hypothesis to be investigated, the founders of the Society seem to have looked upon it rather as a heresy to be written down.'[197] Reddie rarely passed up an opportunity to challenge Darwin and his supporters, and the Institute regularly provided a platform for older, more strident opponents. The 1871 presidential address of Rev. William J. Irons (1812–83) offered characteristic fare: repudiation of human evolution, imputation of Darwin's unsound reasoning and reliance on speculation and not fact, and the fundamental distinction of species.[198] It was a tone which reflected the overwhelming response during the 1860s of those born before 1814, both in its uncompromising hostility, and in the readiness of the participants to identify themselves categorically as opponents.

The Anatomy of Antagonism

For the early Victorians and the dwindling numbers of those from earlier generations, even after the first shock had passed, hostility to Darwinian thought remained visceral. It is noticeable that this opposition was, notwithstanding the Victoria Institute, somewhat less vociferous than it was to become in the 1870s. Pusey's animus was more indirectly expressed than it was later, focussing on the dangers of natural science as part of a broader threat to Christian tradition.[199] Similarly, F.O. Morris, Bree, and Birks, whose responses to the *Origin* have already been noticed, and who together published anti-Darwinian tracts regularly through the 1870s, largely confined their dissent in the 1860s to their private correspondence and interventions in the journals and the scientific societies, although Morris took his outspoken anti-Darwinism to the BA, and published *The Difficulties of Darwin* in 1869.[200] This circumspection was not a sign of complacency or conciliation. Bree, although he seems after one particularly strained exchange with Alfred Newton to have toned down his private challenges to Darwinian correspondents, nevertheless remained highly attuned to any whiff of 'Darwinism' in the naturalists' press.[201]

So there is no doubt that early-Victorian hostility to Darwinian thought was visible during the 1860s. Much of it remains unattributed in the rebuttals of the serious periodicals and the reproaches of the press. Much of the

116 DARWINISM'S GENERATIONS

rest was just the small change of domestic conversation and local discussion, only occasionally leaving a documentary trace, as in the case of E.M. Forster's great aunt, Marianne Thornton (1797–1887), who Forster noted 'usually made fun of [Darwinism], becoming hostile when compelled to be serious'.[202] Even so, the interventions of Morris, Birks, and Bree were merely the tip of the widespread identifiable early-Victorian revulsion, summed up by Disraeli's notorious sneer that he sided with the angels not the apes.[203] Old-school evangelicals continued their uncompromising defiance of what Cardinal Manning (1808–92) called 'the brutal philosophy', convinced not just that it would lead to a loss of faith, but that this was Darwin's deliberate purpose.[204] For this older generation, outside radical circles, there was a pervasive horror at the contempt with which Darwin was thought to have treated the Christian narrative of divine creation and atonement, and the spirituality of humanity. It is 'hard to see how', declared Morris, 'on Mr Darwin's hypothesis, it is possible to ascribe to man any other immortality or any other spiritual existence than that possessed by the brutes'.[205]

During these years Darwin was frequently read through the concurrent debates over biblical scholarship. 'Did you notice that Lyell, Huxley and Darwin are amongst the contributors to [Colenso's] defence fund?', one of his early-Victorian correspondents asked J.H. Balfour in 1864. 'A most significant fact'.[206] But in reality, as 'evolution' was slowly detached from the more long-standing and diffuse notion of 'development', older theological opponents often sought to intervene in the debate without any attempt to engage explicitly with Darwin's works. Indeed, because they were more directly shocked by the fact rather than the mechanism of a single evolutionary chain, early-Victorian commentators were also very likely to comprehend the threat of modern science as primarily geological rather than biological.[207] This sensibility is perhaps one avenue by which some early Victorians sidestepped engagement with Darwinism altogether. As Lightman has shown in the cases of Thomas W. Webb (1806–85) and Ebenezer Brewer (1810–97), a broad focus on the physical sciences, and in Webb's case on the universe rather than earth alone, provided plenty of scope for retaining gratitude for the action of a beneficent Creator.[208]

Where there was direct engagement with Darwin's writings, much early-Victorian opposition continued to be captious and polemical. Particular delight was taken in seizing on Darwin's acknowledgements of the challenges his ideas faced as if these were themselves evidence that his thesis was untenable. There is a pervasive sense of casting around for any vaguely serviceable argument, as in the case of the Oxford astronomer Charles Pritchard (1808–93), whose pamphlet sermon *The Continuity of the Schemes of Nature and Revelation* (1866) attempted to prove the absence of sufficient historical time to effect Darwin's mutations through a convoluted argument that the length of the day was gradually increasing.[209] Many early Victorians fell back on an *a priori* commitment to the essential truth of biblical orthodoxy. The poet Martin Tupper (1810–89), for example, confessed to having no difficulty in reconciling revealed facts (in which he included the dating of Adam, the six days of creation, and the Noachian deluge) with 'whatever I have heard of modern discovery: if I couldn't, I would simply wait for more discovery', he told Richard Owen in 1864.[210] Nor was this hostility merely a product of theological conservatism. Although Richard Bethell, 1st Baron Westbury (1800–73), had defended the authors of *Essays and Reviews*, he retained, it was said, a dislike of Darwinism, against which he would declaim with great eloquence.[211]

It was also not a function of a lack of scientific grounding. Although they showed the sort of relativism Tupper displayed, early-Victorian scientists were as likely as non-scientists to persist in aligning themselves against Darwin. The fiercely sustained enmity of Owen was not merely an idiosyncrasy driven by intellectual jealousy or professional self-interest, even if in his case both burned strong.[212] The roll call of early-Victorian scientific sceptics is far too long to itemise, but it included not just Oxbridge professors and the clerical naturalists from within the established traditions of Anglican science, already discussed, but others still coming into post, most of the Scottish scientific professoriate, including J.H. Balfour, and prominent provincial scientists like James P. Joule (1818–89) at Manchester.[213] For Balfour, even to name Darwin was in his writings of the 1860s occasionally too difficult.[214] The recollection of his son was that the new scientific spirit did not stimulate him as it might have done a younger man. Whilst Darwin's work was incorporated in his teaching, 'the acceptance of Darwin's theory appeared too near the negation of faith. On Balfour indeed, as on others with like views, the immediate effect of the Origin was the opposite of vivifying. It gave a shock'.[215]

Even the intimates of Darwin continued to take refuge in tributes which studiously avoided commitment to his theories themselves. Lyell's fellow geologist, Joseph Prestwich (1812–96), appointed to the Oxford chair in 1874, who had taken a leading role in validating the evidences of prehistoric man, continued until the end of his career to reject the evolutionary hypothesis, arguing that the fossil record showed mass extinctions and

118 DARWINISM'S GENERATIONS

discontinuity.[216] His contemporary George Dickie felt that the idea of descent, even from four or five progenitors, was too great a draft on his belief; he accepted that species were liable to vary, but that it was 'quite another affair when we are called on to believe that a seal for instance has once been a land animal, or the converse'.[217] The same was true of Thomas Campbell Eyton (1809–80), ornithologist, who continued on friendly terms with Darwin, although as Darwin conceded in 1872, Eyton did not believe in evolution.[218] It was said that the gentleman naturalist W.C. Hewitson (1806–78) never lost an opportunity of avowing his hatred of Darwinism, and his commitment to 'the direct and independent creation of each individual species'.[219] Equally, the prominent physiologist Thomas Wharton Jones (1808–91), even if from underlying religious motives, continued to present a sustained scientific critique, questioning the Darwinist argument from analogy, pointing out the speculative nature of the proposed evolutionary sequence, and refusing to abandon Paleyite design arguments.[220] Jones was clear that there could be no truck with the 'weak-minded class [who] commit the absurdity of trying to reconcile the doctrine [evolution] with Belief in a personal FIRST CAUSE'.[221]

This sort of language was much less common amongst the slightly younger group who confronted Darwinism in their forties and fifties. And the birth years around 1812–5 do seem to offer a line of transition, before which sympathy or accommodation are rare, and after which implacable hostility, even from a narrowly theological point of view, is increasingly unusual.[222] There were exceptions. For some mid Victorians, suspicion of science prompted responses more characteristic of the early Victorians. Of Edward Thring (1821–87), Headmaster of Uppingham, described as a stalwart antagonist to the pretensions of science, it was reported that he refused to 'give in' to the *Origin*, and considered popular evolutionism mere 'nursery babble'.[223] Many mid Victorians successfully negotiated the 1860s by deliberately keeping Darwinian debates at arm's length. Interestingly, the biblical scholar Brooke Foss Westcott, although apparently approving of Baden Powell's pre-Darwinian arguments on the transmutation of species, seems to have eschewed direct comment on Darwinism.[224] Despite his intimacy with Huxley, Matthew Arnold's instinctive distaste of science seems also to have served as justification for his complete detachment from evolutionary thought.[225] His important cultural criticism of the decade, eventually collected as *Culture and Anarchy* (1869), made no attempt at engagement; he shuddered rhetorically in his Rede Lecture on 'Literature and Science' at the thought that future students would abandon Homer for Darwin, and

IN THE WAKE OF THE *ORIGIN* 119

later reminiscences suggest that like many classically trained mid Victorians he found it easy to pigeonhole Darwinian science as just another version of the ideas of Lucretius.[226]

But otherwise, the mid Victorians sought less to dismiss than to negotiate degrees of distance. Outright denial was rare amongst a cohort who, even as they looked for weaknesses in Darwin's arguments, seemed instinctively to accept some sort of evolutionary history. Their inclination was against; but they were as likely to suspend judgement as to oppose.[227] John Ruskin (1819–1900) gives a flavour. Although over time his hostility hardened, he was initially reluctant to pronounce Darwinism false, even if ready to note that he had not heard any logical argument in its favour, and had 'heard and read many that were beneath contempt'.[228] Mid Victorians generally refused to accept the full scale of Darwinian evolution by natural selection; but came to concede, albeit with extreme reluctance and only in the face of overwhelming evidence, that evolution of some sort had taken place, and earlier doctrines of the immutability of species were no longer tenable. They were fully aware of the challenge laid down to conventional religious teaching. But they regretted the crude conceptions of the opponents of Darwin and Wallace, feeling that the unbridled hostility of older generations, the 'morbid action of dogmatic theology' as the poet William Allingham (1824–89) put it, harmed religion not science.[229] Both clerics and scientists readily accepted the compatibility and indeed complementarity of science and theology.[230] Hence H.W. Bates' reassurance to Kingsley that even if the evolutionary process was purely natural this did not harm religion, 'for the whole process shows how <u>spiritual</u> are the laws of Nature; & the contemplation therefore spiritualizes the mind instead of materializing it'.[231]

The mid Victorians refused to demonise Darwin or dismiss his achievement. Before the publication of *Descent*, where they appeared to express themselves in more hostile tones, it was generally because they seized upon the unspoken implications of the *Origin* for human evolution. The prevalent tone is well captured by R.H. Hutton's address to the Church Congress held in Liverpool in 1869. Hutton expressed sympathy with those anxious about the implications of Darwinism; but he warned that the struggle for existence was one of the great formative principles in the animal world, and overzealous opposition had 'done far more to give a false impression of the weakness of theology, and to lend an impulse of factitious popularity to the fancied atheism of the theory'.[232] Hence the mid Victorians' suspicion of the early Victoria Institute, with its seemingly crude attacks on Darwin. Charles

120 DARWINISM'S GENERATIONS

James Ellicott (1819–1905), Bishop of Gloucester and Bristol, described the Institute as 'a wretched affair. Second-rate men trying to help religion and so injuring it'.[233] Such mid Victorians were often indeed identified as Darwinians, albeit by less sympathetic early Victorians, and at times applied the identity to themselves, while reserving the elements of Darwinian thought they denied.[234] They had no difficulty in seeing Darwin as a great man; they sought out his autograph and his *carte de visite*.[235]

Under the designation 'Christian Darwinisticism' (the designation is significant in that it rightly emphasises the extent to which the question of evolution was, both for mid-Victorian supporters and opponents, fundamentally a religious question), James Moore has grouped Frederick Temple (1821–1902) (ultimately Archbishop of Canterbury), the Duke of Argyll, and St George Jackson Mivart.[236] We can extend this cluster, as representative of a much broader assembly of non-Darwinian mid Victorians which included the judge Sir Edward Fry (1827–1918), the philologist Max Müller (1823–1900), the physicist William Thomson (Lord Kelvin) (1824–1907), and the Irish Presbyterian businessman, Joseph J. Murphy (1827–94). This group, we might also call them in Adrian Desmond's term, 'providential evolutionists', was responsible for a welter of important interventions in the debates over Darwinism in the 1860s and early 1870s, often based on previous periodical articles with their own wide circulation, including Argyll's *The Reign of Law* (1867) and *Primeval Man: an examination of some recent speculations* (1869), Mivart's *On the Genesis of Species* (1871), Murphy's *Scientific Bases of Faith* (1873), and Fry's defence of Darwin's ideas on evolution in *Darwinism and Theology* (1872).[237] They did not deny some sort of evolution, only, as Murphy put it in 1873, 'that special form...which we call the Darwinian theory'.[238]

It was the strategy of Huxley and his close collaborators to dismiss the leading mid-Victorian 'non-Darwinists' as scientifically incompetent if not intellectually inadequate, and members of the group can come across as blundering dilettantes. Argyll's interests were bewilderingly wide, and his interventions inevitably sometimes superficial if not confused; and there is no doubt that he increasingly found himself involved in controversies in which his misunderstandings were mercilessly exposed. But he was widely read and well-connected, and wasn't so readily dismissed by later generations who were more willing to acknowledge the force of some of his doubts.[239] This was even more so the case for Mivart, a Harrow and King's College educated FRS who held a number of academic posts. For much of the 1860s he was to all intents and purposes a part of the inner Darwinian circle, sufficiently regarded by Newton to be commissioned to prepare one of the volumes in the series of Zoological Manuals he was editing for Macmillan.[240] Although Mivart's slow shift over the decade into a more overtly sceptical frame of mind was influenced by his strong Catholicism, his estrangement from the Darwinians was as much personal as scientific, and his interventions in the 1860s and 1870s provided scientifically credible criticism around which the resistance of mid and early Victorians could coalesce, while highlighting some of the positions which distinguished mid Victorians from the older generations, not least his acceptance of some degree of evolution, a degree of variability of species, and some role for natural selection. But he also offered a full repertoire of the characteristic mid-Victorian attempts at scientific rebuttal of Darwinian science: a teleological defensiveness, a refusal to entirely abandon some role for sudden breaks or saltations, and a commitment to the absolute distinction between animals and humanity.

One of the most powerful commitments shared by the mid Victorians and all previous generations was the absolute distinction between humans and the rest of the animal kingdom. For the generations before the high Victorians, Darwinism frequently came to be thought of primarily as 'the ape connection'.[241] The particular power of the trope of monkey turned into man was a register of the extent to which, in incorporating humanity into the evolutionary process, Darwin was eroding the very foundations of the older generations' worldview. Older commentators often responded to the implication of human evolution with outrage. The Anglican historian Henry Hart Milman (1791–1868) suggested that the very fact that Darwin had produced the *Origin* was proof that mankind could not be descended from the tadpole, and was unmollified by attempts to align geological discoveries with the Mosaic account; for Milman, special pleadings of this sort created the ridiculous notion that the divinely inspired writers had created a riddle that would mislead all readers until the advance of knowledge more than two millennia later would suddenly reveal its hidden meaning.[242] Younger heads were more willing to construct defences by distinguishing the corporeal and spiritual man, allowing for a moment of divine intervention which marked off a bestial pre-Adamite man from his fully human successor. In the wake of Lyell's *Antiquity of Man*, the Scottish preacher George Gilfillan (1813–78) accepted his descent from an ape, confessing 'I do not tremble now, as I used to do'.[243] What Gillian Beer has described as 'the space left by the suppression of man in the argument of *The Origin*', frequently allowed disputes over humanity's place in evolution in the 1860s to

122   DARWINISM'S GENERATIONS

be displaced or conducted by proxy.[244] So the line separating man and animals was contested most vigorously in philology, where high Victorians like Julia Wedgwood, F.W. Farrar, and E.B. Tylor (1832–1907) (along with Hensleigh Wedgwood), opposed by the leading mid-Victorian philologist, Max Müller, sought to construct an evolutionary account of language which encompassed the transition from sound to word.[245]

For the early Victorians, rejection of transmutation and commitment to the constancy of specific forms and multiple creations was all but universal. Individual species might evolve, but not one species into another, or one species into many. 'Bears', as Birks put it in a reference to one of Darwin's most controverted passages, 'cannot have turned themselves into whales, by fishing in the water for insects through successive generations'.[246] John Gould, whose identification of new species of finches had been so influential on the development of Darwin's evolutionary thinking, was himself unconvinced that they offered any indication of transmutation.[247] F.W. Farrar's father, Charles Pinhorn Farrar (1798–1877) remained wedded to the belief that species were immutable.[248] Early Victorians simply reapplied well-established anti-Lamarckian arguments in which they were well versed to Darwin, drawing heavily on the work of figures like Lyell, whose *Principles* in the editions circulating in the early 1860s had explicitly rejected transmutationist theories.[249] Much was made both of the evidence of the stability of species in the fossil record over immense periods, and the lack of intermediate or transitional types, and the suggestion that this was merely because the existing geological record was so incomplete was scoffed at.[250] For William Keddie, the 'great thing to establish is the fixity of specific characters, even though species themselves vary', describing the proven infertility of hybrids as 'one of the fortifying buttresses against the transmutation of species'.[251] Similar significance was attributed to the lack of demonstrability of transmutation in contemporary breeding practices. For early Victorians like Keddie, Darwinian science would overthrow the whole edifice of providential governance for blind chance, inexorable laws, and mere physical force. These positions accepted only a disrupted and discontinuous evolution which had little in common with Darwin's gradualism. Hence James Martineau's *The Place of Mind in Nature and Intuition in Man. A Lecture* (1872) argued for successive stages of evolution, as separate overlying layers indicative of divine intervention. Martineau thought, even more than Wallace, that the evolutionary process needed fresh assistance at certain critical stages, that the highest forms of life were not a mere development of lower.[252] J.F.W. Herschel also enthusiastically greeted the idea of

IN THE WAKE OF THE *ORIGIN* 123

'jumps', or 'discontinuous deviations', amounting 'to the introduction into the process, visibly and perceptibly, of mind, plan, design'.[253]

In contrast, the mid Victorians accepted the evidence of variation of species, and some degree of species evolution, but were very reluctant to go as far as arguing that there was evidence of unrestricted transmutation of species, or that this was the result of chance. And the implications of human evolution were firmly resisted even before *Descent* made them explicit.[254] Some mid Victorians, like Thomas Wollaston, the geologist Samuel P. Woodward (1821–65), and the naturalist and theologian John Duns (1820–1909), were only prepared to accept very limited variation.[255] Woodward took his stand on the border between species variation and transmutation: the former was 'a fact', but 'the transmutation of organic types is a thing past my comprehension!'.[256] Inevitably, it is possible to find numerous interventions from clerics like Edward Duke (1814–95), whose lecture 'What Is Geology Doing for the Bible?', in March 1871, disputed the evolution of animals one from another.[257] W.A. Miller told the Church Congress in 1867 that there might have been some succession of species, but only as result of divine creative intervention; at the 1870 British Association meeting he interrupted Huxley to announce that 'he had been ordained to combat the Heresy of Huxley and Tyndall' and had to be physically restrained.[258] Adaptation to environment was recognised as significant; at the same time much was made of the fact that some varieties were more permanent than others. Many held fast to what Argyll described as 'the *substantial* truth of the Mosaic representation of Creation', even if 'highly "metaphorical"': species were at some point(s) created and then might evolve slowly over time.[259] Others, like Mivart, argued for barriers which limited change at certain points, or in certain directions, which opened up some possibilities for transmutation from one species into another. There was room for some shifting of position here, and it is likely that many mid Victorians exploited the vagueness of the formulations they used to silently shift positions; accepting, like Argyll in the 1890s, that species were not immutable, but arguing for a very high degree of stability and endurance, or like J.J. Murphy, that 'there is no limit to the possible extent of variation acting cumulatively'.[260] We can contrast these sorts of concession with the categorical support for transmutation of someone like Sir W.H. Flower (1831–99).[261]

When it came to mechanisms, few early Victorians were prepared to accept natural selection even as a partial evolutionary mechanism, with all the implications that this would have involved for restricting the role of the

124 DARWINISM'S GENERATIONS

divine. Where they did, their commitment was vague and hedged around with conditions and caveats. James McCosh's *Christianity and Positivism. A Series of Lectures to the Times* (1871) clearly outlines his doubts as to the efficacy of natural selection, and denial of evolution of man from animals without divine intervention.[262] For the majority, natural selection was a mere metaphor, 'a term to be utterly discarded...a verbal deception'.[263] Those prepared to engage in detailed refutation deployed what became a catalogue of familiar arguments, including the extreme stability of some creatures, the sheer improbability of something as complex as the eye emerging by chance, and the unreasonably immense stretches of time required.[264] After 1867, when Fleeming Jenkin (1833–85) published an essay pointing out the mathematical difficulty of any useful variation avoiding being swamped, the danger of reversion back to the mean became another staple of early-Victorian arguments.[265] But for many, natural selection could be rejected simply because it denied the role of an all-wise Creator and His intelligent design, leaving all to chance, and thus contradicting the pervasive evidences of design long-rehearsed by natural theology, not just the eye, but the profligate beauty of nature created for the gratification of man, even the elaborate mechanics of the orchid. '[A]midst all the diversity of nature', as John Phillips put it, 'nothing appears accidental, nothing indefinite, nothing unforeseen'.[266]

The mid Victorians, in contrast, generally recognised the prima facie credibility of natural selection as an explanation of the development of natural life, with all the implications of this in limiting the role of supernatural influences. For many mid Victorians it was one of the elements of the *Origin* which was so compelling. In its varying ways this was as true of Mivart and Argyll as it was of Huxley and Wallace. The mid Victorians deployed a variety of often detailed technical arguments against the *Origin*. They worried over the evidences of geographical distribution, for example the variation of fish species on each side of the central American isthmus, and how to explain on the basis of a single point of creation how almost half were absolutely identical on both sides, but not like species in the more temperate zones.[267] But the mid Victorians' most common point of departure from the Darwinians was in their refusal to accept natural selection not merely as a complete explanation for evolution, but even as a mechanism capable on its own of producing the transmutation of species, and their subsequent use of arguments of insufficiency as a basis for questioning or challenging the status of Darwinism, and scientific naturalism more generally. Hence William Armstrong's 1863 British Association address identified natural selection as

IN THE WAKE OF THE *ORIGIN* 125

capable of determining variations amounting to specific differences of degree, but not 'the production of a new organ not provided for in original creation.'[268] They argued that it was difficult for natural selection to account for the development of rudimentary structures that could not become useful at once, like a bird's wing or the giraffe's neck,[269] and that the improbabilities of the numerous combined changes that would be required to make complicated organs like eyes and ears effective were enormous.[270] They gave short shrift to the ingenious ideas the Darwinists advanced to address these concerns, such as 'correlation of growth', in which one physiological change with clear utility could explain other changes which, initially at least, had no use-value. As Argyll remarked to Lyell in 1865, 'Correlation of Growth' 'is simply one of the many phrases by which men deceive themselves into the belief that they are <u>explaining</u> phenomena, whilst in truth they are merely restating them as fact.'[271] Mid-Victorian critics also contended that natural selection could not explain similarities of organs across different species where they were put to different uses (and so should have adapted in different ways).[272] Argyll stressed he was not antagonistic to natural selection. He was quite prepared to acknowledge its significance 'so far as the preservation and extinction of species <u>once "born"</u> or once "created"'; it was 'just' that he refused to accept that it could explain the creation of new species.[273] There was little to distinguish Mivart and the marine zoologist Charles Wyville Thomson (1830–82) from Friedrich Engels (1820–94), who explicitly rejected any sort of special creation, but described natural selection as '<u>merely</u> a first, provisional, incomplete expression of a newly discovered fact'.[274]

These positions contributed to a widespread unease visible in the correspondence and conversations of the mid Victorians. Even someone like the Liberal academic Goldwin Smith (1823–1910), whose 1877 *Macmillan's Magazine* article suggested a broad acceptance of evolutionary history, confessed in private his scepticism about natural selection, and indeed the whole notion of evolution through gradual variation.[275] Others generally sympathetic to Darwinism, like John Ball (1818–89), naturalist and first president of the Alpine Club, took the stance that natural selection could not explain everything.[276] It was in the face of this weight of argument that Darwin hedged his bets in the 1869 edition of the *Origin*, conceding he might have given too much importance to natural selection.

Both early and mid Victorians clung to notions of breaks in the evolutionary process which could not be explained by natural selection. While accepting the advance across geological ages from lower to higher forms,

McCosh argued that there had been breaks in the series.[277] The mid Victorians read into the palaeontological record strong indications of disrupted development, and they posited instead the need for sudden jumps or saltations. Murphy thought it 'most likely that, in many cases, species have been formed at once by considerable variations...amounting to the sudden formation of new species and new genera'.[278] The figure of the rolling polygon which Mivart borrowed from Galton emphasised his belief in the staccato nature of change.[279]

Mid-Victorian Darwinists like Huxley and Tyndall might have adopted a much more positive register, but they were no more wedded to natural selection as a complete evolutionary mechanism. Unlike many of the younger Darwinists, they were relatively relaxed about the extent to which ancillary factors such as Lamarckian acquired characters might need to be smuggled into service. Huxley continued to play out his own uncertainties with correspondents. Hence his comment to the Manchester palaeobotanist W.C. Williamson in 1871 that 'So far as animals are concerned, I am quite satisfied that evolution is a historical fact. What causes brought it about is another matter'.[280] Joseph Hooker was similarly ambivalent. For Jim Endersby, Hooker's Darwinism was 'more complex and ambiguous than has hitherto been recognised. Hooker strove to reshape natural selection into something working naturalists could use but, far from embracing his friend's theory wholeheartedly, was anxious to distance himself from some of its implications'.[281] Wallace's careful crossing out of 'natural selection' and its replacement with Spencer's formula of the 'survival of the fittest' throughout his copy of the *Origin*, even though it was primarily a matter of language rather than biology, reflects his discomfort, and he was unsettled (as indeed was Darwin) by some of the challenges of the mid-Victorian critics.[282] 'How incipient organs can be useful is a real difficulty', he told Darwin in 1869, 'so is the independent origin of similar complex organs'.[283]

But for the anti-Darwinians, limiting the role of natural selection was not just a marginal issue; it was crucial in opening up spaces for the insertion of arguments of design, and thereby teleological claims. Early and mid Victorians clung tenaciously—and often tautologically—to the proofs of design in nature as demonstration that natural selection could not be their cause.[284] To some mid Victorians, denial of design was the fundamental flaw of the Darwinian scheme. For the astronomer and mathematician William Fishburn Donkin (1814–69), 'the vice of Darwin's theory seems to me to be that he will have his variations to be <u>accidental</u>...the more I try to trace in detail the steps by which an eye or an ear would be developed by

IN THE WAKE OF THE *ORIGIN* 127

accidental variation, and natural selection, the more monstrous does the hypothesis appear. If he would amend his theory by leaving out accident as the main agent, I think it would be almost irresistible.[285] The talk Mordecai Cubbitt Cooke (1825–1914) gave to the Society of Amateur Botanists in 1864 emphatically rejected the notion of accident.[286] (As late as 1884 the early Victorian Charles Pritchard wrote to George Gabriel Stokes in almost identical terms, stating that his 'only quarrel about Evolution is that such an [organ?] as the eye is stated to have been the outcome of—Accidental Variations, Accidental Environment & the survival of the strongest'.)[287] Beauty, in its pure form, divorced from any significant utility, acted as a gloss to design; just look at the extraordinary patterning of a bird's feathers, Argyll told Lyell, in which the markings of each feather have to be so exactly positioned so as to interleave effectively with the adjacent feathers to produce the ultimate effect.[288] In 1871 William Thomson noted that he was convinced that the argument of design had been too much lost sight of in zoological discussions, 'overpoweringly strong proofs of intelligent and benevolent design lie all around us'.[289] There remained a market in the popular religious periodicals for natural history that would 'speak of the perfections of God', and someone like Henry Liddon (1829–90), Oxford cleric and Bampton Lecturer in 1866, could still see the religious functions of nature in fostering the sense of an Invisible World, and in suggesting, however dimly, its 'awful author and ruler'.[290]

For the early Victorians, design could either retain its traditional static meaning, based on the perception of perfect adaptation to nature's exigencies, or it could allow the existing understanding of human history as a story of progress towards moral perfection to be extended to the physical progress of humanity and of nature. In either case, they were not inclined to acknowledge that the *Origin* had reconstituted the terms of debate. Modern science, James Martineau told Henry Sidgwick in 1869, including the Darwinian theory, had not 'in the smallest degree altered the logical weights in the Theistic problem'.[291] For mid Victorians, purpose allowed the evolutionary change they were prepared to accept to be absorbed into religious frameworks. While the logical conclusion of Darwin's thought excluded designed progression, many mid Victorians were willing to accept the fact of development, if shorn of the apparent environmental relativism of natural selection, because they were thereby able to retain a commitment to an organising intelligence, and a belief that evolution served some end. In this sense, first and foremost, as a frustrated opponent exclaimed in the 1890s, the mid-Victorian resistance was a 'revolt of teleology against

128 DARWINISM'S GENERATIONS

Darwinism.[292] The Trinity College geologist Samuel Haughton conceded the possibility of some evolution in his *Animal Mechanics* (1873), but even then only in an entirely teleological way.[293] Both Darwinian sympathisers like Frederick Temple and sceptics like the Hegelian philosopher James Hutchison Stirling (1820–1909) rallied to the cause of purpose in evolution.[294] Hence it was said of the literary critic Henry Morley (1822–94) that '[h]e believed in teleology, and to the end of his life would never admit that it was overthrown by Darwinism.[295] Mivart's *Genesis of Species*, and his later *Lessons from Nature* (1876), advocated a teleological evolution as the only way to explain features like homological relations between diverse species, and their tendency to develop symmetrically.[296] This gave him sustained purchase, across the range of mid-Victorian responses, bolstering those willing to embrace evolution, albeit in a non-Darwinian form, but also (and more tellingly) providing ammunition for those whose doubts loomed larger. Into the 1880s the Free Church minister Edward White (1819–98) was exploring the potential of numerical progressions in nature, and whether they could be seen to offer evidence of an intelligent force; he wondered if it might not be possible to recognise the 'action of "Mind" in the measurement of the curves of the double convex—and in the placing of the lens at [?proper?] distance from the retina?'.[297]

Notably, this progressivism was true of those who in other respects occupied positions closer to Darwin. Hence W.B. Carpenter in 1885, while fully accepting evolution, and discounting the need for special creation as suggesting a lack of foresight on the part of God, was still led back to a beginning and to a subsequent unceasing exertion of developmental energy, and from the 'multitudinous progressions' and the 'complexity of their mutual relations', and the wonderful harmony of the entire scheme, was convinced that such a result could not have come about without an original preordination.[298] Despite his Darwinian sympathies, P.M. Duncan was another who argued for the 'necessary and fundamental truth /to me\ that evolution takes place in accordance with laws which initiated in, and are rendered continuous in their influence by a great first cause. I preach design and foregone conclusions, and the actions and reactions of the material and spiritual in a manner which I expect will make the doctrines of evolution a little less dreadful in the eyes of the ignorant but pious'. Duncan regretted Huxley's efforts to separate off Christian Protestantism from science, and committed himself to oppose 'the Huxleyan dogma'.[299] But here again Huxley himself was not above ambivalence, telling J.S. Blackie in 1876 that his lecture to the Glasgow Science Lectures Association that year had done no more than suggest that

IN THE WAKE OF THE *ORIGIN*   129

'the teleological argument has just as much force (or as little) to those who admit evolution as to those who deny the existence of that process.'[300] In this way, the space available within evolutionary thought for design allowed many mid Victorians to insulate theology from Darwinian biology: as C.A. Row (1816–94) put it, the Darwinian theory was only a special form of the theory of creation by evolution.[301]

These arguments gained fresh force from the late 1860s with the observations of William Thomson that solar physics and the rate of the cooling of the sun meant that the amount of geological time required by the gradualism of natural selection was unlikely to have been available.[302] As early as June 1861, Thomson had written to John Phillips warning that if geologists were to 'give adhesion to Darwin's prodigious durations in geological epochs', they 'must contrive your ancient plants and animals to have lived without a sun', whose age he would not extend beyond twenty million years, a position he publicised in his 'On the Age of the Sun's Heat' article in *Macmillan's Magazine* (1862).[303] Thomson believed that although this might not disprove the transmutation of species, it did disprove the doctrine that transmutation had taken place through descent with modification by natural selection.[304] As late as 1897 he was still affirming that the age of the earth fit for life was around twenty-five million years, certainly within twenty to thirty million years, and was rebuffing calls from high Victorians like T.G. Bonney to 'hurry up' the geological dynamics.[305] Darwin was sorely exercised by the 'odious spectre' of Thomson, and indeed until the discovery of radiation and its implications for the rate of cooling of the earth, his argument presented Darwin's supporters with an almost insurmountable problem. In the meantime, mid Victorians like James Croll (1821–90) and Archibald Geikie 'acknowledged the necessity of fitting geological data into the shorter time supplied by physics', while the older Rev. William Carus (1804–91), Canon of Winchester, welcomed Thomson's 'keeping us *modest* in Geology'.[306]

In this respect natural selection, to the extent that it operated, became, as one mid Victorian put it, 'but a name, and not a power', a description of 'the order and mode according to which Providence works'.[307] By accepting natural selection, but not the randomness of variation, Argyll was able to sidestep the problem of time: the variations on which natural selection worked were not merely accidental variations, 'but processes which maintain and preserve a prepared correlation between the changes of external condition and the changes (if any) in organic life'.[308] The appeal of such approaches was that they slid easily into more diffuse notions of the action

130   DARWINISM'S GENERATIONS

of God via some 'force' or 'will' which were widely deployed by more liberal or progressive religious thinkers. The evolutionary commitments of Temple, as developed in his Bampton Lectures of 1884, illustrated the sort of hollowed-out and largely Spencerian evolutionary sensibility, shorn of any engagement with the actual mechanics of evolution, that could result.[309]

This was a particularly powerful frame, and one that continued to appeal to some high Victorians (usually those closest in age to the mid Victorians) reluctant to embrace Darwinism.[310] But it was distinct from the most common high-Victorian versions of divine transcendence in the operation of natural laws, not least in its much greater commitment to the active presence of the Creator. Mid Victorians had little truck with the notion that having created natural laws, God then just left them to work themselves out. Evolution, they asserted, required the 'direct power of a beneficent Creator'.[311] Argyll's *Reign of Law* (1867) emphasised the need for divine intervention over and above the laws of nature; as he put it, creation was not a single act, done and finished once and for all, but a long series of acts—a work continuously pursued through an inconceivable lapse of time.[312] Kingsley's sympathy with this argument is clear in his private comment that '"a convert to Darwin's views" *could* view the world as being "like an immensely long chapter of accidents", but it was "really…a chapter of special Providences of Him without whom not a sparrow falls to the ground"'.[313] Temple's Bampton Lectures moved further in the direction of transcendence: arguing that God endowed such inherent powers as in the ordinary course of time living creatures were developed. He might not have made things, but He made them make themselves. But even here the impress of God was 'either at the beginning or at some point in the history of his creation', and the argument from design still featured strongly.[314] In effect, as Lightman puts it, mid Victorians retooled the natural theology tradition by clinging to final causes, design, and the law of nature as the work of God, while also retaining a sense of active continued intervention.[315]

Occasionally, the limits of the evidence encouraged mid Victorians, while holding back from direct rejection, to emphasise the highly speculative nature of Darwin's reasoning. Their fundamental position was that Darwinism as not so much wrong (though elements might be disputed) as insufficient,[316] or that, as G.H. Lewes put it, Darwin's theories offered the best current account, but were likely to be superseded in their turn.[317] For R. Payne Smith (1818–95), Regius Professor of Divinity at Oxford, the *Origin* was 'full of interesting facts and ingenious speculations, but the speculations can scarcely be said to have consistency enough to merit the name even of a theory'.[318] Mid-Victorian apologetics in particular sought to treat evolution as 'suggestive theory' rather than established fact.[319] Characteristic of this temper was Henry Acland, whose general sympathy with evolutionary ideas was tempered by a strong scepticism about the stability of scientific knowledge. Disturbed by the bewildering pace of scientific change, Acland remained both acutely sensitive to and grateful for the provisionality of all science.[320] This position was also a useful avoidance mechanism for mid Victorians less inclined to sympathy. 'I pin myself to nothing—geological truths of today are the absurdities of tomorrow', as one put it at the end of the 1870s.[321]

Some of the commonalities of mid-Victorian pro- and anti-Darwinians were disguised by the deliberate efforts of both sides, but more especially the Darwinians, to perform their disagreements as fundamental conflicts. Despite Kingsley's intimacy with Huxley and the generally more conciliatory stance of the mid-Victorian doubters of Darwin, the dominant tone of their exchanges remained confrontational. This was as much an effect of Huxley's determination to exclude any explicit reservation of a role for non-natural forces as it was to the antagonism of the mid Victorians to evolutionary thought in general. Huxley and others developed an emphatic dogmatism that is not visible in the succeeding generation: a mirror for the dogmatic opposition they initially faced from the early Victorians. As Michael Foster observed, 'The reception which [the Origin of Species] met with entered like iron into Huxley's soul; he never forgot it'.[322] The lay sermonising he sustained to the last was a consequence of his overwhelming preoccupation with theological opposition. Even his friends could not wean Huxley from his anticlerical broadsides.[323] Supporters were identified as disciples and opponents as enemies.[324] There is a puritanical zeal to Huxley's treatment of opponents like Mivart, even if Mivart was far from blameless himself. Darwin's disciples, the Liberal Herbert Asquith (1852–1928) later commented, 'almost surpassed the theologians in their arrogance'.[325] Conversely, despite their rejection of the uncompromising dismissals of the early Victorians, mid-Victorian sceptics also continued to think in evolutionary binaries. Edward White's correspondence with Stokes is peppered with such references, figures who are with 'us', the opposing 'party'.[326]

Conclusion

It is not impossible to find contemporary accounts of Darwinian 'victory' as we approach the end of the 1860s, and it would be relatively easy to construct an account of rapid diffusion out of the anxieties of opponents and

132 DARWINISM'S GENERATIONS

the retrofitted reminiscences of supporters. The Church *Guardian* seems to have been especially anxious, suggesting that, at the British Association in Norwich in 1868, Darwin's 'reign was triumphant and almost unopposed'.[327] And it is true that Wallace returned from Norwich lamenting the absence of opposition to evolution from anyone who knew anything about natural history, and so the lack of the 'good discussions we used to have'.[328] Darwin was certainly widely talked about, Darwinian motifs had undoubtedly soaked into the culture; the initial shock of the *Origin* had inevitably dulled, and the tone of the debate had become (to an extent) less hysterical. Evolutionary perspectives were strengthening their hold on the natural sciences.[329] The evidence of the antiquity of man was much more widely accepted.[330] Huxley the controversialist was box office, singlehandedly able to boost periodical sales, giving a profile to evolutionary ideas they might not otherwise have achieved.[331] As the decade progressed, it is easier to find supporters suggesting that the fact of evolution was now accepted, even if debates over the importance of natural selection remained, and to find opponents who felt with Argyll that the question was 'pressing with increasing weight upon men's minds and understandings and emerges continually in all sorts of discussions, Political, Religious, and Philosophical'.[332]

But generally, there was little conviction amongst the Darwinians of the triumph of Darwinism. Efforts to establish a 'house' journal had failed, and the X-Clubbers were not optimistic about the prospects for *Nature*.[333] In the recent distinction of Jane Goodall, Darwinism was still more cultural anxiety than culture shock, a matter for hushed conversation in trembling tones rather than an irresistible intellectual wrecking ball, never mind a stable new foundation.[334] Darwinists were still constrained to offer conditional endorsements of evolution. A liberal intellectual like Mill could produce a work such as *The Subjection of Women* (1869) as if the *Origin* had never appeared. Even in the cultural mainstream of the metropolis, the strength of opposition generated extremes of caution. In 1868, Anthony Trollope (1815–82), then editing the *St Paul's Magazine*, claimed that he was 'afraid of the subject of Darwin' and quite incapable of overseeing an article on the topic.[335] Observing the situation as a young curate in Kensington, Stopford Brooke (1832–1916) regretted not just the 'pitiable' 'cry of dismay' with which Darwin, geology, and the higher criticism were being met, but the 'miserable' 'shifts of argument and endeavours to get around the truth'.[336] In large parts of English and especially Scottish Nonconformity, Darwin was still a dark shadow rather than a vital force. In 1872 the Kendal Literary and Scientific Institution were looking for a popular lecturer in Geology, who

IN THE WAKE OF THE *ORIGIN* 133

would offer a course suited to families of members, 'calculated to interest them in the facts of nature—without going into abstruse theories which partake often of mere wild speculation. We are especially desirous as God-fearing men, not to have any lectures which would teach our young people to indulge in dreams as to creation etc, which are encouraged by many great names at the present day,...We believe however that it is possible to gain lectures on Geology...without broaching the errors of Darwin.'[337] Darwinian entertainments were still largely of the class which presented 'demented heads' and 'marvellous transformations of plants and animals into human beings', as one show at Piccadilly's Egyptian Hall promised in 1868.[338]

Much contemporary comment spoke vaguely about shifts in opinion, but although the tone may have moderated, the reaction to *The Variation of Plants and Animals under Domestication* when it was published early in 1868 offered little sense of a widespread embrace of evolutionary ideas. *Variation* sold slowly (only 5,000 copies by the time of Darwin's death), and although some reviewers took the opportunity to affirm a wider commitment to the arguments of the *Origin*, many simply sidestepped the larger questions, and for those that didn't, doubts that *Variation* could be reconciled to the *Origin* were more common than the endorsement offered by progressive titles such as the *Pall Mall Gazette*. While the tone of the specialist periodicals was usually deferential, and much was made of the book as a store of examples for the practical breeder, its central argument of 'pangenesis', that heredity was to be explained by the intermingling of invisible 'gemmules' which carried with them information about the different cells of the body, was poorly understood and frequently ignored. Bunbury found the whole thing, in both book and subsequent discussion, simply unintelligible; Hooker was equally unconvinced.[339] The mean-spirited review Owen published in the *Athenaeum* trumpeting the anticlimactic retreat from '*Origin*' to '*Variation*' was widely extracted in the press.[340] Even the Darwinians were not enthusiastic. Boyd Dawkins used his *Edinburgh Review* notice to raise doubts about natural selection, and although Galton's *Hereditary Genius* (1869) was initially enthusiastic, his research quickly caused him to question the gemmules theory.[341]

For Huxley, the fate of Darwinism was still 'such a worry to us all'; 'I cannot well be blacker in the general mind' he commented to James Knowles in 1869.[342] In many respects, Huxley's bulldozing style created as much alienation as support, even amongst intimates and allies like Lyell and Rolleston. His 1870 British Association address didn't go down well in early-Victorian circles where he was seen as hectoring the scientific community rather than speaking on its behalf.[343] While increasingly influential in the institutions of science, he was less so in the wider culture. It is telling that as late as 1869, R.H. Hutton, despite the breadth of his intellectual contacts in London, had not actually met either Huxley or Tyndall.[344] This indeed was one of the reasons for the enormous significance of the Metaphysical Society in the years after its establishment in 1868, as a conduit for the X-Clubbers into the heart of Victorian intellectual culture. The celebration of Huxley's embarrassment at the discoveries, or rather non-discoveries, of rudimentary life forms which it had been hoped the Porcupine deep sea exploration expedition of 1869 would dredge up from the ocean floor was indicative of the persistence of anti-Darwinian positions. 'It will shake the theories of some of these Bible revisers and correctors' commented the clergyman and anti-slavery campaigner Horace Waller (1833–96).[345] The care with which Darwin and Spencer publicly downplayed the differences of their evolutionary ideas speaks to the anxiousness of both to enrol even doubtful allies wherever they could.[346]

As the decade progressed, many older opponents became perhaps more inclined to reticence. It is noticeable that J.S. Bowerbank generally avoided direct comment in his correspondence with his friend Henry Lee, even though Lee was also a sceptic; though he did remark cryptically at one point in 1870 that like Lee and Buckland, 'I am an unbeliever'.[347] But quiescence did not indicate acquiescence. The impact of Darwinism was often negative, prompting those born before 1830 especially into accommodations which allowed existing beliefs to coexist with the problems Darwin posed; less out of any sense that his arguments had convinced, and more as a prophylactic against future proofs. These positions were by no means always, or even usually, fixed; but the shifts discernible are generally both grudging and piecemeal. The early Victorians fought an intellectual rearguard; concessions were made at times, but only partially and often not in ways which trenched on the arguments of the *Origin* themselves. So, after pouring over the evidence, the geologist Adam Sedgwick eventually accepted in 1868 that he had to concede the overthrow of the biblical chronology of man, while continuing to deny even more vehemently the evolution of man from animals.[348] The responses of some were thoughtful, prepared to make concession, but cautious and sceptical. Hence, by the end of the 1860s, the Edinburgh advocate Robert Stodart Wyld (1808–93) had recognised that science had overthrown, painfully, some of the 'beliefs of our childhood', and was at one level prepared to announce that he was 'reconciled to the theory'; but while this was not the normal outright hostility

IN THE WAKE OF THE *ORIGIN* 135

of the early Victorians, on closer inspection, the nature of Wyld's concessions was strictly limited: he accepted evolution of existing species (not
least because 'it guarantees the exhilarating prospect of continued progress
and improvement'), but not transmutation, continued to affirm 'the legitimate inference of a designing Cause', rejected a role for chance in the creation of species, and expressly repudiated the position taken by Huxley as
'not only absurd in phil[osph]y but incorrect in science'.[349]

If for the early Victorians Darwinism was still to be resisted, for the mid
Victorians it was to be neutered. The 'hard and hopeless onesidedness' of the
claims of Huxley and Tyndall continued to repel.[350] Acland's concession of
1868 that scientific developments 'helps to lessen the improbability of the
hypothesis of Darwin' was entirely characteristic, not just of him personally,
but of numerous mid-Victorian verdicts from the speaker's table of local scientific and natural history societies.[351] And for many mid-Victorian clerics or
ministers, the distinctions between scientific knowledge and spiritual verities
held firm. Henry Liddon's Bampton Lectures of 1866, which very explicitly
denied that any new scientific knowledge could have any implications for
understandings of Creation, represented a still-powerful strain of thought.
A.P. Stanley, the Broad Church Dean of Westminster, while acknowledging
the power of knowledge to create new truths, never felt that it threatened the
claims of revelation. 'To me', Stanley said to friends, 'a break in scientific order
never makes a difficulty, possibly because I have no science in me'.[352]

The evidence of accommodation to evolutionary ideas in some quarters
(and the tendency of anxieties about the acceptance of Darwinian ideas to
be expressed hyperbolically) must be balanced by indications of a hardening of the doubts of many of Darwin's more sympathetic opponents, and an
erosion of belief amongst many of his original mid-Victorian sympathisers.
Like Henry Tristram and Fenton Hort before them, by the end of the 1860s
Darwinians like Rolleston, Kingsley, and even of course Wallace in respect
of humanity, were expressing their previous doubts even more forcibly. The
efforts of Huxley and others not merely to confront attempts to use religious
ideas and protocols to determine scientific truth, but in fact, it seemed to
many, to mobilise science to attack religious beliefs or institutions was a
dangerously double-edged weapon in the struggle for the commitment of
the majority of contemporaries who retained strong religious beliefs, as the
movement in 1864–5 to collect the signatures for the 'Declaration of
Students of the Natural and Physical Sciences' shows.[353]

Rolleston, although perhaps the most closely aligned with Darwin and
Huxley, signed up to the Declaration (even though Kingsley and many of

136 DARWINISM'S GENERATIONS

the mid-Victorian Darwinian sceptics, including Max Müller, refused).[354] Less than two years later it would seem that he deliberately (and abruptly) detached himself from the Darwinians, his dismay at their increasingly anticlerical tone brought to a head by the tenor of an article on the relations between science and organised religion which had appeared in the *Reader*.[355] His precise position on Darwinian thought thereafter becomes difficult to pin down. Lecturing in Leeds in 1866, he affirmed the single point of the creation of species followed by their migration (noting the links of this to Darwinian theory), but seems from the newspaper reports to have avoided any direct identification with the *Origin* or Darwinism.[356] His address to the BMA in 1868 was similarly obtuse; accepting the adaptability of nature and unsympathetic to cries of horror at suggestions of evolution of man; but also rejecting 'materialism', and avoiding any unequivocal endorsement of Darwinian positions.[357] While expressing 'much pleasure' at *Descent*, he refused to accept the common evolution of man and animals, and opposed what he called Haeckel's 'speculations' on the pedigrees of animals.[358] As late as 1880 he confessed to being 'impressed' (and perhaps worried?) by the imperfection of the geological record.[359] Rolleston was never entirely cast out of the Darwin–Huxley circle, but his stance was recognised as being more and more detached, and he was increasingly seen as part of the problem for younger Oxford Darwinists. '[H]e had the misfortune to be impregnated with the spirit of College Tutorism—which is the exact reverse of the spirit of scientific progress', Huxley commented in 1880.[360]

It was a common pattern. The geologist H.G. Seeley commented in 1867 that although the mid-Victorian physiologist G.M. Humphry was 'more inclined to regard [Darwin], calmly, as a real thinker, than he was [before]', even so, 'most of the abler men that I know are settling to the conviction that though Darwin's process many have produced some final modifications like those called species, they leave the modification of the organisation quite unaccounted for'.[361] By the mid-1860s Tristram was dismissing 'dreaming Darwinists, building up worlds of fact, out of visionary hypotheses', and was offering accounts of how the natural history of Palestine could be reconciled to Mosaic accounts of Creation.[362] After his death, Tristram was often cited as an early clerical convert, but his position—at least the one he maintained in public—is summed up by the version of 'not proven' given in his presidential address to the Newcastle Natural History Society in 1880, along with an extended argument that Darwin's importance lay not in his evolutionary theories but rather in his novel approach to observation and study.[363] Although W.H. Flower always remained a more determined

IN THE WAKE OF THE *ORIGIN* 137

proponent of evolution broadly conceived, he too seems to have drifted away from the Darwinian camp in the later 1860s and 1870s, his comments on Huxley and Tyndall increasingly unsympathetic.[364] The same is true of Wollaston, who, perhaps in part reflecting his discomfort at operating in 'this endless field of speculation', became if anything more convinced of the limits of evolution as the decade proceeded, happy to put the wider debates to one side, and when challenged, digging in his heels, arguing that if anything the fauna of the Atlantic islands 'speaks of deterioration—or (for some cause or other) a dwindling down from a higher type, <u>never</u> of a development from a supposed /lower\ one'.[365]

No doubt greater numbers of high Victorians came to an even more decisive shift in opinion in favour of Darwinism, encouraged by what William Thomson described as the hazardous tone so attractive to young men that Huxley favoured.[366] It is unlikely that Edward Wrench's progress from the antipathy of 1860 to a general ascription to evolution in 1872 was a singular case.[367] Although many high Victorians who eventually came round to acceptance of evolution had not made this accommodation by the end of the decade, the tendency towards acceptance was much stronger. In this sense, such advance in acceptance as occurred before 1871 seems largely to have been constituted by the demographic shift which carried off the later Georgians, pushed the early Victorians into retirement, and brought the high Victorians platforms through which to articulate their views more audibly. Otherwise positions were, if anything, hardening: it was said of the energetic Darwinian doctor Lawson Tait (1845–99), lecturer in physiology and biology at the Midland Institute, that his outspoken advocacy raised 'a storm of abuse'.[368]

The publication of Darwin's *The Descent of Man* in 1871 significantly altered the terms of the debate again. The British Association meeting in Edinburgh in the same year offers a chance to take stock of the position at that moment and provides a balance to any metropolitan sense of Darwinism triumphant. Thomson's presidential address, notorious for its far-fetched suggestions of meteorite-borne life arriving from another planet, for all its apparent fancifulness, did much to encourage the Darwinian doubters.[369] 'Darwin Eclipsed' proclaimed the *Illustrated London News* (Figure 3). Henry Wilkinson Cookson (1810–76), master of Peterhouse, welcomed the tone of the Edinburgh meeting and of many of the attendees, who 'repudiate the simple materialism of the Southern metropolis, acknowledge the evidence of design, and argue for a Designer'.[370] Looking back to life in Cardiff at around the same time, the journalist Clement Shorter

138 DARWINISM'S GENERATIONS

Figure 3 'Darwin Eclipsed', 'The British Association at Edinburgh—Humours of Science', *The Graphic*, 26 August 1871.

IN THE WAKE OF THE *ORIGIN* 139

(1857–1926) suggested that there wasn't a single inhabitant who believed, or who was prepared to publicly avow belief in Darwinism, and that anti-Darwinian feeling 'was such that many a worthy man would have been glad to have brought a faggot to his burning'.[371]

Notes

1. Take as examples, E. Caudill, 'The Bishop-Eaters: The Publicity Campaign for Darwin and *On the Origin of Species*', *Journal of the History of Ideas* 55 (1994), 441–60, and the more cautious M.J.S. Hodge, 'England', in Glick, *Comparative Reception of Darwinism*, 3–4, which suggests that 'the decisive early developments took place in the first ten years after the announcement of Darwin's theory'.
2. Although the minute book recorded explicitly that the *Origin* did not form part of the basis for the award; see discussion in Browne, *Darwin*, II, 244–7. For Sabine's hostility to the *Origin* from the outset, as propounding 'an idea as illusory as that of Lamarck', see W. White, ed., *The Journals of Walter White* (1898), 124–5, and Sabine to John Phillips, 12 November 1863, DCP-LETT-4340F, Darwin Correspondence.
3. Huxley at the Birmingham Midland Institute in October 1867, *Birmingham Daily Post*, 8 October 1867; *Western Morning News*, 29 December 1870.
4. *Charles Kingsley: His Letters and Memories*, II, 249–50; Kingsley to Darwin, 11 December 1867, DCP-LETT-5730, Darwin Correspondence. Compare with the comments of A.W. Bennett in 'Theory of Natural Selection from a Mathematical Point of View', *Nature*, 10 November 1870, which notes that the Darwinian hypothesis has 'so completely taken over the scientific mind' in Britain 'that almost the whole of our rising men of science may be classed as belonging to this school of thought'.
5. Kingsley to Darwin, 8 November 1867, DCP-LETT-5673, Darwin Correspondence; the reference is to Maria Antionetta, Marchioness of Huntly, judging by Kingsley to Frances Kingsley, nd, AM16745, Box 24, Folder 3 Charles Kingsley Papers, M.L. Parrish Collection, PUL.
6. Tuckwell, *Reminiscences of Oxford*, 54.
7. M. Ruse, *From Monad to Man* (1996), 172.
8. Wallace to Darwin, 16 July 1871, DCP-LETT-7868, Darwin Correspondence; Wallace quoted by M. Shermer, *In Darwin's Shadow* (2002), 309–10; Newton, 'Early Days of Darwinism', 249.
9. See B.C. Beddall, *Wallace and Bates in the Tropics: An Introduction to the Theory of Natural Selection* (1969). The *Naturalist* appealed to pre-Darwinian naturalists, including apparently Bunbury, who was reading it in early 1869 with 'much pleasure', Lyell to Wallace, 13 March 1869, ff.101–3, Wallace Papers, Add. MS 46435, BL.
10. Lyell to G. Poulett Thomson Scrope, 29 April 1863, Sir Charles Lyell Papers B L981, APS.
11. Lewes, 'Principles of Success in Literature', *FR* 1 (1865), 85–95.
12. Quoted in James Sully, *My Life and Friends: A Psychologist's Memories* (1918), 132.
13. See Saverio Tomaiuolo, *In Lady Audley's Shadow: Mary Elizabeth Braddon and Victorian Literary Genres* (2010); Isa Knox (1831–1903), reference in her serialised novel 'Deepdale Vicarage', *Quiver*, October 1866, 61.
14. J.V. Nixon, '"Lost in the Vast Worlds of Wonder": Dickens and Science', *Dickens Studies Annual* 35 (2005), 267–333.
15. William Baker, *Wilkie Collins' Library: A Reconstruction* (2002); most obviously the attack in *The Law and the Lady* (1875) at those who took their ideas about religion from science lectures delivered by 'the last new professor, the man who has been behind the scenes at creation... We were all monkeys before we were men, and molecules before we were monkeys' (121).

140   DARWINISM'S GENERATIONS

16. *The Egoist* (1879), I, 3; Jonathan Smith, '"The Cock of Lordly Plume": Sexual Selection and *The Egoist'*, *Nineteenth Century Literature* 50 (1995), 51–77; Holmes, *Darwin's Bards*, 54–62, which traces a broad 'evolutionary philosophy' in collections such as *A Reading of Earth* (1888), and *Poems and Lyrics of the Joy of Earth* (1883). (Holmes' suggestion that because Meredith expressed wholehearted sympathy with Huxley in his confrontation, he was at that point a 'card-carrying Darwinian' isn't defensible, and without this his argument that 'Meredith moves from a Darwinian understanding of evolution to a more catholic evolutionism which incorporates other theories alongside Darwinism' falls, and Meredith's tendency to distance humanity and animals aligns more straightforwardly with other mid-Victorian positions. More apposite is Clodd's observation that the arguments of evolution had 'no cross-examination at his hands; their evidence supplemented to him what were inborn convictions', Edward Clodd, 'George Meredith: Some Recollections', *FR* 92 (1909), 26.)
17. Holmes, *Darwin's Bards*, 246–56.
18. William Knight recorded Tennyson in 1890, speaking 'of Darwin, and of the great truth in Evolution; but it was only one side of a truth that had two sides', W.A. Knight, *Retrospects* (1906), 57; for Emily Tennyson (1813–98) William Allingham reported in 1868, 'She dislikes Darwin's theory', *Diary of William Allingham*, 12 August 1868, quoted in C.Y. Lang and E.F. Shannon, eds., *Letters of Alfred Lord Tennyson 1851–70* (1987), 501. This did not prevent Tennyson becoming a symbol of evolutionary thought; but even when younger evolutionists enthusiastically claimed Tennyson for evolution, they came up against 'Locksley Hall Sixty Years After', which one admitted showed that 'he had never assimilated' evolution as he had assimilated other ideas '...the real man in him protested this time against the very doctrine which had mastered him and had compelled him to give it voice' (see H.W. Clark, 'Tennyson: A Reconsideration and Appreciation', *FR* 92 (1909), 233–5, 235). For discussions of the evolutionary ideas underpinning Tennyson's poetry see Holmes, *Darwin's Bards*, 61–74.
19. Holmes, *Darwin's Bards*, 46–9.
20. Richard Bellon, 'Inspiration', 393–420. For Charles Pritchard, *Fertilization of Orchids* prompted 'an irresistible impulse to uncover and bow my head, as being in the too immediate presence of the wonderful prescience and benevolent contrivance of the *Universal Father'*, *Proceedings of the Church Congress of 1874* (1874), 350.
21. Lightman, *Popularizers*, 81. Lyell, ed., *Life of Sir Charles J.F. Bunbury*, II, 217; compare Roland Trimen (1840–1916) to Darwin, 18 November 1869, DCP-LETT-6995, Darwin Correspondence, and J.O. Westwood to Lubbock, 28 December 1874, ff.75–6, Avebury Papers, Add. MS 49644, BL. Mungo Ponton (1802–80) argued that the book directly contradicted Darwin's theory by showing that modifications would be swamped, *The Beginning: Its When and Its How* (1871), 377–8. Even later generations who accepted evolution still recurred to *Orchids* for its vindication of design, see John Matthews, 'Evolution and the Problem of Evil', in W.F. Muir, ed., *Christianity and Evolution* (1887), 130.
22. See Dallas, 'Mr Darwin's Theories', *Westminster Review* ns 35 (1869), 207–27; Boyd Dawkins, 'Variation of Animals and Plants under Domestication', *Edinburgh Review* 128 (1868), 378–97; *Ibis*, ns 4 (1868), 218.
23. *Gardeners' Chronicle*, 1 July 1871. Of course there were plenty of older horticulturalists prepared to sustain the opposite; see James Buckley (1801/2–83) to the Carmarthenshire Club, *Cambrian News*, 15 May 1873.
24. Brunton, 'Address in Medicine', *The Lancet*, 1 August 1891, 216.
25. Alexander Shewan (1851–1941), *Meminisse Juvat* (1905), 99. Similarly, in the early 1870s students at the Catholic University in Dublin had to remonstrate with the college authorities about the way they were left to 'devour the works of Haeckel, Darwin' and others in private; see J. Tyndall, *Reply to the Critics of the Belfast Address* (1875).
26. Oliver Lodge, *Past Years* (1931, 2012), 68; to this point his only reference point to Darwinism was a memory of the outcry against Darwin at the time of the publication of the *Origin*.
27. As in the case of William Cole (1844–1922), founder and long-time Honorary Secretary of the Essex Field Club, who did not read the *Origin* until after 1871, James Marchant,

Raphael Meldola: Reminiscences of His Worth and Work (1916), 121. See William Woods Smyth (1843–1928), who apparently did not read the *Origin* until 1871, *Evolution Explained and Compared with the Bible* (2nd ed., 1883), viii.

28. Frederick Rogers, *Life, Labour and Literature* (1931), 38. Picton's acceptance both of the outlines of evolution and indeed of Huxley's 'Physical Basis of Life' is clear in *The Mystery of Matter and Other Essays* (1873).

29. Letter, *Daily News*, 1 February 1894.

30. See 'Notebook 1867–69', Jowett Papers/1/H/25, Balliol College, Oxford; see also the amateur Kent geologist Benjamin Harrison (1837–1921), who despite his broad interests only came to read the *Origin* in 1869, Edward R. Harrison, *Harrison of Ightham* (1928), 63 (he was reading the *Beagle* two years earlier, ibid., 55).

31. J. Morlais Jones (1843–1906), *Leeds Mercury*, 13 May 1896.

32. D.A. Stack, *The First Darwinian Left: Socialism and Darwinism, 1859–1914* (2003), 9.

33. See his letters to Alexander Campbell Fraser, Box 13, Dep.208, A. Campbell Fraser Papers, NLS; quote from 8 April 1861.

34. See George A. Denison (1805–96) in *The Churchman*, 29 March 1866; Martineau to Knight, December 1870, ff.11–12, Carpenter MS 6, Harris Manchester, Oxford; J.R. Seeley to Kingsley, March [1869], f.142, Kingsley Papers, Add. MS 41299, BL. For this whole episode, see I. Hesketh, *Victorian Jesus: J.R. Seeley, Religion, and the Cultural Significance of Anonymity* (2017).

35. W. Bagehot, *Physics and Politics* (1873), in *Collected Works* IV, 484. Bagehot 'devoured all Mr Darwin's and Mr Wallace's books and many of a much more technical kind'. He clearly rejected arguments from contrivance and design, but was ultimately more obviously Lamarckian than Darwinian; see R.H. Hutton 'Walter Bagehot', *FR* 22 (1877), 453–84; see discussion in T. Bennett, *Pasts beyond Memory* (2004), 91–8.

36. W. Robertson Nicoll, *James Macdonell, Journalist* (1890), 147.

37. [Peal] 'Experiences of an India Tea Grower', *Gloucestershire Chronicle*, 21 September 1867.

38. See *Gardeners' Chronicle*, 13 June 1863, 24 December 1864; Burkhardt, 'England and Scotland', 47–53.

39. Dawkins to Darwin, 17 July 1869, DCP-LETT-6834, and also 22 August 1867, DCP-LETT-5614, and 17 July 1869, DCP-LETT-6834, Darwin Correspondence.

40. Dawkins to Darwin, 27 August 1867, DCP-LETT-5618, Darwin Correspondence. Dawkins was a complex case. He clearly worked hard to maintain good relations in the 1860s with the Darwinians, perhaps with his ambitions for a chair in Oxford or Cambridge in mind; but from the outset Darwin had his suspicions, and albeit from behind the mask of anonymity, Dawkins' responses to Darwin's subsequent evolutionary works were hostile; see below, p. 216, n.21.

41. Jeffrey Paul von Arx, *Progress and Pessimism: Religion, Politics and History in Late Nineteenth Century Britain* (1985), 91–2. Much the same might be said of E.A. Freeman; see for example Freeman to Edith Thompson, 13 October 1872, *Life*, II, 6–61.

42. For a full study see Ruth Barton, *The X Club: Power and Authority in Victorian Science* (2018).

43. Dallas, 'Mr Darwin's Theories', *Westminster Review* 91 (1869), 207–37; for his relationship with Huxley see W.S. Dallas to Newton, 9 November 1868, MS Add. 9839/1D/27, Newton Papers, CUL. For Sclater's position see letter to Newton, 5 March 1862, MS Add. 9839/1S/594, Newton Papers, CUL. Sclater's letters to Alfred Newton show that he had no time for the hardline anti-Darwinians; but there is limited evidence of direct commitment to Darwinian ideas in his correspondence with Darwin, who noted in 1861 that he was 'heretical' on species, Darwin to Sclater 12 [March 1861], DCP-LETT-3086, Darwin Correspondence.

44. John Henry Newman to J. Walker of Scarborough, 22 May 1868, C.S. Dessain and T. Gornall, eds., *The Letters and Diaries of John Henry Newman*, XXIV (1973), 77–8. For Searles Valentine Wood see E.B. Poulton, 'Convincing Evidence in Favour of Organic Evolution: An Unpublished Essay Written Several Years before the Appearance of the *Origin*', *Proceedings of the Linnean Society* (1932), 114.

45. Letter in *Life of J.D. Hooker*, quoted in Hesketh, *Oxford Debate*, 74.

142 DARWINISM'S GENERATIONS

46. G.H. Lewes, 'Mr Darwin's Hypothesis', *FR*, 1868, quoted in Tjoa, *George Henry Lewes: A Victorian Mind* (1977), 93. Lewes himself thought natural selection a 'formula of speculative beauty', but amounting only to an 'immense probability', ibid., 95; Beer, *Darwin's Plots*, 22.

47. Bates to Hooker, 11 January 1862, JDH/1/2/2/36–40, Hooker Papers, RBGK; Charles Kingsley, 7 December 1859, ff.160–1, Box 19, Huxley Papers, ICL.

48. Carpenter to Frank Darwin, 13 June 1882, MSS.DAR.198/33, Darwin Papers, CUL. On *vera causa* see M.J.S. Hodge, 'Darwin's Book: *On the Origin of Species*', *Science and Education* 22 (2012), 2267–94.

49. [W.H. Hudleston], Obituary, in 'Biographical Notices' in *Ibis: Jubilee Supplement*, 9th ser., II (1908), 115.

50. Augustus Henry Lane-Fox Pitt Rivers, 'Inaugural Address to the Anthropological Section of BAAS', *Journal of the Anthropological Institute* II (1873), 350–61; Pitt Rivers, 'The Evolution of Culture', in his *The Evolution of Culture and Other Essays* (1906), 24; Adam Kuper, 'Darwin and the Anthropologists', in M. Teich et al., *Nature and Society in Historical Context* (1997), 274–94. Note the argument of Robert C. Dunnell, 'Evolutionary Theory and Archaeology', *Advances in Archaeological Method and Theory* 3 (1980), 35–99, that Victorian anthropology was always much more Spencerian than Darwinian.

51. Froude to Huxley, 3 June [1877], f.284, Box 16, Huxley Papers, ICL.

52. F.W. Newman to J.R. Mozley, 22 May 1889, Newman Papers, AL342/23, Senate House Library, UL. Newman's ridicule of both Darwin's logic and his evidentiary base ('never was so vast a structure reared on so narrow a foundation') strongly echoed the responses of other anti-Darwinian early Victorians.

53. Desmond, *Politics of Evolution*; Laycock, 'Reflex, Automatic and Unconscious Cerebration', *Journal of Mental Science* 21 (1875), 477–98, especially 484–5.

54. See Robert Chambers to Alexander Ireland, 31 December 1859, Robert Chambers Correspondence with Alexander Ireland, Dep.341/112, NLS; also Sir James Lamont (1828–1913), in his *Seasons with the Seahorses* (1861). For a flavour of the Edinburgh Philosophical Society circle see the correspondence in the Daniel Wilson Papers, S65 Volume II, Toronto Public Library.

55. See Richard Vyvyan to Lyell, 30 March 1863, COLL 203/1/5863–68, Lyell Papers, Edinburgh. For a similar reading see Edward FitzGerald (1809–83) to Stephen Spring Rice, 2 March 1860, A.M. Terhune and A.B. Terhune, eds., *Letters of Edward FitzGerald* (1980), II, 355, and repeated 359.

56. Lewes, 'Mr Darwin's Hypothesis', *FR*, 1868, quoted in Tjoa, *Lewes*, 93.

57. Jowett, 9 April [1871], J. Prest, ed., *Dear Miss Nightingale: A Selection of Benjamin Jowett's Letters to Florence Nightingale* (1987), 207.

58. Browne, *Darwin*, II, 283.

59. See E.L. Youmans to J.S. Mill, 10 December 1868, and response of Mill, 20 December 1868, Mill Papers, MS350, Yale.

60. Ruse, *Monad to Man*, 181; 'Lux (Divina) in Tenebris: A Personal Retrospect', Vol II, Box 33, Dep.208, A. Campbell Fraser Papers, NLS.

61. James Hinton, *James Hinton: Selections from Manuscripts, III (1860)* (1874), 87, 363, 507. For Hinton, see also Dixon, *Altruism*, 81–9.

62. E.F. Benson, *As We Were: A Victorian Peep-Show* (1930), 149; Tollemache, 'Recollections of Jowett: A Fragment', *Journal of Education* 41 (1895), 306.

63. See L. Stephen, 'Life of Jowett', *National Review* 29 (1897), 446; his sermon 'Darwinism and Faith in God' is discussed in Hinchliff, *Jowett and the Christian Religion*, 182–208; Jowett to Nightingale [11 April 1871], *Dear Miss Nightingale*, 208.

64. A.V. Dicey to W. Sanday, 19 March 1912, f.80, William Sanday Papers, MS. Eng. Misc. d.123(i), Bodleian.

65. Jowett to A.P. Stanley, 17 July 1864, Jowett Papers III/S/115, Balliol College. Oxford.

66. Tollemache, *Benjamin Jowett*, 57. We can see the persistence of this preoccupation in his response to Liddon's sermon in 1884, Mark Pattison to Acland, 3 March 1884, f.189, MSS Acland d.59, Bodleian.

IN THE WAKE OF THE *ORIGIN* 143

67. W.R. Greg, *Enigmas of Life: With Prefatory Memoir* (1891), 221; see M. Francis, *Herbert Spencer and the Invention of Modern Life* (2007), 141–2.
68. See Frederick Arnold, *Reminiscences of a Literary and Clerical Life* (1889), II, 137–8.
69. Julia Wedgwood, 'Frederick Denison Maurice', *British Quarterly Review* 79 (1884), 298.
70. Alexander Crauford, *Recollections of James Martineau* (1903), 98–9.
71. Leslie Stephen, 'Browning's Casuistry', *National Review* 40 (1902–3), 537. This would tally with one of the few direct references to Darwin in Browning's writings, his comment in a letter to F.J. Furnivall, 11 October 1881, in Thomas J. Wise, ed., *Letters from Robert Browning to Various Correspondents* (2 vols, 1895–6), I, 83–5. This is a complex and not entirely straightforward discussion in which Browning distances himself from the 'un-Darwinized', and affirms his general belief in developmentalism (as per various classical texts including Paracelsus), while at the same time apparently incorporating some sense of 'successive acts of creation'. Later critics perceived a distinction between Browning and Meredith, who accepted the results of science, however, melancholy, 'in a way that Mr Browning, of course, refused to do', Le Gallienne, *George Meredith: Some Characteristics* (1905), 12. See the brief discussion of Browning as not a Darwinian, and indeed explicitly in *Caliban* rejecting Darwinian notions, in Holmes, *Darwin's Bards*, 39–40, 84–9.
72. *Life and Letters of Francis Galton* (1924), II, 4–5. In his *Memories*, Galton told Darwin at the end of 1869 that the *Origin* 'drove away the constraint of my old superstition as if it had been a nightmare, and was the first to give me freedom of thought', Galton to Darwin, 24 December 1869, DCP-LETT-7034, Darwin Correspondence. For evidence of his contemporary enthusiasm for the new vistas of the *Origin*, without any sense of intellectual crisis, see John C. Waller, 'Becoming a Darwinian: The Micro-Politics of Sir Francis Galton's Scientific Career 1859–65', *Annals of Science* 61.2 (2004), 141–63.
73. For this see Newton to Vines, 18 May 1901, MS Add. 8580/60, Sydney Vines Papers, CUL.
74. Newton to Tristram, 30 July 1860, copy in MS Add. 9839/1T/209, CUL; see A.F R. Wollaston, *Life of Alfred Newton* (1921), 110–12; Jonathan Smith, 'Alfred Newton', 137–56.
75. An excellent example is James Martineau to W.B. Carpenter, 10 November 1884, f.34, Lant Carpenter Papers, 3, Harris Manchester, Oxford.
76. Lyell, ed., *Life of Sir Charles J.F. Bunbury*, II, 227.
77. Alfred John Jukes Browne (1851–1914) to Wallace, 16 May 1892, ff.265–6, Wallace Papers, Add. MS 46436, BL; in 1860 Darwin commented characteristically that 'Considering his age, his former views and position in society,…his action has been heroic', Darwin to Asa Gray, 22 July [1860], DCP-LETT-2876, Darwin Correspondence.
78. Lyell, *Life of Sir Charles J.F. Bunbury* (1906), II, 161.
79. Bartholomew, 'Lyell and Evolution', 295–6; Lyell to Herschel, 22 August 1861, HS/11/425, Herschel Papers, Royal Society; Darwin to Lyell, 4 October [1867], DCP-LETT-5640, Darwin Correspondence; Moore, 'Deconstructing Darwinism', 371.
80. Argyll to J.E. Denison, 7 January 1862, Os/778, Denison Collection, University of Nottingham; William Carruthers to John H. Balfour, 27 October 1866, C41, Hutton Balfour Papers, RBGE.
81. Lyell to H.W. Bates, 2, 16 September 1866, Sir Charles Lyell Papers B L981, APS; Bates to Lyell, 1, 11 September 1866, COLL 203/1/149–55, Lyell Papers, Edinburgh.
82. Lyell to Wallace, Wallace Papers, 4 March 1867, ff.26–7, also Lyell to Wallace, 14 April 1867, ff.37–9, Wallace Papers, Add. MS 46435, BL; in response to Argyll to Lyell, 25 April 1867, COLL 203/1/85–87, Lyell Papers, Edinburgh. Lyell to H.W. Bates, 13 March [nd] [probably 1872], Sir Charles Lyell Papers B L981, APS.
83. Sedgwick to Livingstone, 16 March 1865, J.W. Clark, ed., *Life and Letters of Adam Sedgwick* (1890), II, 411 (for other references to Lyell's conversion see R.M. Beverley, *The Darwinian Theory of the Transmutation of Species* (1867), 21–3); Lightman, *Popularizers*, 242; Lyell to Argyll, 25 January 1865, in I.A. Campbell, *George Douglas Eighth Duke of Argyll: Autobiography and Memoirs* (1906), II, 484; Bunbury noted that Lyell agreed 'that Darwin is too apt to exaggerate the importance of his hypothesis of Natural Selection, to deify Natural Selection (this was Lyell's expression) to speak as if Natural Selection were a great

144 DARWINISM'S GENERATIONS

primary law of nature, which would explain the real origin of all the diversity of organic forms: instead of being at the utmost, the process by which varieties are segregated into species', Lyell, ed., *Life of Sir Charles J.F. Bunbury*, II, 217.

84. J.W. Judd, *The Coming of Evolution* (1912), refers to Lyell's long resistance and continued reservations, but concluded that 'at the end, (from conversation with both), Lyell was more completely satisfied with Natural Selection than was Huxley' (Judd to Thiselton-Dyer, 13 November 1919, f.204, W.T. Thiselton-Dyer Papers, In-letters II, RBGK). This perhaps says more about Huxley's doubts than Lyell's convictions.

85. George Bentham to Frank Darwin, 30 May 1882, MSS.DAR.198/14, Darwin Papers, CUL.

86. Bentham, 'Revision of the Genus Cassia', *Transactions of the Linnean Society* 27 (1869), 503–93, see Darwin to Hooker, 30 March [1869], DCP-LETT-6688, Darwin Correspondence.

87. Bentham to Gray, [7 June 1859], Asa Gray Correspondence, Harvard.

88. (Accepting there may have been an element of politeness to his correspondent here), Bentham to Gray, 11 April 1860, Asa Gray Correspondence, Harvard.

89. See W.S. Dallas to Bentham, 6 May 1868, BEN/1/2/662, George Bentham Papers, RBGK; extract of letter from Bunbury to Mary Lyell, 30 October 1869, BEN/1/2/451, George Bentham Papers, RBGK. Bentham remained unsurprisingly unwilling to follow all 'Haeckel's German speculations', Bentham to Gray, 1 January 1883, Asa Gray Correspondence, Harvard.

90. Quoted in Green, *History of Botany*, 498. See Richard Bellon, '"The Great Question in Agitation": George Bentham and the Origin of Species', *Archives of Natural History* 30.2 (October 2003), 282–97.

91. See Bentham to Asa Gray, nd but annotated late 1865, and ditto, 15 October 1862, Bentham–Gray, Asa Gray Correspondence, Harvard. In later correspondence it is noticeable that Bentham was studiously non-committal about his Darwinian views; nor does the correspondence offer any direct comment on the death of Darwin, except for a bland acknowledgement of Gray's obituary notice of Darwin, 14 October 1882, Bentham–Gray, Asa Gray Correspondence, Harvard.

92. 'Dr Allen Thomson', *Proceedings of the Philosophical Society of Glasgow* 15 (1883–4), 13. In public lectures in 1877 he indicated that the evolutionary account of the brain was the one he was 'most inclined to adopt', *Dundee Courier*, 16 February 1877. For evidence of his sympathetic stance as early as 1861, see Thomson to Royal Society, MS Gen 1476/A/Series 1/1574, Allen Thomson Papers, University of Glasgow.

93. Allman to Huxley, 9 July 1860, Box 10, Huxley Papers, ICL. In the struggles in the Linnean in the 1870s and 1880s Allman clearly sided with the Darwinians, although he was considered ineffective, and Tyndall was unimpressed with the half-hearted endorsement in his presidential address at the British Association in 1879; see Tyndall to Heinrich Debus, 5 September 1879, *Letters of John Tyndall*, Volume 16 (forthcoming).

94. 'Presidential Address', *Transactions of the Edinburgh Botanic Society* 9 (1867), 206–31.

95. H. Pengelly, *Memoir of William Pengelly* (1897), 22. Moncure Conway recalled that 'Pengelly expressed his amazement that the really religious mind of England had not welcomed Darwin's discovery', Moncure Conway, *Autobiography: Memories and Experiences of Moncure Daniel Conway* (1904), II, 188. There is, perhaps, also the lawyer and railway promoter Charles Austin (1799–1874) who, it was reported, 'inclined to Darwinism', L.A. Tollemache and B.A. Tollemache, *Safe Studies* (4th ed., 1899), 235. (Tollemache's somewhat slapdash approach to ideas makes it difficult to know exactly what beliefs are being described here.)

96. See Charles D. Cashdollar, *The Transformation of Theology* (2014), 194, citing McCosh, *Religious Aspects of Evolution*, viii–ix, and *The Method of Divine Government* (1850). See also McCosh's *Christianity and Positivism: A Series of Lectures to the Times* (1871), whose extensive Appendix clearly outlines the significant limits of his adherence, and also B.J. Gundlach, *Process and Providence: The Evolution Question at Princeton, 1845–1929* (2013).

97. Kingsley to Maurice [1863], *Kingsley: His Letters and Memories*, II, 172.

IN THE WAKE OF THE *ORIGIN*   145

98. Fawcett, quoting Hooker at the BA in 1860, quoted in David Stamos, *Darwin and the Nature of Species* (2007), 202. At the same time, even at the beginning of 1859 it is clear that he was entirely convinced that his work would need to proceed on explicitly Darwinian lines, Hooker to Asa Gray, 6 January 1859, JDH/2/22/1/1/14–15, Hooker Papers, RBGK.

99. Endersby, *Imperial Nature*, 326, see 318–26. Hooker's commitment was primarily to a 'simple, necessary and inevitable' belief in the 'derivative hypothesis' and he had no time for accretions such as pangenesis; see Hooker to Asa Gray, [1868], JDH/2/22/1, Hooker Papers, RBGK.

100. See Carolyn Pennington, *The Modernization of Medical Teaching at Aberdeen* (1994), especially 17–21; Arthur Keith, *Anatomy in Scotland during the Lifetime of Sir John Struthers (1823–1899)* (1911), 17.

101. There was clearly a shift by 1874, when his address brought an anti-Darwinian into the fray, John Struthers to Frank Darwin, 5 July 1882, MSS.DAR.198/198, Darwin Papers, CUL.

102. He was criticised by Romanes for his arguments on the presence of useless organs, *Nature*, 12 March 1874; see *Aberdeen Press and Journal*, 10 February 1883.

103. Cobbold, Presidential Address to the Quekett Society in 1879, *Journal of the Quekett Microscopical Club VI. 1879–81* [1881], 2–9. For his initial opposition see *London City Press*, 12 May 1860. His Swiney lectures in 1868 concluded with both a 'not proven' verdict, but also a rejection of the proposition that all organic beings had descended from one primordial form, see *Edinburgh Evening Courant*, 8 December 1868; while at the BA in 1869 he was prepared to challenge views that the palaeontological record did not demonstrate succession, see *Express and Echo*, 27 August 1869, and in 1870 to argue that the evidence even suggested clear links of man and the other primates, *Liverpool Daily Post*, 20 September 1870. Cobbold is another of those instances where later 'conversion' to Darwinism is claimed on flimsy evidence (in his case some obscure and allusive passages in his *Worms* (1872); see Shang Jen Li, 'British Imperial Medicine in Late Nineteenth Century China and the Early Career of Patrick Manson', unpublished PhD thesis, Imperial College, 1999, 159.

104. W. Huggins, *The Royal Society* (1906), 89.

105. '[D]ogmatic theology', he told H.W. Acland, 'is becoming the Frankenstein of religion', Huggins to H.W. Acland, 31 May 1866, ff.238–9, December 1883 [?], ff.234–6, MSS Acland d.63, Bodleian. 'The clergy seem incapable of understanding the seeking of truth, for truth's sake', he remarked to Acland in a subsequent letter, ibid., 5 May 1884, ff.250–1. For an effective account of Huggins' career and shifting position see Robert W. Smith, 'The "Great Plan of the Visible Universe": William Huggins, Evolutionary Naturalism and the Nature of the Nebulae', in Bernard Lightman and Michael S Reidy, eds., *The Age of Scientific Naturalism: Tyndall and His Contemporaries* (2015).

106. Charles Pritchard to Stokes, 9 February 1884, MS Add. 7656/1P/728, Stokes Papers, CUL; see also his *Occasional Thoughts of an Astronomer on Nature and Revelation* (1890), providing succour to the design beliefs of the reviewer for *John Bull*, 21 June 1890.

107. Lyell to J.D. Hooker, 9 March 1863, in *Life, Letters and Journals of Sir Charles Lyell*, II, 362, cited in Bartholomew, 'Lyell and Evolution', 300; Arthur Milman, *Henry Hart Milman, Dean of St Paul's: A Biographical Sketch* (1900), 271–2.

108. *Life of Frances Power Cobbe as Told by Herself: Posthumous Edition* (1904), 448–9.

109. Bartholomew, 'Lyell and Evolution', 268.

110. T. Jeffrey Parker, *William Kitchen Parker* (1893), 39; for Parker's evolutionism, see his *Mammalian Descent* (1884), where he observes that he would not exchange Darwin's *Origin of Species* for all the 'schoolmen' ever wrote, 204.

111. Parker to Newton, 12, 18 April 1883, MS Add. 9839/1P/78–79, Newton Papers, CUL; Parker to E.T. Newton, 28 May 1889, ff.217–18, Sherborn Autographs, Add. MS 42,585, BL; Parker to A.C. Haddon, 2 Mar? 1890, Envelope 23/1/3, Haddon Papers, CUL; Parker to Stokes, 20 July 1887, MS Add. 7656/1P/63, Stokes Papers, CUL; G. Findlay, *Dr Robert Broom* (1972), 97. The contrast with his son, T. Jeffrey Parker (1850–97), almost whose

146   DARWINISM'S GENERATIONS

first act on taking up his post as Professor of Biology in Otago in 1881 was to use his inaugural lecture to deliver a paean to the *Origin*, whose publication made belief in special creation impossible, and installed evolution as 'the central doctrine of biology', *Nature*, 6 October 1881, could not be starker.

112. 'The Christianity of Scientific Men', MS Add. 5657/PA13111, Stokes Papers, CUL (though Stokes was confident that 'I feel sure that he did not reject the supernatural'); Arabella Buckley to John Stuart Blackie, 14 December 1882, ff.352-3, MS.2634, Blackie Papers, NLS.

113. Letter of Edward L[angridge] Lunn (1856-?), *West Surrey Times*, 10 December 1887; Rev Andrew Henderson (1825?-?), lecture on 'Darwin and Darwinism', *Paisley & Renfrewshire Gazette*, 27 October 1900.

114. Pollock, *For My Grandson* (1933), 99; 24 April 1866, JT/2/32d, Thomas Hirst Journal, 1863–84 tps, RI. Matthew Arnold suggested Kingsley's writing was too lightweight to be treated as serious, Arnold to Max Müller, nd, f.12, Ms. Eng. c.2805 Max Müller Papers, Bodleian. Müller agreed that Kingsley could not be treated seriously as a theological thinker, letter to A.P. Stanley, December 1 [1866], f.77, Ms Eng d.2346, Max Müller Papers, Bodleian.

115. See Kingsley to [H.W. Bates], 13 April 1863, Box 23 Folder 2, Charles Kingsley Papers, M.L. Parrish Collection, PUL; Bates to Kingsley, 27 June 1863, ff.91–4, Kingsley Papers Add. MS 41299, BL. Bates' response sought to encourage Kingsley to distinguish 'the directing hand of a personal providence [which] is seen in the starting of these varieties or sports; [and] the laws of Natural Selection afterwards deciding which shall live and which shall die'. Kingsley recognised that although he became 'most intimate and confidential' with Huxley, they were 'more utterly opposed in thought than he is to the general religious and other public', unaddressed letter/note, ff.42–3, Kingsley Letters, Add. MS 41298, BL. Kingsley found Argyll's *Reign of Law* 'very fair and manly', and an important corrective to Darwin; see Kingsley to Darwin, 6 June 1867, DCP-LETT-5565, Darwin Correspondence.

116. Kingsley to My Dear Sir, 9 July 1872 AM14756, Box 23, Folder 3, Charles Kingsley Papers, M.L. Parrish Collection, PUL.

117. Kingsley to Duke of Argyll, 22 December 1862, Argyll Papers 1209/985, Inveraray Castle; Kingsley, *Scientific Lectures and Addresses* (1890), 313; Kingsley to Helps, 30 March 1871, AM17335, Box 23, Folder 3, Charles Kingsley Papers, M.L. Parrish Collection, PUL; see his praise for Balfour Stewart on energy, Kingsley to Macmillan, 8 April 1870, AM81.56, Box 24, Folder 8, Charles Kingsley Papers, M.L. Parrish Collection, PUL.

118. Ms of Kingsley lecture 'Heroism' [annotated 1874] AM92-13, Box 24, Folder 18, Charles Kingsley Papers, M.L. Parrish Collection, PUL.

119. Conlin, *Evolution*, 213, citing Kingsley to unnamed correspondent, AM14756, Charles Kingsley Papers, M.L. Parrish Collection, PUL.

120. *The Field*, 26 August 1871. Tegetmeier had a first edition of Tennyson's *In Memoriam* which scientific friends, including Huxley and Darwin, had autographed; see 'Mr W.B. Tegetmeier at Home' reprinted from *The World*, in *Hampstead and Highgate Express*, 5 May 1883.

121. E.W. Richardson, *A Veteran Naturalist, Being the Life and Work of W. B. Tegetmeier* (1916), 205.

122. Nicely exemplified by the passages relating to Darwin in *Theology and Science: An Address Delivered to Students Preparing for Ordination at the Clergy School at Leeds, in December, 1880* (1881). See S. Paget, *Memoirs and Letters of Sir James Paget* (1902), 323–6; Francis Paget to WTTD, 3 January 1900, f.92, W.T. Thiselton-Dyer Papers, In-letters III, RBGK.

123. Pausing only at sufficiency of natural selection, see Carpenter, 'On Mind and Will in Nature', *Contemporary Review* 20 (June 1872), 738–62. As a transitional figure, it is not surprising that Carpenter comes to demonstrate many of the approaches of the mid Victorians, including his invocation of 'the controlling and sustaining action of an intelligent mind acting in accordance with a determinate plan', quoted in Boase, *Few Words*, 117.

124. 'Dr Carpenter and His Reviewer', *Athenaeum* (1863), 461, quoted in Dawson, *Darwinism, Literature and Victorian Respectability*, 126.
125. Desmond, *Evolution's High Priest*, 40, 39–43.
126. See the discussion in Desmond, *Evolution's High Priest*, I, 342 etc., Stanley, *Huxley's Church*, 242–63. For the *Reader*, see Pearson, *Life, Letters and Labours of Galton*, II, 67–70; Ruth Barton, '"An Influential Set of Chaps?": The X-Club and Royal Society Politics', *BJHS* 23 (1990), 53–81.
127. James Scott Bowerbank (1797–1877) to Owen, 3 March 1863, Owen Papers, Gen Corr Vol IV, NHM. For another example of men of science feeling excluded by the X-Club, see Albert Gunther (1830–1914) to Newton, 11 January 1865, MS Add. 9839/1G/651, Newton Papers, CUL. For an even earlier sensibility of the hidden forces at work, see John G. Macvicar (1800–84), describing Tyndall as 'one of a set of fellows who are perfect tyrants in Science'. Macvicar to Hutton Balfour, 6 September 1860, M85, Hutton Balfour Papers, RBGE.
128. As reported by Hewett C. Watson to W. Carruthers, Carruthers Papers, U–Z, DF404/1/18, NHM.
129. Josipa Petrunic, 'Evolutionary Mathematics: William Kingdon Clifford's Use of Spencerian Evolutionism', in Lightman and Reidy, *The Age of Scientific Naturalism*, 90, which cites his essay 'Philosophy of the Pure Sciences'. Clifford told Bence at the Royal Institution that his object was to show that the '"greatest <u>development</u> of the greatest number" is a better foundation of ethics than the "greatest happiness"'. He presented his ideas as an attempt 'to popularise the views of Mr Herbert Spencer, though with rather different results', Clifford to Bence, 31 December [1867], Henry Bence Papers, HBJ/A/1/40, RI.
130. See lecture notes and correspondence, Box 48/7, Westwood Papers, OMNH.
131. H.W. Bates to Hooker, 19 March 1861, JDH/1/2/2/33, 12 May [nd], JDH/1/2/2/52, Hooker Papers, RBGK.
132. A. Crawforth, *The Butterfly Hunter* (2018), 232. Bates was certainly alienated later from the Darwinist mainstream, responding to Kropotkin's theories of mutual aid that 'That is true Darwinism. It is a shame to think of what "they" have made of Darwin's ideas', quoted in Crawforth, *Butterfly Hunter*, 235.
133. Pusey to Talbot, 19 February 1880, LBV 126; Pusey to Shaw Stewart, 10 October 1879, LBV 69, Pusey Papers, PHL; for Pusey it was 'the highest honour in the university being applied to Darwinism', Pusey to H.W. Acland, nd, ff.36–7, MSS Acland d.60, Bodleian.
134. M. Wheeler-Barclay, *Science of Religion in Britain, 1860–1915* (2010), 108; Smith, 'Newton', 138, 147, 149–50.
135. Ruse, *Monad to Man*, 219.
136. See the account of Theodore Hook (1838–?), who graduated from Cheshunt College, *c.*1862, *Hampshire Telegraph*, 30 September 1882; or the comments of Ebenezer Griffith-Jones (1860–1924) on the way evolution was presented during his training, 'all mechanical, materialist' with no room for God, *Nonconformist*, 28 October 1897. The tenor of the debate in Scotland can be seen in the correspondence of Alexander Thomson; including of Free Church minister, Hugh Mitchell (1822–94), who noted that 'such books as Darwin's *Origin of Species* and *Essays and Reviews* show in what direction thought is leading, and danger is brooding', Mitchell to Thomson, 5 March 1861, THO 72, Alexander Thomson Papers, New College Edinburgh.
137. Royal Society Letter Books, MMXIX, #47, cited by A.J. Harrison, 'Scientific Naturalists and the Government of the Royal Society, 1850–1900' (unpublished PhD thesis, Open University, 1989), 415.
138. Harrison, 'Scientific Naturalists', 387.
139. See Stokes to Huxley, 16 January 1873, f.87, Box 27, Huxley Papers, ICL.
140. Hugh Falconer to Charles Darwin, 7 November [1864], DCP-LETT-4662, Darwin Correspondence.
141. Tristram to Newton, 9 October 1867, MS Add. 9839/1T/227, Newton Papers, CUL; Frederick Burkhardt, 'England and Scotland'.

148 DARWINISM'S GENERATIONS

142. Jack Morrell and Arnold Thackray, *Gentlemen of Science: Early Years of the British Association for the Advancement of Science* (1981), 449.

143. Desmond, *Huxley: The Devil's Disciple*, 350; Ellegard, *Darwin*, 79; though some newspaper accounts suggest that he carefully hedged his position, contenting himself with saying it was a proper subject of study, but that evolution was 'hardly yet formularised', and despite the high authorities advocating it, 'the preponderance of authority would necessarily be on the other side'; see, for example, *Newcastle Daily Chronicle*, 23 August 1866.

144. 25 August 1864, JT/2/32d, Thomas Hirst Journal, 1863–84 tps, RI. Sir John Evans is an excellent example of a mid Victorian whose sympathies were largely driven by his overriding concern to resist the encroachments of religion, who was really a Huxleyan rather than a Darwinian; see Joan Evans, *Prelude and Fugue: An Autobiography* (1964), 37, 59.

145. Where he proclaimed 'he did not believe a word of the Darwinian theory and was surprised that men of eminent talent should lend themselves to such a belief', *Newcastle Chronicle*, 1 September 1866.

146. Recollections of Professor M.A. Lawson (1840–96), *Morning Post*, 26 August 1882.

147. Lubbock to Darwin, 28 September 1867, DCP-LETT-5635, Darwin Correspondence; *Nottinghamshire Guardian*, 28 August 1868; *Nairnshire Telegraph*, 26 August 1868. Hooker had been most reluctant to occupy the presidential chair, and had given in only after months of resistance; see Hooker to Asa Gray, 22 March 1867, JDH/2/22/1/24, Hooker Papers, RBGK; Alexander Robertson (1825–93) commented that the Darwinians 'had not established a single fact, and it was absurd to rear a system of truth on a mere hypothesis', *Northern Warder*, 21 January 1868. The classicist C.E. Prichard (1820–69) was prepared to see Tyndall's lecture at Norwich as retreating from his previous materialism; see Prichard to J.D. Coleridge, 26 August 1868, ff.142–5, Coleridge Papers, Add. MS 63084, BL.

148. Rolleston to H.W. Acland, 27 August 1868, ff.23–4, MSS Acland d.65, Bodleian.

149. *Daily Telegraph*, 21 August 1869; John Eliot Howard to Joseph Howard, 27 August 1869, in M. Crewdson, *Memorials of John Eliot Howard* (1885), 232–3; see also H.F. [Henry Fletcher] Hance to Sir Joseph Dalton Hooker, 21 January 1870, DC/150/524, Directors' Correspondence, RBGK. Hooker frequently expressed his disillusion with the BA in these years in his correspondence with Asa Gray, highlighting 1874 as exceptional in its usefulness, Hooker to Thiselton-Dyer, [1874], JDH/2/16/25, Hooker Papers, RBGK. Even in 1873 at Bradford, under the presidency of the Comtist chemist A.W. Williamson, the meeting was, according to Hooker, 'simply miserable'; see Hooker to Huxley, f.212, Box 3, Huxley Papers, ICL.

150. See Huxley to Hooker, 23 August 1871, quoted in Harrison, 'Scientific Naturalists'; J.D. Burchfield, *Lord Kelvin and the Age of the Earth* (1975).

151. For Ronald Rainger, 'Race, Politics, and Science: The Anthropological Society of London in the 1860s', *Victorian Studies* 22.1 (1978), 51–70, the dominant character was anti-Darwinian, not least because of the leadership of Hunt. Evelleen Richards, 'The Moral Anatomy of Robert Knox: The Interplay between Biological and Social Thought in Victorian Scientific Naturalism', *Journal of the History of Biology* 22 (1989), 373–436; see also discussion in Barton, *The X-Club*, 3.3, 'Science of Man: Ethnologists against Anthropologists'.

152. Huxley to Lubbock, 18 October 1867, ff.63–5, Avebury Papers, Add. MS 49641, BL; also Huxley to Müller, 7 January [possibly May?] 1869, ff.253–4, Ms. Eng. c.2805 Max Müller Papers, Bodleian. Crawfurd reviewed the *Origin* unfavourably and wrote apologetically to Darwin to explain why he could not approve, Browne, *Darwin*, II, 101.

153. Hunt made explicit his anti-Darwinian position (and that of the Anthropological Society) at their conference at Dundee in 1867, *Dundee Courier*, 7 September 1867, see E. Sera-Shriar, *The Making of British Anthropology, 1813–1871* (2013), 127–32. Davis was a polygenist opponent of human transmutation, and an anti-evolutionist, J. Barnard Davis to Wilson, 16 February 1863, 88c, Daniel Wilson Papers, S65 Volume II, Toronto Public Library. The Society's normal stance is illustrated by the chorus of hostility which met the attempts of C.O. Groom Napier (1839–94) to explore the physiological links of man and

IN THE WAKE OF THE *ORIGIN* 149

animals during the mid-1860s, 'Referees reports 1867–68', Anthropological Society Papers, A8, Royal Anthropological Institute.

154. See A. Lane Fox to Hooker, 10 June 1871, JDH/1/2/8/205, Hooker Papers, RBGK; for a full account see G.W. Stocking, *After Tylor: British Social Anthropology, 1888–1951* (1995); David N. Livingstone, *Adam's Ancestors: Race, Religion, and the Politics of Human Origins* (2008).

155. J.F.N. Clark, *Bugs and the Victorians* (2009), 211.

156. Including Edward Caldwell Rye (1832–85), Henry Guard Knaggs (1832–1908), Thomas Blackburn (1844–1912), and Henry Tibbats Stainton (1822–92). Stainton's position was perhaps fairly traditionally mid-Victorian; he was dismissive of Edward Newman's brand of outright opposition to Darwin, but by no means entirely supportive; see Clark, *Bugs*, 114. *The Entomologists' Weekly Intelligencer*, which he edited, gave only a brief and far from positive notice of the *Origin*, Gowan Dawson, et al., *Science Periodicals in Nineteenth Century Britain* (2020), 223. Certainly in the 1890s Meldola described him as a 'pure systematist' and placed him firmly in opposition to Poulton's pro-Darwinian stance, Meldola to Poulton, 24 June 1892, Meldola Papers, OMNH.

157. Poulton, 'The Influence of Darwin upon Entomology', *Entomologist's Record and Journal of Variation* 13 (1901), 72–6, 74. For some sense of the ongoing turmoil, see the correspondence of Francis Walker (1809–74) in the A.H. Haliday Papers, Royal Entomological Society.

158. Quoted in Clark, *Bugs*, 123; Westwood to Lubbock, 28 December 1874, ff.75–6, Avebury Papers, Add. MS 49644, BL.

159. See his description of himself as a 'factist' in contrast to evolutionist, in his review of Thomas Belt's *The Naturalist in Nicaragua* in *The Zoologist* 2nd series, 9, (1874), 3921.

160. Bates to Meldola, 20 November 1877, #1925, Meldola Papers, ICL; Robert McLachlan (1837–1904) to Bentham, 13 April 1871, BEN/1/7/2555, George Bentham Papers, RBGK. Bates apart, the very occasional apparent support for Darwinian positions came from high Victorians like Murray A. Matthew (1838–1908); see *Zoologist*, ns 2 (1867), 831.

161. A.T. Gage, *History of the Linnean Society of London* (1938), 57. Bowerbank to Lee, 2 March 1867, MS5733, Bowerbank to Lee, MS5377/5, Bowerbank Letters, Wellcome Library.

162. Burkhardt, 'England and Scotland', 47–53. Interestingly, Bunbury felt that it was only in his 1869 address that Bentham 'avowed his entire and complete adhesion to the Darwinian creed', extract of Bunbury to Mary Lyell, 30 October 1869, BEN/1/2/451, George Bentham Papers, RBGK.

163. Richard Spruce to Wallace, 28 December 1873, ff.270–1, Wallace Papers, Add. MS 46435, BL. The affair clearly continued to rankle with Spruce; see G. Stabler to Isaac Bayley Balfour, 13 March 1894, In-correspondence, St–Sz, Bayley Balfour Papers, RBGE.

164. *Gardeners' Chronicle*, 29 July 1871. As E.R. Lankester pointed out in an article in *Nature*, 30 September 1875, the fear of 'speculation' was a feature of naturalists brought up in the years before the *Origin*, becoming 'as much a part of their nervous systems as the fear of precipices…It remains for the present and later generations', brought up not to fear, to turn Darwin's speculations to account.

165. What Frederick Curry called Carruthers' 'noisy vulgarity' is discussed in Gage, *Linnean*, 68–74. For Bentham's account see printed circular dated 2 March 1874, BEN/1/6/2501, and Bentham's draft letter of justification, BEN/1/6/2502–13, George Bentham Papers, RBGK (neither of which suggest any particular alignment to debates over Darwin); also Bentham to Gray, 23 July 1873, Asa Gray Correspondence, Harvard. For Carruthers' evolutionary doubts, see William Carruthers to J.H. Balfour, 27 October 1866, C41, Hutton Balfour Papers, RBGE, H.N. Ridley, 'Life of a Naturalist', 57, HNR/3/3/1–3, Ridley Papers, RBGK, and Carruthers' presidential address to the Geological Society in 1876, *Proceedings of the Geologists Association* 5 (1878), 1–35, 'Evolution and the Vegetable Kingdom', *Contemporary Review* 29 (1877), 397–409, and also his address to the Biological section of the BA in 1886. Carruthers was a dogged resister not just to Darwinian mechanics, but to a single evolutionary chain, as made explicit in his letter to

150 DARWINISM'S GENERATIONS

G.J. Romanes, 29 January 1892 [copy], Carruthers Papers, DF 404/1/6, NHM, which outlines his arguments as to why the fossil record, even setting aside the issue of gaps, does not in his mind offer persuasive support for the evolutionary theory, whether Darwinian or Lamarckian.

166. McLachlan to Bentham, 13 April, 16 July 1871, BEN/1/7/2555–6, George Bentham Papers, RBGK. Other mid-Victorian anti-Darwinists like Henry Lee and J.W. Salter were identified as among the malcontents, George J. Allman to Hooker, 1 June 1876, JDH/1/2/1/89, Hooker Papers, RBGK, and it is significant that the *Globe* newspaper, which the Carruthers party used to circulate their side of the affair, was edited by the Darwinian sceptic Mortimer Collins (1827–74).

167. Hooker to Asa Gray, 24 April 1874, 12 March 1876, 28 October 1880, 23 February 1886, JDH/2/22/1/41–43,56–57,71,85, Hooker Papers, RBGK; Lubbock to Isaac Bayley Balfour, 3 March 1886, Linnean Correspondence Folder, RBGE; also F.J. Bell to Newton, 14 February 1885, MS Add. 9839/1B/462, Newton Papers, CUL. In 1883 George Murray, along with Miller Christy (1861–1928) and E.M. Holmes (1843–1930), were seeking to gain a foothold in the Linnean Committee, squeezing Carruthers out. See Murray to Geddes, 4 July 1883, ff.155–8, MS.10522, Geddes Papers, NLS. See also the later warnings of Thiselton-Dyer, prompted by the comments of Charles Barron Clarke (1832–1906), President, in favour of the ideas of George Henslow in 1894, that such comments damaged the reputation of the Linnean, and suggested that it was falling into the hands of 'uninstructed laymen'; and that a failure to recognise the mixed character of the society, and the need to conciliate all elements was 'the rock on which Mr Bentham came to grief', Thiselton-Dyer to Clarke, 15 November 1894, f.7, A. Gunther Collection 16, NHM; Hooker to WTTD, [nd, 1875], JDH/2/16/46, Hooker Papers, RBGK.

168. Including Vines, Gardiner, Marshall Ward, D.H. Scott, Bower, and Frank Darwin, who found old Botany in the universities in the later 1870s moribund; for their frustration see Isaac Bayley Balfour to Hooker, 21 July 1884, DC/78/42–43, Director's Correspondence, RBGK, and the concerns of Lubbock of the dangers of competition, Lubbock to Isaac Bayley Balfour, 2 December 1886, Linnean Society file, RBGE.

169. S. Pace, 'The Collection of Material for the Study of "Species"', *Nature*, 21 March 1901.

170. Burkhardt, 'England and Scotland', 39, 43–4.

171. Dawson et al., *Science Periodicals*, 161. Murchison to A. Geikie, 6 February 1868, Murchison Papers, LDGSL/789/129a, Geological Society, London. For Murchison's 'younger geologists' comment, and his clear rejection even of evolutionary descent, see Murchison to A. Geikie, 7 January 1868, Murchison Papers, LDGSL/789/125, Geological Society, London.

172. Including Dr Thomas Wright, *Geological Magazine* 4 (1867), 373, Presidential Address, Geological Section of BAAS 1875, *Bath Chronicle*, 4 November 1880; George Twemlow (1795–1877), *Facts and Fossils Adduced to Prove the Deluge of Noah and Modify the Transmutation System of Darwin* (1868), and James Bryce (1806–77), 'Presidential Address', *Proceedings of the Royal Philosophical Society of Glasgow* 7 (1870–1), 382–5.

173. Geikie in Linnean Society, *The Darwin–Wallace Celebration* (1908), 52; Alfred John Jukes Browne to Wallace, 16 May 1892, ff.265–6, Wallace Papers, Add. MS 46436, BL. J.W. Judd remembered talking to many 'of the old generation, who shook their heads gravely over "poor Lyell's fads"', Judd to Meldola, 1 November 1910, #1116, Meldola Papers, ICL.

174. F. DuCane Godman, 'President's Address', *Ibis: Jubilee Supplement*, 9th ser., II (1908).

175. W.P. Pycraft, *History of Birds* (1910), 318.

176. See P. Raby, *Alfred Russel Wallace: A Life* (2020) 173, 197.

177. Robert Chambers to Lee, 30 November 1877, MS5400/133, Henry Lee Papers, Wellcome Library.

178. Herbert Spencer to Lubbock, 12 November 1864, ff.174–7, Avebury Papers, Add. MS 49640, BL; discussion in Barton, *The X-Club*, 3.4, 'The *Reader*: A Liberal Alliance and Its Collapse', Baldwin, *Making Nature*, 24–5.

179. See Lightman, 'Popularizers, Participation and the Transformations of Nineteenth-Century Publishing: From the 1860s to the 1880s', *Notes and Records: The Royal Society Journal of the History of Science* 70.4 (2016), 343–59; Ruth Barton, 'Just before *Nature*:

The Purposes of Science and the Purposes of Popularization in Some English Popular Science Journals of the 1860s', *Annals of Science* 55 (1998), 1–33. For Samuelson's views see his *Views of the Deity* (1871).

180. W.G. Blaikie to J.H. Balfour, 9 August 1864, B337, Hutton Balfour Papers, RBGE.

181. For Murray's complaint of the 'defection of most of our younger anatomists and physiologists' see Murray to Richard Owen, 7 October 1867, f.67, Owen Collection, vol. XX, NHM. For Turner's general commitment to Darwinian evolution, see *Popular Science Monthly* 58 (November 1900), 34–48.

182. Including Wentworth William Buller (1834–83). Lindley himself remained studiously ambivalent, Frederick Keeble, 'John Lindley', in F.S. Oliver, *Makers of British Botany* (1913), 173–4; *Gardener's Chronicle*, 4 June 1864.

183. It included both F.O. Morris and C.R. Bree as correspondents.

184. Bowerbank to Lee, 19 April 1873, MS5377/86, J.S. Bowerbank Papers, Wellcome Library.

185. R. Meldola, identifying as 'A Darwinian', *Land and Water*, 4 February 1871, also his 'Mimicry in the Insect World', *Land and Water*, 6 May 1871, 27 May 1871.

186. Carpenter had initially proposed the Antiquity of Man as the topic of lectures at the Hartley Institute, Southampton, but had to change this to 'The Metamorphoses of the Lower Animals'. The rival Philosophical Institute stepped in to take the rejected lecture; see W.B. Carpenter to Lubbock, 15 June 1864, ff.116–17, Avebury Papers, Add. MS 49640, BL.

187. See the letters of Charles Piazzi Smyth (1819–1900), S114–128, Hutton Balfour Papers, RBGE.

188. See inaugural address of Peter Price (1825–92) to the Cardiff Naturalists' Society, *South Wales Daily News*, 25 January 1884.

189. Register of the Warrington Field Naturalists' Club, f.26, MS29, Warrington Library. Finnegan notes that amateur naturalists 'understood Darwinism as a set of experimental investigations as much as a body of ideas', D. Finnegan, *Natural History Societies and Civic Culture in the Nineteenth Century* (2009), 161.

190. Duncanson's paper was printed in full in the *Alloa Advertiser*, 8, 22 June 1867. In 1871 Duncanson was upbraided for surreptitiously introducing the Darwinian doctrine of the descent of man in a paper on the condition of savagery, *Alloa Advertiser*, 9 December 1871, although the *Origin of Species* had been added to the Society's library earlier in the year, along with works of Lubbock, Bates, and Wallace, balanced by several of Hugh Miller and the Duke of Argyll, *Alloa Advertiser*, 14 January 1871.

191. Including Thomas Wright (1809–84), S.P. Woodward (1821–65), Charles Daubeny (1795–1867). For the Society's initial hostility, see above, p. 52. See W.C. Lucy, *The Origin of the Cotteswold Club* (1888), which gave no indication that debates over Darwin were a feature of the club's history.

192. See recollections of Hugh Robinson (1845–90), *Northern Whig*, 26 November 1888. An echo of this is apparent in the letter of Robert Smith to Darwin, 24 February 1873, DCP-LETT-8784, claiming to be the only member of the club 'who openly avows his belief in your views with regard to life and development'. Although the writer of this letter is listed in the DCP as born 1853, it is most likely to be Robert Smith Snr (1818–79), watchmaker; he did have a son, also Robert Smith, living at the same address, but the son died in 1867, aged twenty-three.

193. J.G. McKendrick, *Story of My Life* (1919), 20.

194. *Huddersfield Chronicle*, 7 September 1867. Likewise George Stewardson Brady (1832–1921), who gave a paper on Crustacea which included a full endorsement of Darwinian evolution, printed in the *Intellectual Observer* 10 (1866–7), 327ff.

195. For a full account see Stuart Mathieson, *Evangelicals and the Philosophy of Science: The Victoria Institute, 1865–1939* (2021).

196. See James Reddie, *On the Credibility of Darwinism: By George Warington [with] a Reply by James Reddie* [1867], and Reddie, *On Geological Chronology, and the Cogency of the Arguments by Which Some Scientific Doctrines are Supported: In Reply to Professor Huxley* (1868), copies at Lambeth Palace Library; Charles Staniland Wake to George Harris, 17 July 1874, George Harris Papers, UCLA. There were some older mid Victorians who

152 DARWINISM'S GENERATIONS

were active, including Rev. Walter Mitchell (1816/7–74) one of the vice-presidents, and Charles Adolphus Row (1816–94).

197. Walter R. Browne, 'The Present Aspect of the Conflict with Atheism', *Churchman* 7 (1883), 107.

198. W.J. Irons, *Annual Address to the Victoria Institute* (1871).

199. Kingsley was glad to see that 'old Pusey hates me as what I am, his thorough-going adversary', Kingsley to Müller, nd [1860s], Ms. Eng. c.2806/1 Max Müller Papers, Bodleian. Most of Pusey's correspondence on Darwinism explicitly is from the 1870s, but see Pusey to Acland, 1 December 1861, Pusey–Acland Correspondence, 'Acland from the Clergy' f.12, MSS Acland d.60, Bodleian.

200. For Morris, see Lightman, *Popularizers*, 43–8. See Morris' 'Difficulties of Darwinism' to the BA in 1868, also later *All the Articles of the Darwinian Faith* (1875), and *the Demands of Darwinism on Credulity* (1890). Morris was encouraged by correspondents like Roundell Palmer (1812–95) and William Wood, Lord Hatherley (1801–81), who praised his attempts to rebut 'the preposterous fictions of Darwin', quoted in *Francis Orpen Morris*, 294. Palmer's own stance is clear from his inaugural address as Rector of St Andrew's in 1878: evolution could not be sustained by experimental verification or deductions therefrom and was therefore unscientific; see *Dundee Courier*, 22 November 1878.

201. See letter of Bree, *The Field*, 2 September 1871; comment of Tegetmeier, *The Field*, 16 September 1871; Bree to Newton, 19 May 1860, MS Add. 9839/1B/969, Newton Papers, CUL. (Bree was apparently challenged by Newton for allowing general issues of evolutionary thought to intrude into their correspondence.)

202. E.M. Forster, *Marianne Thornton* (1956), 239. Of his grandfather, Rev. Charles Forster (1788–1871), Forster suggested that 'Darwinism was of course anathema at the rectory, but my grandfather had become so remote from events that he had not realised its full enormity', ibid., 248.

203. Disraeli's most direct attention to evolutionary thought can be found in *Lothair*; see the 'General Introduction' to the 1882 Hughenden edition, *Volume X. Lothair*, xvii; the character of Lothair denies men of science with their theories of 'nebular hypotheses, development, evolution, the origin of worlds, human ancestry'.

204. Henry Manning, ed., *Essays on Religion and Literature* (1865), 51. For references to Manning's belief that Darwinism was a philosophy which 'augured immeasurable evils', see Cobbe, *Life of Frances Cobbe, Written by Herself: Posthumous Version* (1904), 500.

205. Morris, *All the Articles of the Darwinian Faith* (1875), 33–4; Birks, *Scripture Doctrine of Creation* (1872), 185, 188.

206. William Keddie (1809–77) to J.H. Balfour, 5 March 1864, K68, Hutton Balfour Papers, RBGE.

207. One example is George Anthony Denison (1805–96), see Denison to Right Hon. Sir Robert Phillimore, 21 November 1878, in Louisa Evelyn Denison, *Fifty Years at East Brent: The Letters of George Anthony Denison, 1845–96* (1902), 205–6.

208. Lightman, *Popularizers*, 57–71.

209. See C. Pritchard, *Occasional Thoughts of an Astronomer on Nature & Revelation* (1889), 16–36; *Rev. Charles Pritchard…Life*, 92–4; Pritchard, 'Modern Science and Natural Religion' at the Church Congress, Brighton, 1875, in his *Occasional Thoughts* (1889), 119–42, and his strong defence of the separate creation of man in C. Pritchard, 'Bishop Butler', *The Churchman* 1 (1880), 330–1. Pritchard cautioned Gladstone about his efforts to reconcile science and Genesis in the 1880s, intervened vigorously in the case of Samuel Kinns (see below, pp. 250–51), and offered 'The Creation Poem of Genesis', in *The Guardian*, 10 February 1886, which argued that the Genesis story was not a narrative of facts, but a 'sublime and God-inspired dream'.

210. Tupper to Owen, 13 December 1864, ff.441–2, Richard Owen Papers, Add. MS 39954 (part 2), BL.

211. E. Abbott and L. Campbell, eds., *Letters of Benjamin Jowett* (1899), 186.

212. For Owen's later position see L.A. Tollemache, 'Sir Richard Owen and Old World Memories', *National Review* 21 (1893), 606–19.

IN THE WAKE OF THE *ORIGIN* 153

213. Joule to William Thomson, 13 May 1861, J269, Kelvin Collection, CUL, which welcomed Thomson's exposure of 'some of the rubbish which has been thrust on the public lately', while exonerating Darwin, who 'had no intention of publishing any finished theory but rather to indicate difficulties to be solved', quoted in C. Smith, *The Science of Energy* (1999), 525. For another see George Augustus Rowell (1804–92), who added an anti-Darwinian postscript to the second edition of his *On the Beneficent Distribution of the Sense of Pain* [1862]. For Rowell see also *Mr. Darwin's Theory [on the origin of instincts, a speech by G.A. Rowell at a meeting of the Ashmolean society]* [1860], for which see *Oxford Chronicle*, 8 December 1860.

214. J.H. Balfour, *A Manual of Botany* (1863), xiii. Margarita Hernandez-Laille notes that a reluctance to name Darwin was a common feature of textbooks, 'Darwin in Natural Science School Textbooks in the Nineteenth Century in England and Spain', in Carolyn J. Boulter et al., *Darwin-Inspired Learning* (2015), 310–22.

215. I. Bayley Balfour, 'The Edinburgh Professors', in F.S. Oliver, *Makers of British Botany* (1913), 298.

216. See Prestwich, 'The Position of Geology', *The Nineteenth Century* (1893), in his *Collected Papers on Some Controverted Questions of Geology* (1895), enforced in Prestwich to Wallace, 14 June [nd but after publication of *Collected Papers*], ff.45–6, Wallace Papers, Add. MS 46437, BL; also Sir Henry Holland (1788–1873), whose 'long and intimate friendship' with Darwin brought him to opine, 'Whatever be the fate of his doctrines, he has given to the greatest problem of Natural History a new framework and direction of research', *Recollections of Past Life* (1872), 6.

217. As evidenced by his correspondence with John Hutton Balfour, including 2 January 1868, D118, Hutton Balfour Papers, RBGE.

218. Darwin to Eyton, 4 March [1872], EYT/1/43, Eyton papers, CLB.

219. See the notice in *Nature*, 20 June 1878.

220. Thomas Wharton Jones, *Evolution of the Human Race from Apes and of Apes from Lower Animals a Doctrine Unsanctioned by Science* (1876).

221. Jones, *Evolution…Unsanctioned*, vii. For others offering a similar temper, see Ponton, *The Beginning*, which offered a broad conspectus of anti-Darwinian themes, drawing extensively on J.H. Stirling and Beale.

222. See David Page (1814–79) as a possible transitional figure. His *Chips and Chapters: A Book for Amateur and Young Geologists* (1869) offered a strong endorsement of 'evolution' in a developmental sense, but associated with affirmation of repeated extinctions and creations; see Lightman, *Popularizers*, 223–38. Another would be Alfred Smee (1818–77), who continued to reject transmutation, see *Memoir of Alfred Smee by His Daughter* (1878). A number of figures born in the early 1830s present as slightly out of time mid Victorians, including the anthropologist James Hunt, and John Laidlaw (1832–1906), author of *The Bible Doctrine of Man* (1895).

223. John Huntley Skrine, *A Memory of Edward Thring* (1889), 123.

224. See L.A. Tollemache, *Old and Odd Memories* (1908), 142. (On his death, Westcott was certainly presented as someone who hailed with delight the rapid acceptance of the general doctrine of evolution, albeit more for its inherent fitness to show 'how all creation is one act at once' than because of 'arguments which suggest rather than establish it', 'Bishop Westcott in Relation to Contemporary Thought', *Contemporary Review* 80 (1901), 506. That he did not embrace a fully naturalistic view of the world is clear from his *The Gospel of the Resurrection* (1866).)

225. See reference in his letter of December 1875 to Frances Arnold, in G.W.E. Russell, *Collected Letters of Matthew Arnold* (1901), II, 143, and similar comments in his *Discourses in America* (1885). For a discussion of Arnold see Paul White, 'Ministers of Culture: Arnold, Huxley and Liberal Anglican Reform of Learning', *History of Science* 63 (2005), 115–38.

226. *Cambridge Review*, 21 June 1882; recollection in Judd, *Coming of Evolution*, 3. (This was a position Arnold shared with David Masson, and one articulated by W.H. Mallock, although Frank Turner suggests Mallock was too well informed to actually believe it, see Turner, 'Lucretius among the Victorians', *Victorian Studies* (March 1973), 329–48,

154   DARWINISM'S GENERATIONS

341–2.) Arnold's *Literature and Dogma* (1873) steers well clear of Darwin or natural science generally. (His observation on John D. Coleridge's discussion of Richard Owen in 1884 was that whatever the relative merits of scientific education, 'you can hardly ever read a page of a scientific man's writing on general subjects without feeling that the man has somehow a sense missing', Arnold to Coleridge, 6 March [1884], ff.100–1, Coleridge Papers, Add. MS 86249, BL.)

227. See also the older high Victorian, T. Wickham Tozer (1832–1908), *Ipswich Journal*, 30 January 1888.

228. J. Ruskin, *The Eagle's Nest* (1880), 199. *The Eagle's Nest* displayed Ruskin at his most anti-scientific, denying science's role in the discovery of new facts, arguing its role was the proper grasp of facts already known; see Holmes, *Pre-Raphaelites*, 172.

229. St G. J. Mivart, *Genesis of Species* (1871), 2. [W. Allingham], *Rambles* (1873), 155–6; for similar concerns see George Gilfillan, 'Frederick W. Robertson', in his *Remoter Stars in the Church Sky* (1867), 160.

230. Assertions of this sort were central to F.J.A. Hort's 1871 Hulsean Lectures, and became a persistent refrain of mid-Victorian clerics; see Graham Patrick, *F.J.A. Hort: Eminent Victorian* (2015), 56–9. While Hort welcomed advances in natural science, he accepted the importance of tradition and rejected any sense that only truths that could be proved scientifically were valid. We can see a similar position in Francis Turner Palgrave (1824–97), who hoped Huxley might 'see some day that a true science cannot exclude these regions', Palgrave to Wallace, 15 November 1871, ff.228–9, Wallace Papers, Add. MS 46435, BL.

231. Bates to Kingsley, 27 June 1863, ff.91–4, Kingsley Papers, Add. MS 41299, BL.

232. R.H. Hutton paper to the Church Congress on 'Phases of Unbelief and How to Meet Them', *Authorized Report of the Church Congress Held at Liverpool, October 5th–8th 1869* (1869), 99–106, 101; an approach contested by Rev. Richard Yonge (1809/10–77) who attacked the 'man–monkey unbelief of Darwin'.

233. Fragment of letter from Ellicott [to Tait?] on the Victoria Institute, TAIT/177/7–8, Lambeth Palace Library. Kingsley was appalled at the direct attack on Lyell in the institute's initial circular. Kingsley to General Knollys, 1 June 1865, AM87.14, Box 23 Folder 2, and Kingsley to Lady Bunbury, 9 June 1865, AM16342, Box 23, Folder 6, Charles Kingsley Papers, M.L. Parrish Collection, PUL.

234. For a vivid example of this see Max Müller to Duke of Argyll, 4 February 1875, f.9c, Ms Eng d.2347, Max Müller Papers, Bodleian.

235. H.F. Hance to Sir Joseph Dalton Hooker, 25 July 1867, DC/150/494, Directors' Correspondence, RBGK.

236. See also William Thomson (1819–90), Archbishop of York, discussed in Keith Francis, 'Nineteenth Century British Sermons on Evolution and The Origin of Species: The Dog that Didn't Bark?' in R.H. Ellison, *A New History of the Sermon: The Nineteenth Century* (2010), 269–308, 295–6. For Mivart see M. Artigas, Thomas F. Glick, and R.A. Martinez, *Negotiating Darwin: The Vatican Confronts Evolution, 1877–1902* (2006). Moore's group also (although he does not give him much attention) includes George Henslow (1835–1925), but Henslow was a different beast, an ardent Lamarckian evolutionist, whose attack on natural selection bore no underlying attack on either evolution or naturalism.

237. Fry concluded with confidence that religion had nothing to fear from science, which he was happy to stand by forty years later, see A. Fry, *Memoir of Rt Honourable Sir Edward Fry* (1921), 63–4. Fry rallied to the support of Darwinism at the BA in 1894 in the face of the opposition of Charles D'Arcy, *Guardian*, 14 August 1894. At the same time, he remained ambivalent about natural selection, and his support of Mivart's *Genesis of Species* emphasised the need to consider more the 'law of variability', letter in *Nature*, 27 April 1871. In 1903 he was still puzzling over mimesis, Fry to Beddard, 9 March 1903, Beddard Papers, COLL 623, Edinburgh.

238. Murphy, 'Presidential Address on the Present State of the Darwinian Controversy', *Proceedings of the Belfast Natural History and Philosophical Society, 1873–74* (1874), 23.

239. For Beddard, Argyll was a writer 'who has something to say on most scientific topics—and says it well too', Beddard, Scrapbook, Box 4/7, Evolution Collection, Mss 28, UC-Santa Barbara.

IN THE WAKE OF THE *ORIGIN* 155

240. See Raby, *Wallace*, 169; Newton to Wallace, 4 March 1867, ff.26–7, Wallace Papers, Add. MS 46435, BL; describing himself as 'a thorough-going disciple of the school of Mill, Bain, and H. Spencer', Mivart to Darwin, 10 January 1872, DCP-LETT-8154, Darwin Correspondence.

241. See the account of the views of Professor Neil McMichael (1806–74), Professor of Church History, United Presbyterian Church, whose comment to this effect was said by Charles Jerdan (1843–1926), in his *Scottish Clerical Stories and Reminiscences* (1920), 6, to have 'expressed the view of the theory of evolution which obtained among many middle-aged and elderly people for a time after Charles Darwin's great book appeared'.

242. Lyell to George Ticknor, 9 January 1860, in K.M. Lyell, *Life, Letters and Journals of Sir Charles Lyell* (1881), 329; for later comment see Milman to Charles Lyell, 2 April 1862, in Arthur Milman, *Henry Hart Milman, Dean of St Paul's: A Biographical Sketch* (1900), 270.

243. Gilfillan, 'Notes on the Antiquity of Man', in *Sketches Literary and Theological, Selections from an Unpublished Manuscript*, ed. by F. Henderson (1881), 141. In the mid-1860s Gilfillan argued that Scripture offered only a thickly veiled allegory of the origin of man, and drew attention to the almost complete unanimity in the BA on the broad principle of development; see letter to *Dundee Advertiser*, 1 September 1866, also sermon on the 1867 BA, *Dundee Advertiser*, 16 September 1867.

244. Beer, *Darwin's Plots*, 139.

245. See Radick, *Simian Tongue*, 25–31.

246. Birks, *Modern Physical Fatalism* (1876), 300, 305. This became a serviceable image for other doubters. Beer, *Darwin's Plots*, quotes Sedgwick, 105, and we can see it being deployed by Edwin Paxton Hood (1820–85), in his popular lecture 'Man among the Monkeys', *Merthyr Telegraph*, 5 November 1880. Darwin made a special point of thanking Sir Joseph Lamont for his apparent support for this possibility; see Darwin to Lamont, 25 February [1861], DCP-LETT-3071, Darwin Correspondence.

247. John Gould, *An Introduction to the Trochilidae* (1861), 5 (cited by Beverley, *Darwinian Theory*, 8–9).

248. Reginald Farrar, *The Life of Frederic William Farrar, Sometime Dean of Canterbury* (1904), 108.

249. See citations of J. Phillips, *Life on the Earth* (1860), 185–7.

250. See Phillips, *Life on the Earth*, especially 204–8; Birks, *The Bible and Modern Thought* (1862), 502–3, Birks, *Scripture Doctrine of Creation*, 238–40. These provided resources for mid-Victorian controversialists like Brewin Grant (1821–92); see his debate with Charles Bradlaugh at South Place Chapel, June 1875, and the enduring evolutionary scepticism of Edward White; see White to Stokes, 'Friday 17' [n.d], 30 September 1894, MS Add. 7656/1W/531, 614, Stokes Papers, CUL.

251. William Keddie to Balfour, 5 March 1864, K68, Hutton Balfour Papers, RBGE. Three years later he commented that, despite thinking it over, 'It seems to me morally <u>impossible</u> to harmonise' Darwinism and Revelation, Keddie to Balfour, 9 April 1867, K72, Hutton Balfour Papers, RBGE.

252. Martineau resisted all moves towards scientific naturalism; see Crauford, *Recollections of James Martineau*, 51–2; viz. his comment that 'The miracle question...must be settled by criticism of writings, not by physiological interrogation of nature', Martineau to Anna Swanwick. 14 August 1881, Martineau papers, Bancroft Library BANC MSS 92/754 z Box 6/102; in the 1880s he was commenting that he saw modern thinking 'as acknowledging a limit to the resources of evolution, and a returning suspicion of the absolute character of absolute monism', Martineau to Allen, 29 January 1884, Box 3, Joseph Henry Allen Correspondence, bMS416, Andover Theological College, Harvard.

253. Herschel to Lyell, 13–14 April 1863, Herschel Papers, Royal Society, quoted in Bartholomew, 'Lyell and Evolution', 300. For another advocacy of successive creations see C.A. Row, *Principles of Modern Pantheistic and Atheistic Philosophy* (1874).

254. See the account of a sermon by William Thomson in Westminster Abbey directly repudiating such ideas, as reported in William Grylls Adams to Mary Dingle, 25 April 1866, AM/825, Grylls Adams Papers, Kresen Kernow.

156 DARWINISM'S GENERATIONS

255. Clark, *Bugs*, 112; [John Duns], 'On the Origin of Species', *North British Review* 32 (1860); D. Livingstone, 'Public Spectacle and Scientific Theory: William Robertson Smith and the Reading of Evolution in Victorian Scotland', *Studies in History and Philosophy of Sciences* 35.1 (2004), 6, which notes that Duns kept this stance through to his death. A similar stance is visible for Andrew Whyte Barclay (1817–84); see 'Harveian Oration', *BMJ*, 25 June 1881, and John Plant (1819/20–94), *Papers of the Manchester Literary Club* 24 (1884–5), 411.

256. Woodward to Lyell, 25 November 1862, COLL 203/1/6472–3, and his later description of himself as 'a progressionist and anti-transmutationist', Woodward to Lyell, 21 February 1863, COLL 203/1//6479, Lyell Papers, Edinburgh.

257. *Salisbury and Winchester Journal*, 18 March 1871.

258. See *Authorized Report of the Papers…Church Congress 1867* (1868), Smith, 'William Huggins', 120–1.

259. Argyll to Müller, 2 February 1875, Argyll, *Autobiography and Memoirs*, II, 529–30.

260. Argyll, 'Spencer and Salisbury on Evolution', *TNC* (April 1897), 397; J.J. Murphy, *Habit and Intelligence* (1879), I, 173.

261. Accepting that it is more emphatically stated in his 1883 address 'Recent Advances in Natural Science in Relation to the Christian Faith' to the Reading Church Congress, than in 1870, W.H. Flower, *Essays on Museums* (1898), 123–34.

262. See also Gundlach, *Evolution Question at Princeton, passim*.

263. Beverley, *Darwinian Theory*, 49.

264. Birks, *Modern Fatalism*, 291–7, outlines the various postulates, including no limits to the law of variation, the permanence of slight variation via transmission to offspring, Birks, *Scripture Doctrine of Creation*, 176–7, 243–52; see J.F.W. Herschel's endorsement of Charles Pritchard's comments to this effect, Herschel to Pritchard, [copy], 1866, HS/24/182, Royal Society; and the comments in the Griffith copy of the *Origin* (see above, p. 58).

265. For Jenkin's article see Susan W. Morris, 'Fleeming Jenkin and "The Origin of Species": A Reassessment', *BJHS* 27.3 (1994), 313–43. This argument was taken up readily by mid-Victorian opponents who argued that, without virtually limitless time, natural selection could not have worked without the guiding hand of providence to push it in the right direction; see sermon of C.P. Reichel, *Belfast Newsletter*, 21 May 1883, or the letter of the astronomer James Croll (1821–90), *Nature*, 10 January 1878.

266. Phillips, *Life on Earth*, 2. It was noted of James Martineau that having previously been largely contemptuous of Paleyite natural theology, 'when the Darwinian theory of evolution was advanced, he deliberately turned round, and at the very moment when the argument from design seemed to have completely gone, he started and built it up', *Life and Letters*, I, 435.

267. See letter to Lyell, 19 November 1866, COLL 203/1/1711–12. It is notable that Gunther continued to use multiple creation language, telling Lyell in 1871 that his studies 'must be turned to the predecessors of the present creation', 4 January 1871, COLL 203/1/1715, Lyell Papers, Edinburgh.

268. *Report of the Meeting of the British Association for the Advancement of Science, Held at Newcastle in August and September 1863* (1864), lxiii. For Sir William Armstrong (1810–1900), see DCP-LETT-4284, Darwin Correspondence. Another example is Horace Benge Dobell (1828–1917), author of *Intelligence in the Van* (1887), an argument about the 'wider teleology'; Dobell accepted natural selection but not necessarily the Darwinian argument that it was the dominant mechanism for evolution; see *Lectures on the Germs and the Vestiges of Disease* (1861).

269. Mivart, *Genesis of Species*, 23–62. Mivart's intervention was enthusiastically endorsed as summing his own view by Charles Kingsley, 'The Natural Theology of the Future', *Macmillan's Magazine* (March 1871), 369–78; see also Argyll, including correspondence with Flower in the early 1880s, Argyll, *Autobiography and Memoirs* II, 488–90.

270. The central argument of Murphy, *Habit and Intelligence*, 318–23.

271. Argyll to Lyell, 26 January 1865, reiterated 31 May 1865, COLL 203/1/73–6, 83–4, Lyell Papers, Edinburgh.

272. Mivart, *Genesis of Species*, 63–96.
273. See, for example, Argyll to Lyell, 1,15 May 1867, COLL 203/1/89–92, 93–6, Lyell Papers, Edinburgh.
274. Engels in 1872, quoted in Naomi Beck, 'The *Origin* and Political Thought', in M. Ruse and R.J. Richards, eds., *Cambridge Companion to the Origin of Species* (2009), 311. For Thomson see his *The Depths of the Sea* (1873), 11, 9–11; a point reiterated in his introductory lecture to the natural history class at the University of Edinburgh, *Times*, 9 November 1876; his view was that the results of the *Challenger* expedition gave strong support to the general doctrine of evolution, but no support at all to the theory that it was caused purely by natural selection, *British Quarterly Review* 77 (1883), 106.
275. Goldwin Smith, 'The Ascent of Man', *Macmillan's Magazine* 35 (1887), 194–204; Smith to Coleridge, 5 November 1882, ff.23–4, Coleridge Papers, Add. MS 86307, BL. Smith clearly sought to retain the essence of the mid-Victorian belief about the fundamental distinction of man and animals, while blurring the precise nature of this position as the century progressed; see his 'Is There Another Life?', in his *The Riddle of Existence* (1897), 100–2. Smith was perhaps not the only mid Victorian who was prepared to blow with the scientific orthodoxy as the century progressed, who was, as Lionel Tollemache put it, 'suspicious of Evolution militant, but loyal to Evolution triumphant', Tollemache, *Nuts and Chestnuts*, (1911), 20.
276. *Nature*, 1 February 1872. See John Ball, *Notes of a Naturalist in South America* (1887), and 'On the Origin of the Flora of the European Alps', *Proceedings of the Royal Geographical Society* 1 (1879), 564–89. See also William King, 'On the Origin of Species', *The Geologist* (1862), 254–6.
277. James McCosh, *The Development Hypothesis: Is It Sufficient?* (1876), 9–10.
278. Murphy, *Habit and Intelligence*, I, 344; J.W. Dawson, *The Origin of the World According to Revelation and Science* (1877), 372. Wallace was not unsympathetic, feeling there was much that was unintelligible, but also 'some very acute criticisms and the statement of a few real difficulties', Wallace to Darwin, 20 October 1869, DCP-LETT-6949, Darwin Correspondence.
279. P.J. Bowler, *The Eclipse of Darwin: Anti-Darwinian Evolution Theories in the Decades Around 1900* (1983), 49–50.
280. Huxley to Williamson, 27 June 1871, Huxley Papers, APS. Rupke goes further, noting that Huxley was not above assuring potential allies that he was not a Darwinian, and indeed that 'only in 1868, probably influenced by his twin Darwinian "bulldog" Haeckel, did he explicitly incorporate the concept of evolution into his research papers', Rupke, *Richard Owen*, 206. Huxley's propaganda intervention, 'The Coming of Age of the "Origin of Species"', was another instance of bypassing entirely the question of natural selection. Judd noted that to the end Huxley could not overcome some of his difficulties with the tendency of hybrids to infertility, Judd to WTTD, 13 November 1910, f.204, W.T. Thiselton-Dyer Papers, In-letters II, RBGK. But there is also evidence in Huxley's private notebooks of his personal research work which does provide a bridge between his public advocacy of evolution, and his largely non-evolutionary teaching, with 'research notes littered with phylogenetic tree trunks...evolutionary diagrams [which] came to dominate his notes', Desmond, *Evolution's High Priest*, 110.
281. Endersby, *Imperial Nature*, 5–6.
282. Browne, *Darwin*, II, 312.
283. Wallace to Darwin, 20 October 1869, DCP-LETT-6949, Darwin Correspondence.
284. See J.W. Salter's syllabus of lectures, 'On the Order of Creation', reverse of Salter to A.C. Ramsay, 4 October 1867, GSM/GL/SL/1 British Geological Survey. Or take the observation of the engineer Francis Roubillac Conder (1815–89) that the argument that 'no direct, creative, providential or divine design has...directed or rendered possible, the course of development' was a 'gratuitous proposition', 'Scepticism in Geology', *Edinburgh Review* 147 (1878), 360; compare Eustace R. Conder (1820–92), 'Natural Selection and Natural Theology', *Contemporary Review* 42 (1882), 400–12, and his *Basis of Faith: A Critical Survey of the Grounds of Christian Theism* (1877); discussed by A.F.P. Sell, *Philosophy, Dissent and Nonconformity* (2009), 190–2. In this sense Salisbury's

158   DARWINISM'S GENERATIONS

performance at the 1894 British Association, with its continued appeal to the design argument, see below, p. 327, marks him as a transitional mid Victorian.

285. W.F. Donkin to Acland, 28 February [1865], ff.122–3, MSS Acland d.63, Bodleian. Donkin's limited sympathy is suggested by the withering 'An Imaginary Conversation' skit which he sent to Acland, in which Darwin's justification that his theory 'accounts for the facts' is countered by the observation that so did the theory of the luminous ether, ff.126–31, ibid.

286. Mary English, *Mordecai Cubitt Cooke: Victorian Naturalist, Mycologist, Teacher and Eccentric* (1987), 154–5.

287. Charles Pritchard to Stokes, 9 February 1884, MS Add. 7656/1P/728, Stokes Papers, CUL.

288. Argyll to Lyell, 31 January 1865, COLL 203/1/77–80, Lyell Papers, Edinburgh. For a discussion of the arguments around beauty see Phillip Prodger, 'Ugly Disagreements: Darwin and Ruskin Discuss Sex and Beauty', in Barbara Larson and Fae Brauer, eds., *The Art of Evolution: Darwin, Darwinisms, and Visual Culture* (2009), 40–58. Prodger suggests that by the end of the 1860s 'those who disputed Darwin's account of evolution began to rally around the beauty problem as the Achilles' heel that would finally disprove evolution by natural and sexual selection'.

289. S.P. Thompson, *Life of Lord Kelvin* (1910), II, 608. In 1873 he reported finding in Madeira 'at every turn something to show (if anything were needed to show) the utter futility of [Darwin's] philosophy', 637.

290. W.G. Blaikie to J.H. Balfour, 9 August 1864, B337, Hutton Balfour Papers, RBGE; Liddon to Müller, 14 January 1879, ff.79–80, Ms. Eng. c.2806/1, Max Müller Papers, Bodleian.

291. Martineau to Sidgwick, 6 February 1869, c94/117, Sidgwick Papers, Trinity College, Cambridge.

292. Thiselton-Dyer, *Nature*, 16 January 1890; Moore, *Post-Darwinian Controversies*, 231.

293. See G. Jones, 'Darwinism in Ireland', in David Attis and Charles Mollan, eds., *Science and Irish Culture* 1 (2004), 115–37, 118; he was firmly hostile to Darwinism in a presentation in 1876; see account reprinted in *Monmouthshire Merlin*, 21 April 1876.

294. F. Temple, *The Relations between Science and Religion* (1884), quoted in John C. Greene, *Debating Darwin* (1999), 56. Similarly, in his 1890 Gifford Lectures, *Philosophy and Theology*, and in *Darwinianism: Workmen and Work* (1894), J.H. Stirling was still arraigning Darwin for his denial of purpose in evolution. The Archbishop of York, William Thomson adopted a similar position, rejecting Darwin's denial of design, but accommodating himself to the basics of evolution; see the discussion in Francis, 'Sermons', 295, and his sermon to the Christian Evidence Society in May 1882, printed in full, *Champion of the Faith*, 25 May 1882.

295. H.S. Solly, *Life of Henry Morley* (1898), 185. Morley's exact relationship to Darwinism is less clear-cut; his public stance in 1881 was studiedly ambivalent, Morley, 'The Literature of Today, with a Guess at That of Tomorrow', *General Baptist Magazine* 83 (1881), 18–24, 20. For James Croll the determinations of nature occurred not at random but according to an objective idea, 'Evolution by Force Impossible', *British Quarterly Review* 77 (1883), 25. For other examples see the physician Sir George Johnson (1818–96), 'Address in Medicine', *BMJ*, 12 August 1871; John W. Ogle (1824–1905), 'Herveian Lecture', *BMJ*, 31 July 1880.

296. This was attractive to Lamarckians like Samuel Butler, who ascribed to his reading of Mivart's *The Genesis of Species* his recognition of the 'fallacy of Natural Selection', Butler to Mivart, 27 February 1884, ff.43–4, Samuel Butler Papers, Add. MS 44030, BL. But it also appealed to even evolutionary-inclined mid Victorians like Kingsley, who praised Mivart for exactly supplying his requirement for a system which showed the divine hand at work; see Kingsley's *Scientific Lectures and Addresses* (1890), 313.

297. White to Stokes, 24 August 1882, MS Add. 7656/1W/449, Stokes Papers, CUL.

298. Carpenter, 'On Mind and Will in Nature', Carpenter, *Agnosticism* (1885), 19. His paper 'The Argument from Design in the Organic World' (*Modern Review* 5 (1884), 641–700) provided succour to early Victorians like James Martineau, see Martineau to Carpenter, 10 November 1884, f.34, Lant Carpenter Papers, 3, Harris Manchester, Oxford. Martineau reported a conversation with G.J. Allman (1812–98) twelve months

IN THE WAKE OF THE *ORIGIN* 159

previously, in which Allman indicated that he agreed with Carpenter that 'the teleological [argument] needed only a modified exposition, in order to retain its full force'.

299. P.M. Duncan to A.R. Wallace, 27 February 1871, ff.214–15, Wallace Papers, Add. MS 46435, BL. Tellingly, in his later writings Duncan seems to have been most reluctant to address questions of evolution. His *Heroes of Science* (1882) is notable for its omission of Darwin.

300. Huxley to Blackie, 24 February 1876, ff.33–4, MS.2632, J.S. Blackie Papers, NLS. In his *The Genealogy of Animals* (1896), Huxley, while maintaining that Darwin had confuted traditional teleology, argued for the need to 'remember that there is a wider teleology', a continuity of orderly becoming, according to definite laws', cited by J. Arthur Thomson in Supplement, *Nature*, 9 May 1925.

301. Row, 'Address to the Victoria Institute', in Charles Maurice Davies (1828–1910), *Heterodox London* (1874), II, 357. See also, for example, Edward Henry Bickersteth (1825–1906), quoted by T. Gouldstone, *The Rise and Decline of Anglican Idealism in the Nineteenth Century* (2005), 170.

302. Thomson, 'On Geological Time', *Transactions of Geological Society of Glasgow* 3 (1871), 1–28. Thomson continued to regard this as a formidable obstacle to natural selection, although not to the fact of evolution itself; see William Thomson to Rev. E. Davys, 5 October 1898, MS Kelvin LB5, University of Glasgow.

303. William Thomson to Phillips, 7 June 1861, JP/C/1861/10, John Phillips Papers, OMNH. See the discussion in Smith, *Science of Energy*, 183.

304. W. Thomson, 'Of Geological Dynamics', *Transactions of Geological Society of Glasgow* 3 (1871), 222.

305. Thomson to Duke of Argyll, 14 November 1897, Argyll Papers 1209/1620, Inveraray Castle; Thomson to T.G. Bonney, 27 April 1894, Bonney Correspondence, Royal Geographical Society, London.

306. Browne, *Darwin*, II, 314; Carus to Stokes, 10 November 1871, MS Add. 7656/1C/161, Stokes Papers, CUL. The fifth edition of the *Origin* included various compromises designed to allow for the speedier evolution required by Thomson's attack, Browne, *Darwin*, II, 315.

307. James Harrison Rigg (1821–1909), 'Pantheism', in C.J. Ellicott, ed., *Modern Scepticism: A Course of Lectures* (1871), 61.

308. See Argyll to Lyell, 28 October 1868, COLL 203/1/107–14, Lyell Papers, Edinburgh.

309. Gouldstone, *Anglican Idealism*, 102–3; P.B. Hinchliff, *Frederick Temple, Archbishop of Canterbury* (1997), 179–91. Temple's stance struck a chord with his contemporaries, for example, Francis T. Palgrave (1824–97), see the diary entry for 21 July 1870, in G.F. Palgrave, *Francis Turner Palgrave: His Journals and Memories of His Life* (1899), 134.

310. For example, James Bell Pettigrew (1834–1908), a Scottish anatomist, who was at best 'indifferent' to Darwinism, and followed more teleological approaches; see Rev. A.D. Sloan, 'Professor J. Bell Pettigrew: An Appreciation', *St Andrew's Citizen*, 8 February 1908.

311. Samuel Osborne Habershon (1825–89), 'Herveian Oration', *BMJ*, 30 June 1883. This was very much the position of the Christian Socialist J. Llewellyn Davies (1826–1916), whose sermons and essays in the 1870s and 1880s sought to reassure 'persons trained in [older] modes of apprehending divine action' (as he put it in 'Debts of Theology to Secular Influences' in his *Theology and Morality* (1873), 31), that the reorientations required by Darwinism were productive, *Theology and Morality*, 32–3, 141.

312. Argyll, *Reign of Law*, 218, and his strategic dissection of Darwin's use of terms which implied design in, for example, *Edinburgh Review* 116 (1862), 378–97. This continued to form part of his response through to the end of the century; telling Benjamin Kidd in 1898 that 'I have myself given up using the word "Supernatural" as essentially fallacious', Argyll to Kidd, 27 February 1898, MS Add. 8069/A/50, Kidd Papers, CUL. See also H.P. Liddon, *The Recovery of St Thomas* (1882). See discussion in 'Argyll, Race and Degeneration' in Edward Beesley, *The Victorian Reinvention of Race: New Racisms and the Problem of Grouping in the Human Sciences* (2010), 112–28.

313. Kingsley to Bates, 13 April 1863, AM18153, Charles Kingsley Papers, M.L. Parrish Collection, PUL, also Kingsley to Maurice, nd [1863], f.147, Add. MS 41297, BL; similar

160   DARWINISM'S GENERATIONS

sentiments to Darwin, see *Kingsley: His Letters and Memories*, II, 247–8. For the combination of intelligent creator and 'the actual interfering presence of a personal creator at every stage of its operations', see James Fraser (1818–85), Bishop of Manchester, *Manchester Guardian*, 3 October 1877.

314. See Gouldstone, *Anglican Idealism*, 100–1; see also J. Durant, 'Darwinism and Divinity: A Century of Debate', in Durant, *Darwinism and Divinity* (1985), 20.

315. See Lightman, *Popularizers*, 71–81, which talks about Kingsley 'co-opting Darwin'. For an interesting transitional teleology see Frank H. Hill (1830–1910), 'Gaps in Agnostic Evolution', *National Review* 26 (1895), 97–107.

316. Murphy: 'I regard his theory as not false, but totally insufficient', *Habit and Intelligence*, 332; 'an utterly inadequate account of the world', Jowett to R.D.B. Morier, 16 January 1873, in *Life and Letters of Benjamin Jowett*, II, 89.

317. From the outset Lewes was quite prepared to argue vigorously for Darwinian positions; see the account of a discussion with Owen at a private party at Owen's house where he 'pressed the professor so closely that I saw the (to me) well-known shadow of annoyance in his face', according to the 'Journals of Sir William White Cooper', Vol. II, f.82, MS0303/2, Royal College of Surgeons; yet is clear from his published writings that he was far from endorsing the *Origin* in its entirety, both as to specific components such as unity of origin, especially of man and animals, and natural selection itself, which was 'an imperfect explanation', for all its impetus to research, G.H. Lewes, 'Mr Darwin's Hypotheses', *FR* 3 (1868), 353–73, 364. In like manner, Eliot remarked that 'to me the Development theory and all other explanations of process by which things came to be, produce a feeble impression compared with the mystery that lies under the process', Eliot to Barbara Bodichon, 5 December 1859, quoted in A. Richardson, *After Darwin: Animals, Emotion and the Mind* (2013), 146; although Gillian Beer and others have brought out the strength of Darwinian themes underpinning novels such as *Middlemarch*, Eliot remained sufficiently cautious when revising Lewes' uncompleted *The Study of Psychology*, after his death, to remove many of the explicit references to Darwin, ibid., 157.

318. R. Payne Smith, 'On Science and Revelation', in Ellicott, *Modern Scepticism*, 540.

319. Joseph Lightfoot (1828–89) quoted in Henry Cotterill, *Does Science Aid Faith in Regard to Creation* (1883), 4; or Adam Milroy (1826–99) to the Perthshire Society of Natural History (1884), cited in Finnegan, *Natural History Societies*, 159–60.

320. See his extended correspondence with Gladstone, Gladstone Papers, Add. MS 44091, BL.

321. Matthew Forster Heddle (1828–97) to Geikie, 22 November 1879, Sir Archibald Geikie Papers, Gen 525/2, Edinburgh.

322. Michael Foster, 'Huxley', *National Review* 43 (1904), 436.

323. Foster, 'Huxley', 439.

324. Sir James Paget is an excellent example; see *Memoirs and Letters of Sir James Paget*, especially 407–9.

325. Asquith to Venetia Stanley, 20 February 1915, quoted in Glick, *What about Darwin*, 7.

326. White to Stokes, 4 September nd [probably 1894], 7 October [ny], MS Add. 7656/1W/595, Stokes Papers, CUL.

327. *Guardian*, 2 September 1868, cited in Ellegard, *Reception*, 83. Rolleston was distressed at the tone of the discussions and the lack of sensitivity to the difficulties of clerical scientists, Rolleston to Acland, 27 August 1868, ff.23–4, MSS Acland d.65, Bodleian, Oxford.

328. Browne, *Darwin*, II, 309, citing Marchant, *Raphael Meldola*, I, 221. Add.

329. As in the case of Richard Spruce, whose work on his Amazonian specimens in the 1860s made him 'more Darwinian than ever', see obituary by G. Stabler in *Transactions of the Edinburgh Botanical Society* 20 (1893/4–1895/6), 106.

330. W. Bagehot to Lady Lubbock, December 1866, quoted in J.E. Owen, 'A Significant Friendship: Evans, Lubbock and a Darwinian World Order', in A. MacGregor, ed., *Sir John Evans, 1823–1908* (2008), 216.

331. The appearance of his 'On the Physical Basis of Life' in the *Fortnightly* meant the edition required reprinting six times, Morley to Huxley, 12 November 1869, f.15, Box 23, Huxley Papers, ICL.

IN THE WAKE OF THE *ORIGIN* 161

332. Duke of Argyll to Harriet, Duchess of Sutherland, 18 October 1866, Argyll Papers, Inveraray Castle; George Stewardson Brady (1832–1921), Presidential Address, Tyneside Naturalists' Field Club, March 1871, in *Transactions of the Natural History Society of Northumberland Durham and Newcastle upon Tyne* 4 (1870–2), 279–90.

333. J.D. Hooker to Lockyer, 27 July 1869, Norman Lockyer Papers, EUL110, Exeter University.

334. Goodall, *Performance and Evolution*, 6 and *passim*; Sir Harry Hamilton Johnston (1858–1927) recalled overhearing the teachers of one of his private schools in the mid-1860s 'trembling…discussing the bearings of Darwin's *Origin of Species*', Johnston, *Story of My Life* (1923), 10.

335. See Trollope to John Ellor Taylor, 25 September 1868, in N.J. Hall, ed., *Letters of Anthony Trollope* (1983), I, 447, extracted in J. Secord, ed., *Charles Darwin: Evolutionary Writings* (2008), 227.

336. L.P. Jacks, ed., *Life and Letters of Stopford Brooke* (1917), I, 141. Brooke attended Huxley's lectures at Jermyn Street in 1861, despite their anti-theological animus; but his own pronouncements were also at times studiously vague; see the comments on a lecture in *Liverpool Daily Post*, 14 October 1875.

337. Isaac Whitwell Wilson (1833–81) to Adam Sedgwick, 9 October 1872, MS Add. 7652/II/HH/64, Sedgwick Papers, CUL.

338. Advert, *Morning Post*, 13 May 1867.

339. Lyell, ed., *Life of Sir Charles J.F. Bunbury*, II, 236; Hooker to Asa Gray [1868], JDH/2/22/1/26, Hooker, RBGK.

340. *Athenaeum*, 15 February 1868, widely reproduced in the provincial press.

341. [Dawkins], 'Variation of Animals and Plants under Domestication', 378–97.

342. Huxley to Darwin, 16 July 1869, DCP-LETT-6830, Darwin Correspondence; Huxley to J. Knowles, 15 October 1869, Huxley Papers, APS. In 1890 Huxley was still talking as if acceptance and respectability were a fairly recent development, see [Edward Clayton, son of Nathaniel Pegg Clayton], 'Conversation with Professor Huxley at the Grand Hotel, Eastbourne, Monday September 29, 1890, 6 o'clock to 6.30 pm', ff.220–6, Box 12, Huxley Papers, ICL.

343. See comments on the opinion of his brother J.B. Reade in [George Reade?] to Henry Lee, 15 September 1870, MS5392/17, Wellcome Library.

344. Hutton to Lubbock, 1 January 1869, ff.1–2, Avebury Papers, Add. MS 49642, BL.

345. Horace Waller to Lee, 23 December 1869, MS 5396/11, Wellcome Library; Bentham feared the outcomes of the deep-sea dredging 'showing the actual survival of representatives of organisms elsewhere long since extinct', Bentham to Gray, 27 February 1870, Asa Gray Correspondence, Harvard.

346. B. Lightman, 'The "Greatest Living Philosopher" and the Useful Biologist: How Spencer and Darwin Viewed Each Other's Contribution to Evolutionary Theory', in Hesketh, ed., *Imagining the Darwinian Revolution*, 37–57.

347. Bowerbank to Lee, 24 October 1871, MS5377/70, Wellcome Library. John Hancock (1808–90), Newcastle naturalist, corresponded with Huxley for over twenty years straddling the publication of the *Origin*, without mentioning Darwin once.

348. See *Letters of Adam Sedgwick*, II, 440. John Phillips, in his obituary of Sedgwick, stressed his opposition to Darwinian theory 'against which he not only used a pebble but great use of his heavy hammer', *Nature*, 6 February 1873. To the end in his Cambridge lectures Sedgwick gave 'solemn warning' against the spread of Darwinian doctrines, apparently moving his undergraduate audience to tears with his eloquence on the wonders of creation, see *Medical Times and Gazette*, ns 35 (1867), 689, and George Dawson Rowley to Newton, 30 November 1866, MS Add. 9839/1R/391, Newton Papers, CUL.

349. Wyld to Wallace, 10 May 1870, ff.171–5, Wallace Papers, Add. MS 46435, BL.

350. Response of R.W. Church to a lecture of Tyndall, Anne Mozley, ed., *Letters of J.B. Mozley* (1885), 296.

351. *Oxford Times*, 8 August 1868. For example, Joseph Crompton (1823/4–78), Presidential Address for 1871, *Transactions of the Norfolk and Norwich Naturalists' Society* I (1869–74) (and Frederic Kitton (1827–96) in the same volume).

162 DARWINISM'S GENERATIONS

352. R.E. Prothero, *Life of A.P. Stanley* (1885), II, 496.
353. See Hannah Gay, '"The Declaration of Students of the Natural and Physical Sciences" Revisited: Youth, Science and Religion in mid-Victorian Britain', in Richard Feist and William Sweet, eds., *Religion and the Challenges of Science* (2007), 19–38.
354. Rolleston to W.R. Browne, 3 December 1880, IMS 461/1/43, Institute of Mechanical Engineers.
355. Rolleston to Huxley, 4 January 1865, ff.171–4, Box 25, Huxley Papers, ICL. This was not an entire break, see his rather incoherent letter to Lubbock, 23 June (or January) 1865. ff.53–4, Avebury Papers, Add. MS 49642, BL. See also his concern at the anti-religious tone of the Norwich meeting of the British Association, Rolleston to H.W. Acland, 27 August 1868, ff.23–4, MSS Acland d.65, Bodleian.
356. *Leeds Mercury*, 10 January 1866, although when Rolleston gave this lecture in Newcastle in late 1864, he told Huxley that he 'gave them so much of Darwin that the question was "Was I an Atheist or a Unitarian?"', Rolleston to Huxley, 1 January 1865, ff.166–7, Box 25, Huxley Papers, ICL.
357. *BMJ*, 15 August 1868.
358. Rolleston to Darwin, 20 April 1870, DCP-LETT-7168, Rolleston to Darwin [after 22 February 1871], DCP-LETT-7521, Darwin Correspondence. For another example of his rather oblique positioning around this time, see his letter to the *Eastern Daily Press*, 8 May 1872.
359. For example, see Rolleston to Marsh, 9 July 1880, 1399–49, Box 28, Folder 1183, Othniel C. Marsh Correspondence, Yale. Interestingly, for someone like Bartle Frere (1815–84), Rolleston was aligned with Acland and Ruskin, and potentially even older figures like Sir Benjamin Brodie; see Bartle Frere to Acland, 17 June 1871, ff.227–36, MSS Acland d.77, Bodleian.
360. Huxley to Thiselton-Dyer, 22 November 1880, f.164, W.T. Thiselton-Dyer Papers, In-letters II, RBGK.
361. Seeley to Adam Sedgwick, 24 May 1868, MS Add. 7652/II/GG20, Sedgwick Papers, CUL. For Seeley, who was influenced as a student by Owen, and acted as Sedgwick's assistant, see A. Desmond, *Archetypes and Ancestors* (1982), 135–7, 186–201.
362. Tristram to Newton, 11 February 1865, MS Add. 9839/1T/213, Newton Papers, CUL. This didn't stop Osbert Salvin referring to 'a most Darwinical letter from H.B. T[ristram]…A backslider returned to the faith', Salvin to Newton, 25 March 1865, MS Add. 9839/1S/147, Newton Papers, CUL. (For another suggestion of Tristram as 'accepting the theory outright'. Edward Clodd, 'Darwinism up to Date', *T.P.'s Weekly* 11, 18 March 1910, 301.)
363. Tristram, 'Presidential Address', *Transactions of the Natural History Society of Northumberland, Durham and Newcastle upon Tyne*, 8 (1880–9), 1–22, 18; Newton, 'Henry Baker Tristram', *Proceedings of the Royal Historical Society*, Series B, 80 (1908), xlii–xlv.
364. John Bland-Sutton's later recollection was that Flower 'never entered heartily into the Darwinian controversy', Bland-Sutton, *Story of a Surgeon* (1930), 99. See also Flower to Randall Davidson, 1889, asking that all scientists not be tarred with same brush as Huxley and Tyndall, G.K. Bell, *Randall Davidson* (1935), I, 154.
365. Wollaston to Lyell, 20 March 1867, COLL 203/1/6189–92, Lyell Papers, Edinburgh.
366. See his letter to J.H. Balfour, 15 December 1873, T103, Hutton Balfour Papers, RBGE.
367. See comment in his diary on his amusement at seeing an article in favour of evolution in an SPCK magazine, Diary, 30 October 1872, Wr./D/17, University of Nottingham; compare his earlier approval for the orthodox natural theology of a lecture by Rev. Thomas Hincks (1818–?), Diary, 19 February 1861, Wr./D/6 and welcome of Sir William Armstrong's speech at the BA against the Darwinian theory, 1 September 1863, Wr./D/8, University of Nottingham.
368. Christopher Martin, *Lawson Tait: His Life and Work, with Personal Reminiscences* (1931), 8. See Tait's 'Progress of the Doctrine of Evolution', *Proceedings of the Birmingham Philosophical Society* 5 (1885); in the mid-1890s his wife recalled that pro-Darwinian

IN THE WAKE OF THE *ORIGIN* 163

lecturing at this time was 'neither easy nor safe', Sybil Anne Tait (1845–1909), 'The Rights of Animals', *The Animals' Friend* 2 (1895–6), 195–8.

369. See John Brown (1810–82) to Jessie Crum, 11 August [1871], in J. Brown, ed., *Letters of Dr John Brown* (1907), 204.

370. Cookson to Adam Sedgwick, 9 August 1871, Sedgwick Papers, MS Add. 7652/II/HH/26, CUL; Lubbock who remarked that 'From all accounts our section at Edinburgh was not particularly strong', Lubbock to E.B. Tylor, 16 August 1871, Lubbock Papers, B L961, APS, while Ray Lankester, while prepared to describe the meeting as a whole as 'fair', felt the zoology 'weak', and 'the anthropologists were let loose in the most extraordinary manner', Lankester to Alfred Newton, 8 July [1871], MS Add. 9839/1L/13, Newton Papers, CUL.

371. *The Sphere*, 18 December 1909, in a discussion of how Edwardian practices of library 'censorship' might have affected the reception of the *Origin*; exactly the same point was made by Peter Price (1825–92) in recollections in 1884, *South Wales Daily News*, 25 January 1884; see the similar tone of H.W. Massingham's recollections of Norwich in the 1870s, in which Darwinism was an odd affectation, largely confined to the idiosyncratic manufacturer, F.W. Harmer, 'Norwich as I Remember', *The Sphere*, 2 February 1924.

The Reception of Darwinian Evolution in Britain, 1859–1909: Darwinism's Generations. Martin Hewitt, Oxford University Press. © Martin Hewitt 2024. DOI: 10.1093/9780191982941.003.0003

3

The *Descent of Man* and the High Victorians

One reason for the popularity of the Darwinian theories as generally understood, was that they represented the secular process as a glorified Cup Tie competition with the mammoth and the ichthyosaurus disappearing in the qualifying rounds, and man emerging triumphantly from the final—in contrast with the unsportsmanlike theories of creation, in which man got his post by a job. (A.H. Sidgwick, *Walking Essays* (1912), 114.)

The 1860s were a period of phoney war in the debates over Darwinism and humanity. The great antiquity of man was becoming indisputable, but Darwin's initial reticence in the *Origin* meant it was possible to sidestep the corollary that man evolved with the rest of nature. And it was not until Tyndall's Belfast Address, in 1874, that the extent of the materialist claims of the new biology became inescapable.[1] Tyndall had always been something of a loose cannon, kept away from such prominent platforms for as long as possible. Now he discharged a volley of intellectual grapeshot with predictably bloody effect. Of course, the implications of Darwin's argument had been clear enough to anyone prepared to look carefully. The anthropologist E.B. Tylor's Christmas festivities for 1867 were said to have included an impromptu 'lecture on the Darwinian theory', in which he 'develop[ed] [him]self upward from the gorilla stage, as an illustrative experiment'.[2] Even so, it was disingenuous of Huxley to claim that Darwin's opinions were perfectly obvious, and that, for years after the publication of the *Origin of Species*, one could hardly go to a dinner party without hearing of them. Throughout the 1860s the Darwinians had laboured strenuously to avoid any formal acknowledgement of human evolution, and even Huxley's own *Man's Place in Nature* had dwelt on the fact that humanity alone demonstrated intelligible and rational speech.[3]

The publication of *The Descent of Man and Selection in Relation to Sex* in February 1871, and its companion study *The Expression of the Emotions in*

Man and Animals the following year removed all doubt. There was a sudden intensification of engagement in Darwin's ideas. Indifference was replaced by investigation. Bystanders became disputants. 'Man', Darwin now asserted bluntly, 'is descended from a hairy quadruped, furnished with a tail and pointed ears, probably arboreal in its habits'; humans shared with animals emotions and ways of expressing them, and evolutionary dynamics applied not just to biological man, but also to human culture and intellect. The theory of sexual selection enabled Darwin to meet some of the challenges which had been made to the *Origin* in respect of apparently non-adaptive features, such as the colourful plumage of some birds. But *Descent's* central message, reinforced the following year by his study of emotion in man and animals, made clear Darwin's belief that evolution accounted not just for man's physical frame, but also for rational thought and moral sensibility, denying religious Victorians even the ability to draw a line between body and mind.

If *On the Origin of Species* had sought to effect a scientific revolution, *Descent* and *Expression* initiated an even more far-reaching intellectual one. As Edward Royle has noted, 'Biology itself, and even theories of evolution in the animal kingdom, were compatible with theology; it was the new anthropology, concerned with the evolution of man, which was not'.[4] It is telling that in the early twentieth century, when the labour leader John Bruce Glasier (1859–1920) described the intellectual reading of socialist miners, alongside Ruskin's *Unto this Last*, and Spencer's *Study of Sociology*, it was *Descent of Man* rather than the *Origin* which represented Darwin.[5] Humanity as an entirely evolved organism, especially if its conscience was merely the product of accidents of natural selection and the shifting environment, could hardly be made in the image of God.

At the same time, the empirical evidence for the earth's evolutionary past continued to accumulate. Gaps in the fossil record were being filled in. Evidence of the extraordinary recapitulations of embryological growth were filled out. Discoveries in America in particular enabled for the first time the construction of complete evolutionary sequences, including the horse, an example that Huxley lectured on at the Royal Institution in 1870, and returned to repeatedly as the decade progressed.[6]

As a result, from the early 1870s, the terms of the debate began to shift. Although the short-hand use of 'Darwinism' for the fact of evolution and common descent remained, increasingly it became possible to oppose 'Darwinism' while accepting, admittedly in varying degrees, evolution. In some instances, Darwinism began to signify more specifically a belief in the

166 DARWINISM'S GENERATIONS

entirely natural evolution of humanity and its anthropological connotations, including the natural origins of religion, in ways which squeezed out the careful ambiguities of the *Origin*. At the same time for others, the theological challenge of evolution (and often of 'Darwinism') was neutered by the overlay of some form of providential plan or the insertion of evolutionary processes which undercut the primacy of natural selection. As interest intensified and the issues at stake escalated, we move into a new, more ecclesiological phase, in which alongside the theoretical debates more attention was given to defining more precisely the limits of Darwinian thought, to the policing of evolutionary orthodoxy, and the excommunication of apostates.

Immediate Responses

Inevitably, public interest in *Descent* was intense, if not as febrile as for the *Origin*. Four thousand copies were sold in the first six weeks and 14,000 by the end of the decade. Aided by the novelty of its photographic illustrations, *Expression* did even better, selling 9,000 copies in the first four months alone. There was a noticeable quickening of intellectual interest. Figures hitherto prepared to avoid a close engagement with Darwin's ideas now felt compelled to give them serious attention. Henry Liddon, who had managed not to make a single reference to Darwin in his diaries across the entire decade, or to discuss evolutionary issues in his Sunday evening lectures, subjected *Descent* to close attention and annotation.[7] Benjamin Jowett and Florence Nightingale engaged in an urgent exchange of letters discussing the status and implications of its arguments. It was said that *Descent* competed in the drawing room with the latest novels, encouraging dinner party discussions of whether the shapes of ladies' ears vindicated Darwin's discussion of rudimentary survivals.[8] London's showmen were quickly in on the act. By the end of May 1871 one Piccadilly impresario was recommending his 'Lilliputian Queen' as a specimen by which to consider Darwin's thesis, and knowledge of the arguments of *Descent* had spread sufficiently widely by the autumn for it to be lampooned in the London music halls.[9] Exceptionally large crowds flocked to the Zoological Gardens, and the following year a visitor to the monkey show at the North Woolwich pleasure gardens reported that he heard Darwin's theory very freely discussed.[10] *Descent*'s disturbing power is visible in the extraordinary suicide of Lemuel Howard (1839/40–71), a tramping stonemason, who jumped to his death

THE *DESCENT OF MAN* AND THE HIGH VICTORIANS   167

from the Eden Bridge in Carlisle in October 1871, leaving a journal in which he ruminated on the meaninglessness of life, and the potential horrors of old age, 'Well may Darwin—yes, *Mr. Darwin*—call us naturalised monkeys. You will all be monkeys when you can't call yourselves to recollection.'[11]

Despite this incident, by and large initial responses were a great deal more measured and temperate than had been the case for the *Origin*. One anonymous annotated copy of the first edition of *Descent* griped about the 'absurd and meaningless expressions' concerning the origins of reason: 'How tortuous the way man will travel to avoid giving thanks to the living God.'[12] There was the inevitable rush of pulpit denunciations.[13] A few impatient outbursts were tossed in Darwin's direction, including J.S. Blackie's furious denials of his brutish ancestry at the Edinburgh Philosophical Society, J.H. Stirling's comparison of *Descent* with the 'peculiarity of the mad to tear off their clothes, and cavort their nakedness', and a clever piece of doggerel by Charles, Lord Neaves (1800–76), which appeared in *Blackwood's Magazine*, and was widely reprinted.[14] In general, though, there was little of the visible anguish of 1859. Religious readers were quick to seize not just upon the explicit argument about human evolution, but on Darwin's retreat from the sops to divine responsibility which had been a feature of the *Origin*. But the comic writer F.C. Burnand (1836–1917) saw no dangers in regaling the visitors to the bazaar of the National Orthopaedic Hospital in June 1871 to a lecture on *Descent*, repeated every ten minutes until he was exhausted.[15]

Early reviews were mixed. Predictably enthusiastic summaries were offered by the *Saturday Review*, *Westminster Review*, and the *Pall Mall Gazette*. Otherwise, many reviewers opted to play safe with a non-committal rehearsal of *Descent*'s central arguments (although doubts frequently poked through); but the *Times* offered a thorough survey of the weaknesses of Darwin's position, and a number of other titles followed suit.[16] Not only was there still plenty of evidence of out-and-out rejection, and not just in the religious titles, but the critical balance was markedly against the evolution of man from animals.[17] Unsurprisingly, *Land and Water* offered an unremittingly hostile review, flanked by several squibs from 'Frank Buckland's monkey', but even the *British Medical Journal* concluded that by including man in his evolutionary scheme Darwin had simply gone too far.[18]

Where authorship can be ascertained, the generational pattern offered continuity, but also progression. There were fewer early-Victorian blasts, and the predominating voice was now mid Victorian. Mivart in the *Quarterly Review* spoke of his 'mingled feelings of admiration and

168   DARWINISM'S GENERATIONS

disappointment', seizing on Darwin's various modifications of some of the more speculative elements of the *Origin* and suggesting that the same fate awaited 'sexual selection'. In the *Contemporary Review* the Principal of Edinburgh University, Alexander Grant (1826–84), urged the clearly unprogressive nature of 'savage' humanity as evidence of the necessity of a break in the evolutionary ladder, a point eagerly seized upon by other commentators.[19] Apart from W.S. Dallas (hardly a disinterested voice) in the *Westminster*, the more positive responses seem to have come from high Victorians. John Morley's long review in the *Pall Mall Gazette* presented *Descent* as 'one of those rare and capital achievements of intellect which effect a grave modification throughout the highest departments of the realm of opinion'.[20] For some high Victorians, religious commitments continued to make acceptance of the evolution of man challenging.[21] In general, though, high-Victorian readers were inclined to side with Darwin rather than deploy the sorts of criticism offered by Mivart.[22]

On the provincial platform the register was shriller. YMCA lecturers sneered; older naturalists scoffed. Great play was made of the ludicrous image of monkeys wearing out their tails and turning into men. 'If it could be showed that any human being existing anywhere had ever possessed a tail, it would be a most important fact', mocked one mid-Victorian lecturer.[23] Another warned of the scientific scepticism of which Darwin was the greatest leader, rejoicing that in *Descent* he 'had to retract statements that his followers relied upon'.[24] The writings of Mivart and Müller, distinguishing between reason and instinct, and between animal cries and human language, were exploited extensively to dispute the central pillars of Darwin's argument.[25] Across the country, mid Victorians who had hitherto stood aloof mounted the rostrum to proclaim that they had read *Descent* and found Darwin's case 'not proven'.[26]

A similar generational contrast emerges in private. Early Victorians blenched. Pusey found *Descent* distressing, and expressed the hope that God might raise up naturalists willing to 'destroy the belief in our apedom'.[27] Although fascinated by the discussion of sexual selection, Sir Charles Bunbury confessed himself unconvinced as to the origins of man.[28] The sanitary reformer, John Sutherland (1808–91), exulted in private at the 'squelching of Darwin' by the reviewers.[29] Charles Bree read *Descent* with meticulous care, leaving annotations on more than fifty of the openings of his copy. Many of these use exclamation marks to indicate Bree's dissent, or to draw attention to Darwin's own concessions as to the weakness of the evidence for his argument. But Bree was also prompted to express his

THE *DESCENT OF MAN* AND THE HIGH VICTORIANS   169

frustrations by posing questions to Darwin: where is the man-like animal who could throw with precision that Darwin conjures, how does Darwin's theory of the origin of the Roman numeral VI as a single hand of digits plus one tally with the rendering of IV; and ultimately, and especially in respect of the sections on sexual selection, he was driven to outright dismissal: Darwin's musings on the origins of vocal organs are brushed aside at several points as 'Twaddle', while his discussion of the mating patterns of wild birds and the ways sexual selection operate to privilege the more healthy pairs is dismissed as 'All Bosh'.[30]

Charles Kingsley was mortified, but determined not to oppose publicly: 'I cannot say anything', he lamented, 'if I must say anything against the dear old sage'.[31] The annotations in his copy of *Descent* make it clear that Henry Liddon read it very much against the grain, marking the passages in which Darwin made his characteristic concessions to the strength of the opposing argument, especially of the enormous difference in the mental capacities of humans and animals, and the lack of an animal moral sense.[32] Younger readers were less resistant. The circle of Cambridge philosopher Henry Sidgwick (1838–1900) found *Expression* very entertaining; in private John Morley pronounced it 'very amusing, although also 'awfully daring in its leaps from solid ground'.[33] As for *Descent,* one of Darwin's high-Victorian correspondents told him that it was 'as strong as iron and clear as crystal'.[34] The historian J.R. Green (1837–83) exulted at the wonderful vistas of enquiry the book would open up, while his fellow historian W.E.H. Lecky found it powerful and plausible, its theme 'a most noble one and the promise of a great future for the world'.[35]

The Persistence of Denial: Early Victorians

For the rest of the decade, and indeed beyond, older readers responded to the putative common ancestry of humans and apes with an uncomprehending vitriol which underlined their alienation from the shifting terms of the debate. Their responses were frequently more ones of visceral contempt than argued contestation, happy to evade scientific engagement entirely in favour of *reductiones ad absurdum*. Evolutionary change was rendered ludicrous, as in the off-hand musing of William Howitt (1792–1879) on the potential for flies to develop 'some millions of ages hence,...into men and women'; they 'would have dropped their wings and a lot of legs, and instead of eating one another, have begun to eat sheep'.[36] 'What monkey ever wrote

170   DARWINISM'S GENERATIONS

an epic poem?', or appeared as a lecturer on Aristotle, asked J.S. Blackie.[37] Older readers felt themselves out of time, increasingly forced to endure evolution as a fashionable topic for after-dinner talk in which their denials were dismissed as antiquated.[38] But we cannot write them out of history too quickly. Especially in the provinces, the early-Victorian naturalist, delighting in nature as it was, without enquiring too much about how it came to be, remained a characteristic presence.[39]

Most early Victorians remained steadfast in their denial of Darwin's evolutionary thought. In a series of interventions before the Victoria Institute, the chemist John Eliot Howard (1807–83) affirmed the order, design, and fixedness of nature, unwilling even to countenance the Duke of Argyll's attempts at resolution via the *Reign of Law*.[40] Thomas Cooper's popular lectures, published as *Evolution, The Stone Book, and The Mosaic Record of Creation* (1878), persisted with attempts to reconcile geology and Genesis.[41] In comparison to the trickle of the previous ten years, the 1870s saw a torrent of early-Victorian attempts to refute Darwin in print. Francis Orpen Morris published a number of interventions, including four editions of *All the Articles of the Darwinian Faith* between 1875 and 1882, and was vigorously supported by Thomas Birks and Charles Bree.[42] In their varying ways these works reiterated not just the early Victorians' challenges to natural selection (in the case of Bree's *Exposition of the Fallacies in the Hypothesis of Mr Darwin* (1872) this alone stretched across sixteen consecutive chapters), the struggle for survival, and the common ancestry of all living things, but also their willingness to take on pangenesis and sexual selection, their insistence on acts of special creation, their affirmation of the absolute distinction between man and beast, and the persistence of their characteristic modes of argument. Over and above the lack of evidence, the falseness of its reasoning, it was the inevitable consequences of an acceptance of evolution—the denial of God—which provided the ultimate rationale for its rejection.[43] And these were just the tip of the iceberg: at times it seems that almost every early-Victorian author felt impelled to have their say on the Darwinian question, whether they were cleric or sportsman, whether they were writing about geology or logic; all took advantage of publication to take a swipe at Darwin, rarely progressing beyond *ad hominem* condemnations, or religious rants at Darwin's 'damnable doctrines'.[44]

E.B. Pusey never lost his sense of the scientific weakness of Darwinism 'in that continual "perhaps", probably, possibly, it may be,...[which] lies at the surface of all Darwin's books'.[45] His sermon in the University Church in 1878, published as *Unscience, not Science, Adverse to Faith*

THE *DESCENT OF MAN* AND THE HIGH VICTORIANS 171

created a sensation.[46] Pusey's contemporaries exulted. Bonamy Price (1807–88), Professor of Political Economy at Oxford, described the sermon as following exactly the lines on which he had always opposed the scientific atheists.[47] Younger hearers were dismayed. One undergraduate later recalled 'the sick feeling with which I read that pronouncement. It suffocated one's hopes. It was an ultimatum, on the part of the Church, to a foe which already wore the unmistakable mien of victory.'[48] When the sermon was published, with the addition of notes he described as conveying an entirely erroneous idea of the theory of evolution, another young listener, the botanist Henry Nicholas Ridley (1855–1956), was moved to remonstrate with Pusey in private.[49]

Unlike Pusey, after his public humiliations of the early 1860s, Richard Owen largely withdrew from evolutionary controversies, maintaining his precarious balancing act of denigrating Darwin while seeking space for his own alternative evolution.[50] Owen retained sufficient influence, despite the efforts of Huxley and his allies, to carry to successful conclusion his plans for a new natural history museum at South Kensington, producing a monument, as John Holmes has demonstrated, to a determinedly pre-Darwinian sensibility, with its presentation of species as distinct types, its separate gallery for humanity, and its obstructed movement through the evolutionary series.[51] For the late-Victorian zoologist Frank Beddard (1858–1925), Owen was a man who had lived beyond his time, and whose ideas did not find favour with 'the younger generations of naturalists'.[52] But for older anti-Darwinians he continued to represent an active force. His portrait hung above the mantelpiece at Frank Buckland's Museum of Fisheries in London, a source of inspiration to the end.[53]

Owen's studied ambiguities signal the difficulties of situating the responses of the older generations to Darwinian ideas as the intellectual tide turned against them. W.E. Gladstone (1809–98) offers another case. Gladstone's initial response to the *Origin* seems to have been to circumvent the effort needed to comprehend its biology by focusing instead on the geological challenge to the creation narrative, and on the lack of any direct evidence of species transmutation. As late as 1874 he abjured any opinion on evolution, 'except that the results assigned to it were unwarrantable'.[54] But never one to be accused of lazy rhetoric, he was finally prompted by *Descent* to make a serious effort to understand Darwinism, reading Mivart's interventions of the early 1870s and emerging puzzled at the 'truly portentous' readiness of people to jump to ulterior conclusions neither hinted at nor required by Darwin's writings.[55] Ultimately, like most of his contemporaries,

172 DARWINISM'S GENERATIONS

he remained confident that Darwin would not sweep away that fabric of belief which had stood the test of 1,800 years, but his insurance was to revert to what he described as his own 'loose and inexact way of looking at Evolution'.[56]

For the early Victorians, in the minority of cases where concession was made, it was reluctant and highly conditional. Hence by 1877, the physiologist William Sharpey (1802–80), was prepared to concede to his friend Allen Thomson that 'your Darwinism (greatly limited) may be right'.[57] Significantly, those early Victorians who did move towards acceptance of evolution often did so with Lamarckian rather than Darwinian frames, although frequently the precise nature of the positions being adopted remained mired in spurious semantic niceties.[58] Take the conchologist John Gwyn Jeffreys (1809–85), interesting as someone who retained a relatively prominent position in the institutions of British science, helping to found the Marine Biological Station which later played a not insignificant role in the development of evolutionary biology. In 1873 Jeffreys told friends that he 'believe[d] not in evolution, but in descent with modification', while at the British Association in 1877 he gave a remarkable address in which, while disclaiming any commitment to fixity of species or successive creations, he also described evolution 'in its modern interpretation' as a hasty theorisation, concluding that many centuries would have to pass before the subject could be properly understood.[59]

Any Evolution but Darwin's: Mid Victorians

The move during the 1870s towards a greater weight of mid-Victorian voices, unsympathetic to the fulminations of Blackie or the uncompromising resistance of Birks, might have been expected to produce a decisive shift in the tone of public debate. That it didn't is partly because there was no sudden disappearance of the early Victorians, and partly because the incorporation of humanity into the evolutionary narrative and the spread of materialist approaches intensified mid-Victorian anxieties as well. *Descent*, by shifting the focus of contention towards the narrower questions of human evolution and the barriers between humans and animals, perhaps made the acceptance of some form of non-human evolution easier, but it also made the conflict over 'Darwinism', now applied more particularly to that version of evolutionary thought which treated the origins of humanity as entirely part of the wider evolution of the natural

THE *DESCENT OF MAN* AND THE HIGH VICTORIANS   173

world and by implication repudiated any place for the divine, more bitter. As a result, a considerable number of mid Victorians gravitated into the anti-Darwinian camp even as they came to some sort of acceptance of evolution as a historical process.[60]

If the Duke of Argyll had been the most prominent mid-Victorian voice in the 1860s, St George Jackson Mivart and the group known as the 'North British Physicists', William Thomson, the physicist Balfour Stewart (1828–87), and Peter Guthrie Tait (1831–1901), Professor of Natural Philosophy at Edinburgh from 1860 to 1901, were central to mid-Victorian reshaping of the Darwinian debates of the 1870s. Mivart's overt challenge to 'Darwinism' and the personal animosity of the North British Physicists to the X-Clubbers should not distract attention from the relatively unambiguous commitment to the fact of some sort of evolution which they all came to offer. As Tait and Stewart reiterated in their *Paradoxical Philosophy* (1878), their problem was not with evolution per se, but with 'the abuse of this theory by mechanical bigots'. Their *The Unseen Universe, or Physical Speculations on a Future State* (1875) rehearsed the core mid-Victorian position that Darwinism could not account for the production of the visible universe, of life, or of man. They formed an important nucleus of resistance, especially in Scotland. Rooted in faith and personal rivalry as it might have been, their influence was nonetheless considerable. During the 1870s Thomson and Tait furnished the huge readership of magazines such as *Good Words* with digestible arguments against Darwinism, and introduced cohorts of students, whose notable members included Henry Drummond (see below), to the ways in which energy science could be used to check the rise of scientific naturalism by placing limits on geological time, and by positing the need for a First Cause.[61] This challenge became an important component of the analysis of many mid-Victorian scientists.

Mivart was a far better biologist than Argyll, and steeped in Darwinian science he could not be so easily dismissed by his opponents. His *On the Genesis of Species*, which appeared just before *Descent*, and its follow up *Man and Apes* (1873), which together re-presented many of the established mid-Victorian challenges, made him a standard point of reference for early and mid Victorians seeking authority for their doubts, and the focus of much of the controversial effort of Huxley and his allies.[62] In his *Scientific Bases of Faith* (1873), and a second edition of *Habit and Intelligence* (1879), the Belfast industrialist Joseph John Murphy (1827–94) doubled down on the various conventional arguments against natural selection, not least the idea taken up, especially by the mid Victorians, that not all incipient

174 DARWINISM'S GENERATIONS

variations were useful. Behind them followed a phalanx of mid Victorians who combined varied forms of evolutionary belief with an anti-Darwinian rhetoric, including Max Müller (in his *Introduction to the Science of Religion* (1873), and the final volumes of his *Chips from a German Workshop*) and a fresh wave of religious literature which served up criticism of Darwin in packets of proselytism, such as C.J. Ellicott's *Modern Scepticism* (1871) and Edward White's *Life in Christ* (1878); no longer challenging evolution per se, but equating Darwinism with materialism and chance, and with the absorption of humanity completely into the evolutionary narrative, all of which were strenuously denied.[63] Sales of titles of this sort provide evidence of the continued resonances of anti-Darwinian rhetorics, and they furnished valuable resources for sceptical readers of all stripes. The 'danger of compromise with Evolutionists', as Edward White put it, remained much on the mid Victorians' mind.[64]

For a minority of mid Victorians, these religious anxieties helped to sustain an absolute refusal to concede ground. Of these, perhaps the most prominent figure during the 1870s was the Harley Street doctor Charles Elam, whose anti-evolutionist writings included *Winds of Doctrine; being an Examination of the Modern Theories of Automatism and Evolution* (1876), *The Gospel of Evolution* (1880), and a series of articles in the *Contemporary Review* in the 1870s. Elam did not deny the *possibility* of evolution, and did not fall back on conventional concerns at the fossil record; instead he argued that the lack of evidence of the operation of both natural selection, and of the capacity of even selective crossing to create new species, meant that even detailed phylogenies could only suggest, but not demonstrate, evolution. Elam's obituaries suggested that despite the progress of acceptance of evolution, he still 'succeeded in convincing many'.[65] Otherwise, the most vocal mid-Victorian antipathy both to Darwin and to evolutionary history came, significantly, from outside the metropolis, in the person of the Canadian geologist J.W. Dawson, whose writings were widely circulated in Britain.[66]

We must not allow the shrill special pleadings of a small minority of conservative clerics and medics to disguise the fact that many mid Victorians increasingly simply detached themselves from evolutionary debates, or discussed evolution as a largely unthreatening hypothesis which could be accommodated without any fundamental intellectual upheaval, accepting that it might become a demonstrated scientific fact at some point in the future, even if for the present proof still hung in the balance.[67] For some, this continued to be a matter of indifference rather than denial.

Leslie Stephen's elder brother, James Fitzjames Stephen, was typical of a strand of classically educated mid Victorian who 'cared very little for what may be called the scientific argument. He was indifferent to Darwinism and to theories of evolution. They might be of historical interest, but did not', as his brother put it, 'affect the main argument'.[68] For others, like the popularising naturalist Rev. J.G. Wood, carefully avoiding having to take up a clear position on the topic was an understandable part of his desire to create as wide as possible an audience for his lectures and publications.[69] 'I have read almost all of Darwin's books', Charles Swainson (1820–87), Master of Christ's College, Cambridge, told a leading Darwinian in 1874, and 'his hypothesis or hypotheses as such do not in any degree terrify me'.[70] This didn't make Swainson any the less unconvinced; Darwin's arguments, he went on to observe, were too narrow and inclined to misinterpret the evidence. In January 1878, Caroline Haddon (1827–99) (mother of the anthropologist A.C. Haddon) paid a visit to Mrs Turner, their minister's wife, after one of Haddon's lectures on evolution: 'She did not like the Darwinian theory', Caroline reported, 'and thought from what she had read of his writings that he rather jumped at conclusions. I really do not understand the question sufficiently to argue and I told her so. I may not see my way to Darwin's conclusions, but I cannot see that the principle of evolution drives God from his universe'.[71]

For most mid Victorians, the temporising stance they had adopted to the *Origin* and Darwinian evolution left them open to a gradual shift in position as the evidence of evolution accumulated. But often this emerged as much despite Darwin as because of him, less a biological revelation than a geological recognition of the infilling of the sequences of progression in the fossil record and the extending evidence of the structural similarities of all life. Take the early English feminist Anna Swanwick (1813–99). One of the first female members of the Royal Institution in 1858, her commitment to 'the splendid discoveries known under the general name of Darwinism', despite the initial shock to her firm religious beliefs, was clearly predicated on the antiquity and chains of life demonstrated by geology rather than on any engagement with Darwin's biological mechanisms.[72] Mid Victorians who had sought to uphold the fixity of species found the evidence of change increasingly unanswerable.[73] Some sought to hold the line at belief in a series of successive creations, perhaps recognising limited change within each era.[74] Slowly, and with reluctance, most went further, towards a cautious and contingent acknowledgement of some form of evolutionary past, coupled with a vigorous reservation of a wider teleology, and a refusal to

176 DARWINISM'S GENERATIONS

identify with 'Darwinism.[75] And here again, the concessions were often convoluted and inconsistent.

Although there were a few exceptions, especially amongst the younger mid Victorians such as John Burdon Sanderson (1828–1905), who adopted a Darwinian position of sorts, mid-Victorian medical men, especially outside the universities, showed the greatest reluctance.[76] The prickly Professor of Physiology at King's College, Lionel Beale (1828–1906), illuminates the contortions that could result. Although a dogged critic of the materialist tendencies of the Darwinians, and a prominent Darwinian sceptic in the early 1870s, by the early 1880s Beale had accepted the palaeontological evidence of deep time, and processes of descent and derivation, and indeed the attraction of evolutionary ideas compared to those of sudden creation. But alongside this he retained a commitment to continuity of forms, saltationary changes, a refusal to accept 'endless modification', and a general dismissal of the absurdity of 'evolutionary theory'.[77] The entomologist T.V. Wollaston's work in the 1870s progressed as far as allowing that variation might have 'full play', and species be 'indefinitely plastic', but insisted they still needed to remain true to their type. Wollaston and Thomas Davidson, who spent forty years working on the brachiopods, Palaeozoic marine animals which survived largely unchanged to the contemporary period, both reflected a marked tendency to row back from direct denial to take refuge in dislike of 'speculation', or the inadequacy of evidence 'with respect to "What is Life", "The Origin of Species", "evolution"'. Though careful in his scientific memoirs not to draw directly hostile conclusions from the limits of the palaeontological evidence, Davidson told a correspondent in 1879, 'much time must elapse and much more research must be followed up before we can hope to arrive at any really satisfactory conclusions with respect to these perplexing questions'.[78]

The palaeobotanist W.C. Williamson offers an unusually well-documented case. Williamson was initially unwilling to make any concessions to the arguments of the *Origin*. But lecturing in Chester in 1877, after years of cajoling by Huxley and others, he announced that he accepted Darwinism 'to a large extent', and thereafter his public pronouncements made clear to his audience that be believed it was impossible to deny 'evolution' in the sense that current forms of life were the descendants of earlier forms, having developed from them via external agency over enormous periods.[79] Even so, writing to Owen in the same year, he went only so far as to allow that evolution was 'an admirable working hypothesis and <u>may</u> someday be proved to be true', noting that he himself could not yet accept it in the decided way in which

THE *DESCENT OF MAN* AND THE HIGH VICTORIANS   177

Huxley did, not least because of the paucity of the fossil evidence.[80] It is doubtful if Williamson was ever persuaded that Darwin's evolutionary mechanisms had been satisfactorily evidenced, but twelve years later, in the face of the obstinacy of his long-time correspondent William Carruthers, he warned that '[t]he fact is Botanical science is rapidly assuming a far more philosophical form than it did in the days of [John] Lindley, [John S.] Henslow and [John Hutton] Balfour. You are free, if you prefer it, to lag behind with them. My innumerable facts, <u>all impelling me forwards with the advancing currents of thought</u>, prevent my doing so'.[81] (This did not signify a complete adoption of more advanced views. In 1894 he decided after some internal debate not to make his usual appearance at the British Association meeting at Oxford in 1894; he had been planning 'to utter [his] final protest' against modern tendencies, but concluded in the end that the controversies would be fruitless, and that instead he would 'leave Oxford to the younger men'.[82])

Another symbolically important figure was the physicist (and later President of the Royal Society), Sir George Gabriel Stokes (1819–1903). Stokes had largely held aloof from evolutionary debates in the 1860s. In 1872 he confessed himself open to listening to evidence on the evolution of man, but as yet unconvinced by Darwin's arguments, feeling that he had 'ridden his hobby a great deal too hard'.[83] Stokes became the acknowledged if informal leader of those mid-Victorian scientists who sought to navigate a careful path between denial of evolution and too ready acceptance of Darwinian orthodoxies, who were content to try to establish a position for the Church which abandoned the old lines of defence, and broadly accepted evolution or at least the great likelihood of evolution (even though, as one put it 'You will understand fully that our belief goes well beyond all this').[84] Although Stokes remained suspicious of theological hair-splitting, he was gradually drawn into a more public antipathy, not least through his involvement in the Victoria Institute, to which he gave two widely circulated addresses, *On the Bearing of the Study of Natural Science* (1879), and *On the Absence of Real Opposition between Science and Religion* (1884), while being held up as affirmation that academic science was not exclusively Darwinian.[85]

Stokes' comment to the Church Congress at Derby in 1882, that Darwinism had been accepted 'by many eminent biologists with a readiness which is puzzling to an outsider', alerts us to the hesitancies of these commitments, and the extent to which intellectual assent was combined with emotional distance.[86] This was most obvious beyond the scientific community. William Allingham's collection of poetry *Blackberries picked off many bushes* (1884)

178   DARWINISM'S GENERATIONS

was full of jibes at evolutionary science: not direct repudiations, but dogged, and it must be said often doggerel, resistances. The poetry of Emily Pfeiffer (1827–90), in the 1870s preoccupied with evolutionary questions, offered a conflicted acceptance of the fact of evolution with an at times desperate rejection of the implication of a natural world emptied of meaning.[87]

The partial habilitation of the Victoria Institute hints at such shifts of alignment, and their limits. After 1871, while never entirely throwing off the reputation it had rightly acquired as a crude anti-Darwinian front, the Institute became a more respectable, if not entirely scientifically credible, venue for discussions of science in a Darwinian-sceptic frame. It reassured potential supporters that it sought to avoid any bending of science to suit theological ideas, And having long resisted membership, Stokes agreed to become its president in 1886.[88] But the Institute continued to attract an unrepresentatively large participation from the outspoken fringes. Contemporary description of cadaverous men with long white hair braving the weather to hear anti-Darwinian diatribes speaks vividly of its predominant character.[89] As late as the 1890s the appearance of Samuel Laing (1812–97)'s enthusiastically evolutionary *Modern Science and Modern Thought* prompted a rush of calls from the Institute's membership for some official rebuttal.[90]

The Persistence of Resistance to Ideas of Human Evolution

It was the incorporation of man into the evolutionary schema which provided the mid Victorians with their greatest challenge, and which shifted the grounds of hostility to 'Darwinism'. *Descent*'s more explicit argument for human evolution placed some who had been amenable to the broader evolutionary outline in a quandary which pushed them into a more general or more hysterical hostility.[91] Benjamin Jowett who continued to acknowledge the *Origin* as 'one of the greatest and most far-reaching books' of the century, apparently told a correspondent in the wake of *Descent* that 'I do not believe a word of it'.[92] Frances Power Cobbe, sympathetic to Darwinism in the 1860s, was horrified by *Descent*.[93] Cobbe is a good example of the many mid Victorians who would have liked to go further with Darwin, but who ultimately just couldn't reconcile this with their underlying intellectual commitments. Through the 1870s and 1880s, Cobbe kept up a steady campaign against the corrosive implications of this version of Darwinism: its destruction of religious belief, its crushing of moral ambition, its arrogation

THE *DESCENT OF MAN* AND THE HIGH VICTORIANS   179

of spiritual leadership.[94] One of Cobbe's correspondents, the Greek historian George William Cox (1827–1902), despite his liberalism in religious matters, shared her fear of the collapse of moral and spiritual force which would follow the acceptance of the 'deadly cold negations' of the X-Clubbers.[95] The historian J.A. Froude was another whose post-*Origin* sympathy (admittedly primarily as a weapon of science in its struggle against religious 'superstition') seems steadily to have leached away in the years after 1871, as the implications of the common descent of man and animals loomed larger.[96] The Liverpool clergyman-naturalist Henry Hugh Higgins (1814–93), who overcame his significant qualms about transmutation and the general efficacy of natural selection to acknowledge 'the laws which Mr Darwin has discovered', continued to reject the common evolution of man and animals.[97] In the same way, many who had given a qualified acceptance of Darwinian evolution skirted around the implications of human evolution. But this was to beg the question rather than provide a solution, and for most born before 1830, even for those inclined to accept some version of common descent, the irresistible logic of religious faith was to reject the subsuming of man into the evolutionary sequence.[98]

For some, there was a performative element at work here: a distinction deployed to maintain a psychologically comforting distance from Darwin without having to deny his greatness, or even, perhaps, to come to a precise position on evolution. But for many who had already been struggling to accommodate themselves to evolutionary ideas, any sympathy they had was wiped away by *Descent*'s wider claims. This was certainly true of Ruskin, for whom the measured scepticism of the 1860s became the frenzied condemnations of the 1870s. Although he laboured to avoid toppling over into absolute rejection, Ruskin became more and more outspoken as time passed. In his journalism and writings of the 1870s and 1880s, including *Eagle's Nest* (1872) and *Proserpina* (1875–82), he attacked evolution's 'filthy heraldries' and its appeal for 'every impudent imbecility in Europe'.[99] His regular lectures were often derailed by long anti-evolutionary digressions.[100] By the early 1880s he had tried to resign from the Geological Society, and was confronting Huxley at the end of one of his lectures.[101] Inevitably, when he received a catalogue of John Lubbock's '100 Best Books' in 1886, he deleted the *Origin* from the list.[102] Ruskin's Pre-Raphaelite disciples followed suit: 'Darwinism', declared John Lucas Tupper (1823?–79) in 1872, 'is <u>practical atheism</u>'.[103] Visiting Holman Hunt in 1878, Edward Lear was nonplussed to find that he 'is becoming a literalist about all biblical lore, & has a holy horror of Darwin'.[104]

180 DARWINISM'S GENERATIONS

Figure 4 'Baffled Science Slow Retires', *Punch*, 4 January 1873.

A minority of early and mid Victorians continued to rest their case in part on the long-standing physiological arguments about the relative size of human and non-human brains, or the contrast between human and non-human hands. But the weight of anatomical evidence against was increasingly incontrovertible, and these arguments more and more a matter for ridicule (Figure 4). Instead, (while this did not itself rule out some degree of evolution for humanity as a species) most of the members of the older generations took refuge behind arguments for what Henry Liddon described as three crucial gaps in the evolutionary chain in which the Creative Will must have intervened: mind and intellectual capability, capacity for speech, and morality and conscience.[105]

In the 1870s much of the debate revolved around the extent to which the Darwinians were able to demonstrate that mental distinctions were merely matters of degree and not of kind. Huxley had long recognised that this would be a crucial stumbling block. He went out of his way in his lectures to working men in the early 1860s, while enforcing the kinship of man and ape, to reassure his audience that 'no-one is more convinced than I am of the vastness of the gulf between civilised man and the brutes'.[106] And mid

THE *DESCENT OF MAN* AND THE HIGH VICTORIANS 181

Victorians in the Darwinian camp, like W.B. Carpenter, Lyell, and Newton continued to worry about man's distinctiveness, and the problem of the development of the human mind, happy to resort to rhetorical concessions which created a productive ambiguity which freed them from the need for clear and categorical statements of their position.[107] Those like G.H. Lewes, who accepted that man and animals possibly shared similar mental functions, continued to argue that an impassable barrier existed in respect of language and creative capacity.[108] Even Wallace, despite being one of the first to apply the principle of natural selection to human development, had by the end of the 1860s shifted to a position in which he refused to accept that the mental and moral nature of man developed from lower animals solely by the same processes as the evolution of his physical body, a position he maintained—and indeed insisted on as 'pure Darwinism'—for the rest of his career, to the frustration of Darwin's friends and the delight of Darwin's opponents.[109]

Many of the immediate responses to *Descent* by mid Victorians took up this issue: it was a core concern of Alexander Grant, both in his *Contemporary Review* essay and also in his address to the Edinburgh Philosophical Society in 1871.[110] Grant argued that while the 'extreme sensationalist school' might accept that the mental activity of man and animals was indistinguishable, most philosophers would not. The Edinburgh philosopher Alexander Campbell Fraser (1819–1914) remained convinced that the creation of self-consciousness remained one of the stages of evolution which could not be explained by natural law. This issue was taken up through the first half of the 1870s by Mivart in articles subsequently collected in *Lessons from Nature as manifested in Mind and Matter* (1876), and continued to be a staple of the mid-Victorian critique in the final quarter of the century, as illustrated by the work of the Presbyterian theologian Henry Calderwood (1830–97), not least his *Evolution and Man's Place in Nature* (1893, 1896): Darwin might explain organic evolution, but he could not account for rational thought.[111]

Language was particularly significant in these debates, in that it could embody commitment to broad evolutionary approaches, while also operating as a bulwark against human evolution. Mid-Victorian philologists, such as Max Müller, were happy to trace the descent of language from a common stock, while at the same time offering emphatic arguments that the *faculty* of language had required some form of separate intervention.[112] Müller was another mid-Victorian Darwinist manqué. In many respects his interests and methods should have encouraged Darwinian alignments. As he observed in an 1884 essay, he was an adherent of the historical school, believing that to understand what something is one must examine what it has been.[113] In 1863 he argued that natural selection and the struggle for life were needed to unlock the mysteries of language, and in the aftermath of the publication of *Descent* he told Huxley that he was 'deep in Darwinian speculations'.[114] But ultimately, he refused to take the final step of accepting that capacity for language and the higher mental faculties had evolved. Language, and the concepts and thoughts that were its building blocks were, he argued, 'entirely absent in animals'; it was 'the barrier between beast and man'.[115] Looking back in 1891, Müller recalled that it had required courage at times to stand up against the authority of Darwin, but that it was still the case that 'all serious thinkers...agree that there *is* a specific difference between the human animal and all other animals, and that that difference consists in language'.[116] Müller was supported in this position by the neurologist Frederic Bateman (1824–1904), in papers and lectures gathered in *Darwinism Tested by Language* (1878). Bateman disputed the existence of a specific physical locus of language skills in the brain, and maintained that without it evolutionary accounts could not hold true. These arguments became a standard component of early- and mid-Victorian resistances; creating what Mivart described as an 'impassable limit to evolution'.[117]

The third gambit was to argue that natural development could not encompass moral discretion. The moral sense was the stronghold of those who had made strategic movements of retreat from other defensive positions.[118] Efforts to ascribe moral sensibility to animals were firmly rejected, even by those who campaigned for animal rights, not least Frances Power Cobbe. Even accepting 'as I suppose we all do now' the evolution of animal life from lower to higher forms, Cobbe told Mary Somerville, 'I think all attempts to trace our spiritual nature to such sources have proved signally futile'.[119] Moral sensibility must be the work of the Creator, and faced with the accumulating evidence of the evolutionary history of man, the mid Victorians reworked ideas of special creation to insist that there must have been a moment of the breathing of the soul into humankind—perhaps Adam and Eve themselves—at the point at which the beast became 'man'.[120] Even Hensleigh Wedgwood, whose *On the Origin of Language* (1866) offered an exception to early- and mid-Victorian refusals to accept the evolution of rational thought, disputed Darwin's account in *Descent* of the development of conscience.[121]

The broader epistemological challenge of scientific naturalism was brought into sharp focus by John Tyndall's 'Belfast Address' to the British Association in 1874, with its overt hostility to institutional religion and its provocative ambition to wrest the entire domain of cosmology from theology, but the debate had been raging at least since Huxley's 1868 lecture and article on the 'Physical Basis of Life', which set off a whirl of interest in protoplasm and arguments about the nature of life.[122] It is possible to argue that neither Huxley nor Tyndall espoused a fully materialist agenda, adopting it as scientific methodology rather than general philosophy.[123] And it was a sign of Darwin's own improving reputation that some opponents sought to suggest he would have had little to do with such manifestos.[124] But inevitably, Tyndall's address, and subsequent performances such as his 1877 lecture on 'Science and the Soul', prompted a fierce backlash, much of it in the hysterical register of Carlyle's verdict that it was a 'philosophy fit for dogs', but also in the sort of sustained rebuttal of Tait and Stewart's *The Unseen Universe*, which was primarily concerned to use the principle of the continuity of matter to argue that belief in a future life and the existence of a divine life-giver was entirely compatible with modern science.[125]

The efforts of the North British Physicists were challenged by high Victorians such as W.K. Clifford, who sought instead to unify matter, space and mind into one entity: consciousness was merely molecules stimulating nerve fibres.[126] A non-natural Creator was of course fundamental to Christian beliefs whether reconciled with a version of Darwinism or not, and J.H. Stirling's persistent hostility to Darwinism on the basis of its essential materialism, initially in his *As Regards Protoplasm*, first published in 1869 in response to Huxley's lecture and reissued in 1872, and thereafter in various works, including his 1890 Gifford Lectures, *Philosophy and Theology*, was a characteristic response of early and many mid Victorians.[127] Into the 1890s early and mid Victorians continued to maintain that the Darwinian view of the world was too mechanical.[128] As Mivart told one younger biologist in 1887, 'I quite maintain from the point of view of physical science we have to seek mechanical explanations and that other explanations are <u>thus</u> "out of court"; but that court is not the supreme court, nor is physical science all, or the highest science'.[129]

The Emergence of the High Victorians

Mivart acknowledged in 1871 that Darwinian belief had been steadily gaining ground, and there is little doubt that as the 1870s progressed the mid Victorians felt themselves ever more embattled. A new set of voices were emerging, in the quarterlies and the new monthlies, on the lecture platform and in provincial literary and scientific societies. Some of these voices belonged to the generation of liberal intellectuals who had emerged from the universities in the 1850s and early 1860s, who, as Frank Turner has put it, 'achieved intellectual adulthood' during this decade, including Leslie Stephen, Henry Sidgwick, John Morley, Walter Pater, and W.K. Clifford, but they also came from figures outside this circle, including Samuel Butler, Thomas Hardy, and Edward Clodd.[130] They joined a small number of figures, most significantly John Lubbock, who although of a similar age, had already established themselves as important voices in the 1860s. Their emergence encouraged the impression, as one high-Victorian journalist put it, that an older scientific school was being pushed aside by 'the greater energy and boldness of the newer school' of Darwinians.[131]

In the years after the publication of *Descent*, high Victorians who had been promoted into positions of influence behind the X-Club vanguard strengthened the institutional power base of Darwinism. Despite the disquiet of figures like Stokes and Beale, Huxley became one of the Royal Society secretaries in 1872, and thereafter from 1873–85, Hooker, Spottiswoode and then Huxley himself served as President of the Society (Figure 5). The

Figure 5 'Professor T.H. Huxley, President of the Royal Society', *The Graphic*, 21 July 1883.

THE *DESCENT OF MAN* AND THE HIGH VICTORIANS   185

generational underpinnings of this process were powerful, if not always public. In 1883 Huxley was initially very keen not to have his interim appointment made permanent, and only agreed in the face of strong representations from 'the younger men'.[132] Although control over the British Association was always more distributed because of its system of largely autonomous sections, the greater influence of the X-Club was manifest in the election as presidents of Tyndall, Spottiswoode, Lubbock, along with W.B. Carpenter, Allen Thomson, and Ramsay between 1872 and 1881; and in the slate of evening lecturers, not least in 1874 when Lubbock spoke along with Huxley. High Victorians were no longer just pushing back against previous resistance, but were offering support from positions of influence and authority.[133] The fates of Mivart, and even more so Henry Charlton Bastian (see below), ruthlessly marginalised by the X-Clubbers, offered a stark sign of the times.[134]

After 1870 the universities began to escape from the previous 'torpor' of scientific education. Oxford and Cambridge were being opened up to all denominations and a wider curriculum. Cambridge quickly became the acknowledged centre of British Darwinism. Huxley, firmly established at the School of Mines, was gathering round him a new evolutionary cadre. The oldest high Victorians, including T.G. Bonney, W. Boyd Dawkins, and William Turner, were establishing themselves in academic posts at the end of the 1860s.[135] The younger ones followed on, steadily finding posts during the following decade.[136] By the later 1870s, even in religiously conservative Scotland, students were studying Darwin and being asked to compete for essay prizes on 'Evolution Old and New'.[137]

Emblematic of this process is the career of the physiologist Michael Foster (1836–1907).[138] A protégé of Huxley, Foster was made a fellow of Trinity College, Cambridge, in 1870, and over the following decade built the most important physiological school in Britain there. His *Textbook of Physiology* (1877) helped to establish evolutionary approaches at the centre of the field. From the 1870s he was actively involved in refashioning institutions like the Linnean Society, in the wake of the Bentham affair, and after 1881, when as Huxley's nominee he succeeded him as secretary of the Royal Society, he became a formidable force in shaping British science, using his secretaryship to great effect in the promotion of younger evolutionary scientists.[139] He could be sparkling company, but he was also a ruthless and not always popular operator, with little truck for the sensibilities of older scientists.[140]

186   DARWINISM'S GENERATIONS

Foster was one of the leading actors in the foundation of the Physiological Society in 1876. He was first editor of the *Journal of Physiology* (estd. 1878), one of a number of such specialist academic journals formed in the 1870s with a largely high-Victorian character, including the philosophical journal *Mind*, launched in 1876, and *Brain*, established in 1878.[141] *Mind*, under the editorship of George Croom Robertson (1842–92) revolutionised psychology and philosophy, developing a field in which the works of the mid Victorians like Bain and Mill suddenly seemed 'altogether childlike, old-fashioned, and quaint'.[142] New evolutionary approaches steadily gained ground over the decade as the work of the high Victorians began to take a much greater share of the research being published. The development of *Nature*, despite its initially patchy circulation, into the 'official' organ of academic science helped this generational shift.[143] Its editor, Norman Lockyer, may not always have been sufficiently 'orthodox' for the X-Clubbers, but nonetheless *Nature* came to provide a platform for the opinions of high-Victorian scientists.[144]

A similar transformation was going on in the wider culture, as the new 'higher journalism' was consolidated. Following the path blazed by figures like Morley, Stephen, and Frederic Harrison (1831–1923), undergraduates of the 1850s and early 1860s found voice as leader writers and ultimately as editors for the London press, and as essayists for the serious reviews. The later 1860s saw the establishment of a number of intellectual periodicals with high-Victorian editorial control: the *Fortnightly Review* (1865–), under editorship of John Morley from 1867; the *Contemporary Review* (1866–), owned by Alexander Strahan (1833–1918) and from 1870 edited by James Knowles (1831–1908), who in 1877 established *The Nineteenth Century*, and the *Pall Mall Gazette* (1865–), edited by Frederick Greenwood (1830–1909). Likewise the *Academy*, also launched in 1869 by Charles Appleton (1841–79), established itself as a literary rival to the *Athenaeum*, whose own anti-Darwinian stance was deliberately reversed by Norman MacColl (1843–1904) when he replaced William Hepworth Dixon (1821–79) as editor in December of the same year.[145]

These periodicals were fed by a less structured and less visible ecosystem of informal spaces for conversation and discussion, Lubbock's 'scientific breakfasts', Huxley's 'tall teas', and Lockyer's weekly 'smokers', dinners, billiard room chat, intellectual salons, and weekend house parties.[146] 'High Elms', Lubbock's home near Downe, was an especially significant base from which contemporaries were introduced to Darwin. By the 1870s the Sunday afternoon salon of Eliot and Lewes at The Priory in St John's Wood was

THE *DESCENT OF MAN* AND THE HIGH VICTORIANS   187

functioning primarily as an important meeting ground for high Victorians who came to wait on Eliot and Lewes, and to make new contacts amongst their peers, talking evolution as they did.[147] Likewise, it was the noisy atheism of the Sunday afternoons of the Cliffords, where Robert Louis Stevenson met figures like Andrew Lang and the psychologist James Sully (1842–1923), which formed the backdrop to Stevenson's explorations of Darwinian ideas in 1874, and where Vernon Lee (1856–1935) met Leslie Stephen and Huxley.[148] From the end of the 1870s the fortnightly summer rambles of Stephen's 'Sunday tramps' provided another site of intellectual exchange in which high Victorians predominated.[149] In the provinces, a similar function was performed by the informal groups of like-minded younger men who coalesced around an interest in science, but which rarely left a record. One which achieved unusual visibility was the coterie which emerged from classes associated with the Science and Art Department in Halifax in the mid-1860s, which met in each other's homes and later in a room in the Halifax Museum 'to compare notes, to exchange ideas, and to discuss Evolution, the origin of species, and the work of Darwin, Huxley and Tyndall', as one member later recalled. By the 1870s this group of high Victorians was becoming prominent in West Riding scientific circles, giving lectures and papers, and taking the initiative in the formation of new societies, like the Halifax Scientific Society (established 1875), much more favourably disposed to evolutionary opinions.[150]

The Halifax Scientific Society was one example of the shifts, uneven and uncertain as they were, manifested by the contrast between the continued disengagement from evolutionary topics in the established naturalists' societies (see above, pp. 113–14), and the history of newer foundations. The Norfolk and Norwich Naturalists' Society, whose later establishment in 1869 seems to have relieved it of the drag of early-Victorian influence and allowed much greater prominence to its high Victorian members, offers another.[151] Formed on the initiative of local doctor, Michael Beverley (1841–1930), the society took up the challenge issued by its president in his address to the society's first meeting to 'take our share in the great battle of species', and provided a platform for a number of high Victorian evolutionists, including John Ellor Taylor (1837–95), whose later stint as editor of *Science Gossip* was only part of a long career as advocate for evolutionary approaches, and Frederic W. Harmer (1835–1923), geologist, whose interventions, including two presidential addresses in the later 1870s, offered a wholehearted support for Darwinism.[152] The Nottingham Literary and Philosophical Society offers a slightly different character: formed in 1864–5

188   DARWINISM'S GENERATIONS

it was very much an initiative of the city's cultural elite, and its early committees were dominated by mid Victorians. But the relatively late date of
establishment again allowed for a not insignificant presence of high
Victorians, including Rev. John Ferguson McCallan (1834–83) as first Hon
Secretary, and Hugh Browne, already referenced (see above, p. 85). This,
and the almost complete absence of early Victorians, again facilitated the
raising of Darwinian topics. One of those who offered a firm endorsement
of Darwin in the 1870s was Robert Edmonstone (1842–1914).[153] Edmonstone
illustrates the way in which generational dynamics extended beyond the
public schools and universities. Beginning clerical work at fourteen (eventually rising to be head of the counting house at a Nottingham lace manufacturer), he began a long association with the Nottingham Mechanics'
Institute; having studied Darwin's writings in the later 1860s, he demonstrated his evolutionary commitments in lectures in the 1870s through to
interventions in newspaper debates in the 1890s.[154]

Concurrently, the capital was developing a rich network of radical societies which organised Sunday lectures in which high-Victorian lecturers
and evolutionary topics were prominent.[155] The Sunday Evenings for the
People, organised by the South Place Chapel in Finsbury, and the Sunday
Lecture Societies which spread across the provinces in the years after 1870,
allowed wider audiences to hear high- and then late-Victorian speakers
including Clodd, Clifford, Prince Peter Kropotkin (1842–1921), Karl
Pearson (1857–1936), Grant Allen, Annie Besant (1847–1933), and George
Bernard Shaw (1856–1950) espouse an essentially evolutionary creed.[156]
South Place's Darwinian tone is encapsulated by a cartoon by G.J. Holyoake
(1817–1906) in which, as the minister Moncure Daniel Conway (1832–1907)
described in his memoirs, 'I am in a little tent marked "Conway's Free and
Airy Tabernacle", having a white flag inscribed "We move on". Above all is a
bust of Darwin, beneath being a stairway of geologic strata on which a
gorilla is climbing, and drawing by his tail Huxley and Tyndall. The text
connected with Darwin was Gen. xxvii, 11, "Behold my brother is a hairy
man, and I am a smooth man"'.[157]

High-Victorian Darwinism

If the mid Victorians initially came to Darwinism steeped in widespread
hostility to *Vestiges*, the high Victorians often responded with intellects
already unmoored by the positivism of Auguste Comte, by H.T. Buckle's

THE *DESCENT OF MAN* AND THE HIGH VICTORIANS   189

History of Civilization, and by the controversy over *Essays and Reviews*. For them, the Huxley–Wilberforce debate was 'intellectual sport'.[158] The fun had in Walter Besant's *The Golden Butterfly* (1876) at the expense of its American millionaire anti-hero, who confuses Darwin for a writer of historical romances and Huxley for a preacher, works both because this knowledge can be taken for granted, but also because of his readers' indifference to the fears of their elders. For this generation, these were years of unsettled opinion; of the necessity and duty, suggested Henry Sidgwick, of 'placing ourselves as far as possible outside traditional sentiments and opinions'.[159] They read the *Origin* with mature attention, with a presumption in favour of evolution. They readily adopted Darwinian languages and frames of reference, and had little appetite for convoluted reconciliations of geology and literal biblicalism, never mind for the peremptory anti-Darwinism of a Disraeli, dismissed by F.W. Farrar as a 'claptrap platform appeal to the unfathomable ignorance and unlimited arrogance of a prejudiced assembly'.[160] For the High Church philologist A.H. Sayce (1845–1933), no utterance of science was clearer than that everything existing was the result of evolution; that there was no break, nothing but an unchangeable continuity of progress. This might make him a 'terrible heretic', he confessed to Blackie, 'but it has sometimes happened that the heresy of one generation has become the orthodoxy of another'.[161]

One recollection of an evening at Chatsworth House in 1875 sums up the contrasting generational sensibilities with clarity. At dinner Frederick Temple voiced his puzzlement at how scientific men adopted Darwin's theories, despite the lack of evidence, and almost entirely on his authority. His companions were initially silent, but in the smoking room after dinner, Spencer Compton (1833–1908), later 8th Duke of Devonshire, remarked, after a long silence, 'The Bishop says that scientific men adopt hastily and without sufficient evidence all the modern ideas about evolution! Why, surely, it is only because these ideas explain facts, and because they are based on facts and on evidence, that scientific men do adopt them, and they would not adopt Darwin's explanations if they were not supported by evidence'.[162]

For the high Victorians, acceptance of Darwinism generally involved a very definite break with the views of their parents, and was none the less attractive for that, although at times it produced bitter family divisions of the sort dramatised by Edmund Gosse in *Father and Son*, and Samuel Butler in *The Way of All Flesh*. Justifying his refusal to follow his father's rejection of Darwinism, Edmund Gosse aligned himself not with his

190 DARWINISM'S GENERATIONS

parent's laborious microscopic work but with the bold theorising of Darwin and Huxley.[163] Laura Forster (1839–1924), looking after her septuagenarian father in his Essex rectory, must stand for many for whom, as E.M. Forster put it, 'such Darwinism as they imbibed [was] of the nature of secret drinking'.[164] Even Frederic Harrison, who worked hard to present his father as a sort of unconscious evolutionary before Darwin, failed to bridge the intellectual divide which clearly widened between them in the years after the *Origin*.[165]

The high Victorians did not just identify themselves as 'evolutionists', they accepted that it was Darwin who had for the first time provided a convincing account. Where non-scientific mid Victorians had tended to take refuge in a sort of evolutionary agnosticism, non-scientific high Victorians were inclined to accept as a matter of faith.[166] For Huxley, the positions of the early and mid Victorians were the critical positions to be contested; for the high Victorians, they could be laughed out of court. The butt of this humour was not evolutionary absurdities, but those who continued to peddle them.[167] The rapid accumulation of fossil evidence meant that evolutionists needed no longer adopt their former apologetic tone, and the knotty problems of evolutionary theory were faced as challenges to which solutions needed to be found, rather than as defects which prompted doubt. George J. Romanes (1848–94), part of the younger late-Victorian generation which shared this sensibility, summed up this mentality in an exchange with the evolutionary sceptic, William Carruthers, in 1892; acknowledging the challenging facts that Carruthers had cited as reasons for his doubts, Romanes' response was that 'the only question is whether when taken in relation to all the other large body of facts which now make in favour [sic] of evolution, they ought not to be regarded as exceptions to be explained, rather than as per se destructive of the Darwinian theory'.[168] The compromises of mid Victorians like Argyll were dismissed as nothing more than a comfortable staging post for those unwilling to follow the evidence to its logical conclusion, and the full implications of Darwin's scientific naturalism were embraced.[169]

For many the acceptance of this stance emerged from an undergraduate crisis which undermined or reformulated their beliefs; or it was encouraged by discussions in mutual improvement and young men's societies, or within sibling and friendship networks. It is this generation that furnished most of the true converts, even if the ease with which juvenile beliefs were cast off often left little trace. The youngest high Victorians were only in their teens when the *Origin* appeared, but all had pre-Darwinian learning that needed to be unlearned. Even so, they had little of the investment of their elders.

THE *DESCENT OF MAN* AND THE HIGH VICTORIANS 191

For the oldest of the scientists it was just possible to have, as Sir William Flower recalled, commenced their careers before the appearance of the *Origin*, but most encountered Darwinism while their opinions were still coalescing, at the point at which, as Flower later put it, they were 'in a larval, or rather chrysalis state as regards all things connected with natural history'.[170]

The South American naturalist and novelist William Henry Hudson (1841–1922) offers a vignette of these processes. In his early twenties in the Argentinian outback when his elder brother returned from Europe full of enthusiasm for the *Origin*, on a first reading Hudson was sceptical. Challenged by his brother to read again and put his preconceptions aside, Hudson found that 'my mind, or subconscious mind, like a dog with a bone which it refuses to drop in defiance of its masters' command, went on revolving it', gradually becoming convinced; '[i]nsensibly and inevitably I had become an evolutionist'.[171] In Hudson's case we can trace the impact of this through his scientific work, but also in the environmental consciousness he expressed in his novel *Green Mansions* (1904).

Time and again in the later accounts, the story is of immediate response, enthusiastic adoption, and unapologetic advocacy. The retrospective nature of many of these accounts perhaps masks the very real challenge Darwin posed to the existing beliefs of the younger generation. The initial encounter was no doubt uncomfortable and disturbing for many. The poet Robert W. Buchanan (1841–1901) found the process 'horrible'.[172] The clerical zoologist Thomas R.R. Stebbing (1835–1926) described first reading the *Origin*, 'with the common prejudices strong upon me', and with the expectation that he would find the flaws in the argument, a feeling soon changing to surprise and pleasure. 'I had to undergo the useful pang of giving up some opinions on which I had formerly been very positive; but in return there was laid before me a view of the world's history, so simple, so harmonious,...that I should have felt genuine admiration, even had I not been also convinced'.[173] For many high Victorians, especially those with stronger religious faiths, this was a difficult and long drawn out process which might involve first adopting Darwinism as a working hypothesis, and then finally acceptance in middle age, sometimes but not always experienced as a definite transition to 'becom[ing] a Darwinite'.[174] Take the historian Peter Bayne (1839–96), whose liberal stance on biblical inspiration cost him his job as editor of the Presbyterian *Weekly Review* in 1865, and who was clearly committed to evolutionary positions by the later 1870s, but by implication only after twenty years of debate and deliberation.[175]

Inevitably, not all high Victorians actively embraced Darwinism. It scarcely touched the preoccupations of the romantic socialist William Morris (1834–96). The selection Morris supplied in response to W.T. Stead's request for a list of his hundred most important books was unusual in not including the *Origin* or any other Darwin volume; as his daughter May Morris observed, 'Darwin would of course be in every way outside the scope of Morris's list.'[176] More typical was the urgent searching after insight and understanding visible in the extensive surviving correspondence of Edward S. Talbot (1844–1934), the Warden of Keble College. Talbot was a liberal Anglican who corresponded with leading Darwinists like E.B. Poulton and William Thiselton-Dyer, influential Anglicans including Archbishops Temple and Benson, politician-intellectuals such as Gladstone and Balfour, and also others with an interest in evolutionary questions, including his cousin Victoria, Lady Welby. Talbot was unsympathetic to the combativeness of Huxley and Tyndall, but his letters reveal a mind constantly wrestling with Darwinian debates, conscious of its own scientific limitations but reading widely in the literature, sharing reading aloud of key interventions, seeking advice about technical points; hesitant at times, but ultimately determined to avoid the temptation to *a priori* opposition.[177] Many religious commentators retained a tone of disapproval, but on closer examination this was most often an exercise of displacement in which the implications of Darwin's ideas, or the stridency with which they were proclaimed, were used to mask a reluctant acceptance of the core of Darwin's position. Gradually or more quickly, anxiety was 'transmuted under the rays of truth into wonder and delight and hope.'[178]

Over time, shifts in the balance of religious opinion made such transitions easier. This seems to have been the case for A. Scott Matheson (1836/7–1913), who as a junior United Presbyterian Church minister in Alloa in the 1860s took a leading role in opposing Darwinian ideas in the local Natural History Society, but who over the ensuing thirty years gradually changed his mind. Looking back in 1900, Matheson recalled that in the 1860s evolution was still a daring novelty, but that after the passage of thirty years 'most of us have come round long ago to revere Darwin as a great student of nature, and to accept his theory of evolution.'[179] In the same way the Warrington railway mechanic John Porritt (1834–1904), whose fervent Congregationalism had been shocked by Darwin and the higher criticism, was ultimately able to calmly adjust his faith to the new ideas.[180]

This helps explain the quality of the zeal of the convert displayed by many high Victorians that later generations never fully shared, the sense that their Darwinism was not merely a coolly rational endorsement of an intellectual proposition, and that the argument revolved not around evidence per se, but was as much a leap of faith. In committing to the transmutation of species and the role of natural selection, they accepted that they had no direct proof but were prepared to argue that their plausible hypothesis, which did not contradict any known principles or facts, was better than the entirely supernatural alternative offered by Darwin's opponents. They were not worried about clashes with the Genesis account, accepting that the Bible was not intended as a scientific text and did not need to be reconciled with shifting scientific truth. As the orientalist F.V. Dickins (1838–1915) remarked, the theory might not be susceptible of proof, 'but the more we know it, the more we understand that it is [only?] defect of knowledge [which] stands in the way of its application to any group of cases'.[181] For high Victorians, the critical question, as the German Darwinist Fritz Müller recognised in the title of his influential book, translated into English in 1869, was whether one was *For Darwin* or not. Darwinism was, or quickly became, the bedrock of the high Victorians' world view, 'coeval with thinking man'.[182] Even for the Catholic priest Gerard Manley Hopkins (1844–89), it was clear that 'everything is Darwinism'.[183]

A more sustained example is provided by the publisher and pigeon-fancier Lewis Wright (*c*.1838–1905). Wright was an amateur scientist who later made a minor contribution to the development of cinematography in Britain, whose published opinions on evolution were made almost entirely as an anonymous reviewer for the *Nonconformist* newspaper in the 1880s.[184] During this time Wright was drawn into a number of exchanges with readers sceptical about evolution, often themselves anonymous, but sometimes identifiable as older mid and early Victorians.[185] Wright's own intellectual journey is not easy to untangle, although there is a suggestion that his introduction to Darwin was through an 1859 edition of the *Origin* and that he was initially unconvinced; but by the 1880s he accepted that evolution was 'little short of a divine revelation'; its status steadily strengthened, notwithstanding the evidentiary challenges which still remained, by scientific progress in optics, embryology, and even palaeontology. However, the primary focus of his newspaper interventions was not to controvert the position of those who continued to doubt. Rather he sought to challenge what he saw as their captious and fundamentally dishonest mode of argument, especially the refusal to acknowledge that the overwhelming consensus of current scientific thought was evolutionary, the fixing of an absolute and unreasonable threshold of 'proof' of evolution, and the tendency to cast

194 DARWINISM'S GENERATIONS

doubt on evolution by the parading of a series of 'authorities' who were either lacking in any scientific standing, were the representatives of a pre-Darwinian generation, or who did in fact accept evolution, albeit with specific caveats which did not bear the construction opponents attempted to place on them. He was in no doubt that 'nine tenths' of those who continued to resist 'are either too old to readily accept new views, or have been distinctly committed to controversy on the theologic side'.[186]

For the high Victorians, Darwin achieved an almost cult-like status. Darwinism became a matter of discipleship.[187] It was said of the poet Mathilde Blind (1841–96) that when she first discovered Darwin's works, she retired to a solitary farmhouse for nine months, leaving all her friends and her other work and devoting herself entirely to the study of the new gospel, 'which inflamed her with the ardour of a worshipper'.[188] To the Liberal intelligentsia around Leslie Stephen, Darwin was more than just an author and thinker; he was a revered presence in the round of dinners, soirees, and tramping visits. Darwin's ill health and indifference to the accoutrements of fame meant these circles were not as wide as they might have been, but even beyond his intimates the high Victorians met with, talked to and corresponded with Darwin as a personal presence in their world. He was, reminisced J.W. Judd, the greatest of teachers, and Bonney recalled the combination of genuine humility and unconscious intellectual strength which impressed itself so deeply on 'all younger men'.[189] The traveller and botanical artist Marianne North (1830–90) described Darwin as 'the greatest man living, the most truthful, as well as the most unselfish and modest'.[190] Those drawn into his personal orbit treasured his letters and his praise, used him as a sounding board for their ideas, and basked in his reflected glory, however slight the acquaintance.[191] Those at distance occasionally engaged in gestures of discipleship, naming their sons or their houses or their streets 'Darwin'. It was entirely characteristic that Lawson Tait tried unsuccessfully to organise a 'Darwin Festival' in Birmingham in 1880, and that when this did not come to fruition the Midland Union of Natural History Societies (despite Darwin's embarrassment) instituted a Darwin Medal prize instead.[192]

Where their elders had fretted about methodological flaws, the high Victorians celebrated Darwin's balance of induction and deduction, his systematic observations, and his ability to develop grand conceptualisations from a mass of material. Indeed, Darwin became emblematic of the rigorous scientific method.[193] H.N. Moseley (1844–91) modelled his *Notes by a Naturalist on the Challenger* (1879) on Darwin's own *Journal of Researches*,

THE *DESCENT OF MAN* AND THE HIGH VICTORIANS   195

with fulsome acknowledgement. The transformation in Darwin's status was summed up by the assessment of the journalist and freethinker J.M. Robertson (though himself a late Victorian) that the claim of Darwinian biology to rank as a science rested on the broad ground of the consistent interpretation of innumerable facts, the logical massing of multitudes of phenomena in causal series, the verification of hypotheses by later evidence, and the discovery of fresh evidence by sound hypotheses.[194]

The contours of high-Victorian Darwinism can be synthesised from a range of interventions from the later 1860s onwards. Given Darwin's own works and the energetic advocacy of Huxley, high Victorians were less inclined to publish extended vindications; but a number did appear in the 1870s, including T.R.R. Stebbing's *Essays on Darwinism* (1871), and Benjamin Thompson Lowne (1839–93)'s *Philosophy of Evolution* (1873).[195] High-Victorian scientists, in particular, were much less inclined to pro-grammatic statements of evolutionary science, although their endorsement was frequently articulated in academic and popular lectures. They preferred rather to attend to the painstaking work of filling out the evolutionary sequences, and to the popularisation of evolution. Hence the sorts of art-icles published in the popular science press in the 1870s were expositional rather than controversial. The Darwinism these interventions reveals was a broad church; one in which allegiance was more important than orthodoxy. They were largely uninterested in the distinction between Darwinian and Lamarckian or Spencerian views of evolution, content to absorb them into a more diffuse evolutionary position in which a Darwinian wrapper often papered over a diversity of positions. This allowed for a mostly untroubled acceptance of W.K. Clifford's fairly thorough-going Lamarckism, and for the Christian Darwinism of a figure like James Hurd Keeling (1832–1909), which probably had as much in common with William Thomson's evolu-tionary beliefs as it did with what Keeling called 'the godless Darwinism of Haeckel and others'.[196] It was only in the subsequent generation that the contradictions and inconsistencies of this broad church worked their way to the surface.

High Victorians were rarely attracted by the scientific ecclesiology of Huxley or the defensive lines constructed by the mid Victorians. Almost without exception, the unbroken evolutionary sequence and the creation of new species by the gradual action of secondary causes was accepted as axio-matic.[197] The onus of proof for theories of special creations shifted onto the anti-Darwinians. The serial creations popular with the mid Victorians were dismissed as merely a hypothesis invoked to explain the absence of connecting links, inconsistent with a proper idea of the divine creative power.[198] Of course the palaeontological record remained patchy, but it was filling out all the time, and rather than dwell on the remaining gaps, the high Victorians focused on the jeopardy of those arguing from geological absences: a reliance gently lampooned in Grant Allen's 'Professor Milliter's Dilemma', a short story in his *The Beckoning Hand and Other Stories* (1887), in which the anti-Darwinian professor stumbles across the 'missing link' fossil, half bird/half reptile, and instantly recognises that all his previous evolutionary scepticism has been overborne.

High Victorians accepted the fundamental importance of natural selection without agonising about its complete sufficiency. They marvelled at the compelling simplicity of the combination of Malthusian competition for scarce resources and the survival of the fittest to produce a mechanism whereby tiny favourable modifications could produce adaptation and transmutation. Darwin's own trajectory in successive editions of the *Origin* in the 1860s and 1870s lessened the distance between him and those who believed that the influence of natural selection had been overrated as an element in the evolution of species.[199] Many, like H. Alleyne Nicholson (1844–89), Professor of Natural History at Aberdeen from 1882, were unwilling to give more than partial endorsement to Darwinian mechanisms, and were soon coming across to students as old-fashioned.[200] But unlike mid Victorians such as Argyll and Mivart, the high Victorians did not generally see this as a fundamental flaw, or as preventing them from identifying as Darwinists, even when, like the botanist A.W. Bennett (1833–1902), they were prepared to accept the thrust of Mivart's argument in his *Genesis of Species*, that natural selection was insufficient.[201] The case of Winwood Reade, who desires to write a Darwinian text, models himself on Darwin, gets Darwin's advice, but is also conscious that he might not be 'a good Darwinian' because he sees natural selection as a secondary cause, alerts us to ways in which high-Victorian Darwinism transcended precise alignment to Darwin's own position.[202]

The amount of historic time available for natural selection in prevailing beliefs of the age of the earth remained a stumbling block. Defenders were prepared to accept that the geological record provided evidence of the vast history needed; drawing in, if only as possibilities, a whole series of contributory factors, including changes in the plasticity of species (that is, their propensity to vary from generation to generation), laws of variation which encouraged variation in particular ways (including Spencerian principles of a definite progress from simplicity to complexity), as well as various sorts of

THE *DESCENT OF MAN* AND THE HIGH VICTORIANS   197

Lamarckian use-inheritance. The Methodist microscopist W.H. Dallinger (1839–1909) recognised that other factors of evolution would inevitably also operate and be discovered, but merely as 'added "laws"; supplementary and coordinated methods'.[203]

Despite Darwin's own reluctance to imbue evolution with any directional quality, the high Victorians' Darwinism manifested a distinctly progressive flavour. Evolution, as the historian Sir Edwin Arnold (1832–92) observed, showed 'a continuous and ennobling ascent'.[204] Although a minority of high Victorians including Hardy and the poets Robert Buchanan and Mathilde Blind were oppressed by their sense of the untameable power of nature, the unceasing struggle for existence and its cold-blooded destruction, most were more likely to follow the breezy positivity of Philip Hamerton (1834–1894), whose *Chapters on Animals* (1874) celebrated the health and happiness of creation and the principle of the survival of the fittest as only apparently pitiless, but in reality 'most merciful'.[205] We can see the shift in the substance and texture of argument in the interventions of the scientific populariser Arabella B. Buckley (1840–1929). Buckley had close relations with many of the Darwinian inner circle, including Darwin himself, and was also one of Wallace's closest confidantes, but also retained the impress of her clerical upbringing and decade as amanuensis for Charles Lyell. Her strongly Anglican evolutionary sensibility has most often been noticed because of her powerful arguments for the evolution of morality and indeed altruism, but these were part of a broad religiously inflected progressive evolutionism.[206]

The high Victorians also broadened the range of supporting arguments they could marshal. In particular, advances in investigative techniques and the power of microscopes enabled them to look to embryology as a counterweight to the annoying gaps in the fossil record.[207] In his presidential address to the British Association in 1881, Lubbock pointed to several compelling indicators which more detailed embryological evidence provided, not just that 'ontogeny recapitulated phylogeny' (that is, that the developing embryo appeared to pass through a representation of the various stages of its evolutionary history), but also the widespread existence of rudimentary organs, and the extraordinary parallels in the development of creatures widely distributed across the animal world. Equally effective use was made of the evidence that the human embryo developed and then lost features associated with the lower animals: a moveable tail, facial hair, an outward turned big toe. As Louis Miall (1842–1921), Professor of Biology at Yorkshire College, put it in an 1883 survey, 'every embryologist is accordingly a Darwinian'.[208] This was ground earlier generations avoided if they could,

198 DARWINISM'S GENERATIONS

and where this was impossible, they generally fell back on arguments that affinity did not prove ancestry, that it might demonstrate a serial but not necessarily an evolutional relationship. The absence of any real genetic theory also allowed older Victorians to portray the evolution of embryo into adult as in itself a sign of an underlying directive force.[209]

Notwithstanding the challenge it represented to traditional Christian cosmogenies, high Victorians were also much more ready to accept the evolution of humanity from animals. Some of the very oldest hesitated on the brink. The Edinburgh anatomist William Turner, despite strong evolutionary inclinations, remained troubled by the absence of any clear fossil evidence amidst the 'mists of a bygone era'.[210] But common descent was increasingly taken as a matter of course, not requiring justification. Naturalists, academic and amateur, comfortably espoused homological arguments: the close similarity in structure of the arm of man with the seal's fin, the bird's wing, or the foreleg of the horse or dog, offered powerful evidence of their common origin, as did the seven neck bones shared both with the long-necked giraffe and the squat-necked whale. Samuel Butler told Mivart in 1884 that he had so long and so often approached this subject and found the balance so decidedly in favour of regarding man and the lower animals as descended from a common ancestor that he feared he was 'now incapable of adopting any other conclusion'.[211] Few remained prepared to challenge the 'strange message Darwin brings' that 'We all are one with creeping things;/ And apes and men/ Blood brethren,/And likewise reptile forms with stings', as Hardy's 'Drinking Song' put it.[212] The symbolic message of Darwin's own physiognomy was hard to resist. 'I always feel as if he had been given that grand gorilla-like face on purpose to teach that lesson' commented the mathematician Mary Boole.[213]

Absolute distinctions of mental capacity were also rejected. The language boundary defended by Müller was dismissed by the younger philologists.[214] (This did not stop them also arguing that God's grace created a deep and impassable gulf of separation from all other creatures.) Many argued for the identity of mental capacity in man and animals, including William Lauder Lindsay (1829–80) in his *Mind in the Lower Animals in Health and Disease* (1879).[215] For Benjamin Lowne, mind and language were equally subject to evolution.[216] In the face of suggestions that animal behaviour was purely instinctive rather than rational, high Victorians argued that domesticated animals demonstrated not just memory and capacity to learn, but also the ability to develop this capacity.[217] Lubbock's popular entomology, including his *Ants, Bees and Wasps* (1882), convinced many of his contemporaries of

THE *DESCENT OF MAN* AND THE HIGH VICTORIANS   199

the continuity of mental evolution between animal and man by showing that bees' social organisation was not dissimilar to 'savages', a position that was taken up even more forcibly by late Victorians.[218]

Unlike even Darwin's most enthusiastic mid-Victorian supporters, the high Victorians were largely inclined to follow *Descent* in accepting that ethics developed out of their social utility.[219] They dismissed Cobbe's fears that evolution implied the death knell of morality; the value of morality was not lessened because it was not dropped into men's hearts direct from heaven.[220] Even the more religiously inclined high Victorian Arabella Buckley, could argue that neither nobility of conscience nor expectations of immortality need be in any way compromised by commitment to evolution.[221] This was particularly the case for rationalists like Leslie Stephen, who welcomed the divorce of morality from theology. W.K. Clifford and J.A. Symonds offered widely referenced assertions of ethics and human rationality in evolution: actions were good or bad according as they improved the organism.[222] Even idiosyncratic evolutionists, like Samuel Butler and the Cambridge psychologist James Ward (1843–1925), accepted the need for a place for the development of humankind's moral faculties and intellectual capacities, arguing that nature included realities and experiences that would satisfactorily account for the development of intellect and ideals.[223] Christian high Victorians finessed this notion by employing the idea of humanity being 'endowed' with a spiritual dimension. Early and mid Victorians looked on appalled: 'they are ascribing conscience and intellect to dogs, and even a soul?' lamented Pusey.[224]

Of course, as Thomas Hardy reminds us, high-Victorian attitudes to evolution might have been accepting but were not always celebratory. Hardy's view was particularly pessimistic. He found deep time oppressive. His characters struggled with the realisation that humanity occupied no privileged position and were as vulnerable to ill-adaptation to circumstance as other beings. Generation was for Hardy, as Gillian Beer puts it, 'the law which rides like a juggernaut over and through the individual identity and individual life spans'.[225] But, ministers of religion apart, only for a small minority of high Victorians did such discomfort prompt an explicitly anti-Darwinian rhetoric, most notably the botanist George Henslow (1835–1925) and the writer Samuel Butler, both of whom developed essentially Lamarckian alternatives.[226] What makes both unusual was the way they positioned themselves in opposition to 'Darwinism'. Despite abandoning the special creationism of his youth for a conventional view of evolutionary history, and accepting the importance of natural selection and even pangenesis

200 DARWINISM'S GENERATIONS

(albeit not as complete explanations), Henslow claimed that Darwinism was 'a theory I never accepted from my first reading of the "Origin" in 1859'.[227] Yet his 'True Darwinism' was entirely evolutionary; it challenged the apparently random nature of Darwinian evolution, and the place of Darwinism within evolutionary thought rather than seeking as the mid Victorians did to circumscribe the place of evolution in natural history, and his later writings make much more sense placed in the context of the sectarian debates of the late Victorians of which they became part.[228]

In any case, as the instance of Samuel Butler demonstrates, we should not overestimate the appeal of this sort of dissent. For a decade after the appearance of his *Life and Habit* in December 1877, Butler kept up a steady stream of often vitriolic attacks on Darwin, which included *Evolution Old and New* (1879), *Unconscious Memory* (1880), and *Luck or Cunning* (1887). Yet it is clear that he was a peripheral figure, whose profile in Darwinian debates derives almost entirely from his wider literary presence, and the subsequent influence of his anti-Victorian volumes, *Erewhon* (1872) and *The Way of All Flesh* (1903). Although influenced by Mivart, his views had little congruence with the main lines of early- and mid-Victorian criticism, and even less with the convictions of his contemporaries. His evolutionary writings had a tiny sale and even less intellectual purchase; reviewers were almost universally critical, and it was reported that only 272 copies of *Evolution Old and New* had been sold before the remaining stock was destroyed by a fire in 1899.[229] The appearance of a second edition in the 1880s was merely a desperate attempt to 'galvanise it into a little life'.[230] Indeed, the vehemence of Butler's resentments, shared by other high-Victorian evolutionary mavericks, speaks eloquently of their sense of the power of the Darwinian orthodoxy ranged against them.[231]

One of the incongruities of *Evolution Old and New* was Butler's attempt to reinstall teleology at the heart of evolution, in a way diametrically opposed to the main current of high-Victorian thought. For most high-Victorian scientists, denial of teleology was central to maintaining Darwinism as scientific naturalism, and they continued to resist vigorously attempts to apply theistic glosses to evolution.[232] Religious opinion was also moving strongly in a similar direction. Theological writers like William Knight (1836–1916), in his *Aspects of Theism* (1894), offered a merciless dismantling of teleological arguments.

Design was not repudiated entirely. David Bebbington notes that those who had trained before the appearance of the *Origin* did not abandon their belief design, 'but enlarged it to take into account the fresh evidence'.[233] This

THE *DESCENT OF MAN* AND THE HIGH VICTORIANS   201

allowed religious thinkers to continue to deploy a language of purpose, while in practice retreating to arguments based on God's transcendent responsibility for evolution.[234] For W.H. Dallinger, because variation was accidental, there could be no Paleyean 'instances' and no 'teleological purpose'; instead the universe, its whole progress in time and space, presented one majestic evidence of teleology.[235] The 'evidence of nature' showed that at the outset God 'determined the potency and prevision of all the life, and all the adaptations, that ever emerged or can emerge'.[236] As the Baptist county court judge Henry Mason Bompas (1836–1909) argued, the laws of natural selection did not remove divine responsibility, just as a piece of machine-made lace, even if produced entirely automatically, was no less a proof of intelligence than a piece of lace made by hand.[237]

High-Victorian Thought

Whatever the particular lineaments of their evolutionary commitments, few high Victorians would have disagreed with Leslie Stephen that 'Darwinism...has acted like a leaven affecting the whole development of modern thought', or the belief of economist W. Stanley Jevons (1835–82) that it had revolutionised views of the origin of bodily, mental, moral, and social phenomena.[238] High Victorians frequently retained a strong regard for early influences, and in particular for Carlyle, Ruskin, and Mill; but time and again they described how these gave way to later evolutionary thinkers, just as the socialist illustrator Walter Crane (1845–1915) identified Ruskin as a key influence counter-acted later by Herbert Spencer and Darwin.[239] At the same time, there was a move towards what Henry Sidgwick described as a more general notion of evolution detached from 'the controversial *melee* which has been kept up for half a generation about the "Darwinian Theory"'.[240] Symptomatic is the cautious confession of the Welsh scholar Sir John Rhys (1840–1915) at a school prize-giving in 1881 that he did not know enough about Darwinism to say that he was a Darwinian, or to be quite sure that he was not; but that nonetheless he could not help extending the law of the survival of the fittest to creeds and religious dogmas.[241]

All this makes teasing out specifically Darwinian impacts complicated; but nonetheless it is possible to identify at least four ways in which the implications of the *Origin* and its successors were fundamental for the high Victorians. First, the naturalistic explanation of phenomena, coupled with the inexorable hegemony of laws of nature over the individual which raised

202   DARWINISM'S GENERATIONS

questions about the possibility of free will. Second, the adaptation to environment as a causal force applied broadly, so that the changing nature of ideas, institutions, social mores, even religions, aesthetics, and ethics, all could be seen to be explained by their fitness for the context in which they operated. This (thirdly) reinforced the historical and the historicist method, the imperative to uncover the genealogies of all things, and to recognise the coherence and interrelation of contemporaneous configurations and of history as a record of their successive decline and replacement. And fourthly, Darwin's insight, as Pater put it in *Plato and Platonism* (1893), that '"type" itself properly *is* not but is only always *becoming*', which encouraged the collapse of fixed categories everywhere.[242] Julia Wedgwood spoke of this 'substitution of a world making for a world made', as 'the greatest in our intellectual history'.[243]

Wherever one looks in Victorian intellectual culture in the final three decades of the century it is possible, even allowing for the alternative evolutionary traditions of Comte and Spencer, to see the active influence of the high Victorians' Darwinism. The more thorough-going Darwinian philology of A.H. Sayce (*The Principles of Comparative Philology* (1874), and *Lectures on the Science of Language* (1880)), and of Henry Sweet (1845–1912), squared up to what they deemed the 'drawing room' methods of Max Müller.[244] Müller's theories about early religion were also mercilessly targeted in the 1870s and 1880s by Andrew Lang, one of a new cohort of more overtly evolutionary anthropologists, including Lorimer Fison (1832–1907), John Stuart-Glennie (1841–1910), and especially E.B. Tylor.[245] Under the influence of Michael Foster, British physiology developed a distinctive evolutionary frame,[246] and following James Ward, James Sully, and John Hughlings Jackson, evolutionary approaches came to dominate British psychology.[247] In philosophy, for many high Victorians, Darwinism offered an antidote to the vague metaphysics of Hegelianism: as George Birkbeck Hill (1835–1903) put it in 1892, 'How much rubbish has the mighty Darwin swept away!'.[248] Nor was it just the life sciences. Imbued with evolutionary ideas, it became the overriding ambition of Norman Lockyer to apply Darwin to the inorganic world.[249]

Where the high Victorians paused was at the direct application of evolutionary forces to contemporary culture and society. Despite his underlying enthusiasm for Darwin's evolutionary sweep, having read *Descent*, John Morley confessed that in respect of the social implications of the theory, 'I don't find Darwin at all satisfactory'.[250] In part this was a question of ethics. Like many mid Victorians, Morley felt that 'all that about ethical

THE *DESCENT OF MAN* AND THE HIGH VICTORIANS   203

evolution and the function of Natural Selection in Civilisation is very queer and doubtful.'[251] In part it was a question of scepticism about the applicability of the principle of the survival of the fittest to human history. For Clifford Allbutt (1836–1925), there was no more brutal and hopeless counsel than to apply crude Darwinism to humanity.[252] Henry Sidgwick offered several refutations in the 1870s.[253] Economists including Jevons and Alfred Marshall (1842–1924), although fulsome in their praise of Darwin's biology, questioned its straightforward application to economic behaviour. Marshall criticised Darwin's 'naïve simplicity', rejecting the analogy of biology and psychology.[254] In much the same way, the philosopher and social reformer Jane Clapperton (1832–1914) was clear as to the limits of Darwinian thought in solving late nineteenth century social questions, reportedly telling one correspondent, in respect of the problem of controlling population, 'Prof Huxley, like Darwin, takes us up to a dilemma and leaves us there.'[255] One potential exception, on the basis of an address of 1882, was the New Zealand geologist F.W. Hutton; but it is notable that Hutton's attempts to publish his theories on the application of selection to society were not encouraged by other high Victorians.[256] Otherwise, it is only the very youngest of the generation, like Frederick Pollock, quite happy to conclude 'as an Aryan and Darwinian' that there was 'no reason for preserving inferior and savage races except that the superior ones get demoralised in the process of supplanting them if it is done by violence', who hint at the beginnings of what became a widespread social Darwinism.[257]

It is difficult if not impossible to discuss high-Victorian cultural Darwinism without the sort of inferential readings I have been keen to avoid. As George Levine for literature and Barbara Larson for art have argued, much that has been attributed to specifically Darwinian influences cannot readily be disentangled from Darwinism's own deep roots in earlier traditions.[258] In art, the high Victorian drawn to the medievalism of the arts and crafts movement could find Darwinian sensibilities alienating. Compton Reade (1834–1909), nephew of the novelist Charles Reade, who had been a contemporary of William Morris at Oxford, lamented in 1904 as 'a young man of seventy' that Darwin's gospel was one of sheer materialism, and that Nonconformity and agnosticism in alliance had 'crushed the old cult of the beautiful with its heavenly symbolism'.[259] But it is possible to see some Darwinian engagements in the high Victorians' art of the 1870s, most explicitly in George Bouverie Goddard (1832–86), 'The Struggle for Existence' (1879), but also in some of the compositions of Marianne North (1830–90), including her 'Cluster of Air Roots of a Dragon Tree, Tenerife'

204   DARWINISM'S GENERATIONS

which reflect adaptive strategies in the face of flora's struggle for existence.[260] And as John Holmes has argued, we might see the portrait of the dying anti-Darwinian John Gould by John Everett Millais (1829–96) as a staging of the passing not just of one natural theologian, but 'of the generation'.[261]

High-Victorian Theology

High Victorians recognised that Darwinian evolution required a changed relationship with the world and with God. The incorporation of humanity into the evolutionary process cast doubt on some of the central tenets of Christianity, including the creation of man in God's image, the 'Fall', the incarnation of God into an essentially bestial form, and the Atonement. In the same way, the arguments of high-Victorian anthropology that religion and even ethics were themselves products of evolution undercut doctrines of the origin of human conscience and of divine inspiration. If man evolved from non-human beasts and ethics evolved as well, at what point was humanity endowed with a soul, and what role was left for it to play in the moral sense?

For a significant number of those born in the years after 1830, these challenges completed the destruction of the foundations of Christian belief which geology and biblical criticism had begun. The desperate search for spiritual meaning in the face of encroaching materialism has long been recognised as at the heart of the high-Victorian condition. In their willing embrace of religious scepticism, many high-Victorian intellectuals seized readily on Darwin. As Leslie Stephen put it in 1901, in abandoning Christian belief, he had taken his 'consolations' from Darwin, 'and they do as well as anything else'.[262]

Perhaps ironically, as a result the gap between the secular and the religious, the scientific and the theological, narrowed markedly. While older clerics continued to claim the right to 'sit in judgment' on science, the high Victorians asked what a credible religion could look like, given the broad truth of evolution.[263] One strand of this process can be seen in the circle of correspondents around Victoria, Lady Welby, which sought to reconcile religion and science by establishing a middle ground between religious resistance and what Mary Boole described as a scientific 'passion of disgust', struggling, as Boole put it, to find solutions for an age divided between two half truths it could not put together.[264] Another strand flowed into the Society for Psychical Research, drawing in many high and late Victorians,

THE *DESCENT OF MAN* AND THE HIGH VICTORIANS 205

who sought in varieties of spiritualism a scientifically endorsed assurance of future life. For F.H. Myers (1843–1901), a prominent figure in the Society, the 'first flush of triumphant Darwinism' was part of the 'very flood-tide of Materialism and Agnosticism...when terrene evolution had explained so much that men hardly cared to look beyond'.[265] A third strand emerged in the variety of secular faiths infused as much by Spencerian progressivism as Darwinism per se, which took root in the 1870s and 1880s, including the Positivism of Frederic Harrison, and ultimately theosophy and the occult.[266]

On the other hand, it is perhaps the ability of theologians to reach an effective accommodation with evolutionary challenges which most clearly shows the distinct temper of the high Victorians. Despite the jeremiads of their elders and the loss of faith of many high-profile contemporaries, most high Victorians were able to find sufficient intellectual resources to reconcile their faith even having conceded to Darwinism its enlarged scope. The challenge of the evolution of man was addressed by acknowledging the existence of man before he became a spiritual being.[267] The evolution of ethics could be presented as the inevitable response of man to the presence of the divine will in the evolution of the world, or as part of an emerging response to what Alexander Henry Craufurd (1843–1917) described as 'the supernatural man within'.[268] Ultimately, much of the critical discussion of Darwinism from high Victorians was reframed as a repudiation not so much of Darwin's biology, as of the attempt to raise it into an 'all-embracing cosmic philosophy', as one put it.[269] High-Victorian Methodists and Independents accommodated, if they did not enthusiastically embrace Darwin, in ways seemingly impossible for their congregational elders, so much so that by the 1880s they were increasingly inclined to deprecate the preoccupation of older preachers with wrestling with Darwin in the pulpit.[270]

As the century progressed, successive revisions of Darwin's texts and the debates within Darwinian science yielded opportunities for the hesitant to prise open room for manoeuvre around what exactly Darwin had advanced.[271] One recourse was to draw a hard line between evolutionary dynamics in the physical and in the moral universes (which had the advantage of offering a line of rebuttal to the tendency of scientists to apply mechanical explanations to spiritual life).[272] Darwinism was true, but not part of the essential truths. Theology had strength enough to prevail against a doctrine not directly addressed to it. Hence the recollections of the Congregationalist A.M. Fairbairn (1838–1912), that despite the upheavals which the *Origin* initially brought, the more he pursued the study of theology, 'the

206   DARWINISM'S GENERATIONS

less was Darwin trusted as a safe guide through its intricacies'.[273] The children of the Broad Church Anglican James M. Wilson (1836–1931) recalled that the *Origin* made little difference to him, in that he 'did without the thought of God for a while', but with continued prayer and reading the Bible he eventually came round to acceptance of both God and evolution.[274]

Of course, acceptance was by no means exclusively a matter of separation. The evolutionary record offered many resources for a marine biologist like H.N. Moseley; not least the coexistence of stable as well as evolving forms as evidence of an underlying plan and process.[275] One response was to undercut the significance of Darwin by emphasising the long history of evolutionary perspectives. Another was to sidestep the critical biological details and opt instead for a broad conception which effectively stripped evolution of its naturalism, not denying but effectively effacing natural selection and insisting on the need for a higher power.[276] Often this involved subsuming Darwin into a broader evolutionary synthesis with Spencer, if not also Hegel, in which the vagueness of Spencer's metaphysics could paper over the intractability of Darwin's biology. George Matheson's *Can the Old Faith Live with the New* (1885), enthusiastically welcomed in religious circles, offers a conspicuous example of this approach.[277]

Whatever their particular preference, there is plenty of evidence that high-Victorian divines found it much easier than their older brethren to live with Darwin. Alexander Whyte (1836–1921), the Free Church minister and theologian, was said to have read Darwin and Spencer 'repeatedly and attentively', and was happy to give Darwin's *Life and Letters* as a gift to ministerial friends.[278] Whyte's contemporary, the Free Church minister William Wynne Peyton (1831–1924), described as 'a naturalist, thoroughly versed in modern science', was 'fully convinced of the doctrine of evolution, and appl[ied] it as boldly to the facts of religious life as he would to the phenomena of Nature' (although his evolution was more Spencerian (and so Lamarckian) than Darwinian).[279] For high Victorians like Hugh R. Haweis (1838–1901), popular lecturer and a prominent member of the Broad Church party, evolution was taken as a matter of fact, and Darwinian references liberally scattered into their preaching and platform work.[280]

Prominent amongst the group of high-Victorian theologians alongside Matheson were figures like Bonney, Stebbing, Dallinger, the well-known Nonconformist minister John Clifford (1836–1923), and the Free Church minister James Iverach (1839–1922). Not all offered straightforward endorsements of Darwinism. No doubt mindful of the sensibilities of their very mixed congregations, high-Victorian ministers often sought to diffuse the threat of evolution 'even if' it were eventually scientifically proven. Again, George Matheson offers a sustained example, continuing to deny his own acceptance of Darwinism, while acknowledging the common origin of all species, indeed of all things.[281] Some were prepared to be more emphatic in their endorsement, while retaining space for doubters. For Iverach it was clear that the theory of evolution had come to stay, providing a point of view so fruitful and suggestive that it should be accepted, at least as a working hypothesis.[282] The Unitarian Joseph Estlin Carpenter (1844–1927) was convinced that all questions of doctrine and institution, of creed and ritual, needed to be studied in the light of their progressive development.[283]

The reconciliation of Darwin and Christianity might have been even more straightforward had it not been for Huxley's uncompromising bluntness. James M. Wilson, searching for a basis for his new pastoral role as Archdeacon of Manchester, was one who felt the contradictory push of Huxley's forceful vigour, but also the pull of a gentler approach, 'by parable and illusion, rather than negations and disillusion'.[284] Edward Talbot was less understanding, and became frustrated at Huxley's unwillingness to treat Anglican attempts to incorporate an evolutionary conception into their faith with other than contempt.[285] In a fascinating contribution in 1888, Talbot explained that 'the relief which the Christian may obtain through Darwinism' came from its consideration of the natural world as a whole, pervaded with rationality, and in which, as Darwin explicitly stated, 'happiness decidedly prevails', even if the relations of cat and mouse appeared merely violent and destructive.[286]

Anxiety that evolutionary ideas continued to be used to deny the existence of God preoccupied older theologians.[287] German thinkers, most prominently Haeckel, remained a powerful reminder of the conclusions to which evolutionary thinking could be taken.[288] Given this, it is not surprising that amongst clerical commentators not a few, like the popular Baptist preacher Charles Spurgeon (1834–92), clung to old hostilities.[289] We cannot ignore the persistence of disavowal, although it usually operated within a changed rhetoric in which challenges were presented rather than direct contradiction offered. R.B. Girdlestone (1836–1923), canon of Christ Church Oxford, offers one example of a high Victorian continuing through life to operate within a largely mid-Victorian frame, in which 'the phenomena of growth are in themselves marks of design'.[290] But more typical was someone like the nature writer Alice Bodington (1840–97), who had nothing but disdain for those who tried to resist the Darwinian demolition of natural theology.[291]

208   DARWINISM'S GENERATIONS

This new Christian Darwinism was also encouraged by the increasingly common presentation of evolution as modal rather than causal: the means by which creation was effected, but not of itself an explanation. Bonney was one who regretted the failure of older clerics to recognise the 'oneness of the power at the back of nature', that all creation was 'the outcome of an omniscient mind and what are called secondary causes as only his modes of working'.[292] Even miracles could be accommodated as extraordinary interventions; for the Anglican cleric John James Lias (1834–1923) Darwin had shown that all nature's processes were subject to law, but this did not preclude a creative divine energy, or mean that the ordinary was never subject to extraordinary law.[293]

Providential oversight allowed a greater sense of God at work at all times. For A.M. Fairbairn, the Creator must not just breathe life, but be always acting. This avoided the dangers of simple first cause arguments which allowed the divine a role in an initial act of creation, but thereafter rigorously excluded further action (a position criticised by the early Victorian John Stevens Henslow (1796–1861): '"God", he observed, "did not set the creation going like a clock, wound-up to go by itself, but from time to time interposes and directs things as he sees fit"').[294] In this sense evolution strengthened the divine presence; not as the skilful manufacturer, but as an eternal force or will which worked through and for humankind.[295] As James Orr put it, evolution had become a new name for 'creation', with the creative power working from within, instead of in an external plastic fashion.[296] The dominant frame employed was of the transcendence of God. Matheson pinned Spencer to his transcendentalism before developing his own account of evolution as the workings of a transcendent God; the Methodist W.L. Walker (1845–1930) continued to stress the need to avoid collapsing the Divine into the world in the way younger advocates of the 'New Theology' came to do: the potent force of evolution was at each stage transcendent to that which was evolving.[297] This sort of active but ultimately external presence could appeal even to those who, like the novelist Walter Besant, had abandoned their family evangelicalism, but who were able to discern 'an intelligent Mind who hears, listens, guides, and directs;...which leads a Darwin in the direction of discovery,...which has ordered the evolution of an insect as much as that of a man'.[298] Mid Victorians like Kingsley were unimpressed: 'To talk of its being done by laws impressed on matter', he retorted, 'is...a mere metaphor...useless for exact science.'[299]

Abiogenesis

One aspect of the debate over materialism in the 1870s drew especially on the implications of evolutionary theory explored by Huxley's essay 'Protoplasm: or the Physical Basis of Life' (1869), Tyndall's Belfast address, and the subsequent appearance of the translation of Haeckel's *The History of Creation* (1876): the question of the origin of life. It was an issue on which religious figures could balance their support for evolution by putting distance between themselves and the materialism of Tyndall and Huxley. By stressing the lack of any sort of evidence of process, or real theory of how life might have come about naturally, they were able to maintain space for Divine Creation, irrespective of wider questions of biological progression. For the Darwinians, the origin of life presented a dilemma. It was clearly implicit in claims to scientific naturalism that there was an explanation other than external divine intervention. Even if in practice the starting point of biological evolution was conceived of as a single originating instance, logic required that a natural event once occurred, might occur again, potentially many times. But this was a topic *Descent* steered clear of, so the question remained: was life the product of the act of a Creator, or of natural processes, and in either case was this a single or repeated event? At times it seems that Huxley wanted to commit both to the ancient and also the potentially contemporary creation of life. He championed 'protoplasm' as a sort of primordial soup of organic matter which could form the basis of all life. But he was also determined to distance Darwinism from the less reputable elements of fringe science associated with 'spontaneous generation', a question which was itself complicated by distinctions between those who saw spontaneous generation as the creation of new life forms out of the recombination of existing life (heterogenesis), and those who looked to demonstrate the complete *ab initio* creation of life (abiogenesis).

Despite some earlier dabbling with ideas of heterogenesis, not least on the part of Richard Owen himself, the response in the 1870s of the early Victorians was overwhelmingly hostile.[300] Debate at the Victoria Institute in 1875 linked opposition to abiogenesis with arguments against transmutation, not least that there was no evidence of either creation of life itself or of new species.[301] Blackie commented that he was too good a theist for spontaneous generation to have any terrors for him; 'as pure materialism is simply nonsense.'[302] The mid Victorians also remained unwilling to give up a divine role in the critical transition from inert matter to life. Hence William

210   DARWINISM'S GENERATIONS

Thomson's pronouncement to the British Association in 1871 that he was 'ready to accept as an article of faith in science, valid for all time and in all space. That Life is produced by Life, and only by life'; and hence also his bizarre musings about the possibility of life arriving on meteors.[303] Mid Victorians continued to draw solace from the fact that the researches of the scientists had never shown that life had been evolved from non-life, and that all scientific investigation had confirmed that every living creature had come from a pre-existing living creature.[304]

In the years after 1859 a lively debate swirled around this question. By far the most vocal opponent of ideas of abiogenesis was the King's College physiologist, Lionel Beale.[305] Beale was an obdurate adversary, prepared, as he told Acland, to 'debate the matter inch by inch...until I am utterly beaten or the physical philosophers beat a retreat'.[306] He was supported by J.H. Stirling, most forthrightly in his *As Regards Protoplasm in relation to Professor Huxley's essay on the Physical Basis of Life* (1869, 2nd ed. 1872), which took Huxley to task for professing to reject materialism while deliberately offering an argument which in reality only strengthened it.[307] Stirling's arguments were enthusiastically welcomed by other anti-Darwinians; John Herschel apparently commented that he could not imagine '[a]nything more complete and final in the way of refutation than this Essay'.[308] Despite occasional efforts to inoculate Christian belief against even the possibility of the natural creation of life, this absolute rejection was a fairly consistent position for mid Victorians to the end of the century.[309]

In the 1860s their main antagonist was the brilliant young Darwinian physician, Henry Charlton Bastian (1837–1915), who undertook extensive experiments in which sealed and apparently sterilised test tubes produced life. Bastian's claims of abiogenesis were supported by a number of other high Victorians including Clifford Allbutt and T.R.R. Stebbing.[310] The problem with Bastian's approach was the difficulty of demonstrating that all existing life in his test tubes had indeed been destroyed with no subsequent contamination; and Bastian's refusal to defer to counselled caution in publishing his experimental results, and his increasingly reckless attempts to press the necessity of spontaneous generation, drove a wedge between him and the leading Darwinists.[311] It was a claim, F.W. Farrar argued, which he derived not from Darwin, who 'has distinctly declared himself against an atheistic materialism', but from Darwin's 'violent and reckless followers'.[312]

Although Bastian's *The Beginnings of Life* (1872) was not without its champions, the X-Clubbers and their high-Victorian followers closed ranks.[313] Huxley dismissed him as 'a clumsy experimenter & an uncritical reasoner'.[314] Advances in the knowledge of microbes only served to confirm the likelihood that his experiments were capturing previously undestroyed life.[315] Bastian continued to champion his researches in the press and on the platform through the 1870s, though by 1880 in the face of the powerful critique offered by John Tyndall's work on germs he had withdrawn from the fray.[316] Thereafter, this was something of a backwater in Darwinian debates. But the inability to demonstrate abiogenesis, and the lack of appetite for discussion of the topic which resulted, left Darwinian evolution vulnerable to criticism, and opened up particular spaces for high-Victorian doubt. For Frederick Temple in his Bampton Lectures of 1884, it was not possible to accept life as a 'mere evolution from organic matter'.[317] From this perspective, all that evolutionary theory had done was to locate the moment of divine action more precisely in the initial creation of life.[318]

For at least some late Victorians this seemed an increasingly pointless debate. George J. Romanes, who became one of the most controversial of Darwin's late-century disciples, dismissed it as irrelevant to the fundamental proofs of evolutionary theory, although some younger readers could not entirely dismiss the theory.[319] By the end of the century most high Victorians with religious beliefs were confident that there was no scientific basis for contemporary abiogenesis; others were happy with a relatively instinctual belief in what one was wont to describe as the 'lucky thunderstorm' theory.[320]

Conclusion

The extent to which Darwinism was normalised across Victorian culture in the 1870s can easily be exaggerated. Careful examination of the fiction of the 1870s and 1880s suggests that it was only in the latter decade that Darwinian references began to proliferate. Notwithstanding Eliot's *Middlemarch* (1871–2) and *Daniel Deronda* (1876) (not least its chapter on 'Sexual Selection' in direct address to *Descent*), much of the flavour of the fiction of the 1870s was supplied by mid-Victorian satires of evolutionary belief.[321] It is true that the sort of playfulness which was generationally uncharacteristic in Kingsley's *The Water Babies* becomes more common, not least in the *Alice* books of Lewis Carroll (1832–98), *Alice in Wonderland* (1865), *Through the Looking Glass* (1872), and in *The Hunting of the Snark* (1876). But for many high Victorians, most strikingly M.E. Braddon, but also Walter Besant and George Du Maurier (1834–93), explicit Darwinian

212   DARWINISM'S GENERATIONS

references are notably scarce in their novels of the 1870s, in contrast to their later writings.[322] There were exceptions, including Samuel Butler's *Erewhon* (1872), in part an explicit attempt to explore the threats that might be posed to humanity by natural selection, Winwood Reade's *The Outcast* (1875), and perhaps Thomas Hardy's exploration of the relationship of personality and environment in *The Return of the Native* (1878), but these have largely assumed significance only in retrospect, or involved only the most fleeting of references. Perhaps to a lesser extent, the same is true for poetry, where volumes such as Swinburne's *Songs Before Sunrise* (1871), and less prominent contributions by William Canton (1845–1926), Robert Buchanan, and Louisa Sarah Bevington (1845–95) (whose *Key Notes* (1876) offers several evolutionary poems), need to be balanced against the many who only developed their evolutionary preoccupations in the 1880s and 1890s, including Mathilde Blind, whose *Ascent of Man* did not appear until 1888.[323]

But by the start of the 1880s there is no doubt that contemporaries felt a tilting of the balance. Self-confessed old fogeys found themselves challenged by the conversational gambits of 'young ladies' who 'adored' Darwin and dissected the latest pronouncement at the British Association.[324] Huxley's celebration of the coming of age of the *Origin* in a lecture to the Royal Institution in April 1880 was characteristically bullish, while largely content with celebrating the transformation in the tone of the debate from the outraged dismissals of the 1860s to the respectful discussions of the 1880s. Even Argyll, so confident in the response to his writings a decade earlier, now found the state of philosophy and belief 'a matter of grave anxiety'.[325] Viewed from a high-Victorian perspective 'The generation of taper-headed youths who airily express their conviction that "Darwin's theory is all humbug"' had not yet died out, but the debates of the scientific societies featured 'quite a different and much more qualified set of young men'.[326] The balance of comment was much more favourable, and it was much less common to find a dismissive response to Darwin. At the Edinburgh University Philosophical Society in November 1881, the high-Victorian philosopher Edward Caird (1835–1908) launched a (thinly veiled) attack on older figures like Blackie, who he accused of 'repeating against men like Mill and Darwin the old watchwords with which Plato attacked the Sophists'. Those who want to develop philosophy he was reported as warning, must renounce the 'questionable luxury of contempt'.[327] Henry Liddon saw rapid change, even in conservative Oxford.[328] The rather belated decision of Cambridge University in 1877 to award Darwin an honorary degree (and to commission a memorial portrait) was perhaps less telling than the proceedings

THE *DESCENT OF MAN* AND THE HIGH VICTORIANS 213

themselves. The accommodation for students had to be increased to meet the enormous crowds who wanted to be present. The atmosphere was boisterous but overwhelmingly sympathetic (and certainly without the bitterness which might have been in evidence a decade earlier). Just as Darwin presented himself to the Vice-Chancellor, a large hairy imitation of an ape dressed in academic costume and dangling a rusty link of chain labelled 'Missing' was lowered down a thin cord which had been strung from one gallery to the other and left dangling just above Darwin's head. Some accounts speak of various monkeys also being thrown down from the galleries, and although after a desperate intervention from an official the effigy was dislodged, the 'missing link' remained suspended above Darwin for the rest of the proceedings.

Even so, for many this felt more like the beginning of a significant shift than its completion. For the historian Peter Bayne in 1879, after twenty years of discussion, the meaning of evolution was 'at length beginning to be understood'.[329] Significantly, the Cambridge nomination oration hedged its bets, focussing primarily on Darwin's non-evolutionary texts, its concluding notice of the *Origin* placing it firmly in the context of the works of Lucretius.[330] It still seemed credible for the mid-Victorian novelist Eustace Murray (1824–81), in his 1879 novel *The Artful Vicar*, to include a set piece in which the young vicar is bated by proponents of Darwinism, their leader 'threatening' to donate a copy of the *Descent of Man* to the local mechanics' institute.[331] At the same time audiences for Frank Frankfort Moore's farce 'Moth and Flame' were being invited to laugh at the veteran entomologist Dr Dugdale, whose mania for Darwinian theories drove him to constantly parade them at the most inappropriate moments.[332] Despite Lubbock offering yet another presidential celebration of the acceptance of Darwinism in 1881, the meetings of the British Association offer plenty of signs of resistance. The 'mad' fringe of F.O. Morris and his ilk was largely excluded, but as late as 1878 there was talk of audience frustration at not being given an opportunity to take issue with Romanes' pro-Darwinian lecture, and in 1879 at Sheffield the X-Clubbers were lamenting 'a reactionary meeting', with anti-Darwinian papers, and Samuel Haughton, 'rabid against Darwinism in an after dinner speech'.[333] Four years later the geologist Thomas Wright (1809–84) was still reiterating his outright denial of evolution.[334] There is little sign that the tastes of provincial lecture audiences had shifted decisively against expressions of Darwinian scepticism. The veteran anti-Darwinian Benjamin Waterhouse Hawkins was still finding a place on the programmes of local natural history societies into his seventies.[335] Thomas

214   DARWINISM'S GENERATIONS

Cooper continued to draw large crowds well into the 1880s (although for the young Todmorden Methodist John Naylor (1868–1945), hearing Cooper 'only increased his early evolutionary ardour').[336] Darwin the man was still something of a cipher, judging by the frequency with which coverage in the provincial press identified him as Dr or even 'Professor'. The *Origin* and *Descent* were still selling steadily, but there remained significant gaps in their penetration. In the Sixth Report of the Royal Commission on Scientific Instruction and Advancement of Science, Manchester Grammar School was the only school to mention the availability of a Darwin text (*Descent of Man*) as a reference book to which scholars had access.

Many high Victorians were themselves still only working through their responses to Darwin at the end of the decade. At Glasgow, John Clelland (1835–1924), Professor of Anatomy at Glasgow, continued to regale students with denunciations of Huxley and a strange non-Darwinian version of evolution.[337] In 1883 the Leeds obstetrician James Braithwaite (1838–1919) purchased his own copy of *Descent*. The surviving volume, extensively annotated, offers an insight into some of the limits to the spread of Darwinism, even at this stage. Braithwaite should have been a natural convert. Admittedly Anglican and Conservative, but with no obvious sign of militancy in either; a respected provincial man of science, member of the Leeds Literary and Philosophical Society, editor of the twice-yearly *Retrospect of Medicine* and occasional contributor of research papers to the medical press, a medical student at the moment the *Origin* had originally appeared, he appears ideally placed to absorb and reflect the shifts of scientific temper of the previous twenty-five years. Yet Braithwaite's reading was fiercely sceptical. His marginalia disputed Darwin's comparative anatomy and argument from homology, challenged his inability to produce an evolutionary sequence of fossils from simian ancestor to man, and remained deeply wary of his evolutionary conclusions.[338]

In the same way, Charles Booth (1840–1916), shipping magnate and later pioneering social investigator, has left notebooks which demonstrate an extensive effort to clarify his thinking about Darwin and evolution in the period from 1879 to 1882. By the end of this period, Booth clearly thought of himself as a evolutionist and even as committed to a form of scientific naturalism; but he also recognised that this was in part a matter of faith, and that there remained some fundamental challenges to the argument that natural selection was the dominant mechanism of evolution.[339] Booth was also doubtful about Darwin's attempts to apply the survival of the fittest principles to questions of instincts and ethics, and was clearly troubled by

THE *DESCENT OF MAN* AND THE HIGH VICTORIANS   215

the role that might be ascribed to some ultimate purpose, whether this was of divine origin or not. There remained an unresolved tension between nature 'perfecting itself in relation to the forces by which it is surrounded' or 'reaching out to an inherent perfection'.[340]

Booth's and Braithwaite's readings offer a clear warning against accelerating the acceptance of Darwinian ideas. Resistance was clearly widespread. The Argyle Mutual Improvement Society in Bath voted in 1881 by a substantial majority that the theory enunciated by Darwin in the *Origin* was unworthy of belief, because it denied the existence of a Creator, and because the arguments were not sufficiently conclusive.[341] Much of this was the persistence of early- and mid-Victorian opposition which continued to generate friction with younger supporters. Tracing the generational patternings of local debates is complicated by the difficulties of establishing the age of many of the participants, but the overall picture is clear: younger members advocating Darwinian positions, often in the face of a solid phalanx of disbelieving older members. And so, for example, a pro-Darwinian lecture at the Stroud Natural History and Philosophical Society in 1879 was met by resistance from Joseph Timbrell Fisher Jnr, solicitor and insurance agent (1829–83), and Edwin Witchell, FGS (1823–87), solicitor, treasurer of the Cotteswold Naturalists' Field Club, even if Fisher was prepared to accept that the responses of many in his position were coloured by 'preconceived notions taught in our childhood and which grow up with us'.[342] Evolution as a topic was still sufficiently sensitive in 1880 for W.H. Dallinger to be hastily replaced as the annual Fernley Lecturer at the Wesleyan Conference, in the face both of the stern hostility of older Methodist leaders like George Osborn (1808–91) and a general anxiety about controversy; and even more tellingly for George du Maurier to have been instructed by the editor of *Punch* in 1880 to take out a reference to Darwin from one of his squibs, for fear of antagonising the paper's readership.[343] As the *Origin* came of age in 1880, it continued to produce not a consensus in its favour but a generationally profuse cacophony of controversy.

Notes

1. Foreshadowed by Huxley's lecture 'On the Physical Basis of Life' in 1868.
2. E.B. Tylor to Juliet Tylor Morse, 12 January 1868, Tylor Correspondence, Box 3, Folder 10, Evolution Collection, Mss 28, UC-Santa Barbara.
3. *Academy*, 2 January 1875, Hooker to Darwin, 6 October 1865, in *Life of Hooker*, II, 54.
4. E. Royle, *Radicals, Secularists and Republicans* (1980), 172.

216 DARWINISM'S GENERATIONS

5. J. Bruce Glasier, *The Meaning of Socialism* (1920), 238.
6. 'On the Pedigree of the Horse', *Proceedings of the Royal Institution* 6 (1870), 129.
7. Nixon, *Hopkins*, 109–12.
8. 'The Descent of Man', *Edinburgh Review* 134 (1871), 99, 99–120.
9. *Era*, 5 November 1871.
10. An account by Frank Buckland, in *Land and Water*, reprinted *Newcastle Guardian*, 10 August 1872.
11. *Westmorland Gazette*, 28 October 1871. See also Winwood Reade's novel *The Outcast* (1875), which has a character who commits suicide having lost his faith reading Malthus and the *Origin*.
12. Copy sold by Broli.com, ebay number 144583967571.
13. William Connor Magee (1821–91)'s Easter Day sermon at St Paul's in 1871 included 'a masterpiece of scathing sarcasm directed against Mr Darwin's book', *Manchester Times*, 15 April 1871.
14. William Gresley (1801–76), *Thoughts on the Bible* (1872); J.H. Stirling to C.M. Ingleby, 1871, quoted in A.H. Stirling, *James Hutchison Stirling, His Life and Work* (1912), 338 (the published version has 'contort', but in the context it seems very likely the original was cavort); Harvey Goodwin, university sermon at Cambridge, *Cambridge Chronicle*, 3 June 1871; and the similar tone of Rev. Osborne Gordon (1813–83), George Marshall, *Osborne Gordon: A Memoir with a Selection of His Writings* (1885), 129.
15. *Standard*, 13 June 1871.
16. *Times*, 8 April 1871.
17. *Gardeners' Chronicle*, 20 May 1871, offers a good example, agreeing that there is only a difference of degree not kind between intelligence of man and animals, but still taking refuge in the idea of man's 'higher faculties' and denying that there is compelling evidence of evolution of man from beasts; also Alexander Goss (1814–72), *Drogheda Argus*, 15 April 1871.
18. *All the Year Round*, 8 April 1871, 445–50; *BMJ*, 15 April 1871.
19. *Nonconformist*, 4 May 1871, quoted by Ellegard, *Reception*, 314.
20. [J. Morley], 'Descent of Man', *PMG*, 20, 21 March 1871; see also T.R.R. Stebbing in *Nature*, 20 April 1871.
21. See the response of W. Boyd Dawkins, analysed in detail in Meiring, 'Scientific Patronage'; Frederick Arnold (1833–98), 'Piccadilly Papers', *London Society* (April 1871), 371.
22. Neville Goodman (1831–90), review of *Descent*, and Mivarts' *Genesis of Species*, *Journal of Anatomy and Physiology* 5 (1871), 362–72.
23. The geologist Peter Bellinger Brodie (1815–97), *Warwick and Warwickshire Advertiser*, 11 March 1871; the Anglican cleric Alfred Gatty (1813–1903) condemned Darwin at the Yorkshire Episcopal Examinations, distribution of prizes, see *Sheffield Daily Telegraph*, 21 April 1871.
24. William Guest (1818–91), congregational minister and geologist, *Islington Gazette*, 3 March 1871.
25. Thomas Francis L'Anson (1825–98), surgeon to the Whitehaven and West Cumberland Infirmary, paper to Whitehaven Scientific Association, *Whitehaven News*, 21 March 1872. See also William Francis Wilkinson (*c.*1813–*c.*1890), *Leicester Guardian*, 27 March 1872.
26. Frederick Thompson Mott (1825–1908), *Leicester Journal*, 16 February 1872; Rev. William Kennedy Moore (*c.*1828–1905), *Cornubian and Redruth Times*, 15 September 1871; John Vertue (1826–1900), *Hampshire Telegraph*, 22 March 1873.
27. Elder, *Chronic*, 36.
28. Lyell, ed, *Life of Sir Charles J.F. Bunbury* II, 284. He later told Katherine Lyell that although he had still not finished *Descent*, and 'I am not a convert to his main theory', it was still a 'remarkable book', ibid., 287.
29. Sutherland to Florence Nightingale, 14 April 1871, f.206, Add. MS 45744, Nightingale Papers, BL. The mid-Victorian Bishop of Bombay felt the grandeur and simplicity of the underlying argument, and marvelled at the details, while retaining a generationally characteristic scepticism that the theory was without foundation in the facts, Maria Milman, *Memoir of the Rt. Rev. Robert Milman, D.D.* (1879), 183–4.

THE *DESCENT OF MAN* AND THE HIGH VICTORIANS 217

30. Christie's, New York, Lot 36 Online Auction, Fine Printed Books, 1–15 October 2021; my thanks to Rhiannon Knol at Christie's for images of the annotations in this volume.

31. Kingsley to Lockyer, 8 November 1872, Norman Lockyer Papers, EUL110, Exeter University. This had not stopped him telling the Devonshire Association of Science, Literature and Art in August 1871 that he deprecated *Descent's* attempt to mix arguments about the origins of language, morality, and beauty up with the theory of evolution, and would recommend the counter-arguments of Mivart's *Genesis of Species*; see *Cambridge Chronicle*, 19 August 1871. For the Baptist philosopher and journalist Thomas Spencer Baynes (1823–87) *Expression* abounded in the 'eager word catching...which characterise the lower forms of religious controversy', [Baynes], 'Darwin on Expression', *Edinburgh Review* 137 (1873), 492–528, discussed in Dawson, *Darwinism, Literature and Victorian Respectability*, 47–8.

32. Passages at pp. 34, 49, 62, 65, 70; Liddon also marked a few passages more aligned to Darwin, including the observation of Carl Vogt that 'no-one in Europe any longer supported the idea of separate creation in all cases', 1, Henry Liddon's copy of *Descent*, courtesy of Pryor-Johnson Rare Books, New York, on abebooks.co.uk [accessed 4 January 2021].

33. Henry Sidgwick to his mother in November 1872, Sidgwick Collection c.99/165, Trinity College, Cambridge; Morley to Harrison [1872], quoted in F.W. Hirst, *Early Life and Letters of John Morley* (1927), I, 224.

34. Secord, *Evolutionary Writings*, quoting James Crichton-Browne (1840–1938) to Darwin, 19 February 1871, DCP-LETT-7492, Darwin Correspondence. The psychiatric doctor Henry Maudsley (1835–1918) cautiously accepted the strength of the argument, Maudsley, *Body and Mind* (1871), 51. See also the review of *Descent* in *Nature*, 6, 13 April 1871, by P.H. Pye-Smith (1839–1914).

35. Stephen, *Letters of J.R. Green*, 292; Letter, 4 March 1871, printed in E. Lecky, *Memoir of W.E.H. Lecky* (1909), 78; Lecky had a childhood interest in geology; see W.E.H. Lecky to Sir W.H. Flower, 22 November 1896, Flower Papers, IC WHF 17/3, JIC.

36. William Howitt to William Oldham, 16 October 1874, Brigg Collection, Bg 11, University of Nottingham. See also the extended discussion of the first gorilla to learn how to use a flint tool, in W. Gresley (1801–76), *The Scepticism of the Nineteenth Century* (1879), 59–61; and also W. Gresley, 'Darwinism', the postscript to his *Thoughts on the Bible* (1872), 211–20. Of course, comedy was also a response of the mid Victorians, as in Robert Kemp Philp's comic poem about a monkey who, on hearing of Darwin's theories, gets ideas above his station, *Sunday Times*, 25 August 1872.

37. Blackie, *Natural History of Atheism* (1878), 5–6; Blackie to the Edinburgh Philosophical Society, *Dundee Courier*, 7 March 1871.

38. Henry Boase, *A Few Words on Evolution and Creation* (1882), v.

39. For an example, see Oliver Lodge's appreciation of Robert Garner (1808–90), reprinted in the *Staffordshire Advertiser*, 7 September 1940.

40. That is, Argyll's attempt to present the Bible as using clearly developmental languages in describing creation, *Reign of Law* (1867); Howard, *Creation and Providence*, *passim*. Other early Victorians expressing their opposition at the Institute include James Challis (1803–82), Derwent Coleridge (1800–83), and Thomas Griffith (c.1797–1878).

41. For a discussion of Cooper see T. Larsen, *Crisis of Doubt: Honest Faith in Nineteenth Century England* (2006), 72–108.

42. Morris also published *A Double Dilemma for Darwinism* (1877), and *Opinions of Men of Light and Leading...on the Darwinian Craze* (1880), and numerous articles in popular journals like the *Leisure Hour*. Thomas Birks followed his tract *The Scripture Doctrine of Creation, with Reference to Religious Rihilism and Modern Theories of Development* (1872), with *Modern Physical Fatalism and the Doctrine of Evolution* (1876); Birks' publications were a frequent resource for opponents of Darwinism in the 1860s and 1870s; for one attributable case, see John Evans (1840–97), *Wrexham Advertiser*, 29 May 1880. For Bree, see *An Exposition of Fallacies in the Hypothesis of Mr. Darwin* (1872), specifically rejecting the ideas of Mivart, 80–1; and also 'On Darwinism and Its Effects upon Religious Thought' to the Victoria Institute in 1872–3.

218 DARWINISM'S GENERATIONS

43. For Bree's continuing insistence on special creation see his letter, *Champion of Faith*, 3 August 1882, 126.
44. Grantley Fitzhardinge Berkeley (1800–81), *Facts against Fiction...with Some Remarks on Darwin* (1874), 5; see also as examples, John Benn Walsh, Baron Ormathwaite (1798–1881), *Astronomy and Geology Compared* (1872), Francis Close (1797–1882), *Our Family Likeness, Illustrative of Our Origin and Descent* (1871).
45. Pusey to Acland, 1 May (nd but *c*.1881-2), 1881–2 file, Pusey Papers, PHL; also Pusey to Sayce, 3 August 1879, ff.30–1, A.H. Sayce Correspondence, MS Eng lett d.62, Bodleian.
46. Ridley to Frank Darwin, July 1882 MSS.DAR.198/168, Darwin Papers, CUL; copy of Ridley to Pusey, nd, HNR/2/1/5/123–8, Ridley Papers, RBGK.
47. Bonamy Price to Liddon, 3 November 1878, Pusey to Liddon, 1871–9 volume, Pusey Papers, PHL.
48. Hubert Handley (1854–1943), 'Anglican Starvation', *TNC* 57 (1905), 987, describing Wilberforce's earlier BA intervention as 'ill-informed raillery in unsurpassable bad taste'.
49. See Ridley's unpublished autobiography, and other materials, HNR/3/3/1, HNR/2/1/5/ 113–127, Ridley Papers, RBGK.
50. Rupke, *Richard Owen*, 175. See, for example, Owen to C.O. Groom Napier, 19 March 1878, Misc MSS/Q(xxii), Royal College of Surgeons, and Owen to George Burrows, 15 October 1881, Temple University Miscellaneous Collections, PC5 (1040), cited in Rupke, *Richard Owen*, 177–9.
51. Holmes, *Pre-Raphaelites*, 232. For a determinedly pro-Owen version of this history see Rupke, *Richard Owen*, 36–43.
52. *Saturday Review*, 24 December 1894, in Beddard Scrapbook, Box 4, Folder 7, Evolution Collection, Mss 28, UC-Santa Barbara.
53. T. Douglas Murray to Richard Owen, 29 December 1880, Richard Owen Papers, B Ow2, APS.
54. Gladstone to W.S. Jevons, 10 May 1874, in D.C. Lathbury, *Correspondence of Gladstone on Church and Religion* (1910), II, 100–1.
55. James Knowles to Herbert Spencer, 24 November 1873, MS791/91, Herbert Spencer Papers, Senate House Library, UL; Gladstone, *Diary*, quoted in R. Shannon, *Gladstone: God and Politics* (2008), 247.
56. Gladstone to Moore, 10 December 1885, Moore Autograph Collection, B M781, APS. P.C. Mitchell recalled a conversation in 1889 or 1890 in which Gladstone seemed content to conclude that the debate was at present 'in a somewhat equivocal state', Mitchell, *My Fill of Days* (1937), 118.
57. William Sharpey to Allen Thomson, 17 February 1877, MS Gen 1476/A/Series 1/8890, Allen Thomson Papers, University of Glasgow.
58. See the account of George Robert Waterhouse (1810–88), naturalist, *Hardwicke's Science Gossip* 28 (1892), 118.
59. See Searles Valentine Wood, *Supplement to a Monograph of the Crag Mollusca. III. Univalves and Bivalves* (1874), 191. For reference to Jeffrey's paper at the Plymouth British Association, which strongly opposed the evolutionary position of Allen Thomson's presidential address, see *Proceedings of the Belfast Naturalists' Field Club 1877–78* (1878), 311–12. In 1880 he was able to quote his opinion of 1862 that the argument of the *Origin* was 'inconclusive and unsatisfactory', noting that he saw 'no reason to change his views', 'Presidential Address', *Transactions of the Hertfordshire Natural History Society and Field Club* 1 (1879–81), 89; see *Biograph and Review* 6 (1881), 377, and the obituary notice in *Western Mail*, 29 January 1885, which remarked that 'He was not a Darwinian in the full sense of that term'.
60. In this I take issue with Hale, 'Darwinian Revolution', 119–20: it was not that Mivart represented a 'few non-Darwinians' in contrast to those like Kingsley, Cobbe, etc., who continued to identify as Darwinians; both because it is only a small part of the story to suggest Cobbe and many other mid Victorians like her still thought of themselves as Darwinian, but also because the implications of *Descent* allowed both a reformulation of the basis of opposition to Darwin and shored up antagonistic positions just at the point at which rejection of some form of evolution was becoming increasingly difficult.

THE *DESCENT OF MAN* AND THE HIGH VICTORIANS   219

61. Smith, *Science of Energy*, 183. See the account of a conversation he had with Fleeming Jenkin in the later 1860s about how this argument was 'fatal' to Darwinism, in James Collier (1847–1925), 'Science up to Date', *Lyttleton Times*, 1 March 1913. Not everyone was equally impressed. D'Arcy Thompson recalled that his father 'firmly believed that [*Unseen Universe*] was a gigantic piece of fun on the part of Tait and Balfour Stewart', Thompson to C.G. Knott, tps, 4 June 1906, ms16120, D'Arcy Wentworth Thompson Papers, St Andrews.

62. Mivart, 'Mr Darwin and His Critics', *Contemporary Review* 18 (1871), 443–76. Mivart's key contributions included various articles in his *Essays and Criticisms*, as well as 'Primitive Man: Tylor and Lubbock', *Quarterly Review* 147 (1874), ibid., 'Reply to George Darwin', also his *Lessons from Nature* (1876) and two essays, 'The Meaning of Life' and 'The Government of Life', published in *TNC* in 1879. Mivart was strongly enjoined to take this stance by William Walter Roberts (1830–1911), who 'again and again' urged the impossibility of Darwinism explaining human intellect, R.J. Richards, *Darwin and the Emergence of Evolutionary Theories of the Mind* (2014), 353–63.

63. Jones, 'Darwinism in Ireland', 11–12. For Murphy's distinction between evolution, which he would support, and Darwinism, which he continued to oppose, see his 'Presidential Address on the Present State of the Darwinian Controversy', *Proceedings of the Belfast Natural History and Philosophical Society 1873–4* (1874), 1–24.

64. White to Stokes, 23 November 1887, MS Add. 7656/1W/406, Stokes Papers, CUL.

65. *BMJ*, 23 July 1889. Other examples include Margaret Bell Alder (1828/9–1902), Belfast naturalist, whose opposition is clear in her correspondence with Stokes, MS Add. 7656/1A/640–1, Stokes Papers, CUL.

66. Dawson's books included *The Story of the Earth and Man* (1873), which went through six editions by 1880, *The Dawn of Life* (1875), *Fossil Men and Their Modern Representatives* (1880), and *Facts and Fancies in Modern Science* (1882). For a flavour of the place Dawson occupied within religious resistance to Darwinism, see the review of *Fossil Men* in the *Literary Churchman*, 9 July 1880, and W.E. Gladstone, *The Impregnable Rock of Holy Scripture* (1890). For others, see for example the paleoanthropologist Thomas Hood Cockburn-Hood (1820–89), whose stance can been gleaned from his support for Owen and J.W. Dawson versus the 'wild theorist[s]', Scrapbook, MLMSS 9843/1–2, SLNSW.

67. Comment of clergyman, astronomer, and geologist Henry Cooper Key (1819–79), see *Transactions of the Woolhope Naturalist's Field Club for 1870* (1871), vi.

68. Leslie Stephen, *Life of James Fitzjames Stephen* (1895), 374–5. It was later reported that after Darwin's death Stephen wrote an article, never published, in which, provoked by a sermon of Liddon, he argued forcibly that anyone who believed Darwinism was a fool if he was a theist and a knave if he proclaimed Christianity; see *Review of Reviews* 39 (1909), 24, a position endorsed by Rev. Joseph Crompton to the Norfolk and Norwich Naturalists' Society, see *Transactions 1870–71* (1871), 8, and Acland's position was not dissimilar; see Acland to Owen (1862), quoted in Altay, *Memoir*, 305.

69. See Lightman, *Popularizers*, 169–96.

70. Swainson to G.J. Romanes, 21 August 1874, ff.39–40, Eng Mss d.3824, Romanes Papers, Bodleian.

71. C. Haddon to A.C. Haddon, 1 February 1878, Envelope 22, Haddon Papers, CUL. The Haddon papers also contain letters of his sisters discussing spontaneous generation and human evolution.

72. Anna Swanwick, *Evolution and the Religion of the Future* (1893, 2nd ed. 1894), 8–9; Swanwick, *Poets, the Interpreters of their Age* (1892), 341. For Swanwick see Mary L. Bruce, *Anna Swanwick: A Memoir and Recollections* (1903).

73. Fragmentary notes on topics relating to 'Spiritual Evolution of the Universe', in commonplace book, Box 33, Dep.208, A. Campbell Fraser Papers, NLS.

74. For example, Rev. Edward Duke (1814–95); see *JTVI* 16 (1883), 191–2.

75. Drysdale authored articles against spontaneous generation with W.H. Dallinger; see Dawson, *Darwin, Literature and Victorian Respectability*, 147–8.

76. Sanderson fully accepted the Darwinian revolution; see his presidential address to the BA in 1893: 'there was no true philosophy of living nature until Darwin', *Life of John Burdon*

220   DARWINISM'S GENERATIONS

Sanderson, 242. See also Richard D. French, 'Darwin and the Physiologists', *Journal of the History of Biology* 3.2 (1970), 253–74. But he also offers interesting idiosyncrasies which betray something of his generational position, not least the strongly Lamarckian undertones to the priority he gave to adaptation over descent, and his consequent rejection of Weismann; see J. Burdon Sanderson to Spencer, 2 November [1894], MS791/229, Herbert Spencer Papers, Senate House Library, UL; and his preference for the terminology of 'development' rather than 'evolution', see Burdon Sanderson to Galton, 25 October 1904, Galton/2/4/13/1/3, UCL; and Burdon Sanderson to Geddes, 5 March 1888, T-GED/9/29, Geddes Papers, Strathclyde University.

77. See Lionel Beale, 'Address of the President of the Royal Microscopical Society', *Science*, 25 June, 2 July 1881. For his earlier doubts, see, for example, *Nature*, 23 November 1871. His work was frequently used by more uncompromising opponnets, and Flower's suspicion in 1895 that Beale might have been the author of the critical review of Huxley's contribution to Owen's biography indicates the suspicion with which he continued to the end to be regarded, W.H. Flower to Huxley, 9 February 1895, f.136, Box 16, Huxley Papers, ICL.

78. Thomas Davidson to Marsh, 15 October 1879 Box 8, Folder 331, Othniel C. Marsh Correspondence, Yale; see also Davidson to 'Mr Jeffreys', 25 April 1880, Davidson folder, #7, Charles Moore Papers, Geological Society, London. There is evidence that this was not without significant 'pressure' from Darwin and Huxley for him to reconsider; see Davidson to Lyell, 21 November 1862, COLL 203/1/712–13, Lyell Papers, Edinburgh.

79. See *Leeds Times*, 11 March 1882, *Dundee Advertiser*, 14 February 1879.

80. Williamson's caveats can be traced through scattered references in the correspondence of his contemporaries and coverage of his various lectures; see *The Guardian*, 3 October 1877, *Cheshire Observer*, 10 November 1877; Huxley to Williamson, 18 November 1869, Huxley Papers, APS; Williamson to Owen, 28 August [18]77?, ff.47–8, Owen Collection, Gen Corr XXVII, NHM, which rehearsed the problems of evidence of the descent of the vertebrates from the molluscs through the ascidians and amphioxus; Williamson to WTTD, 12 January 1875, f.193, W.T. Thiselton-Dyer Papers, In-letters IV, RBGK. Williamson (increasingly conscious of his status as a 'regular old stager' as he described himself in 1886 [see Williamson to L.F. Ward, 16 June 1886, Lester F. Ward Papers, Brown University]) continued to emphasise the partiality of the fossil record and understandings of it; see Williamson, 'The Evolution of Paleozoic Vegetation', *Nature*, 27 October 1881, and the extract of his 1894 letter to Lester F. Ward printed in 'Saporta and Williamson and their work in Paleobotany', *Science*, 9 August 1895, 148. This was a reflection of the position of many palaeontologists, and particularly palaeobotanists, whose subject matter offered several intractable challenges to evolutionary ideas, including the angiosperms, whose unity, sudden appearance without apparent antecedent, and clear differentiation from all other orders, caused even Darwin to describe them as 'an abominable mystery'.

81. W.C. Williamson to W. Carruthers, 19 March 1889, Carruthers Papers, DF404/1/8, NHM.

82. Williamson to Ward, 10 June 1894, Autograph letter books, Subseries E, Lester F. Ward Papers, Brown University.

83. While indicating he was not afraid to listen to any satisfactory evidence, he confessed he had 'not paid much attention to the Darwinian theory and hardly any to biology', Stokes to J.H. Gladstone, 26 July 1872, MS/743/1/80, John Hall Gladstone Papers, Royal Society. For Stokes see Mark McCartney, Andrew Whitaker, and Alastair Wood, eds., *George Gabriel Stokes: Life, Science and Faith* (2019).

84. H.C. Sorby (1826–1908) to Stokes, 25 December 1880, MS Add. 7656/1S/1192, Stokes Papers, CUL. I'm grateful to Ian Hesketh for providing me with full details of this letter.

85. White to Stokes, 13 October 1882, MS Add. 7656/1W/451, Stokes Papers, CUL.

86. Stokes to Lubbock, 25 April 1889, ff.72–3, Add. MS 49678C, BL; Stokes, 'On the Absence of Real Opposition between Science and Revelation', *JTVI* 17 (1884), 200. Stokes' position was maintained in his Gifford Lectures on *Natural Theology* (1891–3), which suggested (somewhat enigmatically), that evolutionary thinking was helpful as long as it was only 'held so far as real evidence, or even probable evidence, fairly conducts us to it', quoted in Livingstone, *Forgotten Defenders*, 98.

THE *DESCENT OF MAN* AND THE HIGH VICTORIANS 221

87. See Karen Dieleman, 'Evolution and the Struggle of Love in Emily Pfeiffer's Sonnets', *Victorian Poetry* 54.3 (2016), 297–324.

88. F. Petrie to Tait, 23 May 1871, TAIT/177/1-3, Lambeth Palace Library.

89. L. Forbes Winslow, *Recollections of Forty Years* (1910), 358–9.

90. F. Petrie to Stokes, 12 September 1895, MS Add. 7656/1V/167, Stokes Papers, CUL.

91. See the account of the Anglican Rev. Edmund Botelier Chalmer (1828–83), who regaled his congregation during the Sheffield meeting of the BA with his own tongue-in-cheek version of evolutionary development, *Sheffield Independent*, 29 August 1879.

92. Correspondence with Dean Fremantle, reported in Tollemache, *Benjamin Jowett*, 68.

93. Cobbe, 'Darwinism in Morals', *Theological Review* 8 (1871), 175; Hale, *Political Descent*, 122.

94. Frances Power Cobbe, 'Agnostic Morality', *Contemporary Review* 43 (1883), 783–94, Cobbe, 'The Scientific Spirit of the Age', *Contemporary Review* 54 (1888), 126–39, Cobbe, 'The Lord Was Not in the Earthquake', *Contemporary Review* 53 (1888), 70–83.

95. G.W. Cox to Cobbe, 11 October 1875, Frances Power Cobbe Papers, Box 4, 1873–7, Huntington Library, San Marino, CA.

96. According to C. Brady, *James Anthony Froude: An Intellectual Biography of a Victorian Prophet* (2014), 382, citing 'Science and Theology, Ancient and Modern', in *The International Review* 5 (1878), 289–302; see also J.A. Froude, *Short Studies on Great Subjects* (1872), II, 'Calvinism', 53, Froude to Hooker, 13 May nd, JDH/1/2/8/287, Hooker Papers, RBGK; but see comments on Darwin in Knight, ed., *Rectorial Addresses* (1894), 126; writing to Huxley in 1877, Froude observed that he had long waited to see if Darwinians could 'elevate Evolution into a theory which will satisfy the eagerness of the imagination': this done 'then science has the world in its hands. If not, I cannot shake off the fear that we may have another era of…superstition before us', Froude to Huxley, quoted in Desmond, *Evolution's High Priest*, 126. This lack of commitment is visible in the language Froude continued to use, of Darwin having 'thrown doubts upon our supernatural origin', 'The Templars', *Good Words* 27 (1886), 378. Ultimately, as the literary critic Herbert Paul later commented, 'Froude cared no more for Darwin than Carlyle did', Paul, *Life of Froude* (1905), 391.

97. See his contributions to the *Proceedings of the Liverpool Literary and Philosophical Society* 14 (1859), 12–28, and 15 (1860), 42–8, contrasted with his *Sermons Broad and Short* (1883), 28, and *Notes by a Field Naturalist* (1877), 45–6; 'Rev. Henry Hughes Higgins, MA', *Geological Magazine* 10 (1893), 380–4.

98. For a very good example see H.F. [Henry Fletcher] Hance (1827–86) to Sir Joseph Dalton Hooker, 20 February and 21 June 1869, Directors' Correspondence, DC/150/504, 508, RBGK; also Robert Arthington (1823–1900) to Stokes, 19 February 1881, MS Add. 7656/1A/826, Stokes Papers, CUL.

99. Ruskin, 'The Choice of Books', *PMG*, 15 February 1886, quoted in Finnegan, *Natural History Societies*, 96. See Jonathan Smith, *Charles Darwin and Visual Culture*, 166–79. Ruskin, *Love's Meinie* (1873), 59, quoted in Beer, *Darwin's Plots*, 9.

100. George William Kitchin (1827–1912), 'Ruskin at Oxford', *Saint George* 4 (1901), 31–43.

101. See the letters in the Bonney Correspondence, Royal Geographical Society, London.

102. See the account in Ruskin to Geddes, 9 August 1884, f.172a–b, MS.10523, Geddes Papers, NLS.

103. Letter to Holman Hunt, quoted in J.H. Coombs, ed., *PreRaphaelite Friendship: The Correspondence of William Holman Hunt and John Lucas Tupper* (1986), 185.

104. Edward Lear to Thomas Woolner, 1 May 1878, in Amy Woolner, *Thomas Woolner* (1917), 284. For another example see the Melbourne palaeontologist Frederick McCoy (1817–99).

105. Liddon, *St Thomas*, 29; the other two were between life and inorganic matter; between matter and nothing. Liddon apparently annotated the passages of his first edition copy of *Descent*, including volume I, 62 and 65, where Darwin spoke of the 'ennobling' belief in an omnipotent God, and acknowledged the presence of a number of 'savage' races which had not developed any concept of divinity (see listing of this copy by Pryor-Johnson Rare Books, New York, on abebooks.co.uk [accessed 4 January 2021]). See also Alexander

222 DARWINISM'S GENERATIONS

Grant, 'Philosophy and Mr. Darwin', *Contemporary Review* 17 (1871), 281, and Henry Reynolds (1825–96), *Light and Peace: Sermons and Addresses* (1892).

106. See the discussion in J. Rachels, *Created from Animals: The Moral Implications of Darwinism* (1990), 81–3.

107. Radick, *Simian Tongue*, 34. For Carpenter see 'Darwin on the Origin of Species', *National Review*, reprinted in D.L. Hull, *Darwin and His Critics* (1973); similarly Alfred Newton continued to resist attempts to read intellectual capacity into animals; see Newton to Romanes, 13 January 1882, ff.252–3, Romanes Papers, Ms Eng, d.3823, Bodleian.

108. Lewes, *Problems of Life and Mind* (1879), 142–3. Lewes of course had little time for religious requirements for a soul, but even that, he conceded, was better than a 'crude materialist hypothesis', quoted in Tjoa, *George Henry Lewes*, 96.

109. See J. Costa, *Radical by Nature: The Revolutionary Life of Alfred Russel Wallace* (2023), 305–6; Wallace, *Darwinism*; I. Hesketh, 'The First Darwinian: Alfred Russel Wallace and the Meaning of Darwinism', *Journal of Victorian Culture*, 25 (2020), 171–84; also 'A Visit to Dr Alfred Russel Wallace, FRS', *Bookman* 8 (1897–8), 121–4.

110. See *Dundee Courier*, 7 April 1871.

111. Livingstone, 'Situating Evangelical Responses', 200–3; this allowed Calderwood to break the link between animal life and human life—rational life was not subject to natural laws; see W.L. Calderwood, *Life of Henry Calderwood* (1900), 189–93, 299–305.

112. This stance puzzled other mid Victorians, including the Duke of Argyll; see correspondence with Müller in Müller Papers, Bodleian, and Daniel Wilson (1816–92), who rejected the miracle explanation of language; see Wilson to Blackie, 2 April 1863, ff.17–18, MS.2625, J.S. Blackie Papers, NLS.

113. Max Müller, 'Forgotten Bibles', *TNC* 15 (1884), 1004–22. For Müller see the discussion in Radick, *Simian Tongue*, 15–49.

114. As reported by Lyell to Lubbock, 20 February 1863, ff.25–6, Avebury Papers, Add. MS 49640, BL; Müller to Huxley, 9 May 1873, f.114, Box 23, Huxley Papers, ICL.

115. Müller to Duke of Argyll, 4 February 1875, and Müller to Darwin, 7 January 1875, in *Life and Letters of the Rt Hon Friedrich Max Müller* (1902), 476, 509, Müller to Blackie, 24 November 1875, ff.360–1, MS.2631, J.S. Blackie Papers, NLS. Also his 'Mr Darwin's Philosophy of Language', *Fraser's Magazine* 7 (1873), 525–41, 659–78, 8 (1874), 1–24. For Müller's sustained hostility to what he later described as 'the thoughtless extravagances of the so-called Darwinian School', see Müller, 'Literary Recollections', *Cosmopolis* 4 (1897), 341; 'language', Müller continued to assert, 'is something that puts an impassable barrier between beast and man'. For high-Victorian criticism of this see A. Bain and T. Whittaker, eds., *Philosophical Remains of George Croom Robertson* (1894), 434–5, which described Müller's as a rambling discourse that 'bristles in detail with points of questionable statement'.

116. Radick, *Simian Tongue*, 49, quoting *Chips* 1894 ed., I.

117. John Laws Milton (1820–98), *The Stream of Life on Our Globe* (1864), Thomas Wharton Jones, *Evolution of the Human Race from Apes and of Apes from Lower Animals a Doctrine Unsanctioned by Science* (1876), J.J. Murphy, *Habit and Intelligence* (2nd ed., 1879); Mivart, 'A Limit to Evolution', *TNC* 16 (1884), 263–80, 280.

118. Henry Maudsley, 'An Address on Medical Psychology', *Journal of Mental Science* 17 (1872), 402.

119. Cobbe to Mary Somerville, 7 May [nd, but probably 1870 or 1871], MSFP-18, Somerville Papers, dep c.358, Bodleian; Cobbe confessed that 'I personally like and respect him so warmly that it was altogether a most trying task', before going on to recommend Mivart's imminent *Genesis of Species*.

120. Argyll's *Primeval Man*, Mivart's *Genesis of Species*, Rev. Thomas Smith (1817–97), *The Bible Not Inconsistent with Science* (1866), 28–9; Temple, *Relation between Religion and Science* (1885); see Gouldstone, *Anglican Idealism*, 101. Liddon claimed that God must have imbued primitive man with some 'feeling or instinct about Himself', Liddon to Müller, 8 April 1878, ff.71–5, Ms. Eng. c.2806/1, Max Müller Papers, Bodleian.

121. See Richards, *Evolutionary Theory*, 220–1.

THE *DESCENT OF MAN* AND THE HIGH VICTORIANS   223

122. For the Belfast Address, see, for example, Ruth Barton, 'John Tyndall, Pantheist: A Rereading of the Belfast Address', *Osiris* 3 (1987), 111–34. Clodd describes the storm created by Huxley's 1868 lecture on the 'Physical Basis of Life', Supplement, *Nature*, 9 May 1925.
123. See the distinction of Ruth Barton, 'John Tyndall, Pantheist', 111–34.
124. For example, John Tulloch, *Philosophy and Religion* (1884): Darwin 'eminently deserving of all due honour', and his 'good sense can hardly have welcomed this outburst', 145.
125. For others see James Michell Winn (1808–1900), 'Materialism', *Journal of Psychological Medicine and Mental Pathology* ns 1 (1875), 1–20, Winn, *Modern Pseudo-Philosophy* (1878), and his 'Darwin', *Journal of Psychological Medicine and Mental Pathology* ns 9 (1883), 163–80, and C.M. Ingleby to C.J. Munro, 8, 12 October 1877, ACC/1063/2629A/2630, Monro Family Papers, LMA. See also Samuel Wainwright (1824–99), *Scientific Sophisms: A Review of Current Theories Concerning Atoms, Apes, and Men* (1886), George Sexton (1825–98), *Scientific Materialism Calmly Considered* (1874).
126. See, J. Petrunic, 'Evolutionary Mathematics: William Kingdon Clifford's Use of Spencerian Evolutionism', in Lightman and Reidy, eds., *Scientific Naturalism*, 105–28.
127. See the manuscript volume including 'Philosophical Systems', a paper to the Glasgow Philosophical Society (date unclear), and various other notes on evolution, Box 33, Dep.208, A. Campbell Fraser Papers, NLS.
128. H.P. Liddon, diary entry 4 May 1882, in J.O. Johnston, *Life and Letters of Henry Parry Liddon* (1904), 275–6, quoted in Glick, *What about Darwin*.
129. Mivart to Patrick Geddes, 29 April 1887, T-GED/9/11, Geddes Papers, Strathclyde University.
130. Turner, *Science and Religion*, 39. Although Stephen's most significant interventions in the evolutionary debates, in particular *The Science of Ethics* (1882) and *An Agnostic's Apology and Other Essays* (1893), did not appear until later, his influence was being established by the appearance of his essays through the 1870s, not least 'Darwinism and Divinity', *Fraser's Magazine* ns 5 (1872), 409–21. See S. Collini, *Public Moralists, Political Thought and Intellectual Life in Britain, 1850–1930* (1991), 170–96.
131. J. McCarthy, *Reminiscences* (1899), II, 266–7.
132. See the letters to Michael Foster and William Flower, quoted in Marie Boas Hall, *All Scientists Now: The Royal Society in the Nineteenth Century* (1984), 119.
133. Examples to include Dr Charles Egerton Fitzgerald (1833–98) as President of the Folkestone Natural History Society, *Folkstone Express*, 11 November 1871, 15 February 1873.
134. The full account of the manoeuvrings by which Bastian was ruthlessly marginalised is given in James E. Strick, *Evolution and the Spontaneous Generation Debate* (2001), especially 157–82.
135. Turner was appointed Professor of Anatomy at Edinburgh in 1867, T.G. Bonney in Cambridge in 1868, and then from 1877 Professor of Geology at University College London, and Boyd Dawkins at Manchester in 1870 (Professor in 1874). Also Joseph Reay Greene (1836–1903), Professor of Biology at Queen's College, Cork and E. Perceval Wright (1834–1904), lecturer in Zoology and later Professor of Botany at Trinity from 1869.
136. Sidney H. Vines (1849–1934), influenced by Huxley's laboratory methods, being appointed at Cambridge in 1876, and E.R. Lankester (1847–1929), Professor of Zoology at UCL from 1874.
137. *Dundee Evening Telegraph*, 18 April 1878.
138. For Foster see G.L. Geison, *Michael Foster and the Cambridge School of Physiology: The Scientific Enterprise in Late Victorian Society* (1978), and B.J. Hawgood, 'Sir Michael Foster MD FRS (1836–1907): The Rise of the British School of Physiology', *Journal of Medical Biography* 16.4 (2008), 221–6.
139. See his correspondence with William Thiselton-Dyer, W.T. Thiselton-Dyer Papers, In-letters II, RBGK. Foster later made clear both his affection for Huxley, and his fundamental lack of sympathy with the anticlericalist focus of his Darwinian advocacy, in Michael Foster, 'Huxley', *National Review* 43 (1904), 421–39.

224 DARWINISM'S GENERATIONS

140. For the view of him from one undergraduate in the early 1880s see letter of DWT, 2 March 1882, ms47014, and DWT to Joseph Gamgee, 30 November 1884, ms47126, D'Arcy Wentworth Thompson Papers, St Andrews.

141. *Brain* was established in 1878 by Hughlings Jackson (1835–1911), David Ferrier (1843–1928), and James Crichton-Browne (1840–1928), along with Sir John Charles Bucknill (1817–97).

142. *The Nation*, 22 March 1888, quoted in I. Skrupkelis and E.M. Berkeley, eds., *Correspondence of William James Volume 6 (1885–89)* (1998).

143. Kingsley's enthusiasm for the journal in 1872 was tempered by his lack of understanding of much of the content and the limited readership this brought, Kingsley to Lockyer, 8 November 1872, Alexander Macmillan to Lockyer, 10 November 1871, Norman Lockyer Papers, EUL110, Exeter University. In the late 1890s neither the Brighton public library nor the Brighton Medical Library subscribed; see R.J. Ryle to K. Pearson, 21 July 1898, Pearson/11/1/17/102, UCL.

144. See Baldwin, *Making Nature*, 39–73 (Baldwin describes this generation more broadly as those born in the 1840s and 1850s).

145. N[orman]. MacColl (1843–1904) to Nora Sidgwick, 25 April [1904], Sidgwick Papers c.103/71, Trinity College, Cambridge; MacColl noted that previously J.E. Gray of the British Museum had done most of the *Athenaeum*'s reviewing in this area, but he turned to Henry Sidgwick to review J.R. Liefchild's *Origins of Man*, and got 'an antagonistic but quite courteous criticism…There was never any doubt, afterwards, of the paper's attitude towards Darwin's views'. (This was part of a more general shift to a younger group of reviewers; see M. Demoor, *Their Fair Share: Women, Power and Criticism in the Athenaeum, 1870–1900* (2000).) MacColl's appointment was itself the consequence of the passing of ownership of the paper to Sir Charles Dilke (1843–1911), on the death of his father. The editorial shift was, of course, a gradual process, still working its way through in the 1880s, as in the case of the appointment of D.C. Lathbury (1831–1922), as editor of the [Church] *Guardian* in 1883, at which point he scouted his contacts to find someone to review science who was 'not a violent Anti-Evolutionist'; see E.S. Talbot to E.B. Poulton, 11 December 1883, Box 11, Poulton Papers, OMNH. During the later 1870s, albeit under the new editorship of a mid Victorian, Thomas Chenery (1826–84), the *Times* on more than one occasion looked to Huxley to offer its readers a guide to developments in evolutionary thought, Thomas Chenery to Huxley, 3 November 1878, ff.176–7, Box 12, Huxley Papers, ICL. Chenery indicated that 'my own belief and opinion entirely inclines to the doctrine'.

146. For a flavour of Edward Clodd's house parties at Aldburgh, see Clement Shorter's reminiscences in *The Bookman* 10 (1899–1900), 76–8.

147. James Sully, 'Sir Leslie Stephen', *Atlantic Monthly* 95 (1905), 347–56, 350. Visiting in 1877, Reginald Brett, later Viscount Esher (1852–1930), reported that Eliot 'adore[d] Charles Darwin, because of his humility', O. Brett, ed., *Journals and Letters of Reginald, Viscount Esher* (1934), I, 40.

148. See J. Reid, *Robert Louis Stevenson: Science and the Fin de Siecle* (2006), 115–16; Violet Paget to Matilda Paget, 25–27 June 1881, https://libguides.colby.edu/ld.php?content_id=1059875 [accessed 23 February 2022]. For brief account of Sunday afternoons at the Cliffords, and Huxley's early evening 'tall teas', see Clodd, *Memories*, 37–8, 42–4.

149. See William Whyte, 'Sunday Tramps', *Oxford Dictionary of National Biography*, https://doi.org/10.1093/ref:odnb/96363; key members included Sir Frederick Pollock (1845–1937), George Croom Robertson (1842–92), Norman MacColl (1843–1904), James Cotter Morrison (1832–88), Sir Robert Romer (1840–1910), Shadworth Hodgson (1832–1912), James Sully (1842–1923), Francis Ysidro Edgeworth (1845–1926), the psychologist George Henry Savage (1842–1921), the explorer Douglas William Freshfield (1845–1934), the novelist Richard Ashe King (1839–1932), along with a smattering of younger figures, including the artist John Collier (1850–1934), and R.B. Haldane (1856–1928).

150. In addition to Charles Hobkirk (see above, p. 114), the leading members of this group were the palaeobotanist James Spencer (1834–?), microscopist Thomas Hick (1840–96), William Cash (1843–1914), George Henry Parke (1844/5–1913), the geologists James

THE *DESCENT OF MAN* AND THE HIGH VICTORIANS 225

William Davis (1846–93) and George Brook (*c*.1851–?), and the marine biologist Walter Percy Sladen (1849–1900) (of whom Cash recalls specifically that his masters were 'Darwin, Huxley, Lyell and Tyndall', William Cash, 'Walter Percy Sladen', *Proceedings of the Yorkshire Geological and Polytechnical Society* 14 (1901), 268). See also William Cash, 'James William Davis: An Obituary', *Proceedings of the Yorkshire Geological and Polytechnical Society* ns 12 (1895), 319–34. A similar group was attended by the later school teacher and lecturer Jonas Bradley (1858–1943), who recalled that it was only when 'I read Darwin, Huxley, Tyndall, Mill and Boyd Dawkins [in this group, that] light came to me', 'Men with a Mission: Jonas Bradley', *Nelson Leader*, 21 June 1929.

151. The active figures in the society in the 1870s were with the exception of Crompton nearly all either very young mid Victorians, or high Victorians, including Thomas R. Pinder (*c*.1828–1902), Treasurer, Octavius Conder (1828/9–1910), Secretary, James Reeve (1833–1920), Curator, Rev. Jonathan Bates (1829–79), Charles Golding Barrett (1836–1904), Herbert D. Geldart (1831–1902), Thomas Southwell (1831–1909), and Rev. John Alfred Lawrence (1836–1928). Geldart's presidential address in 1875 accepted the instability of species but continued to express doubts as to 'the Darwinian theory'; see Geldart, 'Presidential Address', *Transactions of the Norfolk and Norwich Naturalists' Society, 1874–79* 2 (1874–9), 1–11.

152. See *Transactions of the Norfolk and Norwich Naturalists' Society, 1874–79* 2 (1879), 355–76. By 1872 Harmer was a powerful pro-Darwinian voice in the Norwich press; see his responses to Bateman's ideas on language, *Eastern Daily Press*, 27, 30 April, 24 May, 13 August, 3 December 1872.

153. *Nottinghamshire Guardian*, 14 May 1875.

154. *Nottinghamshire Guardian*, 2 July 1875, *Nottingham Journal*, 12, 23 April 1898, 26 August 1909, *Nottingham Evening Post*, 17 July 1914.

155. See the lists which appeared weekly in newspapers such as *The Radical* in 1880. Significantly, as David Stack has noted, this culture embraced Darwinism, but was suspicious of those like Edward Aveling (1849–98), whose *Gospel of Evolution* (1881) appeared to erect Darwinism into a new religion, with scientists as a new priesthood, Stack, *Darwinian Left*, 171–2.

156. For example, Thomas Okey (1852–1935), *A Basketful of Memories* (1930), 42–3; Henry Snell, *Movements, Men and Myself* (1936), 76–7, 156–7. An excellent impression of this world is provided by Roy Macleod, *Archibald Liversidge, FRS: Imperial Science under the Southern Cross* (2009).

157. Quoted in I.D. MacKillop, *The British Ethical Societies* (1986), 53, which notes it was published as a broadside under the name Ion in the 1870s, and reissued in 1883; see endpapers of W.S. Smith's *London Heretics* (1967). For the Darwinian character of the Sunday Lecture Societies see Ruth Barton, 'Sunday Lecture Societies: Naturalistic Scientists, Unitarians, and Secularists Unite against Sabbatarian Legislation', in Lightman and Dawson, *Victorian Scientific Naturalism*, 189–219.

158. Leslie Stephen, 'Thomas Henry Huxley', *TNC*, 40 (1900), 905, and also his reflections on the two great intellectual shocks received by 'The generation which was growing to maturity in the decade 1850—60' in Leslie Stephen, 'An Attempted Philosophy of History', *Fortnightly Review* 27 (1880), 672–95; Hirst, *Life and Letters of John Morley*, I, 18, notes the impact of Darwin, Spencer and Mill. Similar intellectual obligations were testified to by Frederic Harrison in his *The Creed of a Layman: Apologia Pro Fide Mea* (1907).

159. J.B. Harford, *Life of Handley Carr Glyn Moule (1841–1920), Bishop of Durham* (1922), 33–4.

160. Julia Wedgwood, 'Ethics and Science', *Contemporary Review* 72 (1897), 220–33, 227; F.W. Farrar, *Chapters on Language* (1865), 50. It was said of Farrar that he 'never fully accepted the Darwinian theory of evolution in the animal kingdom, inclining to the belief that species were immutable', Farrar, *Life of F.W. Farrar*, 108.

161. Sayce to Blackie, 21 January 1876, ff.21–2, MS.2632, J.S. Blackie Papers, NLS; A.H. Sayce, *The Principles of Comparative Philology* (1874), and *Lectures on the Science of Language* (1880).

226 DARWINISM'S GENERATIONS

162. B.H. Holland, *Life of Spencer Compton, Eighth Earl of Devonshire* (1911), I, 133.
163. E. Charteris, *Life and Letters of Sir Edmund Gosse* (1931), 48.
164. E.M. Forster, *Marianne Thornton* (1956), 248.
165. See the interesting discussion in Harrison, *Thoughts and Memories* (1926), 12–14, 106, 175.
166. For examples of this unstudied acceptance, see the suffragist and marriage reformer Jane Hume Clapperton (1832–1914), in her *Scientific Meliorism and the Evolution of Happiness* (1885), 89, 334–5.
167. See the skit on 'Evolution' by the Scottish lawyer J.H. Balfour Browne (1845–1921), 'Casual Conversations', Acc.4223, volume 15, NLS.
168. G.J. Romanes to W. Carruthers, 31 January 1892, Carruthers Papers, DF 404/1/6, NHM.
169. F.W. Hutton, *Lesson of Evolution* (1907), xvi.
170. Flower to Hooker, 31 May 1870, JDH/1/2/8/141, Hooker Papers, RBGK.
171. W.H. Hudson, *Far Away and Long Ago* (1918), 327–30.
172. H. Jay, *Robert Buchanan: Some Account of His Life* (1903), 21. This can be followed up in the interesting correspondence with Wallace, Wallace Papers, Add. MS 46441, BL, including his observation that he could not accept pain as beneficent, 4 August 1899, ff.195–6.
173. T.R.R. Stebbing to Darwin, 5 March 1869, DCP-LETT-6643, Darwin Correspondence. Falconer Larkworthy (1833–1924) was 'captivated at once by Darwin's ideas of creation of species via evolution', but 'wholehearted acceptance' took 'some time' because of the way it overthrew the belief in special creation he had 'held since infancy', Harold Begbie, ed., *Ninety-One Years: Being the Reminiscences of Falconer Larkworthy* (1924), 157; Alfred Comyn Lyall (1835–1911) was 'haunted' by *In Memoriam* and 'by Darwin's struggle for existence' in 1864 when in his late twenties, Henry M. Durand, ed., *Life and Letters of Alfred Comyn Lyall* (1913), 109.
174. [Charles Warren Adams (1833–1903)] to President and Vice-President of the Zoological Society, 26 October 1901, in Frank E. Beddard Papers, COLL 623, Edinburgh; for other cases, see Sidney Biddell (or Biddle) (1830–1911) to Ellen Nussey, 20 April 1892, BC MS 19c Bronte C13, no. 127, Bronte Parsonage Museum; W.G. Wheatcroft (1834–93) in his presidential address to the Postal Microscopy Society, *Journal of Microscopy and Natural Science* 9 (1892), 50–1.
175. See Peter Bayne, *Lessons from My Masters—Carlyle, Tennyson and Ruskin* (1879); and his letter to Blackie, 1 January 1878, ff.1–2, MS.2633, J.S. Blackie Papers, NLS.
176. See her Biographical Introduction, in volume 22 of *The Collected Works of William Morris*, xii. For an account which places Morris more centrally within evolutionary thought see Hale, *Political Descent*, 256–64.
177. See in general the Talbot to Poulton correspondence, Box 11, Poulton Papers, OMNH, but in particular letters of 31 July 1887, 19 August 1889, 7 September 1889, 17 April 1896; also the very extensive correspondence with Lady Welby, 1970–010/016(06)–/017(15), Lady Victoria Welby Papers, York University, Canada. For Talbot see Gwendolyn Stephenson, *Edward Stuart Talbot, 1844–1934* (1936).
178. Marcus Dods, *Books Which Have Influenced Me* (1887), 100.
179. See *Alloa Advertiser*, 18 January 1868, 24 November 1900.
180. Arthur Porritt, *More and More Memories* (1947), 21.
181. F.V. Dickins to Thiselton-Dyer, 5 October 1888, f.198, W.T. Thiselton-Dyer Papers, In-letters I, RBGK.
182. William Lucas Distant (1845–1922), editorial address for his first edition of the *Zoologist* (1897), 2–3; Distant recognised a marked contrast in his approaches from the mid-1860s when he was in the Malay peninsula to South Africa in 1890, Distant to Meldola, 28 August 1890, #1952. Meldola Papers, ICL.
183. Although it is clear that Hopkins was at the same time sufficiently revolted by some of the implications to reject what he thought of as 'downright Darwinism', somewhat squeamishly telling his mother in 1874 that 'I do not think that Darwinism implies necessarily that man is descended from any ape or ascidian or maggot, and so on; these common ancestors, if lower animals, need not have been repulsive animals', Nixon, *Hopkins*, 122, 126, 132.

THE *DESCENT OF MAN* AND THE HIGH VICTORIANS 227

184. For Wright see *Optical Lantern and Cinematograph Journal*, 15 January 1906. Wright identifies himself as the *Nonconformist* reviewer in a letter to the paper, *Nonconformist*, 19 January 1888, from this and previous passing comments it is possible to reconstruct at least some of his activity reviewing and participating in correspondence debates back to 1882; he also identifies himself as the 'Layman' who published *Conversations on the Creation: Chapters on Genesis and Evolution* (1881), initially published in the *Sunday School Chronicle*. For a signed contribution see Lewis Wright, 'The New Dogmatism', *Contemporary Review* 54 (1888), 192–213. Wright also gave occasional public lectures; see for example, lecture on the eye, *Nonconformist*, 19 January 1888.

185. It has to be acknowledged that perhaps Wright's most pertinacious critic was actually a high Victorian, Rev. Henry Sturt (1832–1922), Congregationalist minister of the Ebenezer Chapel, Dewsbury, 1865–95, who continued his resistance to evolutionary thought to the end of the century; see *Leeds Mercury*, 16 October, 6 November 1896.

186. *Nonconformist*, 25 February 1888. For a similar figure, see the Edinburgh stationer William Durham (1834–93), and his *Evolution, Antiquity of Man, Bacteria, etc* (1890).

187. See Sybil Anne Tait (1845–1909)'s description of her husband at the time they married as 'a devoted follower of Charles Darwin', 'The Rights of Animals', *The Animals' Friend* 2 (1895–6), 195.

188. Ludwig Mond to Wallace, 13 July 1899, f.47, Wallace Papers, Add. MS 46437, BL; and Arthur Symons, 'Introduction', to his *A Selection from the Poems of Mathilde Blind* (1897), vi. Blind listed the *Origin* as one of the two books which most influenced her, and confessed that 'Darwin naturally became the chief mental factor in my development', F.J. Gould, *Chats with Pioneers of Modern Thought* (1898), 54.

189. J.W. Judd to Seward, 29 March 1908, #123, A.C. Seward, Correspondence re *Darwin and Modern Science* 1909, MS Add. 7733, CUL. Review of *Further Letters of Darwin, Nature*, 9 April 1903. Judd's correspondence with Lyell makes it clear that he saw Darwin and Lyell as mentors; see reference to 'my guides and philosophers', and also one of Darwin's 'disciples', Judd to Lyell, 27 September 1874, 27 December 1873, COLL 203/1/3093–6, 3035–8, Lyell Papers, Edinburgh; also Judd to J.D. Dana, 12 August 1889, Dana Family Papers, MS164 Box 2, Yale.

190. M. North, *Recollections of a Happy Life* (1892), II, 87; see also Julia Wedgwood to T.H. Farrer, 14 September 1876, 9606/4/1/2, Farrer Family Papers, Surrey History Centre.

191. As in James Geikie to Patrick Geddes, 14 December 1881, f.208, MS.10521, Geddes Papers, NLS. See also the comment of J.F. McLennan, that 'Writing to Darwin has been a recovery so far of slipping-away points of view', J.F. McLennan to Alexander Gibson, 1 February 1874, MS Add. 7449/D/453, Robertson Smith Papers, CUL. Returning from Algiers in 1877, McLennan planned to take a cottage near Down, on the basis that he would have access to Darwin's books, McLennan to Gibson, 11 May 1877, MS Add. 7449/D/457, Robertson Smith Papers, CUL.

192. Clark, *Bugs*, 83, quoting Lady Avebury's diary from the 1880s; for Tait, see Browne, *Darwin*, II, 389, quoting Darwin to Tait, 13 January 1880.

193. In hankering after 'these Darwin-like men that we want here', quoted in G.W. Prothero, *A Memoir of Henry Bradshaw* (1888), 293. Frank Maitland Balfour was explicitly the model here, someone who showed this 'Darwin-like singleness of aim'.

194. J.M. Robertson, *Buckle and His Critics* (1895), 316–17.

195. For Stebbing see Alison Wood, 'Responding to Darwin: The Reverend Thomas Stebbing (1835–1926), Clergyman, Naturalist, Apologist', unpublished PhD thesis, King's College, London (2011). Also T.G. Bonney, *Evolution: Two Sermons* (1877), James Ross (1837–92), *On Protoplasm: Being an Examination of Dr J.H. Stirling's Criticism of Professor Huxley's Views* (1874).

196. For Clifford see Petrunic, 'Evolutionary Mathematics', 92–110; James H. Keeling to William Thomson, 12 May 1903, MS Kelvin K1, University of Glasgow.

197. See Arthur Roope Hunt (1843–1914), 'A Vindication of Bacon, Huxley, Darwin and Lyell', *Geological Magazine* (1902), 265–74; which notes Huxley's tendency to describe Darwinian theory as a 'hypothesis'. Thomas Whiteside Hime (1841/2–1920), see *BMJ*,

228 DARWINISM'S GENERATIONS

17 November 1920, 843. Lecturing to the Sheffield Lit and Phil in 1883, he gave nine key propositions of Darwin, 'each of which he considered to be incontrovertibly true'.

198. William Knight (1836–1916), 'Ethical Philosophy and Evolution', *TNC* 4 (1878), 432–56, 442.

199. A.W. Bennett (1833–1902), in a paper at the British Association in 1870, reprinted in *Nature*; see Ellegard, *Reception*, 249–50.

200. Nicholson, 'On the Bearing of Certain Palaeontological Facts upon the Darwinian Theory of the Origin of Species and of Evolution in General', *JTVI* 9 (1875), 373–74, a position he maintained in his *Ancient Life History of the Earth* (1897); see H.R. Mill, *Life Interests of a Geographer* (1945), 25–2.

201. *Nature*, 2 February 1871; see also Bennett's review of the sixth (1872) edition of the *Origin* in *Nature*, 22 February 1872.

202. Ian Hesketh, 'A Good Darwinian? Winwood Reade and the Making of a Late Victorian Evolutionary Epic', *Studies in History and Philosophy of the Biological and Biomedical Sciences* 51 (2015), 44–52.

203. Dallinger, *The Creator, and What We May Know of the Methods of Creation* (1887), 67, 70. Dallinger retained an underlying scepticism, which perhaps becomes stronger again into the 1890s; see the account of a sermon at Brixton Independent Church, DA49/1/2/239 File 268/9, Willoughby Collection, CLB. For Dallinger see J.W. Haas, Jnr. 'The Rev. Dr. William H. Dallinger F.R.S: Early Advocate of Theistic Evolution and Foe of Spontaneous Generation', *Perspectives on Science and Christian Faith* 52 (June 2000), 107–17. For John Clifford, see 'Charles Darwin or Evolution and Christianity', in his *Typical Christian Leaders* (1898). For another example see the entomologist W[illiam] F[orsell] Kirby (1844–1912), 'What Is Darwinism', *Hardwicke's Science Gossip*, 1 November 1868, 241–4.

204. E. Arnold, *Seas and Lands* (1891), 282.

205. P.G. Hamerton, *Chapters on Animals* (1874), 3–4; Robert Buchanan, *The Coming Terror and Other Essays and Letters* (1891), x–xi; James Diedrick, *Mathilde Blind: Late-Victorian Culture and the Woman of Letters* (2018), including Blind to John Todhunter, 1 September 1880, quoted 175.

206. See Dixon, *Altruism*, 153–8, and Gates, *Kindred Nature*, 50–61.

207. For full discussion see Ron Amundson, *The Changing Role of the Embryo in Evolutionary Thought* (2005).

208. Louis Compton Miall (1842–1921), *Life and Works of Charles Darwin: A Lecture to the Leeds Literary and Philosophical Society* (1883), 29.

209. As in W.J. Irons, *Annual Address to Victoria Institute* (1871), 15.

210. A. Logan Turner, *Sir William Turner: A Chapter in Medical History* (1919), 239–41.

211. Butler to Mivart, 29 February 1884, ff.51–2, Samuel Butler Papers, Add. MS 44030, BL.

212. Hardy, 'Drinking Song', M. Irwin, ed., *Collected Poems of Thomas Hardy* (1994), 906–7, discussed in R. Ebbatson, *The Evolutionary Self: Hardy, Forster, Darwin* (1982).

213. Mary Boole to Welby, 7 May 1884, 1970–010/001(47), Lady Victoria Welby Papers, York University, Canada.

214. Farrar, *Chapters on Language*; A.H. Keane, *Ethnology: In Two Parts* (1896), x.

215. Also Lawson Tait, *Has the Law of Natural Selection by Survival of the Fittest Failed in the Case of Man?* (1869), reprinted from the *Dublin Quarterly Journal of Medical Science* 48 (1869), 102–13.

216. Lowne, *Philosophy of Evolution*, 139–50, 158; [Morley], 'Descent of Man'; see Buckman, 'Evolution of Language', tps, GSM/GX/BM/3/24, British Geological Survey, arguing that animals do have their own vocabulary.

217. Stebbing, *Essays on Darwinism*, 62–81, 80. For another example, see A.J. Dadson (1843–1908), *Evolution and Religion* (1893).

218. Clark, *Bugs*, 90. See the discussion in John R. Morss, *The Biologising of Childhood: Developmental Psychology and the Darwinian Myth* (1990), 19–20.

219. See P.L. Farber, *The Temptations of Evolutionary Ethics* (1994).

220. A.H. Crauford (1843–1917), *The Religion of H.G. Wells and Other Essays* (1909), 174.

221. [Arabella Buckley], 'Darwinism and Religion', *Macmillan's Magazine* 24 (1871), 45–51, which argued that, after all, the law of nature was the law of God, the law of what Buckley

THE *DESCENT OF MAN* AND THE HIGH VICTORIANS 229

terms 'the ever-present action of the Infinite and All-Perfect first cause', 47. For Buckley see Dixon, *Altruism*, 152–7, Gates, *Kindred Nature*, 50–61, and Lightman, *Popularizers*, 239–53; as Gates makes clear, discussing *Life and Her Children* (1881) and *Winners in Life's Race* (1883), Buckley's was a Darwinian evolution, not one that had 'taken place by special guidance along certain beneficent lines', 'but that the overwhelming preponderance of healthy, happy, and varied existence has been brought about by the steady working of natural laws among which the struggle to survive and the constant action of natural selection are the most important', 54.

222. Clifford, 'Cosmic Emotion', cited in J. Paradis and G.C. Williams, eds., *Evolution and Ethics: T.H. Huxley's Evolution and Ethics (1893)* (1989), 38; Clifford, 'On the Scientific Basis of Morals' (1875), reprinted in his *Lectures and Addresses* (1879); J.A. Symonds, *Studies of the Greek Poets*, Second Series (1879). For Clifford, see Timothy J. Madigan, *W.K. Clifford and 'The Ethics of Belief'* (2009), and Lindsay Wilhelm, 'The Utopian Evolutionary Aestheticism of W.K. Clifford, Walter Pater and Mathilde Blind', *Victorian Studies* 59.1 (2016), 9–34.

223. Turner, *Science and Religion*, 30.

224. Pusey to Liddon [4 November 1878], LBV 68/202, Pusey Papers, PHL. Similarly R.H. Hutton's positive appreciation in the *Spectator* at Darwin's death offered its only regret at Darwin's misguided attempts to derive 'an "ought" and "a conscience" out of mere victorious sympathy', Dixon, *Altruism*, 179.

225. Beer, *Darwin's Plots*, 241.

226. Discussion of William Salmond (1832/3–?) United Presbyterian minister. See Peter Matheson, 'Transforming the Creed', in S.J. Brown, G.M. Newlands, and A.C. Cheyne, eds., *Scottish Christianity in the Modern World* (2000), 123–31. Salmond adopted an anti-Darwinian tone, but in many respects was steeped in evolutionary ideas and accepted that God did not intervene in the natural law.

227. G. Henslow to Darwin, 22 February 1869, DCP-LETT-6626, Darwin Correspondence; *Acton Gazette*, 18 November 1893.

228. G. Henslow, *The Theory of the Evolution of Living Things* (1873), e.g. 17–19, Henslow, *Heredity of Acquired Characters*, 4, and 'The Origin of Species without the Aid of Natural Selection: A Reply', *Natural Science* 5 (1894), 257–65, 'The True Darwinism', *TNC* 60 (1906), 795–801. For Henslow as a Lamarckian, see Bowler, *Eclipse*, 85–6, Lightman, *Popularizers*, 87–94.

229. Henry Festing Jones to Frank Darwin, 29 July 1910, copy, VIII/18/2, Butler Papers, St John's College, Cambridge. Butler apparently told friends that his books before *Erewhon Revisited* had lost him a combined total of £1100, O.T.J. Alpers, *Cheerful Yesterdays* (1929), 71.

230. Butler to J.F. Fuller, 22 November 1886 [draft], ff.189–90, Butler Papers, Add. MS 44031, BL.

231. Another is Patrick William Stuart-Menteath (1845–1925), who studied at the University of Edinburgh and then the Royal School of Mines, and lived for most of his life near Saint Jean-de-Luz in the Pyrenees. The origins of his antipathy to the Darwinians is not entirely clear, but in 1906 he was 'writing a huge work, which I will send you, to prove that Darwin and Huxley were blackguards and that the scientific tyranny of their successors is a worse intellectual bondage than that of the church ever was'; see Royall Tyler to Mildred Barnes, 12 January 1906, https://www.doaks.org/resources/bliss-tyler-correspondence/letters/12jan1906, and also letter at ms38515/11/36/16, St Andrews.

232. See the controversy in the *Times* prompted by a report on a lecture by Kelvin in May 1903, I.S. Ruddock, 'Lord Kelvin', in M.W. Collins et al., eds., *Kelvin, Thermodynamics and the Natural World* (2016), 32–4.

233. D.W. Bebbington, *The Dominance of Evangelicalism* (2005), 132. Similarly, James Iverach makes natural selection into 'design'. For Iverach see A.P.F. Sell, *Defending and Declaring the Faith: Some Scottish Examples, 1860–1920* (1987), 118–36.

234. See the works of Alexander Balmain Bruce (1831–99), the Free Church minister, and the Presbyterian James Orr (1844–1913), A.M. Fairbairn (1838–1912); see W.B. Selbie, *Life of Andrew Martin Fairbairn* (1914), 188–20. Also Charles Barnes Upton (1831–1929); see for example, 'Evolution and Religion', *Theological Review* 9 (1872), 561–74, Upton,

230 DARWINISM'S GENERATIONS

'Some Recent Signs of Convergence in Scientific and Theological Thought', *Theological Review* 11 (1874), 407–31. For this question see Livingstone, *Dealing with Darwin* (2014).

235. Dallinger, 'Popular Notes on Science', *Wesleyan Methodist Magazine* 117 (1894), 696. 'We profoundly believe still that it is no part of Nature's mission to man to become his moral teacher', W.H. Dallinger, 'Popular Notes on Science', ibid., 700.

236. Dallinger, *The Creator*, 57, 74. In this text Dallinger comes very close to endorsing a notion of design through the effects of the 'law of evolution'. Certainly this is the lesson that the popular religious magazine *The Quiver* sought to draw in its interview with Dallinger, *Quiver*, 24 (1889), 351–5. Dallinger was clear that this involved a clear repudiation of old-fashioned natural theology, 'The Religion of a Scientist: An Interview with the Rev. W.H. Dallinger', *The Young Man* 8 (1894), 363–70. We can see very much the same sort of reading in the notes of W.S. Jevons' reading of the *Origin* dated August 1863, JA/6/36/1, Jevons Papers, MJRL.

237. Bompas, 'The Argument for Belief', *National Review* 21 (1893), 634. See also George St Clair (1836–1908), *Darwinism by Design*, 2. Although St Clair's lecture 'The Great Globe', *Oxford Times*, 24 March 1866, ridiculed anthropocentricism, without directly endorsing Darwinism; by the 1890s, he fully accepted Darwinism, *Bridport News*, 19 July 1895. For others see William Powell James (1837–85), *On the Argument from Design in Nature* (1882). Rev. Robert Flint (1838–1910) was another who argued that Darwinian laws of natural selection and progressive development, 'must imply belief in an all-originating, all-foreseeing, all-fore-ordaining, all-regulative intelligence'. Flint, *Theism*, cited in D. Macmillan, *Life of Robert Flint* (1914), 347. See Sell, *Defending and Declaring*, 39–63.

238. Leslie Stephen, 'An Attempted Philosophy of History', *FR* 27 (1880), 672–95. See also his comment in the Preface to his *Science of Ethics* (1882), v–vi, that after the early influence of Mill, his 'mind was stirred by the great impulse conveyed through Mr Darwin's *Origin of Species*', and the 'great intellectual debt' he was under; *Nature*, 6 March 1873, W.S. Jevons, *The Principles of Science* (1877), 762; Jevons, 'Evolution and the Doctrine of Design', *Popular Science Monthly* 3 (1873–4), 98–100; see also his 'A Deduction from Darwin's Theory', *Nature*, 30 December 1869; in 1873 Jevons told Spencer that his ideas 'afford a complete solution' to problems in the theory of morals, Jevons to Spencer, 27 June 1873, MS791/85, Herbert Spencer Papers, Senate House Library, UL. If Jevons had doubts it would seem to have been Darwin's reluctance to impute progress to biology, Margaret Schabas, *The Natural Origins of Economics* (2006), citing Jevons' essay on 'Utilitarianism' (1890), 273–4.

239. Isobel Spencer, *Walter Crane* (1975), 66–7; Ralph E. Moreland, 'The Art of Walter Crane', *Brush and Pencil* 10 (1902), 257–71; see also D. Donoghue, *Walter Pater* (1995), 28; E. Clodd, *Memories* (1916); see also the architect Henry Heathcote Statham (1839–1924), *Building News and Engineering Journal*, 8 January 1869.

240. Henry Sidgwick, 'The Theory of Evolution in Its Application to Practice', *Mind* 1 (1876), 53.

241. *Cambrian News*, 5 August 1881.

242. Quoted in Nixon, *Hopkins*, 157; for Pater Darwinism was 'the very law of change', Holmes, *Pre-Raphaelites*, 201–2.

243. Wedgwood, 'Ethics and Science', 222.

244. See Henry Sweet 'Phonetics, General Philology and Germanic and English Philology' (1882), in H.C. Wyld, ed., *Collected Papers of Henry Sweet* (1913), 148–66, 154.

245. Robert Alun Jones, *The Secret of the Totem: Religion and Society from McLennan to Freud* (2005); Tylor, *Primitive Culture*, quoted in Kuklick, *Savage Within*, 75. Lang was also *sui generis*; as much a late Victorian as a high Victorian—of a very different breed to Lubbock, for example. At the same time his works continued into the Edwardian period to use Darwin, and particularly *Descent* as a fundamental point of reference, and to remain acutely sensitive to any suggestions of divergence from Darwinian positions. See his various contributions to discussions in *Man*, including 3 (1903), 180–1, and 6 (1906). For Stuart-Glennie see Eugene Halton, 'John Stuart-Glennie's Lost Legacy', in Christopher T. Connor et al., *Forgotten Founders and Other Neglected Social Theorists*

(2019), 11–26, and his long correspondence with Patrick Geddes, T-GED/9/190 etc., Geddes Papers, Strathclyde University.

246. Geison, *Foster*, 334. See, for example, the works of Thomas S. Clouston (1840–1915).

247. L.S. Jacyna, 'Reckoning with the Emotions: Neurological Responses to the Theory of Evolution, 1870–1930', in Richardson, *After Darwin*, 215–35. Morss, *The Biologising of Childhood*, notes that Sully's *Studies of Childhood* (1895), is systematically recapitulationary.

248. Hill to Mrs Ashley, Hampstead, 6 February 1892, in Lucy Crump, ed., *Letters of George Birkbeck Hill* (1906), 205.

249. A. Meadows, *Science and Controversy: A Biography of Sir Norman Lockyer* (2008), 172.

250. Morley to Frederic Harrison, 4 March 1871, quoted in Hirst, *Life and Letters of John Morley*, I, 180.

251. Morley's rejection of the struggle for survival as the primary driver of human progress is clear in his review of Hobhouse's 'Democracy and Reaction', *TNC* 57 (1905), 361–72. Harrison was sympathetic to this point of view; see his comments in *The Creed of a Layman: Apologia Pro Fide Mea* (1907), 245.

252. Sir Clifford Allbutt, 'Nervous Diseases and Modern Life', *Contemporary Review* 67, (1895), 228.

253. Sidgwick, 'The Theory of Evolution', 64–65, Sidgwick, 'The Scope and Method of Economic Science', in his *Miscellaneous Essays and Addresses* (1904), 170–99. For some sense of the way this continued to inform Sidgwick's teaching into the 1890s, see G.E. Moore's notebook of Sidgwick's lectures on Ethical Systems from 1894, Add. MSS 8875/10/2/1, Moore Papers, CUL.

254. Schabas, *Natural Origins*, 148, 273–4, citing Marshall's manuscript 'The Law of Parcimony [sic]' (1867).

255. See G.A. Gaskell to Huxley, 23 January 1891, ff.20–1, Box 17, Huxley Papers, ICL; for Gaskell, see Dixon, *Altruism*, 169.

256. See F.W. Hutton to W.E.H. Lecky, 25 September 1882, #271a, Lecky Papers, TCD.

257. Pollock to E.A. Freeman, 26 August 1876, Freeman Papers, MJRL, cited by C.W.J. Parker, 'Freeman and Liberal Radicalism: The Racial Ideas of E.A. Freeman', *Historical Journal* 24 (1981), 825–46, 835.

258. Barbara Larson, 'Evolution and Victorian Art', in Lightman and Zon, *Evolution and Victorian Culture*, 121–48.

259. Letter, *Morning Post*, 1 June 1904.

260. And also in conjunction his 'The Fall of Man' (1877), an illustration of *Paradise Lost*; see Diana Donald, 'The "Struggle for Existence" in Nature and Human Society', in D. Donald and J. Munro, eds., *Endless Forms: Charles Darwin, Natural Science and the Visual Arts* (2009), 81–97. For North, see Philip Kerrigan, 'Marianne North: Painting a Darwinian Vision', *Visual Culture in Britain* 11 (2010), 1–24.

261. Holmes, *Pre-Raphaelites*, 245.

262. Stephen to C.E. Norton, 20 June 1901, F.W. Maitland, *Life and Letters of Leslie Stephen* (1906), 463.

263. See Rev. John Kennedy (1813–1900), *Nonconformist*, 7 October 1874.

264. Mary Boole to Welby, 20 June 1886, 1970-010/001(47), Lady Victoria Welby Papers, York University, Canada. See also her letter to Darwin in 1866 in E.M. Cobham, *Mary Everest Boole: A Memoir with Some Letters, etc.* (1951), 27–8, discussing her views of the compatibility of Darwinism and her religious faith. For Welby and her circle, see A. Stone, *Women Philosophers in Nineteenth Century Britain* (2023).

265. Eveleen Myers, ed., *Fragments of Poetry and Prose by Frederick W.H. Myers* (1904), 33.

266. See Joy Dixon, *Divine Feminine: Theosophy and Feminism in England* (2001); Janet Oppenheim, *The Other World: Spiritualism and Psychical Research in England, 1850–1914* (1985). In general, the overwhelmingly progressive evolutionary ideas of these marginal religious groups, while they borrowed from notions of the survival of the fittest, had little if any Darwinian content.

267. A good example is Edmund Symes-Thompson (1837–1906). Symes-Thompson told the 1903 Church Congress that 'There was…only one theory of creation in the field, and

232 DARWINISM'S GENERATIONS

that was evolution, and there was only one theory of origin, and that was evolution... The God of evolution was infinitely grander than the occasional wonder worker of the old theologians', *Sheffield Daily Telegraph*, 16 October 1903. See also Sabine Baring-Gould (1834–1924), 'Primeval Man', in his *Some Modern Difficulties* (1875), 87–9; Francis Peek (1834–99), 'Science and Revelation', *Contemporary Review* 41 (1882), 1025–38; Thomas Hodgkin (1831–1913), L. Creighton, *Life and Letters of Thomas Hodgkin* (1917), 71, 341.

268. Alexander Henry Craufurd, *The Religion of H.G. Wells and Other Essays* (1909), 161; Craufurd explicitly distances himself from the anxieties of someone like Frances Power Cobbe, at the same time recognising that Darwinism could not address the sources of man's spiritual life; see his *Enigmas of the Spiritual Life* (1888).

269. Wright, 'The New Dogmatism', 194.

270. See comments of Joseph Parker, *Christian World*, 6 November 1890; recalling the 1860s, George Slayter Barrett (1839–1916) suggested that the moral was that faith would be wise to keep out of the fray, 'The Certitude of Christianity', *Nonconformist*, 25 March 1897.

271. See the various expositional works of a clergyman such as James Gurnhill (1836–1928), ordained in 1862, including *Some Thoughts on God* (1911), *The Spiritual Philosophy* (1914). For Methodists, see the popular writer and lecturer J.E. Taylor, S.J. Punkett, 'Dr John Ellor Taylor: Guide, Philosopher, Friend', *Proceedings of the Suffolk Institute of Archaeology* 40 (2002), 164–200.

272. Well expressed by the Scottish Congregationalist John Brown Paton (1830–1911) to P. Geddes, 9 February 1905, T-GED/9/590, Geddes Papers, Strathclyde University.

273. Fairbairn, 'Why I Am a Christian', *FR* 86 (1909), 417–29, 427. As he told a congregation at Mansfield in 1894, as a young theologian he had 'pleaded for the acceptance of evolution', *Eastern Daily Times*, 13 August 1894.

274. James M. Wilson, *An Autobiography* (1932), 248.

275. *Oxford Journal*, 2 September 1882.

276. W.L. Walker (1845–1930), *Christian Theism and a Spiritual Monism* (2nd ed., 1907).

277. E.H. Talbot read *Can the Old Faith Live* aloud with Henry Scott Holland, welcoming its powerful treatment of the compatibility of evolutionary thought with Christian theology, Talbot to Poulton, 21 July 1887, Box 11, Poulton Papers, OMNH.

278. George Freeland Barbour, *Life of Alexander Whyte* (1925), 68, 382. Whyte wrote of 'the warm exclamation of wonder and of worship' that arose 'as we lay down... *The Origin of Species*', Whyte, *Bible Characters: Adam to Achnan* (1896), 10.

279. See Walter Smith's review of his *The Memorabilia of Jesus, Commonly Called the Gospel of John* in *Critical Review* 3 (1893), 30; W.W. Peyton, 'Anthropology and the Evolution of Religion', *Contemporary Review* 80 (1901), 213–30, 434–46.

280. *South London Chronicle*, 2 May 1874; Haweis, *Arrows in the Air* (1878); see also William Page Roberts (1836–1928), Vicar of Eye in the 1870s, especially his 'Evolution', in his *Reasonable Service* (1876), and the debate in *Ipswich Journal*, 12, 16 December 1876.

281. Matheson, 'Freedom of the City of God', *The Expositor*, 5th ser., 6 (1897), 208–17.

282. See his review of Griffith-Jones's *The Ascent through Christ*, in *Critical Review of Theological and Philosophical Literature* 10 (1900), 43–8.

283. J.E. Carpenter and P.H. Wicksteed, *Studies in Theology* (1903), 60.

284. J.M. Wilson to Max Müller, 18 May 1891, ff.361–2, Ms. Eng. c.2806/2 Max Müller Papers, Bodleian.

285. For example, Talbot to Welby, 8 July 1890, 1970–010/016(18), Lady Victoria Welby Papers, York University, Canada.

286. [E.S. Talbot], *PMG*, 18 January 1888.

287. See comments in 1892 of the Methodist minister Rev. William Fiddian Moulton (1835–98), W.F. Moulton, *Memoir of W.F. Moulton* (1899), 218–19.

288. [G.H. Curteis (1824–94)], 'Dr Strauss's Confession', *Edinburgh Review* 138 (1873), 536–68. For another see Rev. Osmond Fisher (1817–1914) to James Croll, 1 September 1871, 'I think Darwin has upset some not over strong minds, and I cannot but fear that the moral effects of his doctrines will be injurious', J.C. Irons, *Autobiographical Sketch of James Croll* (1896), 262.

THE *DESCENT OF MAN* AND THE HIGH VICTORIANS   233

289. Rev. M.B. Moorhouse (1840–1925), quoted in *Life of Walter Pater* (1907), 1, 203. The popular preacher Charles Spurgeon (1834–92) defended the immutability of species in the face of the fractured geological record, *The Sword and the Trowel*, referencing belief in evolution as 'asinine', quoted in *Dundee Evening Telegraph*, 5 February 1897. Another opponent was the Rev. Henry Sturt (1832–1922), pastor of Ebenezer Chapel, Dewsbury 1865–95 (see above, p. 227, and his intervention in opposition to J.M. Wilson's support for evolution at the Church Congress in 1896, deploying citations of Mivart, J.W. Dawson, E.R. Conder, and Davidson's doubts over the Brachiopoda, *Leeds Mercury*, 16 October 1896), also his long letter in *Nonconformist*, 1 December 1887. The Baptist John Urquhart (*c*.1836/9–1914) noted in 1895 that the lack of the discovery of the missing link was 'slowly, but surely, killing Darwinism', *The Inspiration and Accuracy of the Holy Scriptures* (1895), 243. Such attitudes were also visible in lay evangelicals like the purity campaigner Samuel Smith (1836–1906); see his *The Ascent of Man; by Henry Drummond: A Review by Samuel Smith, M.P. Reprinted from The Churchman* [1894], and his *My Life-Work* (1902), 474.

290. See Robert Baker Girdlestone (1836–1923), *Old Testament Theology and Modern Ideas* (1909), 53–6.

291. Lightman, *Victorian Popularizers*, 462, and wider discussion, 462–70.

292. Bonney to Stokes, 16 October 1895, MS Add. 7656/2B/482, Stokes Papers, CUL; and the very similar tenor of Brooke Herford (1830–1903) in 'Man's Part in Evolution', in his *Anchors of the Soul* (1905), 78.

293. Lias, *Can Miracles Happen* (1883), 91.

294. Leonard Jenyns, *Memoir of the Rev. John Stevens Henslow* (1862), 213, quoted in Glick, *What about Darwin*. In exactly the same way Pusey at the end of the 1870s resolutely opposed any real prospect of reconciliation, utterly unconvinced that the shred of a first cause 'who, all those aeons ago, infused the breath of life into some primeval forms, and has remained inactive...ever since' could offer a reconciliation of religion and Darwinian science, Liddon, *Life of Edward Bouverie Pusey*, IV, 336.

295. Fairbairn, 'Theism and Science' (1881), in his *City of God* (1883), 17, 72. For another high-Victorian example, see Alexander Macalister (1844–1919) in his 'The Personal Religion of an Evolutionist', Murtle Lecture at Aberdeen in 1909, which outlines a view of evolution as a process not a power, 'only intelligible to him on the hypothesis that behind it there was a continuing agent in whose thought all the actors and their several parts were perfectly present', and that this 'great causal force' was God, *Aberdeen Journal*, 8 November 1909. Also Sydney Thelwall (1834–1922), English clergyman and Christian scholar, to Edward Blakeney, 1 October 1889, MS Add. 8553/3/4, Blakeney Papers, CUL: 'That there may be a measure of truth in the theory of Evolution I am not at all in a position to deny. But I am in a position to deny an Epicurean god, who simply impresses certain properties upon matter, and then leaves this matter, thus endowed, to work out its end with no interference or supervision on his part.'

296. Livingstone, 'Situating Evangelical Responses', quoting 'Science and the Christian Faith', in R.A. Torrey, ed., *The Fundamentals: A Testimony to the Truth* (1910–15), IV, 103. For Orr, see also Sell, *Defending and Declaring*, 154–5.

297. This is clearest in W.L. Watkin, *What about a New Theology* (1907), but see also his *Christian Theism and a Spiritual Monism* (1907), with its sustained engagement with Darwin.

298. Annie Besant, *An Autobiography* (1893), 280–1.

299. *Kingsley: His Letters and Memories*, II, 174, quoting Kingsley to Bates, 13 April 1863, AM18153, Charles Kingsley Papers, M.L. Parrish Collection, PUL.

300. For this and rest of this section see J. Strick, *Sparks of Life: Darwinism and the Victorian Debates over Spontaneous Generation* (2000), *passim*.

301. John Eliot Howard, 'The Contrast between Crystallisation and Life', *JTVI* 8 (1875), 173–201; see also Brooke's 'The Evidence Afforded by the Order and Adaptations of Nature to the Existence of a God', in S. Wilberforce, ed., *Faith and Freethought: A Second Course of Lectures Delivered at the Request of the Christian Evidence Society* (1872).

234 DARWINISM'S GENERATIONS

302. J.S. Blackie to Drummond [Dear Sir], 22 October [1884], Drummond Papers, Acc.5890/2 (not foliated), NLS.
303. *Proceedings of the British Association 1871.*
304. Sandford, *Memoirs of Archbishop Temple*, II, 686, quoted in Gouldstone, *Anglican Idealism*, 208.
305. Strick, *Sparks of Life*, describes him as a 'more extreme vitalist than many of his contemporaries', 58; see his *Life Theories: Their Influence on Religious Thought* (1871), *The New Materialism: Dictatorial Scientific Utterances and the Decline of Thought* (c.1882), and *Protoplasm* (various edn, inc. 1892). See the series of letters in *BMJ* (1863), 109–10, 157–8, 235–6, 365–6, and also correspondence with Acland in the 1860s, MSS Acland d.62, Bodleian, including Beales' comment that 'the whole race of modern Darwinistic and physical origin of living bodies people you may be quite certain are a set who look a very little way into things living', Beale to Acland, 12 January 1861, ff.40–3, MSS Acland d.62, Bodleian.
306. Beale to Acland, 2 February 1865, ff.48–9, MSS Acland d.62, Bodleian; and in 1871 apparently berating Acland for 'tolerating Darwin' (as reported in Acland to Bentham, 28 June 1871, BEN/1/1/4, George Bentham Papers, RBGK). True to his word, Beale was still arguing for the fundamental distinction of organic and inorganic matter at the Victoria Institute in 1899, cited in P.N. Waggett, *Religion and Science: Some Suggestions for the Study of the Relations between Them* (1904), 110.
307. See also Stirling to Bastian, 6 November 1869, 7220/15, Wellcome Library, and Elam, 'Automatism and Evolution', *Contemporary Review* 28 (1876), 537–61, developed into *Winds of Doctrine, Being an Examination of the Modern Theories of Automatism and Evolution* (1876).
308. Stirling, *James Hutchison Stirling*, 220; reporting that 'this opinion was shared by such men as Dr Lionel Beale..., Dr John Brown [1810–82]..., Professor Masson, and Dr Hodge of Princeton'. For Kelvin's verdict that protoplasm was 'a mythical affair', see William Thomson to Rev. E. Davys, 5 October 1898, MS Kelvin LB5, University of Glasgow.
309. Stokes told the Church Congress in 1883 that 'As for the origin of life itself, it was not intended on this theory to account for it, and the experimental researches of our foremost scientific men are adverse to the supposition of its production by spontaneous generation', *The Churchman* 7 (1883), 158.
310. See Strick, *Spontaneous Generation*, 71–2. A similar stance is visible in James Ross (1837–92), *On Protoplasm: An Examination of J.H. Stirling's Criticism of Professor Huxley's Views* (1874).
311. See the text of his 'Introductory Lecture to Course of Lectures on Physiology 1874–5', ff.73–85, J. Burdon Sanderson Papers, MS.20211, NLS; it is significant that in notes on an article by Haldane, Burdon Sanderson should finish with a reference to 'the ever impenetrable mystery of Organic Evolution', f.187, J. Burdon Sanderson Papers, MS.20212, NLS.
312. Farrar, 'The Voice of Conscience', *The Silence and the Voices of God, with Other Sermons* (1875), 40. Even Allen Thomson conceded in his pro-evolutionary 1877 BA presidential address that there was no evidence in favour of abiogenesis.
313. For cautious support see Lowne, *Philosophy of Evolution*, 17–24. Published objectors included William Thiselton-Dyer (see his 'On Spontaneous Generation and Evolution', *Quarterly Journal of Microscopical Science* ns 10 (1870), 333–54), Lankester, who Strick describes as 'a very eager proponent of abiogenesis in the distant past producing *Bathybius*', and Henry Lawson (1841–77), Strick, *Spontaneous Generation*, 101–2.
314. Desmond, *Evolution's High Priest*, 10. The attack on Bastian commenced with Tyndall's 'Dust and Disease' January 1870 lecture at the Royal Institution and continued with Huxley's September 1870 BA address 'Biogenesis and Abiogenesis'.
315. See discussion in Andrew Lang, 'The Genesis of Life', *Gentlemen's Magazine* 241 (1877), 541–62.
316. He confessed that 'in the present phase of the question I do not think lectures best fit it to advance the subject', Bastian to Dr (Sir) Thomas Oliver (1853–1942), 17 July 1880, quoted in 'From the Archives', https://academic.oup.com/brain/article/138/11/3449/332796 [accessed

THE *DESCENT OF MAN* AND THE HIGH VICTORIANS   235

22 February 2022]. Bastian retired in 1897 to renew his experiments into abiogenesis; attracting indifference and then further hostility, but continuing to his death in 1915, Strick, *Spontaneous Generation*, especially 190–4.

317. F. Temple, *The Relations between Religion and Science* (1885), 168.

318. See Matheson, *Can the Old Faith Live with the New?* (1885). Hence Sir Robert Anderson (1841–1914) challenged Huxley and the pro-Darwinians that they were reduced to arguing that abiogenesis 'must have been', Anderson, *A Doubter's Doubts* (1889), 17. See for further examples, the review of *Descent* in *Nature*, 6, 13 April 1871, by P.H. Pye-Smith (1839–1914), [William B.] Pope (1822–1903), *Compendium of Christian Philosophy* (1880), Upton, 'Human Automatism', *Theological Review* 14 (1877), 397–422. Others who firmly rejected abiogenesis included the literary critic Theodore Watts-Dunton (1832–1914); see comments of W. Robertson Nicholl, in 'The Significance of Aylwin', *Contemporary Review* 74 (1898), 798–809.

319. See Romanes, 'A Reply to the "Fallacies of Evolution"', *FR* 26 (1879), 492–504; in contrast, the response of Haddon's sisters Laura and Pussy, to Tyndall's article in *The Nineteenth Century* in 1878, letters 1 March and 1 April 1878, Envelope 22, Haddon Papers, CUL.

320. The reported opinion of Marcus Hartog (1851–1924); see Baldwin Spencer to W.S.B. Goulty, 5 March 1883, W.R. Sorley on 'the Interpretation of Evolution' paper to British Academy (1909); see *Athenaeum*, 4 December 1909; for Hartog see Baldwin Spencer (1860–1929) to W.S.B. Goulty, 5 March 1883, f.37, W. Baldwin Spencer Collection, Pitt Rivers Museum. Edward Kay Robinson (1855–1928), 'The Man of the Past', *TNC* 52 (1902), 789–805, is scathing about the search for abiogenesis.

321. For the late Victorian Joseph Jacobs (1854–1916), 'She alone, we thought, possessed the message of the New Spirit that Darwinism was to breathe into the inner life of man', Joseph Jacobs, *George Eliot, Matthew Arnold, Browning, Newman* (1891), xx. For mid-Victorian satires, see for example Julia Kavanagh's *John Dorrien* (1875), Mortimer Collins, *The British Birds: A Communication from the Ghost of Aristophanes* (1872), and his *Transmigration* (1874), and Wilkie Collins, *The Lady and the Law* (1875). Mortimer Collins was dismissive of Darwinism, observing that 'there is more science in the first chapter of Genesis than in all Tyndall and Huxley and Darwin', F. Collins, *Mortimer Collins, His Letters and Friendships, with Some Account of His Life* (1877), II, 181.

322. See below, p. 244; for Du Maurier see *Peter Ibbetson* (1891) and *Trilby* (1894), discussed in L. Vorachek, 'Mesmerists and other Meddlers: Social Darwinism, Degeneration and Eugenics in *Trilby*', *Victorian Literature and Culture* 37.1 (2009), 197–215.

323. See also Bevington's 'The Moral and Religious Bearings of the Evolution Theory', in South Place Chapel, *Religious Systems of the World* (1905). For discussion of Canton's 'The Latter Law' (1879) as well as even more explicitly evolutionary poems of the 1880s, see Holmes, *Darwin's Bards*, 121–2.

324. See the bewildered account of 'Benedick' originally in *The Standard*, 14 October 1879.

325. Huxley, 'The Coming of Age of *On the Origin of Species*', in *Darwiniana*; Argyll to James McCosh, 10 September 1878, AM11601, Box 2, General Manuscript Miscellaneous Collection, PUL.

326. J.E. Taylor, *Nature's Byepaths* (1880), 220.

327. *Edinburgh Evening News*, 5 November 1881.

328. Liddon to H.W. Acland, 17 October 1881, ff.88–9, MSS Acland d.59, Bodleian.

329. Peter Bayne, *Lessons From My Masters, Carlyle, Tennyson and Ruskin* (1879), 326–7.

330. *Oxford and Cambridge Undergraduate's Journal*, 22 November 1877.

331. E. Murray, *Sidelights on English Society* (1881), 147–8.

332. See *Leicester Journal*, 31 May 1878, *Belfast Morning News*, 3 May 1879.

333. *Christian World*, 23 August 1878; John Tyndall to Heinrich Debus, 5 September 1879, Tyndall Papers, volume 16 (forthcoming); for an account, see *Sheffield Independent*, 25 August 1879. For evidence that Huxley continued to feel himself embattled even before the British Association, see the gratitude expressed to Henri Milne-Edwards, 4 June 1879, B H891, Huxley Papers, APS.

334. Recollection of E.B. Poulton, *Western Morning News*, 2 September 1937.

236 DARWINISM'S GENERATIONS

335. *Falkirk Herald*, 14 December 1878. *Cheshire Observer*, 26 February 1881, 'The Facts of Natural History Compared with Darwinism', *Wigan Observer*, 6 December 1878.
336. See recollections in *Todmorden & District News*, 2 November 1906.
337. Mavor, *Windows*, 213–14; J. Clelland, *Evolution, Expression and Sensation* (1881). In the later 1880s, students would deliberately provoke Clelland by asking him his opinion of Huxley; see G.H. Findlay, *Dr Robert Broom, F.R.S. Palaeontologist and Physician* (1982), 8.
338. Braithwaite's copy of *Descent* (1883, 17th thousand), annotations to 22–3, 25, 35, author's personal collection. Braithwaite's annotations suggest that even at this stage he was unfamiliar with the concept of pangenesis.
339. 'Organic Evolution' (June 1879), MS797/III/26/9, Booth Papers, Senate House Library, UL.
340. 'Notes on Darwin', MS797/III/26/2, Booth Papers, Senate House Library, UL.
341. *Bath Chronicle*, 17 March 1881.
342. *Stroud Journal*, 14 February 1879; the presenter, 'J. Sibree, MA', cannot be reliably identified, but is very likely John Sibree (1823–1909), schoolmaster, who gave a paper to the Cotteswold Naturalists in January 1879, *Stroud Journal*, 18 January 1879. This earlier account makes it clear that Sibree rejected pre-*Origin* catastrophism/special creation, but was more inclined to align with Mivart, Murphy, and others who accepted evolution, but were mindful of the considerable obstacles in the way of accepting Darwinian ideas of natural selection as its key dynamic.
343. J.S. Lidgett, *My Guided Life* (1936), 87–8 (Dallinger eventually gave the lecture in 1887); Leonee Ormond, *George Du Maurier* (1969), 242, citing a letter of 1880 in the Charles Roberts Autograph Collection, Haverford College.

The Reception of Darwinian Evolution in Britain, 1859–1909: Darwinism's Generations. Martin Hewitt, Oxford University Press. © Martin Hewitt 2024. DOI: 10.1093/9780191982941.003.0004

4

The Death of Darwin and After

The 1880s and the Late Victorians

> The purity of our faith, as it was in the early seventies, has become streaked by harassing doubts; heresy and schism are rampant; we are no longer certain whether to pin our faith to the *Origin of Species* or to the gospel as preached and published by Dr August Weismann; some of us are members of that by no means obscure sect, the 'Neo-Lamarckians'. ([Frank Evers Beddard], review of Ernst Haeckel's *History of Creation* 2nd ed., *Saturday Review*, 18 February 1893.)

By the end of 1881 Darwin, long notorious for his poor health, was showing increasing signs of heart disease. In the night of the 18th April 1882 he had a final serious attack, and died the following afternoon. With his death, Darwinism crossed a Rubicon, further distancing the days before the *Origin*. All too quickly, W.H. Hudson observed, modern naturalists were looking on those who lived in those 'pre-Darwin days' as 'one of the ancients'.[1]

Watching Darwin's funeral, the labour leader George Howell (1833–1910) reflected on how his teachings would now be 'sanctified, to use a theological term, in the eyes of the public'.[2] It was a prescient observation. The biographical engagements that Darwin's death encouraged salved some of the sensitivities to his evolutionary theories. There was a noticeable growth in public attention, suddenly a greater willingness to explore his life and ideas on the lecture platform, in the provincial natural history associations, and in local debating and mutual improvement societies. Obituary notices and popular biographies made Darwin's ideas available in a more easily digestible form, and served as reminders of the range of Darwin's less controversial geological, zoological, and botanical work. But most importantly, while Darwin was alive, even as a partial recluse, his work had remained contingent, open to modification and revision, and he remained, as the sensation novelist M.E. Braddon had one of the characters in her novel *The Day*

238   DARWINISM'S GENERATIONS

Will Come (1889) remark, a 'perpetual court of appeal against arrogant smatterers'.[3] With his death, his thought became a fixed body of work, a matter for exegesis and interpretation, a battleground for contending opinions. While rival groups of scientists fought over Darwin's implied sanction, preachers, safe from direct denial, reclaimed Darwin for Christianity.

In the years before Darwin's death the older early-Victorian generation was also passing away, and even the ranks of the mid Victorians were thinning out. E.B. Pusey, the scourge of Darwinism at Oxford, died within six months of his adversary. 'How our generation are disappearing', lamented John C. Shairp (1819–85) later in 1882, 'each new blank gives a blow which it is hard to stand up against'.[4] A new generation had already begun to shape Darwinian debates: the late Victorians, born roughly between 1846 and 1859, the generation of Oscar Wilde and Vernon Lee, Arthur Balfour and Herbert Asquith, Robert Blatchford, and Annie Besant, all of whom experienced the initial debates over the *Origin* as children, and over *Descent* as adolescents or young adults, a generation which had little direct sense of and even less intellectual investment in pre-Darwinian modes of thinking.

In the final years of the 1870s two lectures, the first on 'Animal Intelligence' by George J. Romanes during the Dublin meeting of the British Association in 1878, the second on 'Degeneration' by E. Ray Lankester (1847–1929) at Sheffield the following year, announced two significant new voices in the debates over evolution. Within a few years they were being joined by younger late Victorians, including the entomologist Edward Bagnall Poulton (1856–1943), the comparative psychologist Conwy Lloyd Morgan (1854–1936), the botanist and later town planner Patrick Geddes (1854–1932), and the evolutionary statistician Karl Pearson (1857–1936). After Darwin's death, Romanes, who had worked as his assistant in his later years and was seen by many as Darwin's most committed disciple, took an increasingly prominent role in the discussion of the mechanisms of evolution. His three volumes with the general title *Darwin and After Darwin* (1892–94) sought to gather up some of Darwin's unpublished work and develop his ideas. But it was Lankester, rumbustious, charismatic, and outspoken, who was the more influential. Lankester held a number of important academic posts and for many years contributed a regular column to the *Daily Telegraph*. His British Association lecture, published as *Degeneration* (1880), although most often cited by scholars as a foundational text of *fin de siècle* pessimism, is better understood as an early salvo in the disputes within the evolutionary camp which dominated evolutionary debates for the following fifteen years, between orthodox Darwinians, fellow travellers with Romanes, and those influenced by the more Lamarckian ideas of Herbert Spencer.[5]

At first glance, reactions to Darwin's death seemed finally to demonstrate the achievement of the Darwinian revolution. From every quarter came praise for his character and his science with none of the antipathy that might have been apparent twenty years earlier. Editorials gushed with glowing tributes. The *Times* marvelled at the changed tone of the responses when compared with the confrontation of Huxley and Wilberforce at Oxford two decades before, dryly concluding that evolution's theological opponents had demonstrated 'an adaptation to their environment'.[6] A great deal of attention was given to Darwin's personality, which allowed even those still fearful of the implications of his ideas to separate the man from his theories and especially the claims of his more pugnacious followers.[7] He was transfigured, as the sexologist and social reformer Havelock Ellis (1859–1939) put it, 'by virtue of his method and spirit, his immense patience, his keen observation, his modesty and allegiance to truth'.[8] From mid Victorians like George Prothero (1818–94), Canon of Westminster, and Harvey Goodwin (1818–91), the Bishop of Carlisle, to late Victorians like Francis Paget (1851–1911), son of Sir James Paget and later Bishop of Oxford, assessments praised Darwin for his pure and earnest love of truth, and 'the simplicity and humility and self-forgetfulness which made him mighty and noble as the *minister et interpres Naturae*'.[9] Howell wrote of the way Darwin's 'vast labours, his giant intellect, his noble life, with its lofty aims, have impressed themselves upon his generation…'.[10] If Darwin's ideas were still to be questioned, it could no longer be on the basis of his status or methods as a scientist, although at times it almost felt as though justification of Darwin's ideas was being transferred from the credibility of his reasonings to the character of their author. The thought, of course, was not ignored. It was celebrated as a decisive intellectual epoch, the creation of a new nomenclature and a new way of understanding, albeit largely in respect of the existence of evolution rather than the explanation of its causation. Several prominent waverers, including John Clifford, publicly affirmed their acceptance of evolution.[11] Horror, it seemed, had given way to hagiography. It was all too much even for a loyal Darwinian like Julia Wedgwood, who found the obituary notices over-effusive, and longed for something a little more temperate.[12]

Huxley and Lubbock quickly stepped in to ensure a state funeral and a burial in Westminster Abbey. A week after his death, the original plain local coffin replaced by a grand, velvet-draped casket, Darwin's body processed

240   DARWINISM'S GENERATIONS

from his home to the Chapel of Faith at Westminster Abbey. The following day more than 3,000 mourners crammed into the Abbey, representatives from across British culture and spanning the generations. The Sunday after, the prominent Birmingham Nonconformist R.W. Dale (1829–95) preached in Carr's Lane Chapel on Darwin with a copy of the *Origin* in his hand, and in the afternoon another enormous crowd gathered in the Abbey to hear Harvey Goodwin's funeral sermon, which, steering clear of endorsing any of Darwin's evolutionary propositions, praised his strength of purpose, pertinacity, honesty and ingenuity, and paid tribute to a 'brave, simple-hearted, truth-loving man' who never attacked religious truths himself.[13] These sentiments were echoed from pulpits up and down the country. As Ruth Barton has observed, the irony was that Darwin's funeral, perhaps the greatest symbolic success of the X-Clubbers, affirmed not the independence of science from religion, but the conflation of science, church and state.[14]

The overwriting of the radical evolutionist by the virtuous man of science was reinforced when Darwin's statue in the Natural History Museum was unveiled in 1885. The Prince of Wales received the statue, with Edward White Benson (1829–96), the Archbishop of Canterbury, also in attendance. Presenting the statue on behalf of the Memorial Committee, Huxley pointedly observed that its creation was not designed to seek any official imprimatur for Darwin's ideas (remarking, not without unconscious irony, that 'science does not recognise such sanctions, and commits suicide when it adopts a creed'), but was instead an attempt to ensure that 'as generation after generation of students of Nature' entered the museum they should be reminded of the ideal by which they should shape their lives.[15]

This was Darwin the scientific totem, shorn of the controversial associations of his name. At the British Association a bare mention was enough to raise a hurricane of cheers and the waving of handkerchiefs.[16] The context in which Darwin was encountered changed. The weekend parties at Down House, the visits, the occasional public appearances in London, the exchanges of correspondence, were all no more. A few childhood memories apart, the relationship of the late Victorians had to be not with Darwin the person, but instead through his books, relics, and sites. True, monumental commemorations were few in number, and struggled to generate enthusiasm. In Shrewsbury there was a proposal to buy the old Shrewsbury School building to establish a Darwin Institute, but this came to nothing and even a statue had to be funded by the local Horticultural Society in the face of more general apathy; and the Hope Museum at Oxford acquired another in 1899, through the personal initiative of E.B. Poulton as Hope Professor.

THE DEATH OF DARWIN AND AFTER   241

Picture essays of Down House, initially preserved as Darwin had left it, turned it into something of a site of pilgrimage.[17] One visitor, tentatively identifiable as William Lee Martin (1856–1911), a London manufacturing chemist, described approaching Down in 1910, 'with feelings of reverence closely akin to those of the devotee towards his shrine,…rejoic[ing] in the thought that *here* the master carried out his great life work which emancipated the spirit of men from the thraldom of ancient dogmas and flooded the whole world of life and thought with new light'.[18] In the 1890s after sitting in Darwin's chair, examining the brass nails he had hammered into the adjacent wall, and exploring the plant houses, visitors could also see Darwin's original plain wooden coffin, displayed in an outbuilding behind the village pub.[19]

The years after Darwin's death saw a clutch of biographies, almost all by late-Victorian authors: Grant Allen, E.B. Poulton, J.T. Cunningham (1859–1935), and George T. Bettany (1850–91), as well as primers and introductory essays like *Darwin Made Easy* (1893) by the socialist Edward B. Aveling (1849–98).[20] This wave of lives opened up Darwin to new biographical interpretations. After twenty years of circumspection, lectures and papers on Darwin the man suddenly became a staple of rational recreation.[21] The way in which the transformation of Darwin's personal reputation was prising open spaces for a more positive assessment of his ideas is illuminated by the outcome of a discussion at the Saltcoats Literary Society in 1889. 'Whilst the majority of the meeting were not prepared to endorse Darwin's theories', reported the local newspaper, 'there was a general opinion that he deserved to be regarded as one of the great pioneers of scientific truth'.[22]

In death Darwin became an increasingly pervasive cultural presence. Certainly in the years around his death there was a significant shift in the circulation of Darwinian texts. Sales of Darwin's writings accelerated. By the late 1870s as well as issuing revised editions and new volumes, Murrays were selling nine of Darwin's works in their Standard Works series; total profits in the year ending April 1882 were over £900.[23] As public libraries proliferated in the 1880s and 1890s, so Darwin's books and other popular treatments of evolution became a staple of their circulations. In 1897 the Sunderland Free Library reported that *Descent* had been taken out more times than any book bar the most popular novels.[24] In 1882 the set of books given as a testimonial to the departing secretary of the Irish Association for the Prevention of Intemperance included all the evolutionary standards including Darwin, Huxley, Tyndall, and Spencer, as well as Lubbock and Arabella Buckley, along with many of the other central figures of Victorian

242 DARWINISM'S GENERATIONS

liberal thought; suggesting that these books were now considered sufficiently uncontroversial to be used in this way.[25] In truth, no substantial private library was now complete without its Darwinian texts. It was perhaps not surprising that the study of the radical journalist and novelist Eliza Lynn Linton (1822–98) was lined with Spencer, Darwin, Tyndall, and Huxley, as well as Haeckel's *Evolution of Man*, but more revealing that the bookshelves of the dramatist Henry Arthur Jones (1851–1929) included two shelves of Huxley, Darwin, Spencer, and Ibsen, and that Jones described his careful reading of Darwin and Spencer as 'especially useful to the dramatist'.[26] References to copies of Darwin on drawing room tables and library shelves begin to appear with greater frequency in popular fiction.[27] Even the high society drawing room revolving bookcase could display Darwin's *Voyages* at least, without embarrassment.[28] Grant Allen kept his inscribed copy of the *Origin* in the bookcases in his little dining room, and had his prized letters from Darwin ready to show to visitors.[29] Darwin's piano decorated the Positivists' Newton Hall.[30] Gardeners were encouraged to plant Darwin tulips, or *Berberis Darwini*, its attractive coral-like blooms appearing twice a year, and by 1885 they could also buy a 'Charles Darwin' buttonhole rose.[31]

As Janet Browne has shown, Darwin was a presence not only in his books, but also in the pages of *Punch*, and in the form of photographic portraits and *cartes de visite*.[32] Darwin's letters, even his autograph, were eagerly sought and proudly displayed. Already by 1882 W.C. Williamson had given away virtually all the correspondence he had received from Darwin to his large circle of autograph-hunting friends; by the start of the following decade Darwin letters commanded a good price in the collectors market—not perhaps yet rivalling Dickens, but matching Browning.[33] In an era when the market for celebrity photographs was dominated by the living, demand for Darwin's continued unabated after his death.[34] By the start of 1883, the engraving of John Collier's portrait of Darwin which had been exhibited at the Royal Academy in 1882 and issued by the Fine Art Society was being eagerly snapped up by and for his admirers.[35] Portraits of Darwin became part of the accoutrements of respectable science. Two loaned etchings decorated the rooms of the Sheffield Medical School for its opening ceremonies in 1888.[36] W.B. Tegetmeier's Finchley cottage had several portraits and photographs of Darwin on the walls; while he took his place in the portraits hanging in the sitting room of the radical chartist George Julian Harney, along with Ruskin, Tyndall, Julius Caesar, and Constance Naden.[37] Those with more limited budgets collected the images of Darwin appearing in the new illustrated magazines, like the series of Down House, including

Figure 6 'Darwin's Study', *Illustrated London News*, 10 December 1887.

Darwin's study, sandwalk, and greenhouse produced by Alfred Parsons and used to illustrate an article by Wallace in the *New Century Magazine* in 1883.[38] The young John Bland-Sutton (1855–1936), later a prominent surgeon, collected the portraits of Darwin, Huxley and Tyndall from *Nature* in the mid-1870s and mounted them together on the wall of his study (Figure 6).[39]

And if the novels of M.E. Braddon (1835–1915) are anything to go by, Darwin suddenly became an inescapable presence in polite conversation. In the 1860s and 1870s, Braddon's extensive output, including her popular sensation novels, had offered only oblique and occasional references to Darwinian motifs. In contrast, for about a decade after 1882, Darwinian books and discussions seemed suddenly everywhere. From the 'strong-minded Miss MacAllisters' of *One Thing Needful Thing* (1886), 'who scorned accomplishments as futile, [and] sat in different corners of the drawing-room, one reading Herbert Spenser [sic], while the other devoured Darwin', to the Dalbrook sisters in *The Day Will Come* (1889), who 'quoted *Essays and Reviews*, and talked of Darwin and Spencer, Huxley and Comte', Darwin was the topic of Braddon's drawing rooms. In the same spirit,

244 DARWINISM'S GENERATIONS

Robert Buchanan's picture in his *Foxglove Manor* (1885) of the Rev. Charles Santley, whose study included a bookcase of scientific and philosophical works, including *Descent of Man*, which was kept permanently locked, was designed to portray a fear that was increasingly anachronistic.

Not everyone shared the newspapers' sense of Darwin's death as an epochal event. The Fabian socialist Edward R. Pease (1857–1955) recalled it as passing almost unnoticed, and suggested that notwithstanding Darwin's burial in Westminster Abbey, even in 1882, evolution was regarded as 'a somewhat dubious theorem which respectable people were wise to ignore'.[40] The diaries of Alexander Campbell Fraser and Edward Wrench, despite their long-standing interest in evolutionary issues, pass without noticing Darwin's death. Nor does it have the footprint in private correspondence that might be expected. In one exception, W.B. Carpenter exulted at the apparent transformation of Darwin's standing, indulging in a private fantasy in which the hostile *Quarterly Review* reviewer of the *Origin* returned and was forced to recognise how decisively his view had been proved wrong. Yet at the same time, Carpenter was clearly taken aback. Before Darwin's death he had not noticed any signs that this change was taking place. And perhaps this is because it hadn't been; at least not in the revolutionary terms in which it was being described. Certainly, the change did not seem so dramatic for younger observers, one of whom recalled a period in which the weight of orthodox theology seemed hopelessly at variance with the growing testimony of physical science, leaving many, especially amongst the clergy, stranded: 'They had been trained in the old ideas, and had not come into contact with the new ones, except in the form of rumours and stray reports, which they found themselves able to put aside with ease as mere "surmises" or casual expressions of aberrant intellects'.[41] Carpenter conceded that it was probably to works like Darwin's recent and popular *Formation of Vegetable Mould through the Action of Worms*, rather than to his much more technically and ideologically challenging works on evolution, that the shift in reactions might be attributed (Figure 7).[42] (And he was not entirely comfortable with the perspectives of contemporary science, suspecting that 'the present generation, who have been brought up in the light', did not 'quite apprehend…the utter darkness in which we were groping, or fully recognise the deserts of those who helped them to what they now enjoy'.)[43]

Whatever the extent of the shift, it was as much a case of artificial selection than the survival of the fittest. Gowan Dawson has demonstrated how assiduously the group around Darwin managed his public *persona* from the

Figure 7 'Punch's Fancy Portraits. Charles Robert Darwin', *Punch*, 22 October 1881.

246   DARWINISM'S GENERATIONS

very first publication of the *Origin*, and the posthumous canonisation of Darwin was not achieved without further concerted effort. Hooker took a leading role. He was part of a select group which convened to raise money for a bust in Westminster Abbey and a Darwin Fund to be administered by the Royal Society, leading eventually to the institution of the Darwin Medal, awarded for the first time in 1890.[44] He worked energetically if unsuccessfully to rein in plans for an unsanctioned series of tributes by Norman Lockyer and Romanes, which eventually appeared in *Nature*.[45] Meanwhile, the mantle of biographer fell to Darwin's third son, Frank, a botanist and collaborator in his father's later work. Frank and his siblings resisted pressure to rush into print, and moved quickly to forestall the publication of any of Darwin's letters by 'literary hangers on'.[46] Instead a comprehensive two-volume *Life and Letters* was slowly and methodically prepared. Proofs were sent out to leading participants and great care was taken to downplay the initial opposition to the *Origin*.[47] Details of Darwin's loss of Christian faith were carefully elided (and Haeckel's decision to publish a letter which Darwin had sent to a German admirer led to an acrimonious exchange with Frank, who was incensed at the potential damage to Darwin's new-found status).[48]

When it was eventually published in 1887, the *Life and Letters of Charles Darwin* became a minor publishing sensation of its own, selling out its first edition of 4,000 copies in days, squeezing the circulation of all other books on the shelves of Mudie's library, and being greeted with almost universal praise. It was followed by a condensed single volume *Life* with additional material in 1892, and three further volumes of letters in 1903. For many readers the Darwin of the *Life* was a revelation; used to the harsh tones of the public debate they were astonished to discover 'the most gracious, tender-hearted, modest, humble-minded, unaggressive of men'.[49] The Darwins made strenuous efforts to ensure that reviews highlighted the book's appeal to a general readership, and were no doubt pleased at the way in which reactions focused above all on Darwin's character.[50] Echoing the obituaries, the reviewers offered Darwin more as celebrity than as scientist, and paid remarkably little attention to the substance of his discoveries; one widely reprinted review suggested that for most readers the first volume (which provided the bulk of the biographical detail—the detailed discussion of his science being left largely to the second) would be enough.[51] Alfred Newton's review in the *Times*, more than a full page long, managed to avoid any summary of Darwin's arguments at all.[52] A.V. Dicey bubbled with enthusiasm after a morning with the *Life* and the *Origin*, pronouncing

THE DEATH OF DARWIN AND AFTER 247

them the most edifying writings he knew.[53] The *Life* had an ethical as well as a scientific value, Bonney later remarked; it provided a personal pattern that could be invoked to justify the possibility of noble character without Christianity.[54]

In some ways Darwin's correspondence suddenly became as important as his published writings, offering new opportunities to dip into his thought and cherry pick amenable observations. Sceptics picked up on Darwin's characteristic private expressions of despondency or doubt, and presented them as evidence of the underlying weakness of his ideas; believers seized on his observation that there was nothing contradictory in evolution and Christianity.[55] Darwin's amazement at the progress that Christian missionaries had been able to achieve with the 'primitive' Fuegans of Tierra del Fuego became a staple of the religious platform. In the pulpit, Darwin could be domesticated as the man who supported missionary work, was on friendly terms with his parish priest, and confessed that his lack of faith was a sort of colour-blindness.[56] It is noticeable how often the discussions which flared up in the newspaper columns over the ensuing thirty years were informed as much, if not more, by materials culled from the *Life* as from Darwin's own works.

Reprintings of the *Origin* and *Descent* and the supplemental publication of Darwin's correspondence after the initial tranche printed in the *Life and Letters* continued to be carefully supervised into the 1900s.[57] Even impeccable Darwinists like Poulton could find their requests to publish letters refused.[58] Revised editions of Darwin's other works were issued with carefully commissioned introductions. Controversy was to be avoided: Bonney was engaged to provide a preamble to a new edition of *Coral Reefs*, on the basis that 'it should not be polemical'.[59] More popular studies, at least those produced by Darwinians like Poulton, were sent to the Darwins for comment and approval; in Poulton's case with the result that pressure was brought to bear on him to offer a less equivocal account of Huxley's commitment to Darwin's core ideas.[60] Meanwhile, the influence Huxley wielded as the main editor of Macmillans' Science Primer series enabled him to ensure the refusal of proposals from less orthodox younger men.[61] In turn, figures like Poulton themselves took on informal responsibilities for safeguarding Darwin's reputation.[62]

This newly discovered virtuous man of science, fit to be proposed in 1888 by Henrietta Maria, Lady Stanley (1807–95) as a topic for the annual essay competition she organised for girls' schools, was taken up especially by mid Victorians.[63] Even the 79-year-old Thomas Cooper adjusted his

248 DARWINISM'S GENERATIONS

anti-Darwinian rhetoric. By 1884 his lecture on Darwin strove to separate the man and his love of truth, his humility, the great patience exemplified by works like *The Action of Worms*, and even hearsay about his religious beliefs, from his advocacy of evolution.[64] At the Free Church Congress in 1896 the Congregationalist R.F. Horton (1855–1934) commented that a more perfect idea of holiness for the churches would include the work of Darwin, and the great naturalist, 'instead of being reckoned among the unbelievers would become one of the saints'.[65] It was this sort of response which was gently caricatured in *The House of Quiet. An Autobiography* (1904) by the Edwardian novelist A.C. Benson (1862–1925), in which Darwin's *Life* was described by one character, an elderly parish priest, as 'a wonderful book...from end to end nothing but a cry for the Nicene Creed! The man walks along, doing his duty so splendidly and nobly, with such single-heartedness and simplicity, and just misses the way all the time; the gospel he wanted is just the other side of the wall....Whenever I go to the Abbey, I always go straight to his grave, and kneel down close beside it, and pray that his eyes may be opened. Very foolish and wrong, I dare say, but I can't help it!'.[66]

The Persistence of Older Attitudes

The tone of public debate was changing, in part, as F.W. Farrar remarked, because those who twenty years previously would have offered denunciation and derision had 'at least learned modesty'.[67] Nevertheless, the attention prompted by Darwin's death inevitably raised new tensions. The banker and historian Thomas Hodgkin (1831–1913) had feared, as soon as he saw the reviews of Darwin's *Life*, 'that the old trouble of those days would be to some extent renewed',[68] and the outpouring of admiration was too much for some older anti-Darwinians, who were roused once again to protest. Nicholas Whitley (1810–91), secretary of Royal Cornwall Society, while proclaiming the pleasure he had drawn from reading Darwin's *Beagle Journal*, muttered that the hero-worship that Darwin's death had elicited was a nefarious strategy to validate his theory by the back door.[69] William MacIlwaine, who had rallied opposition in the Belfast Naturalists' Field Club in the 1860s, was stirred into print again.[70] J.S. Blackie held tight to his evolutionary scepticism, drawing succour from the renewed interventions of early and mid Victorians like the Duke of Argyll, who was in turn privately encouraged by correspondents including Richard Owen.[71] Religious

THE DEATH OF DARWIN AND AFTER   249

magazines like *Good Words* continued to offer a vehicle for Argyll to criticise the 'bungle' of natural selection. In a series of articles on 'Darwinism as a Philosophy' later in the decade, he suggested that the correspondence published in the *Life* would accelerate the reaction against Darwin's ideas, by revealing the contrast between his greatness as an observer and his weakness as an interpreter.[72]

Before his death, Pusey's hostility probably kept anti-Darwinism more vigorous in Oxford than anywhere else. His close friend Henry Liddon confessed to coming to preach his own memorial sermon for Darwin 'with discomfort and misgiving'.[73] In the 1880s Henry Liddell (1811–98), Dean of Christ Church (and father of Lewis Carroll's 'Alice'), was still presenting science as less significant than medieval thought, and scientific men as 'infidels',[74] and modernisers like Lankester and Foster still needed to conciliate or circumvent the older professors, Burdon Sanderson and Rolleston, notwithstanding their ostensible Darwinian sympathies.[75] One Oxford undergraduate of the time recalled an afternoon in Max Müller's garden, Müller talking with great vivacity of Darwin and Haeckel: 'It ended by my host turning upon me, and looking with that flash through the glasses: "If you say that all this is not made by Design, by Love"—waving an arm towards the Parks—"then you may be in the same house, but you are not in the same world with me"'.[76] Müller's 1892 Gifford Lectures, *Theosophy or Psychological Religion* restated his limited evolutionism, conceding the evolution of the horse from earlier horse-like animals, but not the entire evolution of the natural world from protoplasm.

In the 1880s the episcopal bench was still well-furnished with old anti-Darwinians. The Bishop of Lincoln, Christopher Wordsworth (1807–85), was still actively combatting evolutionary thought amongst his clergy, denying a temporary curate's licence in April 1882 to one candidate on the basis of an essay he had published on 'Geology and Evolution'.[77] In 1883 William Huggins reported a recent conversation with a 'distinguished professor' who observed that he had lost all chance of patronage in the church because he had spoken in favour of evolution.[78] Even in the universities and science institutions, younger Darwinists continued to encounter barriers to professional progression. Especially in Scotland, where Darwinism was still treated in some quarters as synonymous with atheism.[79] Patrick Geddes was warned that his application for a chair in Biology at Dundee in 1884 was likely to be unsuccessful because of fear 'that you are a belligerent Darwinian'.[80] His friend George Murray (1858–1911), assistant in the Botany department at the Natural History Museum, found his path to promotion

250 DARWINISM'S GENERATIONS

blocked by the hostility of William Carruthers, whom Murray felt had eventually realised that 'I am in the tents of the enemy as to evolution'.[81]

Striking evidence of the persistence of anti-Darwinian attitudes is provided by the Kinns' mission episode of 1884–5. Samuel Kinns (1826–1903) was the author of *Moses and Geology* (1882), a conventional attempt to reconcile the geological record and Genesis. In 1883 a committee of predominantly early-Victorian bishops engaged Kinns to give lectures and sermons promoting the view that the recognised sequence of evolutionary development was exactly mirrored by the Bible account.[82] This position was not of itself anti-Darwinian or even anti-evolutionary, but part of Kinns' activity seems to have been to push sales of *Moses and Geology*, in which he indicated he was prepared to accept 'so much of the Development theory as implies a progression of species from a lower to a higher organic condition', but nevertheless argued for special creation, variation only within species limits and entirely denied the evolution of man from animals.

There certainly remained a constituency for such ideas. In the early months of his mission Kinns was apparently selling 100 copies of *Moses and Geology* a week, and it continued to find purchasers into the 1890s.[83] The first thousand copies of his follow-up, *Graven in the Rocks* (1891), sold out before publication.[84] Significantly, the early-Victorian geologist Joseph Prestwich, although he was ultimately pressed into condemning Kinns' book, repeatedly expressed his reluctance, arguing that at least Kinns' stance was better than the old biblical literalism and would encourage greater attention to modern science.[85] Nor do Kinns' activities seem to have concerned most high Victorians, for whom perhaps, as James Knowles suggested, the whole affair just seemed outdated and irrelevant.[86] But Kinns and his sponsors galvanised vigorous opposition from a largely mid-Victorian group, both Darwinists like the radical-Christian Charles Voysey, William Huggins and T.G. Bonney, and also doubters like Charles Pritchard, Frederick Temple, H.W. Acland, and William Carruthers, concerned that Kinns' stance would just play into the hands of religious sceptics by giving the impression that Christianity had only Kinns' inadequate arguments for its defence.[87] Despite his own evolutionary hesitancies, Carruthers wrote to the *Times* accusing Kinns of mangling both geological facts and the Mosaic account, while Acland drummed up supporting letters and raised his concerns with E.W. Benson, the new Archbishop of Canterbury.[88]

Part of Kinns' difficulty was that although the previous Archbishop, the early-Victorian A.C. Tait (1811–82), had hankered after just the sort of stable reconciliation between scientific and religious positions that Kinns was

THE DEATH OF DARWIN AND AFTER   251

peddling, his mid-Victorian successor Benson was much less interested, agreeing that such 'forced reconciliations...even if they are dextrously done, are...most dangerous.'[89] Kinns' incautious claims to have the endorsement of leading figures in the religious as well as the scientific establishment only intensified Benson's ire. Initially undaunted, Kinns gave a series of lectures in the first half of 1884, from large evangelical venues like Exeter Hall down to drawing room gatherings, while pushing back against those he described as 'bitter opponents amongst scientific men who are disbelievers in Revelation', but ultimately the campaign against him bore fruit, the mission was quietly allowed to peter out, and Kinns diverted into regular parish work.[90]

One footnote to the affair was the report that Kinns' brief notoriety did not dissuade Gladstone from going to hear him preach.[91] And Gladstone's essay wars with Huxley in the 1880s and 1890s further illustrate the persistences and intergenerational tensions of late-Victorian debates. In 1885 the Prime Minister attempted to assert the congruence of Genesis with contemporary palaeontology, in doing so giving Huxley the opportunity to weigh in and demolish his crude renditions of contemporary science in a series of characteristically trenchant rebuttals. Once again, the controversy played out the differences between early and mid Victorians. Acland, although he shared much of Gladstone's anxiety about the theological implications of evolutionary thought, consistently urged caution, counselling vigorously against any attempts at a literal reconciliation of Genesis and science, and warning that on such matters jousting with Huxley was not only futile but self-defeating.[92] Although by this point Gladstone had read relatively widely in the literature of evolution, he drew inspiration for his continued resistance from a network of predominantly early-Victorian non-Darwinians, including Owen, Prestwich, and the philologist Robert Scott (1811–87),[93] collectively determined to give ground to modern scientific interpretations only when and insofar as it was absolutely essential.[94]

Confrontations of this sort were deprecated by more secular high Victorians, and the subsequent controversy whipped up in 1890 over the New Testament account of the Gadarene swine, although it was only tangentially implicated in the wider evolutionary debates, helped reinforce both the whiff of obsolescence and the rout of early-Victorian attempts at reconciliation.[95] Argyll warned Gladstone that many would be indifferent to such trifles, but Gladstone countered that for him, as for Huxley, it was a serious question.[96] It wasn't a serious question for young students like Arthur Conan Doyle (1859–1930), who later recalled that 'when a Gladstone

252 DARWINISM'S GENERATIONS

wrote to uphold the Gadarene swine, or the six days of Creation, the youngest student rightly tittered over his arguments, and it did not need a Huxley to demolish them', nor for May Kendall (1861–1943), who smirked at the clash in her poem 'Nirvana'.[97] Gladstone was still championing the historical veracity of Genesis in the mid-1890s,[98] but by then he could be dismissed as 'the mind of the Oxford of 1829 accommodated to the political circumstances of today'; Edward Clodd told his publisher in January 1897 not to send his new book to Gladstone; 'it will look like a direct challenge', he observed, 'and one cannot meet an old man on equal terms in controversy'.[99] Not that Huxley fared much better. Responses to the reissue of many of his polemical defences of Darwin from the 1860s and 1870s in *Darwiniana* (1893) exposed his distance from contemporary concerns, and Huxley in turn recognised himself as more and more a man of the past.[100] The younger generation 'shut their eyes to the obstacles clericalism raises', he lamented.[101] By the 1890s the world of 'dear old Huxley' (as Lankester condescendingly described him in 1889) seemed increasingly distant to the younger scientists.[102]

Even metropolitan science remained riven with these sorts of discords, as the *fin de siècle* history of the Entomological Society shows. There is no doubt that there were plenty of younger entomologists for whom entomology offered opportunities to explore Darwinian themes, and it is tempting to regard anti-Darwinian polemics like Francis Polkinghorne Pascoe's (1813–93) fiercely hostile *The Darwinian Theory of the Origin of Species* (1890), which combined a general acceptance of evolution with a wide-ranging attack on Darwinian dynamics, as anachronistic recapitulations of long-discredited positions. Yet even a dyed-in-the-wool Darwinist such as Raphael Meldola conceded that Pascoe was not just a credible scientific authority in his specialist field, but also representative of a large class of systematic entomologists in the country who held similar views.[103] In the early 1890s a dispute over the candidature for president of the Society of the naturalist and traveller Henry John Elwes (1846–1922) laid bare tensions which although often couched in terms of scientific expertise also expressed generational differences over Darwinism.[104] Meldola warned Poulton in the early stages of the affair that 'the secession of all the older fossils would...only help to crystallise out a rabid anti-Darwinian, anti-biological faction'; at least with the mixed membership as it then existed, Meldola felt, 'we can leaven them a little'.[105] Elwes was probably more unsympathetic towards the Darwinians than actually hostile to Darwinism.[106] Nevertheless, for the late Victorians like Poulton and J.W. Tutt (1858–1911), he was one of the collector types of the old school; and he certainly stressed the continued importance of old-fashioned systematic work, which enamoured him to anti-evolutionary members.[107] Poulton, who lamented that he just wanted a proper biological entomologist for once, campaigned hard for Meldola to be elected president instead, assembling a list of supporters in which Darwinists like Wallace, Dixey, and Arthur Sidgwick were prominent, but found himself in the clear minority.[108]

It became much more common for early and mid Victorians, unable to sustain their initial outright opposition, to resort to a tactical retreat from the exposed positions into which their initial antagonism had led them. Many over the years will have occupied the intermediate world of the Free Church cleric David Brown (1803–97), of whom it was said towards the end of his life, that while he was not to be drawn into direct condemnation of evolutionary thought, 'he had a certain scepticism as to the permanence of all such speculations, a certain disinclination to make terms with them'.[109] Frequently such withdrawal involved falling back on a suspension of judgement, accepting that evolutionary theory was becoming more probable but insisting it was still not proven. But Huxley's claim in his 1880 address on the *Origin*'s coming of age, that evolution had been established as an undoubted fact threatened this position, forcing mid Victorians who had previously taken refuge in the room for manoeuvre provided by its status as persuasive hypothesis into a more nuanced distinction between the evolutionary past and its causes.[110] More and more mid Victorians were prepared to offer a limited acceptance of evolution, their intellectual history undocumented, and the nature and timing of their intellectual journey unclear.[111] Some, like Sir Richard Strachey (1817–1908) were happy to frame this in Darwinian terms.[112] Most continued to distance themselves from 'Darwinism' and its explanations, seeking to construct a distinction between the 'facts' of evolution and its 'pseudo-facts', or the theories and inferences drawn from the facts, or to console themselves that even if it were proved, it was an irrelevance for Christian belief.[113]

Many mid Victorians probably came to occupy the sort of intellectual no-man's-land typified by the botanist and collector James Backhouse (1825–90), who told Asa Gray in 1882 that Darwinian theory did not alarm him in the way it did some, denied any sharp line of separation between scientific and religious truth, agreed that Darwin's 'marvellous contribution' would tend to advance truth in its highest sense, all without feeling the need to clarify his own position on the fundamental question of whether he could accept evolution.[114] William Alexander (1824–1911), Archbishop of

254 DARWINISM'S GENERATIONS

Armagh and Primate of Ireland, gave a vivid vignette of this sort of process in a sermon preached on the occasion of the British Association meeting in Belfast in 1902, relating how some time before, while sailing across the Atlantic, he 'took up Darwin's *Descent of Man*, and read, not for the first time, the sixth chapter of the first part' ('On the Affinities and Genealogy of Man'). 'I felt for a while pained and dismayed', Alexander recalled, until the sound of singing from a Sunday evening service helped suddenly to bring the realisation that 'the question of questions for the soul was not what man may have been in outward form at one time, but what he is;...not what organisms may have been employed in moulding his body, but what they have become.'[115]

The shifting terms of debate between the generations is symbolised by the petering out of the Metaphysical Society and its partial reincarnation in the Synthetic Society. In the 1870s the Metaphysical Society had offered a welcome forum for discussion of controverted issues of science and religion, and had helped soften the edges of the debate by encouraging friendship among some of the antagonists.[116] But by 1880 it had taken rapprochement as far as it could go and was wound up. Its successor, the Synthetic Society, was described by James Martineau as a resuscitation of the old Metaphysical, 'with the simple omission of the confirmed Agnostic class, for whom no religious problem exists',[117] and it effected a shift away from debates over evolutionary science to its philosophical and theological implications. Indeed, the first meetings discussed Arthur Balfour's paper on 'Can Order and Design in Nature be accounted for without a prior reason working through evolution?'[118]

The Emergence of the Late Victorians

The idea for the Synthetic Society took shape at a dinner at the Junior Carlton Club in January 1896 attended by Balfour, Wilfrid Ward, Edward Talbot, and Charles Gore (1853–1932) (for whom see below), and although the Society eventually included some older survivors from the Metaphysical, and a smattering of older Edwardians, its centre of gravity was firmly among the late Victorians who in the previous decade had been increasingly setting the tone of public debate.[119] That the Synthetic was in its turn quickly mired in doubts as to its efficacy, and a sense that, as the Indian civil servant and literary historian A.C. Lyall (1835–1911) put it, 'we have been at cross purposes, starting from different points, and each keeping straight on to his own ends', was in part a reflection of the intellectual differences opening up between the late Victorians and their predecessors.[120] The terms of the evolutionary debates were being set by a new generation.

Lankester, who had swept into University College London as Professor of Zoology and Comparative Anatomy in 1874, restoring its position as a leading centre of evolutionary research, was only the first of a new cadre of scientists, who were establishing themselves in academic posts in the 1870s and 1880s, (including Meldola, the embryologist Francis Maitland Balfour (1851–82), James Cossar Ewart (1851–1933), Arthur Milnes Marshall (1852–93), and Frank E. Beddard (1858–1925)), and presenting themselves, as Cossar Ewart did at Edinburgh, as fervent disciples of Darwin's law of evolution.[121] The School of Mines under Huxley produced not just Lankester and Conwy Lloyd Morgan, but 'a whole academic generation [which] passed through this South Kensington filter'.[122] In addition, a surprising number of late-Victorian natural scientists and philosophers, including Sydney Vines (1849–1934), R.B. Haldane, his brother John S. Haldane (1860–1936), Andrew Seth (1856–1931) (later Pringle-Pattison), and John H. Muirhead (1855–1940) spent periods studying with the leading German Darwinists.

The institutional landscape was changing. The curricula of the public schools had been overtaken by the lure of evolutionary thought. Religious orthodoxy became impossible to sustain where the elder boys read Spencer, Tyndall, and Darwin.[123] By this point Darwin's works were prominent on the reading list for the examinations of the newly established Diploma of Special Lecturer of the previously ambivalent Secularist Society.[124] By the mid-1880s the Congregationalist training colleges were expanding their curriculum to include aspects of Darwinian theory, including practical work with the microscope.[125] Even the Victoria Institute was losing some of the narrow anti-Darwinian bigotry which had characterised its early years, and while continuing to offer a rallying ground for older anti-Darwinians, it gradually extended its constituency.[126]

From the mid-1870s the Darwinian challenge to religious orthodoxy along with the Idealism of T.H. Green and his disciples dominated the intellectual atmosphere of the universities. Religious undergraduates no longer worried over whether God created the universe, and instead debated how.[127] Biology, and indeed comparative embryology, much more than geology, now seemed to be addressing the really crucial questions facing the natural sciences, and new student societies, like the Cambridge University Natural Science Club, founded in 1872, and the Oxford Junior Scientific Club, formed in 1882, offered spaces for research students to share their interest in

256   DARWINISM'S GENERATIONS

evolutionary questions.[128] Figures like F.M. Balfour and Hans Freidrich Gadow (1855–1928) at Cambridge were developing Darwinian curricula, and founding a new school of embryologists.[129] Alfred Newton now struck undergraduates as a cautious old-fashioned zoologist, lecturing on Darwinism to tiny audiences, but uninterested in, although not actively hostile to, their new preoccupations.[130]

The social psychologist Graham Wallas (1858–1932) read Darwin's works while at Oxford at the end of the 1870s. His wife later noted that Wallas had acquired a personal relationship with Darwin's mind 'of a kind generally only friendship gave'.[131] The recollection of the geneticist William Bateson (1861–1926) of the early 1880s was that every aspiring zoologist was an embryologist, and the one topic of professional conversation was evolution.[132] Darwinian beliefs could all too easily prompt irate response from older tutors, as Karl Pearson recalled of an incident during his tripos examinations, when a chance remark roused the physicist Clerk Maxwell to 'the highest pitch of anger'.[133] But the sway of evolutionary ideas amongst the student body was universal. At Cambridge the first groups of students at Girton debated the implications of evolutionary biology for religious belief, and at Oxford in the mid-1870s, even the unscientific Oscar Wilde (1854–1900) was reading the full range of evolutionary writers and extracting them in his commonplace book.[134]

Simultaneously, there was a noticeable freeing up of provincial intellectual cultures. This was especially visible in the newer societies, whose formation was often driven by a core of younger late Victorians. At Northampton, where a Natural History Society was established in 1876, apparently on the explicit understanding that members would ascribe to evolutionary theory, the lead was taken by a small group of young teachers and professional men, supported by carefully chosen allies with established Darwinian sympathies.[135] At Dundee, another locality where the Naturalists' Society emerged out of a group associated with the local science classes, the generational impulse and its Darwinian agenda was even more overt. Recollections spoke of 'heavy skirmishes' over evolution, but placed the main lines of division not between pro- and anti-evolutionary camps, but between disciples of Darwin and of Spencer, and the society's public lectures provided a vehicle for a succession of leading evolutionary popularisers, including Andrew Wilson, James Geikie, Geddes, Romanes, and Dallinger.[136]

Longer-established local philosophical and scientific societies were also more willing to hear papers and lectures on Darwin and Darwinian biology. Many of these were delivered by high Victorians, but an emerging presence of late Victorians is also apparent.[137] The influence of Linnaeus Greening (1855–1927) in the Warrington Field Naturalists' Society from the late 1870s is one instance of their impact. Greening, a local wire manufacturer and committed arachnophile, lectured almost annually to the Society for more than forty years, helping to break down its previous nervousness about tackling evolutionary questions.[138] What the Gilchrist Trust lectures had done for the high Victorians, the spread of the various forms of University Extension lecture in the final quarter of the century provided for a younger generation of academics, bringing a succession of late-Victorian graduates to the provincial lecture platform: figures like Albert W. Brown (*c*.1874–?), and Conwy Lloyd Morgan, who became a regular performer of Darwin and evolution lectures in the south-west from the mid-1880s, with a consistent line promoting evolution as one of the greatest scientific theories since Newton's theory of universal gravitation.[139]

The Extension movement was just one of the forces broadening intellectual culture in the 1880s and 1890s. The houses of Murray, Macmillan, and Longman had long provided a path to publication for evolutionary studies, but a publishing expansion in the 1880s and 1890s greatly extended the availability of evolutionary literature. The University Extension Manuals launched by John Murray in the 1890s included a number of evolutionary titles, such as J.A. Thomson's *The Study of Animal Life* (1892), and Hugh R. Mill's *The Realm of Nature* (1897). This was just one of several innovative series and new publishers opening up opportunities for the late Victorians, including the Walter Scott Company, which published the Contemporary Science Series edited by Havelock Ellis, with a clear mission for authors to take up 'as advanced a position' as they liked,[140] and whose list included several important evolutionary texts.[141] W. Swan Sonnenschein & Co (initially 1878, restyled 1882), formed by William Swan Sonnenschein (1855–1931), a publishing vehicle used by the Darwinian inner circle, as well as being close to J.H. Muirhead and the Idealists, was joined in the 1890s by Edward Arnold, and by Patrick Geddes and Colleagues, who saw the publication of books bearing particularly on evolution as 'the essential and central purpose of our work'.[142]

At the same time in the late 1880s and early 1890s, the emergence of new cut-price book series gave opportunities for the reissue of evolutionary classics. George Bettany, as editor of the 'ridiculously cheap' two-shilling Minerva Library of Famous Books for Ward, Lock, and Co., created a

258 DARWINISM'S GENERATIONS

whole new market for some of the classics of Darwinian natural history. Darwin's *Beagle* journals, the first title in the Library, went through ten Minerva editions between 1889 and 1891.[143] Examination prizes provided another avenue for late Victorians to acquire their own copies of the *Origin*: the rationalist Frederick J. Gould (1855–1938) obtained his from the science exams administered from South Kensington; the pathologist Sir John Bland-Sutton (1855–1936) chose the *Origin*, along with Dante's *Inferno*, as his class prize at the Middlesex Hospital Medical School in 1878.[144] Popular natural history texts, even where they skirted round the question of evolution, increasingly incorporated the results of Darwin's empirical investigations into orchids, insectivorous plants, or earthworms.[145] There were exceptions, as in the case of the industrialist Edward Davies of Llandinam (1852–98), objecting at a prize-giving for a Methodist Sunday School in Montgomeryshire around 1890 to being asked to give out a *Life of Darwin*, but these were more and more reported as exceptions.[146]

Around this time, as Melanie Baldwin has shown, science's periodical culture shifted. *Nature*, for all its early friction with the X-Club, gradually established itself as the late-Victorian scientists' primary means of communicating with their peers.[147] While the mid and high Victorians carried on looking to the general intellectual periodicals to voice their opinions, the younger generation published much less frequently in these journals. Some of the long-running debates, for example around agnosticism, or the relations of science and religion, had much less appeal to the late Victorians, who while they were increasingly likely to repudiate organised religion, often took up not the belligerent agnosticism of the high Victorians, but a more relaxed theism.

Museums were another space where we can see a new generation of late Victorians progressing into positions of responsibility, with implications for the evolutionary content and design of displays. Until the 1880s the history of museums is in many respects another case study (literally) of evolutionary absence, even though their collections of natural history and especially fossils might have been expected to make them crucial sites for the dissemination as well as the development of evolutionary thinking. But while collections at the Hunterian Museum and at the Museum of Practical Geology continued to play an important role in palaeontological research, museum organisation and display continued to be dominated by dioramas of stuffed animals, trays of specimens organised on taxonomic lines, and the occasional curiosity. In the 1860s and 1870s many of the established museum curators were, like John Plant (1819/20–94) at Salford, old-fashioned

THE DEATH OF DARWIN AND AFTER 259

naturalists with staunchly anti-Darwinian views. As late as 1887 it was suggested by one observer that local museums still remained almost entirely in a pre-Darwinian condition.[148]

The Natural History Museum's ponderous progress, even after the retirement of Owen in 1884 and his replacement as director by the impeccably evolutionary W.H. Flower, and then from 1898 by the belligerently Darwinian Lankester, is an object lesson in the conflicted progress of evolutionary ideas across late-Victorian and Edwardian culture. Externally, Flower and Lankester both became mired in power struggles with their political masters (and in Lankester's case with his formal boss at the British Museum), struggles which their mobilisation of the Darwinian scientific leadership only exacerbated. Internally, both found themselves impotent to override the evolutionary scepticisms of the Museum's powerful department heads, not least William Carruthers, Keeper of Botany, and Albert Gunther (1830–1914), Keeper of Zoology, neither of whom retired until 1895. Flower used his greater control of the Museum's central hall to install Darwin's statue, and to institute an introductory 'Index Museum', which offered illustrations of mimicry, adaptation, and other components of evolutionary biology.[149] Otherwise, he was reduced to personal Sunday afternoon tours for invited guests, where he delighted in pointing out 'points of thrilling interest and mysterious evolution'.[150] His successor Lankester had the advantage of more sympathetic keepers, and indeed personal control of zoology, which opened up opportunities to supplement existing displays along the lines of the materials illustrative of the evolution of the horse, which had been added to the domestic animal displays by 1904.[151] But it is doubtful that he ever prioritised the evolutionary emphasis of the displays over his wider advocacy for Darwinism, even before his tenure descended into the bitter stand-off which led to his forced retirement in 1907.

The only significant exception to this marginalisation before the 1890s was the approach taken by Augustus Pitt-Rivers to the vast anthropological collections he accumulated from the 1850s to the 1880s. In many respects Pitt-Rivers' approach, like much of the evolutionary anthropology of the 1860s and 1870s, was more loosely developmental than Darwinian, although he was more explicit than most in his use of the dynamic of natural selection, and emphasis on change being the outcome of finely gradated sequences of minor alterations.[152] And once his collections found a home in Oxford they became an influential exemplar of evolutionary organisation. The most well-publicised and perhaps most sophisticated evolutionary displays of the Edwardian period, those of the Horniman Museum in Bethnal

260 DARWINISM'S GENERATIONS

Green, were developed under the guidance of A.C. Haddon, himself inspired as an Oxford student by the Pitt-Rivers collections.[153]

In the mid-1880s Thomas Jeffrey Parker (1850–97), the son of W.K. Parker, who had emigrated to New Zealand in 1880, championed phylogenetic displays in *Nature*.[154] Otherwise, moves towards more explicit evolutionary museum design only began to coalesce in the later 1880s and 1890s, with the municipalisation of many local museums and the associated drive to make the collections more educationally useful. By the early 1900s, in addition to the Horniman, significant reorganisations had also been effected at a number of museums outside London.[155] Trophies were replaced with types, curiosities with links in the evolutionary chain. New guides were published which encouraged visitors to reflect on the evolutionary implications of the specimens, and ancillary programmes of lectures reinforced the message.[156] Some of these initiatives were pushed through by younger Edwardians, but most derived from the emergence of a cadre of late-Victorian provincial curators, who also took the lead in the formation of the Museums Association in 1889 and in the discussions at its annual meetings in the following decade, where the case for evolutionary reorganisation was repeatedly made.[157] Darwin was invoked in these discussions as the necessary foundation of reform, the ideal modern museum curator portrayed as working through the new lines of exhibition with a copy of *On the Origin of Species* open on the desk before him.[158]

Late-Victorian Darwinist Identities

The career of Grant Allen illustrates many of the late-Victorian crosscurrents. Although he came to be known primarily as a novelist, and in particular for his 'new woman' novel, *The Woman Who Did* (1895), Allen started out as a jobbing journalist of science, relying heavily on short natural history articles, some of which were collected in a series of volumes with explicitly evolutionary themes published in the 1870s and early 1880s, and periodically thereafter. Although Allen's brand of broad evolutionism ultimately owed more to Spencer than to Darwin, he 'grew up a Darwinian, never anything else but an evolutionist'.[159] He was conscious that the multiplication of academic biologists was leaving him marginalised professionally, forced by the exigencies of journalism into 'quill driving from morning to night', to 'turn a thing into a Cornhill article which ought rather to have been worked up into a scientific paper'.[160] But he clung to a sense of the

THE DEATH OF DARWIN AND AFTER   261

significance of his work, that it was 'better to write evolutionary papers, however undigested, for the Pall Mall, than to write ghost stories for Christmas annuals',[161] and he was quite prepared to engage in controversy with academic scientists, even in the pages of *Nature*. For all this, the appeal of his brand of popular evolutionism seems to have remained fairly limited before the death of Darwin. Despite energetic promotion and favourable reviews, sales of his *An Evolutionist at Large* (1881) were disappointing. The publisher and Allen himself both worried that the prominence of Evolutionist in the title might have harmed the book's prospects.[162] Yet after 1882 he was able to carve out a reasonably successful career as novelist, writer of short fiction, and scientific populariser, publishing short essays on evolutionary themes throughout the late 1880s and 1890s.[163] He never fully escaped the whiff of dilettantism, but as the lines of high-Victorian division became more defined, so his Lamarckism brought him supporters and readers.[164]

The generation to which Allen belonged were generally 'evolutionist down to the tips of [their] toes'.[165] The physician and later secretary of the Research Defence Society, Stephen Paget (1855–1926), recalled his youth as a time 'when the doctrine of evolution so possessed us that we applied it to everything'.[166] If they did not grow up surrounded by the names of Darwin and Herbert Spencer, they experienced the debates over the *Origin* and *Descent* during their formative years. The journalist John Ferguson Nesbit (1851–99), brought up in a Calvinist family in Scotland, 'a precocious boy with a bent for science', remembered 'thinking of Darwin as a very wicked man, with whom, nevertheless, I had some secret sympathy'.[167] The late Victorians retained a sense of evolution as a conflicted doctrine, but a conflict experienced as stirring victories, and they developed a natural assumption of its status as orthodoxy alien even to high-Victorian supporters.[168] They installed Darwin in personal pantheons of heroes.[169] The linguistic landscape of their childhood had been transformed. The poet William Watson (1858–1935), whose verse frequently explored evolutionary themes, recalled that in his youth 'natural selection' and 'the survival of the fittest' were household words.[170] Adolescent and early adult reading was often embodied in personal copies which remained as physical mementoes of youthful enthusiasms.[171] A.C. Haddon was able, after leaving school at fifteen, to immerse himself in botany and zoology, reading Tyndall, attending evening classes at King's College, throwing himself into a number of societies and associations, and getting up various 'talks' on the writings of Darwin and Huxley at home.[172] Their Darwinism was often presented as a

262   DARWINISM'S GENERATIONS

matter of course: the *Origin* simply explained the facts of nature more effectively than any other theory. We see some sense of this in William Robertson Nicoll's (1851–1923) characteristic dismissal of Tennyson's May Queen as a 'silly girl' who 'lived before the times of evolution and the higher criticism'.[173] As Grant Allen told the readers of *The Academy*, 'Man, his origin and nature, his future hopes and reliable ideals, all seem something different to the present generation from their seeming to the generations that lie behind us in the field of time'.[174]

Even those late Victorians who claimed no scientific knowledge felt the fascination of evolutionary ideas. Edward Lyttelton (1855–1942), who became Headmaster of Eton, cited the *Origin* as perhaps the only great book of western civilisation that as a young Philistine he would have known.[175] The essayist Constance Plumptre (1848–1929) described Darwin as 'naturally...the chief mental factor in my development'.[176] Autobiographies and reminiscences expressed pride in the conjunction of birth and the beginning of Darwinian science. George Bernard Shaw (1856–1950), noting that the year of his birth was just three years before the publication of the *Origin*, identified himself as 'belong[ing] to a generation which, I think, began life by hoping more from Science than perhaps any generation ever hoped before'.[177] Many would have shared the straightforward enthusiasm of the 17-year-old Helen Norman in George Gissing's *Workers of the Dawn*, whose experience of reading the *Origin* in 1868 was that it 'created an enthusiasm in me such as perhaps no other book, except the "Leben Jesu" ever did....What immense labour, what a wonderful intellect does it represent! Yes, yes, this is real solid facts, facts one can grasp, handle, examine with the eye or the microscope'.[178]

The intellectual shock of Darwinism did not entirely disappear. Mrs Humphry Ward's Robert Elsmere was a study in this sensibility, aware of Darwinian ideas, but only reading the *Origin* in early manhood, reflecting that even knowing the substance, 'to drive the mind through all the details of the evidence, to force one's self to understand the whole hypothesis and the grounds for it, is a very different matter. It is a revelation'.[179] In fact, the writings of the late Victorians are, like those of the previous generation, studded with the intellectual wrestlings and emotional trauma associated with the abandonment even of their relatively unformed and shallow-rooted beliefs. Frank Harris (1855–1931), later editor of the *Fortnightly Review*, Havelock Ellis, and the banker and classicist Walter Leaf (1852–1927) were all amongst those for whom reading evolutionary literature as teenagers created an emotional and intellectual crisis and a painful loss of faith.[180]

The recollections of Mary Emily Dowson (1848–1941), who wrote various tracts on agnosticism and theology in the early twentieth century, convey the resulting sense of tumult: 'Vaguely I felt that earth shook under my feet [sic]. The world was in flux, and had come of transformation.'[181] The socialist J. Bruce Glasier (1859–1920), destined for the Presbyterian ministry, was said to have endured a year of agonising doubt, reading Darwin and scientific literature late into the night, and then tossing 'sleeplessly on his bed in dreadful mental suffering and unrest', before emerging into scientific agnosticism.[182] This was a common experience for the first generation of Labour politicians, whose ethical socialism was often rooted in strong Nonconformist upbringings, rocked by an early adult crisis of faith. Now, though, the loss of faith from reading Darwin no longer had the power to shock contemporaries.

These encounters were often framed genealogically, both literally and metaphorically. Parents might be themselves unenthusiastic (Poulton later recalled as a youth receiving a gentle parental warning against committing himself too entirely to a belief in evolution), but they were generally not dogmatically opposed, and sometimes positive at least about the stimulus evolutionary ideas might provide.[183] The palaeontologist D.H. Scott recalled that his parents were strongly anti-Darwinian on religious grounds, and that when he first heard of Darwinism at the age of sixteen or seventeen, he was hostile at first, 'from my bringing up, but soon came round to Evolution.'[184] Similarly, although the father of Edward Pease was a 'blind worshipper of the Bible', who subscribed to special creation, Pease's own reading was uncensored and for him too, 'Darwin in those days was a revelation.'[185] Our parents, recalled Pease, 'who read neither Spencer nor Huxley, lived in an intellectual world which bore no relation to our own'; he and his contemporaries were 'cut adrift…from the intellectual moorings of our upbringings, recognising,…that the older men were useless as guides in religion, in science, in philosophy because they knew not evolution.'[186] Arthur Conan Doyle recognised in a similar vein that a 'gap had opened between our fathers and ourselves…suddenly and completely.'[187]

Although there were exceptions (notably, Marie Corelli (1855–1924), several of whose novels, including *Ardath* (1889) offered an unsympathetic account of the corrosive effects of Darwinian thought, or George Moore (1852–1933), whose *Evelyn Innes* (1898) offers 'Darwin and Huxley' or 'Darwin and Spencer' as merely a bogey), endorsements of Darwinism were the norm for late-Victorian fiction, in the guise, for example, of the eponymous heroine of *Dr Janet of Harley Street* (1893), by Arabella Kenealy

(1859–1938), or the self-educated labour mistress, Jane Hardy, in *In Darkest London* (1891) by Margaret Harkness (1854–1923), who went to lectures, read Darwin, and sighed for the time when men would have become belittled by their vices and women become rulers of the world.[188] Anti-Darwinian ideas were increasingly caricatured as the province of the ignorantly opinionated (as in the character of Mr Gresley in Mary Cholmondeley (1859–1925)'s enormously successful *Red Pottage* (1899), or the Anglican curate vainly trying to demolish the entire Darwinian system, to the indifference of the listening scientific squire, who features in *Cherryfield Hall* (1895) by F.H. Balfour (1846–1909)).[189] The plays of Henry Arthur Jones, without trenching explicitly on Darwinism, were experienced by contemporaries as a reflection of his readings in Darwin and Spencer, and an influential representation of contemporary philosophies.[190] Reviewing one expression of this situation in 1883, the *Glasgow Herald* emphasised the stark contrast between contemporary attitudes and the initial response to the *Origin*, 'received like a lighted bombshell in a powder magazine, [when] high hopes of pseudo-scientific Victoria Crosses fired the bosoms of scores of valiant old-world naturalists who rushed recklessly forward to toss it overboard'.[191]

Educated in an academic world increasingly influenced by the early champions of evolution, the late Victorians situated themselves on the modern side of a stark division between the authority of pre-Darwinian prejudice and Darwinian law.[192] Darwin was the starting point, rather than an intervention in a long-running debate. And it was often the works of Darwin which inspired their scientific interests.[193] They recurred to his writings, they appropriated his authority, and they invoked him as final arbiter. This was particularly so for the amateurs and enthusiasts conscious of the condescension of professional scientists.[194] Through the work of Darwin, evolution was no longer theory but 'law'; the facts of transmutation were taken as a given, requiring no demonstration.[195] Their fundamental acceptance of this perhaps contributed to a sense that, for the late Victorians, debate about Darwinism was by the by, and thereby to the less frequent appearance in the general periodicals of late-Victorian scientists that Melanie Baldwin has noted. The high Victorians began to feel that the great struggles they had fought to make Darwinism respectable were being diminished or dismissed. J.W. Judd mused anxiously in 1891 on the growing danger of overlooking the value of the labours of the pioneering Darwinists, whose work was 'often carried on amidst difficulties and discouragements of which the younger generation know nothing'.[196] Yet the late Victorians continued to contribute to the wider discussion; it was just that they either, like Grant Allen, offered popularising evolutionary naturalism for a lay readership, or they focused on debates amongst themselves which intersected with the common culture in application more than essentials.

Late-Victorian natural science, across geology, palaeontology, biology, and comparative anatomy became evolutionary in essence, taking up lines of investigation suggested by Darwin, preoccupied with the phylogenetic identification of species and their evolutionary sequences. The zoologist Alfred Garrod (1846–79) envisaged his whole career as extending the work of Darwin.[197] Although he was himself becoming sceptical about this approach, the young biologist William Bateson recognised in 1890 that it was important that he reassured the electors for the Linacre Professorship at Cambridge that he intended to pursue 'Darwin's problems using Darwin's methods'.[198] Late Victorians threw themselves into embryology as the most promising field of evolutionary progress. If ontogeny recapitulated phylogeny, then the history of the individual could reveal the evolutionary history of the species. In the years before his early death in July 1882, F.M. Balfour and his students had shown convincingly that embryology could provide important clues—and possibly even answers—to the phylogeny of species.[199] Amateur naturalists like the Brighton palaeontologist Agnes Crane (1852–1939) celebrated the way embryology and the phylogeny of fossil fauna provided the ultimate demonstration of Darwinian evolution. 'I have made one diagram showing the ordinal evolution which ought to convince the Archbishop of Canterbury! Anyone who denies the validity of the various stages of growth and their correlations is better fitted for a lunatic asylum than office in a Museum now-a-days', Crane exclaimed in 1893.[200] For all his abhorrence of what he saw as the monism of the scientific establishment, even the conservative intellectual W.H. Mallock (1849–1923) readily conceded that embryology had made Darwinism a demonstrable, indeed a visible, fact.[201] Even high Victorians occasionally found the ubiquity of this approach suffocating. By the early 1890s the entomologist David Sharp (1840–1922) was regretting 'Biology' as 'a vile word', and condemning the habit of '"Biologists" [to] look upon themselves as greatly the superiors of the old "workers"', and to 'occupy themselves specially with phylogeny, natural selection, & all that sort of thing'.[202]

In parallel, the late Victorians focussed their attention on the question of the place of humanity in evolution, and the implications of the evolutionary origins of reason and the intellect.[203] They accepted without pause the congruence of man and animals in nature; as the philosopher F.H. Bradley

266 DARWINISM'S GENERATIONS

(1846–1924) remarked, 'I could never see any difference at bottom between my dogs & me, though some of our ways were certainly a little different'.[204] In his three volumes of the 1880s, which culminated in *Mental Evolution in Man* (1888), Romanes and collaborators like Lloyd Morgan demolished the continued attempts of the mid Victorians to sustain the distinction between the physical and mental development of man and animals, by demonstrating the close similarity of their mental life.[205] Romanes rejected Müller's language barrier; language did not start with given roots, he argued, but with what the historian Greg Radick has described as 'an undifferentiated speech-protoplasm'.[206] At least some late Victorians accepted that conscience was a product of evolution and therefore was as visible in animals as in humans.[207] Romanes was prepared to go even as far as arguing that the mental development of the child rehearsed the history of mental evolution, identifying stages at which the infant was equivalent to insect, larvae, fish, and reptiles.

Darwinism was transformed from theory into creed. In the words of the Idealist philosopher D.G. Ritchie (1853–1903) it became 'not merely a conception by which to understand the universe, but a guide to direct us how to order our lives'.[208] For the new generation of radical freethinkers like Annie Besant (1847–1933) and Karl Pearson, evolution was an article of faith, and Darwin one of its 'high priests'.[209] One characteristic figure is Constance Naden, for whom evolution was less of a scientific theory and more a philosophy, one which she hoped to promote by the formation of an Evolution Society.[210] The economic historian William Ashley (1860–1927) provides another. Despite attendance at some South Kensington science classes as a youth, Ashley remained almost entirely ignorant of natural sciences, but nevertheless he made an exception for the *Origin*, and abandoned Christianity for an amorphous evolutionary creed, defining himself as an 'evolutionary socialist' to the end of his life.[211] Such attitudes found institutional expression in the Ethical movement of the 1880s and 1890s, whose core personal identity was often 'evolutionist', although its specifically Darwinian impulses were only part of an evolutionary cocktail with dashes of Comte, Spencer, and Mill.

Late-Victorian Evolutionary Sectarianism

If the high Victorians had seen their role as one of vindication, the late Victorians were more concerned with exegesis and adjustment, especially of the relative importance of the different components of Darwin's account of how evolution occurred. Hitherto, the critical division had been between those for and against the fact of Darwinian evolution, broadly conceived. The new alignments were distinguished by their understandings of its operation. '[W]e', observed the nature journalist Edward Kay Robinson (1855–1928), 'who have accepted the doctrine of evolution as the main-spring of the machinery of existence, are more concerned with discovering the checks and limitations which modify its workings'.[212] The crucial question was now to which version of Darwinism the late Victorians subscribed, what were the essential causes of the gradual modification of organic forms.[213] To the delight of mid Victorians, who saw in the combat of rival theories the exposure of the weaknesses of all, and to the dismay of high Victorians who looked at the hardening sectarianism with distaste, the pages of the scientific press became clogged with rival claims to be the true heirs of Darwin.[214]

During the 1880s and 1890s a group headed by Lankester, Meldola, and Poulton took up the challenge of defending what they saw as the orthodox position of Darwin and his followers. In doing so, they adopted an increasingly narrow purist interpretation, in which the explanatory force of natural selection and of adaptation were privileged, and Darwin's own hesitant concessions to the possible role of Lamarckian acquired characteristics and other subordinate factors such as sexual selection were dismissed as unnecessary and unhelpful. In the 1880s Lankester was the leading representative of this position. As a spokesman for Darwinism, in many respects he outdid even Huxley, championing natural selection and the struggle for existence as a complete explanation, albeit one that was continuously misrepresented and misunderstood.[215] He praised those evolutionists doing 'truly Darwinian' work, and anyone found wandering from the true path was quickly called back to the Darwinian scriptures, and urged to 'read again and again their Darwin'.[216] 'I am somewhat concerned by what you say about Lamarckism', he warned Patrick Geddes in 1882, 'you must get cured before you write'.[217]

By the turn of the century Lankester's role as chief guardian of the Darwinian heritage was passing to the younger Darwinian entomologist Poulton, who had succeeded J.O. Westwood as Hope Professor at Oxford in 1893. Supported by Meldola and allies such as his Oxford colleague F.A. Dixey (1855–1935), Poulton intervened regularly to reiterate Darwinism's ability to meet earlier critiques, in ways which left less and less scope for mechanisms ancillary to natural selection.[218] A considerable group of late-Victorian

268   DARWINISM'S GENERATIONS

scientists aligned themselves to this position.[219] The younger high-Victorian William Thiselton-Dyer came to share much of the group's sensibility, confessing in the early 1890s of being 'much depressed with the way in which Darwinianism is being messed about', and regretting that 'Darwin's writings are not read or studied'.[220] Commenting on debates at the British Association in 1894, the journal *Natural Science* remarked that the impression left on the minds of the broader scientific public was that the defenders of natural selection were more concerned to maintain an orthodox interpretation of the 'Darwinian scripture' than to investigate the actual facts.[221]

This group, increasingly identified as 'neo-Darwinians', championed the ultra-Darwinism of the German evolutionary biologist August Weismann, whose *Studies in the Theory of Descent* (1875) had taken up the question of heredity which had vexed Darwin in his final years, arguing that the mechanism of heredity was an absolutely stable 'germ plasm' which remained entirely unaffected by any modifications to an organism during its life. Lankester had already organised the translation of Ernst Haeckel's popular Darwinist primer *The History of Creation* (1876), rejoicing to its author that 'The parsons here at Oxford will be furious when they read it'.[222] But in the wake of the translation of the *Theory of Descent* in 1882 (overseen by Poulton and Meldola) and then Weismann's *Essays in Heredity* in 1889, evolutionary debate in Britain moved into a phase in which response to Lamarckian theories of use inheritance became a litmus test of evolutionary affiliation.[223] In the pages of *Nature*, Lankester assiduously championed Weismann, who was apparently greeted with glee at Oxford as affording a prospect of 'dishing Spencer', whose Lamarckian inclinations had contributed to his dubious reputation in university circles.[224]

Darwinism, as Thomas Huxley's grandson Julian Huxley later suggested, began 'to resemble the early nineteenth-century school of Natural Theology *redivivus*,...but philosophically upside down, with Natural Selection instead of a Divine Artificer as the *Deus ex machina*'.[225] All this was, not surprisingly, often too much even for the most Darwinian of the high Victorians, who protested against the arrogations and the arrogance of the neo-Darwinists.[226] Those with Lamarckian instincts were suddenly conscious of their potential 'heresy'.[227] Issuing a clear rebuke to the neo-Darwinians, W.H. Flower expressed regret in his presidential address to the BA in 1889 at the intolerant dogmatism into which some proponents of Darwinism had fallen; a dogmatism which he suggested was alien to Wallace, Darwin, and Huxley.[228] Michael Foster complained likewise that 'a certain school seems to think Weissmann [sic] a sort of second Darwin', and expressed concerns at the unwarranted attempts of 'the ultra-Darwinists who form the bulk of our rising naturalists' to present natural selection as sufficient cause for the whole history of biological evolution.[229] Leslie Stephen, exultant that 'we are all evolutionists now' was happy to confess that he had 'no pretensions to take part in a discussion of Weismannism'.[230] Samuel Butler and others of his generation, like the political economist William Graham (1839–1911), often directed their complaints of dogmatism at the elder statesmen of scientific naturalism, Huxley and Tyndall, but the impulse they were responding to was more essentially late-Victorian.[231]

Amongst the late Victorians, the neo-Darwinians were increasingly in tension with a group (we might call them 'broad Darwinians') who sought to take seriously Darwin's own hesitancies about the adequacy of natural selection, arguing that although it was the most important, it was not the exclusive explanation of evolutionary change (and in doing so often partially conceding the typical mid-Victorian critiques of the instances where natural selection seemed insufficient). In part this involved acknowledgement of Darwin's dalliance with Lamarckian explanations in the later editions of the *Origin*, but more fundamentally it involved the discussion of other secondary factors, and in particular the role of geographic and fertility barriers in the progress from variation to species differentiation.[232]

The key figure here was Romanes, who rejected the narrowness of what he asserted (in an allusion to A.R. Wallace's tight focus on natural selection) was not so much the pure Darwinianism as the pure Wallaceism of the neo-Darwinians.[233] Romanes was a complex figure with a highly conflicted place in the history of evolutionary thought in later-Victorian Britain.[234] A physiologist, he held academic posts at Edinburgh and Oxford, but much of his work was done as a private scholar, and his position was complicated by profound religious convictions. Romanes' prominence derived in no small part from the close relationship he had established with Darwin in the final years of Darwin's life, a relationship he used after Darwin's death to claim the mantle of chief disciple and defender of Darwin's evolutionary legacy. As he told readers of the *Times* in 1889, 'I have ever been an upholder of the Darwinian doctrine as this was left by the matured judgement of Darwin himself', a judgement which he claimed in a series of widely noted lectures at the Royal Institution in early 1890 would have included opposition to Weismann and certainly have incorporated some degree of Lamarckism.[235] Along with figures like the zoologist Frank Evers Beddard (1858–1925),[236] the palaeobiologist D.H. Scott (1854–1934),[237] and the naturalist F.W. Headley

270 DARWINISM'S GENERATIONS

(1856–1919),[238] and many non-scientific late Victorians, Romanes advocated a broad approach, incorporating subsidiary factors to natural selection, including the inheritance of acquired habits, all of which he argued were sanctioned by Darwin's own later works and correspondence.[239] In public these accretions were defended as developments of Darwin's ideas, but they were significant enough to be described in private as a new theory of the origin of species.[240] In particular, Romanes questioned the ability of small modifications to initiate the sort of divergent evolution which could create new species in contexts where continued interbreeding would swamp the initial variations. He also emphasised the danger of 'panmixia', whereby any cessation of natural selection would lead to species degeneration.[241] To counter these, he stressed the importance of reproductive or geographical isolation as a necessary correlate to the operation of natural selection in creating new species.[242] His argument that the mutual sterility of variants ('physiological selection') was 'the <u>cause</u> (or at least the chief <u>condition</u>) instead of the <u>result</u> of specific differentiation' was, he accepted, the opposite view to that held by most evolutionists.[243]

These arguments embroiled Romanes and those aligned to him in a long-running debate with the neo-Darwinians and also older Darwinians, who had no time for those who sought to proliferate evolutionary mechanisms. H.W. Bates' dismissal of what he thought of as a 'factless theory' was typical of the response of the mid-Victorian Darwinians. Wallace was appalled at Romanes' suggestion that natural selection was not a theory of the origin of species, and at his failure to address the question of the role of natural selection in the development of mental capacity from animals to man.[244] His *Darwinism* (1889) launched a direct assault on Romanes, much to the delight of Meldola, who felt it would 'do much good to the true Darwinian theory and help to counteract the influence of "Romanian atrocities".[245] Romanes' theory of physiological selection was condemned as meaningless: 'the whole thing is so frothy that I propose to rename it the theory of <u>fizziological selection</u>', remarked Meldola, who clearly enjoyed a pun or two, and who in one moment of frustration accused Romanes of having done more to harm Darwinism than any other writer.[246] As Piers Hale has demonstrated, Romanes' ideas were rebutted in the pages of *Nature* by late-Victorian neo-Darwinians such as Raphael Weldon and Karl Pearson.[247]

But above all, what stuck in the craw was Romanes' penchant for appropriating Darwin's posthumous imprimatur, what Bates described as his claims to be Darwin's prophet, a 'pseudo-Elisha',[248] and to pontificate to

THE DEATH OF DARWIN AND AFTER   271

other biologists on what Darwin's stance would have been, buttressing his arguments by claiming Darwin's previous private approval. Romanes' almost Tridentine approach to appropriating the authority of Darwin prompted the Scots medical professor William Tennant Gairdner (1824–1907) to describe his work as 'building up a Darwinism *in excelsis*'.[249] The older Darwinians seem to have found this particularly irritating. Lankester was enraged at the 'humbugging piece of foolery' of Romanes' 'attempt to say "Darwin-and-I" and "the Darwin-Romanes theory"'.[250] Romanes was also often annoyingly assiduous in promulgating his views; and (at least as far as his opponents were concerned) in constantly shifting his ground in the face of criticism. Thiselton-Dyer described him as 'elusive as an eel'.[251] Even generalist readers picked up on this sense of underlying lack of solidity: F.V. Dickins dismissed him as 'a very clever word monger and nothing more—too clever by half'.[252]

The heat of the private exchanges around Romanes' work illustrates the extent to which, for the late Victorians, the intellectual and emotional charge previously generated by debates between Darwinists and their opponents had been displaced by the charge arising from divisions within Darwinism itself. Lockyer found himself struggling to prevent the vitriol spilling over into the pages of *Nature*.[253] Meanwhile, like Darwin before him, Romanes came to feel as though his efforts and those of his collaborators were constrained by generational barriers. His initial response to Wallace's attempt at rebuttal at the British Association in 1889 was that the biologists present had 'nearly all come to support what they regarded as the orthodox Darwinian dogma'.[254] '[T]he leading English biologists are not hopeful material to convert', he lamented, 'no doubt the rising generation will prove better able to distinguish the fact that 2 and 2 equals four'.[255]

Beyond Romanes and the broad Darwinians was a flanking group of Lamarckians, or perhaps more properly Spencerians, who while sharing the broad Darwinians' rejection of natural selection as the sole evolutionary dynamic, were distinguished by the priority they gave to the importance of the inheritance of acquired characteristics as an alternative mechanism, and in particular in the extent to which they envisaged evolution as a progressive or developmental process.[256] In their tendency to question the sufficiency of natural selection, these critics often deployed the sorts of arguments advanced by the mid Victorians. But it does not make sense to think of them as 'anti-Darwinian', not least because they operated in a context of a post-*Origin* acceptance of the fundamentals of evolution across the natural world, but also because they, like the broad Darwinians, presented

272 DARWINISM'S GENERATIONS

themselves almost exclusively as working to extend the Darwinian position from within, rather than attempting to assail it from without. Not that this stopped some orthodox Darwinians from blunt vituperation of what Huxley called 'paper philosophers' 'who are trying to stand on Darwin's shoulders and look bigger than he, when in point of real knowledge they are not fit to black his shoes'.[257] Nevertheless, as the debate between advocates of more or less pure 'natural selection' and those who espoused broader views spilled out beyond the confines of biological science, Lamarckian positions proved increasingly popular.

For a while at least in the 1890s and 1900s (before he moved towards a vitalist position which shared much in common with the Edwardians) one of the most prominent examples of this Lamarckism was the socialist and playwright George Bernard Shaw. One of the Fabian Society's 'old gang' along with Beatrice and Sidney Webb, Shaw was as much influenced by Samuel Butler as by Spencer, while also absorbing an essentially Idealist preoccupation with the importance of 'mind' as manifest in nature.[258] Hence his opposition to Darwinian natural selection on the grounds that it had 'banished mind' from the world, and in doing so had obliterated the 'true theory' of evolution.[259] As Shaw put it in a speech in 1906, Darwinism had played a vital role in showing that the cruelties of nature had no moral significance, but it needed to be overthrown in its turn if society was to recover its spiritual energy, for a doctrine that reduced men to flies on the wheel of natural selection was deadly to the human mind. 'An active Socialism' he claimed, 'would be absurd if we did not believe that humanity would develop by its own will even if the operation of Natural Selection were entirely suspended. In fact, our work is largely to defeat Natural Selection, and give free play to Lamarckian processes'.[260]

The popularity of Lamarckian ideas was itself a reflection of the attraction of Herbert Spencer's grand conception of universal evolution for the generation as a whole. Spencer is both a complication and a conundrum in the history of the reception of Darwinism. He claimed not to have read any of Darwin's volumes after the *Origin*,[261] and although his work attempted to accommodate, if not to absorb, Darwin's key arguments, his position was always both different (in that on the central question of the biological mechanisms of evolutionary change he was always at least as much Lamarckian), and much broader (in that for him biological evolution was only a part of a comprehensive edifice of evolution which encompassed the inorganic as much as the organic, and offered a progressive process quite at variance with Darwin's random adjustments to temporary contingencies).

THE DEATH OF DARWIN AND AFTER   273

There was at least an undercurrent of Spencerian thought in all the debates over evolution from the early 1850s, and it is possible to find individuals whose conceptions were particularly moulded by his views in every generation from the early Victorians onwards.[262] His influence in the coterie around John Chapman which included George Eliot and George Henry Lewes has already been signalled. For the high Victorians, Spencer was frequently part of the ferment of ideas in which they were immersed as young adults.[263] He was a formative influence on the individualist thinker Auberon Herbert (1838–1906), who became his executor. Alfred Marshall (1842–1924), who found Darwin's natural science hard going and Spencer's larger generalisations more readily appealing, located a similar influence on his peers in the later 1860s and 1870s.[264] Many were happy to assume the neo-Lamarckian label without abandoning their deference to Darwin, and W.K. Clifford's mathematical evolutionism, the chemical researches of Sir William Crookes (1832–1919), and Hughlings Jackson's psychological theories were all primarily underpinned by a Spencerian view of change operating across organic and inorganic nature.[265]

But these were mostly younger high Victorians, and by and large for the generations before the late Victorians, Spencer's standing could not compare with Darwin's. Where Darwin inspired veneration, Spencer could aspire only to respect. Though few doubted his significance, fewer were able to put it into words. While Darwin's burial in Westminster Abbey was accepted with scarcely a murmur of dissension, efforts to get even a memorial plaque in the Abbey for Spencer were stymied.[266] Unlike the works of Darwin, Huxley, and others, most of Spencer's writings were not broadly circulated in cheap popular editions. Spencer was reported to have distributed a hundred complementary copies of his *Principles of Psychology* but the rest of the 750 print run took another twelve and half years to sell.[267] Academic scientists and philosophers were generally unimpressed. The theologian Hastings Rashdall (1858–1924) once claimed to have been the only person in Oxford ever to have read the *Principles of Sociology* from beginning to end.[268] Even in the 1900s, the philosopher W.R. Sorley (1855–1935) groaned at the 'portentous dullness' of Spencer's *Autobiography*.[269] Social and intellectual snobbery no doubt played a part, as did Spencer's mannered hypochondria and self-importance. He was always an easy target for those who refused to persevere with his jargon, or who suspected him of selective marshalling of the evidence. Fair or not, Huxley's famous barb—that for Spencer tragedy was a beautiful deduction killed by a fact—neatly summed up a widespread reaction to his work. Even R.J. Ryle (1854–1922), one of Spencer's GPs during his final years, commented that in Spencer's 'philosophical talk', 'his theory generally gets a good start of his facts [sic] and the facts come panting after just in time to act as telling illustrations'.[270]

Spencer was much more significant for the late Victorians. Undergraduates arriving at university in the mid-1880s found the philosophical extension of evolutionary theory by Spencer in the heyday of its influence.[271] Recalling his university days in St Andrews and Edinburgh, James Collier, who later became one of Spencer's assistants, described Spencer as a closed book to his teachers, but an acknowledged leader of the younger generation; 'Our fathers and grandfathers conceived of the Cosmos as a creation', Collier recalled, but 'for us it is not unfinished product but unending process'.[272] Havelock Ellis' journals reveal that while he was also reading Darwin, it was Spencer who effected the spiritual revolution he underwent in his late teens.[273] In part this was merely a function of the widening of the canon of evolutionary thinkers which was encouraged by a greater willingness to distinguish evolution in general from Darwinism in particular.[274] But it also reflected the attractiveness of Spencer's progressivism, and the role his ideas gave to individual and collective betterment, which made him of more immediate appeal to those wrestling with the application of evolution as a personal creed or as foundational to a set of political principles.[275] Edward Carpenter's socialism was just one example of a growing tendency to subscribe to Spencerian progressivism.[276] Even as his increasingly anti-state individualism came to alienate Idealists like D.G. Ritchie, so Spencer's Lamarckism attracted many reformers suspicious of the potential fatalism of neo-Darwinism.

Robert Louis Stevenson was not the only one for whom the reading of Spencer led to a loss of faith, nor was the forensic psychiatrist Charles Mercier (1851–1919) alone in attributing the Lamarckian slant of his evolutionary beliefs to Spencer's influence.[277] The Anglican bishop Charles D'Arcy (1859–1928), Conwy Lloyd Morgan, and J.M. Robertson (1856–1923) were three other late Victorians who acknowledged Spencer as a key formative influence.[278] 'We are all Spencerians to-day, whether we like it or not', remarked the 'new Liberal' J.A. Hobson (1858–1940) in 1904, observing that while Darwin exercised a more direct, immediate, and distinct impression on his age, Spencer was the more important, because the largest interpretation and application of the new scientific principles came from him.[279] An even fuller example of the sort of influence Spencer could exert is offered by Hector Macpherson (1851–1924), editor of the *Edinburgh Evening News*, whose numerous popular studies provided a guide through the maze of

THE DEATH OF DARWIN AND AFTER   275

Spencerian thought for the middlebrow public.[280] For Macpherson, '[i]n the history of Evolution there was one great man, Herbert Spencer', and Darwin's natural selection was 'at most, a brilliant confirmation of Spencer's cosmical generalisation'.[281]

For many late Victorians, Spencer's influence was indirect or temporary. Others, like Shaw and the Fabian socialist Sydney Olivier (1859–1943), seem to have come to their Lamarckism via the critiques of Samuel Butler.[282] Lloyd Morgan, for example, moved steadily out of the Spencerian camp and towards acceptance of Weismann's views during the early 1890s.[283] Olive Schreiner read Spencer assiduously in early adulthood, but later observed that 'he has nothing else to give men now'.[284] For others, including the anthropologist and folklorist, James G. Frazer (1854–1941), he was always a more significant influence, and there persisted a group of Spencerian natural scientists, of whom Grant Allen, J.T. Cunningham, and Patrick Geddes were perhaps the most prominent.[285] Edward Clodd described Grant Allen as 'a whole-souled disciple', and Cunningham professed his own discipleship.[286] Patrick Geddes encapsulates some of the intellectual itinerance of this group. Beginning as a student of Huxley's, he moved steadily away from the mainstream of Darwinian ideas and towards an idiosyncratic version of Spencerism in the 1880s and 1890s, joking after spending time with Weismann in 1891 that he found 'Weismann more Darwinian than ever, and we did not attempt to fight much'.[287] As early as 1888 he was arguing for the importance of vitality (and 'vegetativeness' in plants) as a force for evolutionary change independent of natural selection.[288] A decade later he was offering fairly forthright criticisms of Darwin and the arrogations of the Darwinians.[289] In Britain in the 1880s these neo-Lamarckians remained a rather small and embattled group in scientific circles, unlike in America, where neo-Lamarckism became the evolutionary orthodoxy, but their writings, along with Spencer's themselves, obtained very substantial purchase in the wider culture.[290] Even among the numerous late Victorians who were strongly critical of Spencer's ideas, many retained the sense that his ideas were more important, more worthy of the effort of rebuttal, than did the older generations.[291] Instead of remaining semi-detached, Spencer became increasingly central to divisions around the most pressing evolutionary questions.

The disagreements between the three groups perhaps started as a specialised debate within the scientific community around questions of inheritance, and played out in the pages of *Nature* and at the British Association, without the accessibility of the early controversies around the *Origin*. It is notable that despite the endorsement of Darwin, initial sales of the translation of Weismann's *Studies* were poor.[292] But the interventions of Lankester and Weismann at the 1887 British Association meeting, and the subsequent publication of the translation of Weismann's *Essays on Heredity* in 1888, prompted a much wider discussion, not just in the popular science magazines, but also in the literary newspapers like the *Academy* and in the serious reviews. Controversy reached a crescendo in 1893 in the exchanges between the Weismann and Spencer camps in the *Contemporary Review*.[293] For the neo-Lamarckians, Cunningham stressed that what was needed was a theory of variation, and that adaptation and recapitulation could not be explained without acquired characteristics, while Annie Besant continued to argue that society as an evolving organism must advance to greater and greater integration or it would fall apart.[294] The controversy spilled out into other less obvious forums, including the radical *National Reformer* newspaper, which gave the issue sustained consideration between April and July 1893.[295] It also seems to have generated a considerable correspondence from various late-Victorian pathologists, and the geologist W.J. Sollas (1849–1936), assuring Spencer that they endorsed his position.[296]

The Weismann–Spencer debates illustrated the impasse into which evolutionary thinking appeared to have boxed itself by the early 1890s. Understandings of key components of cell biology, such as chromosomes (still being described as 'histological slang' in 1896) were only just emerging. The pangenesis theory, which Darwin had offered in *Variation*, had been effectively disproved by Galton in the 1870s. Even so, it is noticeable that Darwin continued to be a frequent reference point in the discussions of the 1890s.[297] Weismann's work marked an advance in cytology, but his 'germ plasm' theory was just as speculative; hence the scepticism of Romanes' *An Examination of Weismannism* (1893). If Darwinism was to fulfil its potential, and especially if it was to inform interventions in social policy, then greater clarity was required as to the precise nature of the mechanics of evolutionary change in general, and of heredity in particular.

Questions of heredity were garnering more attention. But the rumbling debates in the scientific and medical press during the mid-1890s regarding the science of telegony (the idea that the features of the sire of one offspring could be transmitted to subsequent offspring bred to different sires) and the attention it was given in scientific correspondence at the time, only reinforces how wide the gap was between practical lore and solid scientific knowledge. Even at this stage James Cossar Ewart at Edinburgh was, with Weismann's encouragement, devoting considerable resources to experiments

THE DEATH OF DARWIN AND AFTER 277

crossing zebras with ponies to test the theory.[298] The distance is clear in a long correspondence at the end of 1893 between Charles Herbert Hurst (1856–98), a young ex-student of Huxley's, and the palaeontologist S.S. Buckman (1860–1929), which raised not just the inadequacies of cytology, but also wider definitional problems, including the lack of clear distinction between variability (i.e. diversity of offspring), and variation (i.e. changes in populations over time). Hurst went as far as to suggest that Darwin's theory was 'actually based upon the confusion of these two meanings of the term "variation".[299]

In the early 1890s momentum was developing most obviously around the scientific and statistical study of heredity being pioneered by a small group born in the later 1850s, and so in their early thirties, including Havelock Ellis and Karl Pearson, who began to look for new avenues to address what Ellis later described as the 'howling wilderness of facts which could often only by courtesy be called "facts".[300] Galton's *Natural Inheritance* (1889) had extended his statistical application of Darwinian theory to the mathematics of inheritance. Admittedly, the initial response was muted; notably, Pearson's important *Grammar of Science* (1892) did not address Galton's work at all, while Ellis' *Man and Woman: A Study of Secondary and Tertiary Sexual Characteristics* (1894) took a quite different approach. But, in a number of lectures and essays eventually collected in *The Chances of Death and Other Studies in Evolution* (1897), Pearson did gradually set out the potential of a new statistical biology or 'biometry', which might help uncover the dynamics of inheritance.[301] The challenge was creating a stable agenda. The investigations of the mid-1890s were still marked by a bewildering array of questions: did the father have more hereditary influence on younger children than older, did males tend to vary more than females, was the interaction of organs influenced by the selection of one? This incoherence (and the mathematical promiscuity it encouraged) alienated even Pearson's closest allies. 'You are like H.G. Wells' martians', his close colleague Raphael Weldon (1860–1906) told him after one exchange, 'you stalk upon stilts over plane space, and discharge a poisonous cloud of formulae every time you come to your cross roads.[302]

R.J. Ryle certainly saw the new developments in the study of heredity in part in generational terms, welcoming the way in which something like the very existence of variation as the material for natural selection, which had been an open question for the generation of Asa Gray, should now be a basic explanandum for his generation.[303] Weldon was contemptuous of Lankester and Adam Sedgwick (1854–1913), (nephew to his fierce

278 DARWINISM'S GENERATIONS

anti-Darwinian namesake), who 'hasn't the first idea about variation', very conscious that the work he and Pearson were attempting was a long way from the morphological and developmental approaches of Darwin or the Darwinists.[304] But for reasons discussed in the next chapter, biometrics never attracted a substantial Edwardian following.

Applied Darwinism

As Havelock Ellis observed in his *The New Spirit* (1890), the shift from those he called 'the men of 1859' (by which he meant not those born that year, but those who defined the culture of the year, and so for us the mid Victorians) to their successors was a move from theology to sociology.[305] Greater attention to human evolution encouraged work not just in anthropology but also psychology, economics, and human physiology, and the application of the principles of the survival of the fittest and natural selection to social groups as well as to individuals. As Mrs Humphry Ward later recalled, 'Darwinian debate in the realm of natural science was practically over. The spread of evolutionary ideas in the fields of history and criticism was the real point of interest'.[306] By 1895 the librarian at the Redland Public Library in Bristol was classifying both the *Origin* and the *Descent* in the 'Social Questions and Theology' category, rather than under Scientific works.[307] We can see a similar shift in the trajectories of individual careers: Lloyd Morgan's shift from zoologist and geologist to psychologist, A.C. Haddon's abandonment of biology for anthropology, Patrick Geddes' transformation from botanist to urban reformer, and Annie Besant's journey from Huxley's laboratory to socialism and theosophy. As he told the Dunfermline Naturalists' Society in 1902, Geddes was convinced Darwinism should not remain an abstract doctrine of evolutionary philosophy, but must become an urgent matter for concrete application.[308] Spencer's theories in particular lent themselves to applications to ethics and social development, whether by evolutionary Christians like Henry Drummond (1851–97), Idealists such as Ritchie and Geddes, sociological Darwinians in the mode of Benjamin Kidd (1858–1916), or socialists like Keir Hardie (1856–1915) and Edward Aveling. But this turn to what we may term 'social-biology' was almost equally true of those sceptical of Spencer's views, like the philosopher and social reformer Bernard Bosanquet (1848–1923), who ridiculed the fashion for applying biological terms to social phenomena, and yet at the same time sought to explore the role of natural

THE DEATH OF DARWIN AND AFTER **279**

selection in the development of society; or of W.H. Mallock, whose *Aristocracy and Evolution* (1898) was a direct challenge to the sociological turn, but which largely deprecated the results it had achieved rather than the legitimacy of the analogies it had mobilised.[309]

There was a danger, Karl Pearson grumbled in the 1890s, of Darwinism becoming 'a cant term to cover any muddle-headed reasoning'.[310] Grant Allen in art, Frederick York Powell in history, and many of the contributors to J.H. Muirhead's Library of Philosophy, all sought to remould their field around a Darwinian method. The parapsychologist Edmund Gurney, reputed to be one model for George Eliot's Daniel Deronda, applied his early biological studies to an examination of musical appreciation. His *The Power of Sound* (1880), with its argument that musical sense was an evolution from the resonances of primitive sexual passion, brought the accusation, as he himself put it, that his ideas rested on a 'superstitious belief in the exhaustiveness and omnipresence of Darwinism'.[311] In fact, his correspondence and writings nicely reveal the cross-currents and tensions of evolutionary debates in these years. He explored but was sceptical of Spencer (he spoke of the tendency of 'poor old' Spencer's *First Principles* to break down on careful investigation), but he also had little time for the materialism of the high-Victorian Darwinists and some of their successors, 'the modern *cock-sure* school of Empiricists', as he described them.[312] The art critic and essayist Vernon Lee (Violet Paget) (1856–1935), who described herself as 'one who believes in scientific method, in human development, and in evolutional morality', was also characteristic of the sorts of engagements which resulted.[313] Lee's intellectual interests stretched from travel to biology (where she undertook experiments on the physiological manifestations of aesthetic responses); in 1894 she was writing enthusiastically about her reading in Darwinism and her fascination with natural sciences, and her writings on art and philosophy offer a telling example of the ways Darwinian biologism was applied in domains beyond science.[314]

The question of how far survival of the fittest was a matter of struggle between groups rather than individuals became another of the characteristic debates of the late Victorians. The political permutations which could result are bewildering, and the extent to which they derived from the Malthusian ideas which underpinned Darwin, rather than Darwin's own additions, is not easy to disentangle, which is why the concept of 'social Darwinism' has always been so difficult to pin down in any stable or useful way. The permutations could include the socially conservative common sense philosophy proclaimed by the literary critic W.E. Henley (1849–1903)

280   DARWINISM'S GENERATIONS

and his *Observer* group, or the championing of international strife implicit in Karl Pearson's 1901 lecture 'National Life from the Standpoint of Science'.[315] Such interpretations encouraged the imperialist expansionism of figures like Cecil Rhodes (1853–1902), and the diffuse cultural racism which underpinned British policies of colonial subjugation and appropriation, and also the aggressive nationalism and militarism which were a feature of British jingoism.[316] In this form, social Darwinism could be used to justify various forms of state intervention designed to improve the health of the nation, a position which often had much in common, at least in its implications, with the quite distinct organicism of the left, and of the mindset of groups like the Independent Labour Party.[317]

Simultaneously, social Darwinism could also be associated with extreme individualism based on the argument that welfare provision would only enervate the nation by blunting the ability of natural selection to weed out the weak, justifying the laissez-faire, small government mentality often associated with Manchester School liberalism. Spencer himself increasingly lined up in the small government camp. His *Man versus the State* (1884) warned that state intervention, especially in social welfare, risked neutralising the progressive action of natural selection and facilitating the survival of the unfittest. For Muirhead, Spencer, under the sway of the analytic atomising spirit of the physical sciences for which society was ultimately a mere aggregate of individuals endowed with rights, had produced 'the textbook of all the reactionaries of those days'.[318] Nonetheless, it was a position widely adopted, by libertarian liberals like the secularist Arthur B. Moss (1855–1937), and also by social conservatives like the Ulster-born medical penologist Sir James Barr (1849–1938). The logic was difficult to escape even for maverick radicals like George Bernard Shaw, whose response to a lecture on eugenics from Galton to the Sociological Society in 1904 was that the blunting of the struggle for existence, especially without positive reforms, might 'without countenancing for a moment the crudities of neo-Darwinism… do more harm than good'.[319]

What few late Victorians would have questioned was that societies and nations should be studied as if they were species, applying the methods that evolutionary science had previously developed for the study of nature. In the late 1880s Geddes produced a series of papers urging the application of 'the laws of physiology and evolution' to economic and social policy, urging attention not merely to the cheapening of food and multiplication of the number of eaters, but to the quality of social life.[320] The early Fabian movement, the Men and Women's Club of the group around Pearson, and the

THE DEATH OF DARWIN AND AFTER   281

'New Liberalism' of the Rainbow Circle (a collection of predominantly late-Victorian progressives, including Wallas, Hobson, Sidney Webb, Percival Chubb (1860–1960), and J.M. Robertson, who met for debate and discussion on political issues from 1894 into the 1920s), all demonstrated the powerful lure of organicist approaches for the late Victorians.[321] Grant Allen was taken by the contrast between 'all the men of that first generation [in which he included Tyndall, Huxley and Spencer] who spread the evolutionary doctrine among us [who] are now reactionary in politics', and 'the younger brood whom they trained [who] have gone on to be Radicals, Fabians, Socialists'.[322] The commitment to an evolutionary view of history and of social regeneration distinguished the Fabians from more mainstream Marxist socialists, and attracted, if only temporarily, many other late-Victorian fellow travellers, including Ritchie and Bosanquet. The acknowledgement of a relationship between society and biology was always strained and contingent, and it is easy to multiply instances of late Victorians warning, as J.M. Robertson did, that 'mere biology is a quite blind guide to polity'.[323] But whereas for the Edwardians such observations registered fundamental doubts, for the late Victorians they indicated the need to set limits to this process, and were often a rhetorical device for rejecting one particular version of the crude invocation of Darwinism, rather than social-biological approaches in their entirety.[324]

Although this socially progressive evolutionism drew eclectically from a range of intellectual sources, of which Comte was perhaps the most significant, its Darwinian roots are fully visible in the *Fabian Essays in Socialism* published in 1889. Sydney Olivier's essay on 'Morals' in this volume was explicitly an intervention in debates around the proper application of ideas of the survival of the fittest. For Olivier, insofar as man had attained freedom to do and be as he desired, he had attained it only through the evolution of society.[325] Aware of the various conflicting appropriations of Darwin, the Fabians were anxious to reclaim him for their brand of technocratic ameliorism, unwilling to wait passively for the impersonal forces of social evolution, but also resistant to 'might is right' arguments. Wallas later argued that it was the intellectual tragedy of the nineteenth century that Darwinism had been used to justify capitalist and imperial violence; but he was also convinced that any reformer unprepared to study humanity 'as evolution has for the moment left it' would 'find his life one of constant and cruel disillusion'.[326] Sidney Webb agreed, telling Wallas in 1908 that even if political science could not yet produce its equivalent of the *Origin of Species*, it should take Darwin's *Barnacle* monograph as its model.[327]

282 DARWINISM'S GENERATIONS

There was little in this to distinguish Fabianism from either the socialist or the labourist mainstreams. As the newspaper of the Social Democratic Federation, *Justice*, had put it on the unveiling of Darwin's statue in the Natural History Museum in 1885, 'To us scientific Socialists whose sociological theories are based upon the recorded facts of evolution and development, Darwin's work serves as a sure foundation'.[328] David Stack has demonstrated how a Darwinian lexicon provided both structure and limits to socialist discourse, but promiscuously, drawing equally on Lamarckian or Spencerian ideas, individualism and organicism.[329] He points out that the socialism of Keir Hardie and Tom Mann (1856–1941) was more evolutionary than Marxist; during his period in Australia in the 1900s Mann presented himself as an 'evolutionary socialist', celebrating Darwin's 'epoch-making' contribution, and justifying socialism as an imperative driven by evolutionary forces in nature, industry, and politics.[330] Both Hardie and Mann reflected the ethical socialist and rationalist sub-culture associated with labourist figures like Robert Blatchford (1851–1943), Ben Tillett (1860–1943),[331] John Trevor (1855–1930), and the Scottish anarchist-socialist James Tochatti (1852–1928),[332] sustained by a melange of works, including the popular evolutionary writings of Edward Clodd, continental Darwinist texts such as Haeckel's *Riddle of the Universe* (as translated by Joseph McCabe), Spencer, and comparative studies of primitive religion, of which J.G. Frazer's *The Golden Bough* (1890, etc.) was the most influential.[333] Others sought to exploit more explicitly biological analogies. The articles on mutual aid of the émigré anarchist Prince Kropotkin in the 1890s sought to deploy the evidence of social insects such as ants to challenge blunt social Darwinism, while Edward Carpenter sought to mobilise Lamarckian notions of evolution through willed development to justify what Ruth Livesey has called his 'biological idealism', a celebration of the primary force of evolution as art not struggle.[334]

There was a strong strand of this anti-competitive approach in the thought of Benjamin Kidd, perhaps the fullest development of late-Victorian applied Darwinism.[335] Kidd was a self-educated clerk to the Inland Revenue, entirely unknown when his book *Social Evolution* was published in 1894; but this did not prevent it from becoming the best selling book of non-fiction in the period 1891–1909.[336] *Social Evolution* offered a confection of Darwin, Spencer, and Huxley, filtered through Kidd's own religious sensibility. His rejection of what he saw as the crude individualism of previous applications of Darwinian ideas to society, and his argument that religion provided a vital spur to the altruism needed to mitigate the

THE DEATH OF DARWIN AND AFTER   283

destructive tendencies of self-interest, resonated with the reading public. *Social Evolution* reached its eleventh thousand printing within twelve months, and was still selling 140 copies a week in 1907.[337] The book was avidly sought by the members of the provincial Lit and Phils; at Newcastle it was reported that although the society's library had several copies, they were insufficient to meet the demand.[338] Charles Booth found that his presentation copy was so much requested that he had to buy a second to lend out to his friends.[339] Although Kidd had no formal biological training and was frequently dismissed as a scientific neophyte, he took his Darwinism seriously, and was not afraid to spar with Wallace about the operation of natural selection.[340] Kidd sought to separate two forms of social competition, an 'external' one between nations for markets and resources, and an 'internal' one for jobs, rewards, status; vital for the efficiency of the nation. He supported democratisation and welfare as ways of widening the domestic competition for success; but opposed socialism as working to suspend it, and individualism as antagonistic to the success of the nation in its external conflicts.

It has been suggested that the difference between Kidd and his critics was not a distinction between conservative and reform Darwinism, but 'between a generation who minimized Darwin's significance, and one who translated the new biology into a call for positive social action,'[341] but in truth the generational alignments of the responses to *Social Evolution* were more complicated. Of the mid Victorians, Argyll was highly critical, but the old Christian socialist Thomas Hughes (1822–96) conceded that the book had given him 'a respect for evolution which I no doubt wanted.'[342] Religiously minded high Victorians, always on the lookout for accommodations of faith and evolution, embraced Kidd's argument: 'one of the greatest books we have had since Darwin's "Origin of Species",' suggested the biblical scholar Marcus Dods (1834–1909).[343] But otherwise, high Victorians rejected Kidd's attempt to extend evolutionary dynamics to social conditions; a clearer example of the false method in sociology could hardly be found, commented the Positivist J.H. Bridges (1832–1906).[344] For the Edwardians, it was harder to find an obvious hook. It might appeal to amateur scientists, especially those looking for the extension of evolution into society, but it was often met with disdain; the philosopher John Stuart Mackenzie (1860–1935) told Kidd that although he had evidently said what a great number of people wanted to hear, '[s]uch things are seldom the right things to say.'[345] In contrast, the Liberal-Imperialist John Saxon Mills (1863–1929) recalled the way *Social Evolution* lifted the tyranny of

284 DARWINISM'S GENERATIONS

Darwinism as applied to the problems of life by disengaging it from questions of social and spiritual development.[346]

Despite the presumption of many radicals that evolutionary progressivism was naturally aligned to liberal and against conservative instincts, Darwinism's loss of cultural challenge encouraged the parallel emergence of powerful strands of social Darwinian thought which contributed to new imperialist world views, ideas of the superiority of the white races and the survival of the racially fittest. Darwinian metaphors were suddenly popular in conservative circles. The implications drawn were far from consistent. For some, Darwinism was appropriated as a great argument that progress was necessarily slow and that nothing constructive could come of violent change.[347] In Karl Pearson, such ideas created an elitist socialist with strong imperialist impulses. In Cecil Rhodes, described in one assessment as 'a Darwinian rather than a Christian', and in others nurtured by the evolutionary imperialist imaginaries of novelists like Rider Haggard (1856–1925), including the African adventurer Frederick Courtney Selous (1851–1917) and the colonial administrator Sir Everard im Thurn (1852–1932), they encouraged belief in the hierarchy of races and the right of imperial force.[348] The lure of this line of thinking was particularly compelling for the leaders of Britain's white settler colonies, for whom racial hierarchies were powerfully self-serving.[349] The force of this rhetoric strengthened steadily during the 1880s and 1890s, peaking during the Boer War and its aftermath. By this point high Victorians like Wilfrid Scawen Blunt as well as Liberal Edwardians were looking on sickened at the exploitation of Darwin for crude imperialist purposes.[350]

Although it is not easy to write directly onto the generational scheme proposed here, the emergence of eugenicist ideas at least partially expressed this generational alignment. Primarily a twentieth-century force, feeding off Edwardian anxieties about national decline, eugenics in Britain was built on ideas articulated first by Francis Galton, whose term it was, and whose *Hereditary Genius* (1869) provided many of the foundational assumptions, but more vigorously by late Victorians, not least the youngest of Darwin's sons, Leonard Darwin (1850–1943), who during the 1900s transferred his broad interests in geography militant to eugenics.[351] Part of the problem of making sense of the constituency of eugenics is the different ways in which its central anxieties about hereditary traits and breeding could be interpreted. There was a positive impulse which looked to the introduction of measures to improve social and economic conditions, and a negative impulse which deprecated any effort to protect the weak from the pressures

THE DEATH OF DARWIN AND AFTER   285

of a competitive environment, and sought improvement rather from the management of reproduction, and especially preventing the 'unfit' from having children. Both could lead to eugenic positions, but especially under the anti-Lamarckian influence of Weismann, the mainstream impulses of eugenics derived from the belief that natural selection had to be allowed to carry on operating at the level of each member of society. But it remained possible to deploy eugenic ideas to press for environmental improvement while being opposed to interference in procreation, or for someone like Wallace to express sympathy with the broad notion of attention to better breeding, but to be fiercely hostile to controls on fertility.[352]

Both mid and high Victorians generally had little confidence that any sort of intervention would do other than discourage the respectable but leave the unrespectable unaffected.[353] Many, like Andrew Lang, claimed ignorance to avoid the issue.[354] But by the mid-1890s, there was a growing cadre of predominantly late Victorians who were embracing the cause, including the physiologist John Berry Haycraft (1857–1922), Alexander Graham Bell (1847–1922),[355] the social reformer Mary Higgs (1854–1937),[356] Mary Dendy (1855–1933), who was a leading figure in the movement to provide for children with learning disabilities,[357] and the judge Sir John Macdonell (1846–1921). Poulton stood aloof, but did concede that Darwin was 'bound up with the eugenic idea'.[358] Haycraft advocated an active policy of preventing the 'lower types' from reproducing to avoid the degeneration of the race.[359] Macdonell was even more blunt, suggesting that the teaching of Darwin had placed on a scientific basis the pretensions of 'civilised races' to dominate 'the black race'.[360] Concern for racial health certainly attracted many of the youngest of the late Victorians born from the mid-1850s onwards, including Sidney Webb, and the Scots physician George Archdall Reid (1860–1929), while New Liberals, while they shared much of the underlying diagnosis, rejected the crude individualism and the apparent indifference to premature death of many eugenicists (Geddes caustically annotated one piece of eugenic literature as being intended 'to encourage the Herodians').[361]

Idealism

A similar Darwinisation is visible in the discipline of philosophy, where the Idealism of the disciples of T.H. Green[362] was one of the most powerful currents of late-Victorian thought. J.H. Muirhead, later Professor of Philosophy

286 DARWINISM'S GENERATIONS

and Political Economy at Birmingham University, and one of British Idealism's organising forces, was explicit that its intention should be to take the broad concept of evolution and apply it more directly to the history of philosophical thinking.[363] In part, this meant developing the evolutionary philosophies of Kant, Hegel, and Spencer. But what the late Victorians also did, especially in the later 1880s and 1890s, rather than downplaying the relevance of Darwin as Green and his contemporary Edward Caird had been wont to do, was to filter these approaches through an explicit Darwinism.[364] In *Essays in Philosophical Criticism* (1883), a tribute collection to Green which was intended as a manifesto of 'some of the younger bloods', many of this second generation of British Idealism, who had almost all been immersed in discussions of the implications of Darwinism for philosophy as university students in Glasgow and Edinburgh in the 1870s,[365] staked out their claim to deploy the tools of evolutionary science, facing down a scientific establishment which would, as James Bonar, one of their number, put it, 'tolerate no evolution except of the earth, earthy. Darwin may steal a horse, while Idealists may not look over a hedge'.[366]

The Idealists resisted both the collapsing of human life into social laws in the way Spencer had attempted, and also the narrowly determinist Darwinism of Weismann, described by one of their number as 'the physiological nakedness of evolutionary science', but they nonetheless believed that the categories of evolution, not just natural selection and the struggle for existence, but heredity and inheritance, had to be incorporated into philosophical thinking.[367] The most explicit development of these approaches comes in the writings of David Ritchie, whose 'Idealist Evolutionism', developed in *Darwinism and Politics* (1889, 1891), and the essays collected in *Darwin and Hegel, with other Philosophical Studies* (1893), strongly defended the operation of natural selection in the development of morality and ethics, while simultaneously challenging crude applications of evolutionary natural laws to social relations, and suggesting that ultimately natural selection passed into a higher form of itself, in which the conflict of ideas and institutions took the place of the struggle for existence between individuals and races, so that 'rational selection' replaced natural selection.[368] Ritchie had no sympathy with those who responded to the excesses of some versions of social Darwinism by rejecting it entirely, arguing not just that the biological metaphor remained a serviceable means of freeing discussion from old assumptions while simultaneously guarding against expectations of rapid change, but also insisting that society needed

THE DEATH OF DARWIN AND AFTER   287

to be understood as a living organism, more than the arithmetic sum of its individual members.[369]

Here again, this was not just a question of the assimilation of Darwinism, or even the appropriation of Darwinian themes, but of an ongoing debate about its 'true meaning'. 'Let us ask', F.H. Bradley wrote in his 'Some Remarks on Punishment' (1894), 'what Darwinism teaches and what it does not teach'.[370] Bernard Bosanquet is interesting in this regard because his fascination with Darwin clearly extended beyond the core evolutionary texts—in 1877 he was discussing Darwin's *Form of Flowers* with a friend—and because he was seen as an important interpreter of social evolutionary thought in the 1890s.[371] In his 'Socialism and Natural Selection' (1895), Bosanquet acknowledged the dangers of abolishing the struggle for survival, accepting Kidd's distinction between good and bad socialism, and stressing the importance of producing the right sort of family.[372] But like all the late-Victorian Idealists, he saw the state as embodying the moral will of the people, which meant that the crucial task was ensuring the right frame for state action: contemporary problems were not to be solved by administrative nihilism, he argued, but by care, analytic experience and patient well-doing. Bosanquet's detailed response to James Ward's Gifford Lectures in 1906, in which he challenged what he saw as Ward's implicit Lamarckism, and reaffirmed his own commitment to the struggle for existence as the meeting point of natural selection and social morality, illustrates the pervasive exegetical imperative of the late Victorians.[373]

The Idealists were particularly exercised by the relationship of evolution and morality. Some were only prepared to go part of the way. Muirhead accepted that biological evolution was closely connected with ethics, but not identical to it,[374] while Bradley argued that while Darwinism could have nothing to say about moral ends, it demanded a revolution in thinking about means.[375] But in general the late-Victorian Idealists insisted that natural selection was the mechanism of moral as well as physical progress, and that once attention moved beyond a narrow biological focus to the rational action of humanity, this became incontrovertible.[376] Samuel Alexander (1859–1938) was one of this group who did conceive of evolution as a process whereby the good gradually drove out the bad. Alexander is illuminating as a transitional figure, in some respects representative of the next generation, starting with Idealism, moving through a Darwinian naturalism in the 1890s and then feeling the need to break away from it, while standing aloof from Edwardian pragmatism.[377] Searching from the late 1880s for the solution to what Ritchie called his 'metaphysical applications of Darwin',

288 DARWINISM'S GENERATIONS

Alexander offered a forthright justification of the role of natural selection in morals, and his *Moral Order and Progress* (1889) was very much within the 'evolution and society' tradition.[378] Alexander never entirely threw off this naturalism, but in the years after 1914 he was identified with Lloyd Morgan and A.N. Whitehead (1861–1947) as one of the advocates of 'emergent evolution', a reformulation of earlier saltationist approaches which again questioned the significance of natural selection.[379]

Idealism was a powerful component of the wider late-Victorian sensibility. In the writings of a figure like Hector Macpherson, it contributed to a rejection of high-Victorian monism, and a commitment to an underlying intelligence as the universal driving force.[380] We can see its influence in the persistent teleology of late-Victorian poesy, most notably Agnes Mary Frances Robinson's (1857–1944) 'Darwinism' (1888), and the diffuse impulses which might at a push be described as the late-Victorian 'Ethical Movement', which drew in high-Victorian agnostics like Stephen, Sidgwick, and Seeley as mentors, but whose dynamics were rooted in the following generation, and which became an enduring locus of evolutionist culture. It is visible in the animal rights activism of Henry S. Salt (1851–1939) and the Humanitarian League, and in John Trevor and the Labour Church movement.[381]

Late-Victorian Theology

Despite the naturalisation of Darwinism, for many late Victorians science continued to present a formidable challenge to conventional Christian belief. Notwithstanding efforts such as Oliver Lodge's alternative evolutionary catechism, *The Substance of Faith Allied with Science* (1907), many never fully escaped from the suspicion of science that the tenor of these conflicts often engendered. Darwinism continued to be emptied of its biological substance and presented entirely as a matter of religious scepticism. We can see this in the fiction of Braddon, where Darwin's coadjutors were Buckle, Comte, Spencer, and Schopenhauer, and Darwinian beliefs were generally the prerogative of shallow agnostics like Gerard Hillersdon in *Gerard* (1895), or Brian Walford in *The Golden Calf* (1883). In the memories of Girton in the 1880s of Constance Maynard (1849–1935), Darwinism appears essentially as a materialistic philosophy which subverted religion.[382] The forces of institutional suspicion remained strong. Praise for Darwin in the presidential address at the Glasgow meeting of the British Medical

Association in 1888 prompted denunciations from several pulpits on the succeeding Sunday.[383] Presbyterianism in particular continued to nurture versions of resistance, in which the vehemence of the surface rhetoric often disguised a degree of underlying accommodation, as in the case of the Irish Presbyterian Samuel Biggar Giffen M'Kinney (1848–1908), whose *The Origin and Nature of Man* (1907) launched a wholesale assault on natural selection as a blind unconscious force, while conceding the possibility of non-human evolution.[384] It was certainly possible for one late-Victorian botanist to think of his identity as scientist as one who 'dares to look into matters denied him by the priesthood', and to despise religious belief as requiring belief in science for six days and theology on the seventh.[385]

But increasingly such antitheses did not reflect the balance of opinion within the churches. The example of figures like Aubrey Moore encouraged believers that it was possible to accept Darwinism and Christianity.[386] Late-Victorian preachers often offered a deliberately non-confrontational assessment, which left space open for those who clung to the creation narrative, but clearly inclined to evolution.[387] W.H. Dallinger's Fernley lecture, blocked in 1880, was listened to without undue anxiety in 1887, and in the mid-1880s the London Welsh Calvinistic Methodists were shaken by the popularity of Darwin and Spencer amongst the younger members of the congregation.[388] By the end of the decade the Congregational Union was being told 'we are all evolutionists now', and even the National Council of Evangelical Free Churches was listening with equanimity to the suggestion that they should not be afraid to let the whole of humanity be brought within the pale of evolutionary law.[389]

It is conventional to suggest that for the Anglican Church, it was Frederick Temple's Bampton Lectures of 1884 that made Darwin's evolutionary hypothesis respectable; and the same might be said of the importance within Dissent of the acknowledgement of the American Congregationalist, Henry Ward Beecher, that that he was a 'cordial Christian evolutionist'.[390] But perhaps it makes more sense to identify a new generation of clerics shifting the timbre of the debate around this date. One striking example is Hugh Price Hughes (1847–1902), whose youthful interest in natural history left him with a particularly intense sense of the confusion and chaos wrought in many minds by the idea of evolution.[391] Hughes was the leader of progressive Methodism at the *fin de siècle*, and he clearly accepted the fact of evolution, and deprecated the frame of mind of many older Methodists, who 'slept in fear of waking up next morning to find that either Darwin or Huxley had MADE SOME TERRIBLE DISCOVERY which would provide an

290 DARWINISM'S GENERATIONS

absolute contradiction to the truth of the Gospel'.[392] But he made little attempt to discuss Darwinian or even evolutionary themes in his preaching or addresses, and was just as likely to attack men of science, or what he saw as the ethical nihilism of the pitiless struggle of existence.[393]

The Manchester lay preacher Richard Wilkinson (1857–?) had no such qualms, embracing Darwin, Huxley, and even Haeckel, all of whom he later read with his precocious daughter Ellen, adding in lectures at the Free Trade Hall, and later Bergson's *Evolution Creatrice*.[394] Late-Victorian apologists more fully assimilated evolutionary ideas into theological thinking. Forthright evolutionism had long been a feature of the progressive fringe congregations, but within the main denominations the shift is associated with the high and late Victorians, although pulpit language, mindful of the range of congregational views, was often chary about identifying too directly with a particular evolutionary position.[395] There was a move from guarded acceptance to positive advocacy. For A.W. Momerie (1848–1900), Professor of Logic and Mental Philosophy at King's College, London, evolution saved God from the picture of wanton destructiveness implicit in pre-evolutionary ideas of successive creations; while the popular Congregationalist preacher, William Hardy Harwood (1857?–1924), argued that Darwin had helped the nineteenth century mind to realise how orderly and definite were the purposes of God running through history.[396] Much of the theological discussion avoided any effort to engage with specific Darwinian evolutionary mechanisms, and even prominent evolutionist preachers retained a residual providentialism.[397] Evolution became the gradual growth of the divine purpose which featured prominently in the Bible.[398] This often amounted, as John Kent has noted, to little more than the adoption of a pre-Darwinian, Hegelian belief that history, now extended to the biological pre-history of man and other creatures, manifested a process of development in which Spirit progressively dominated Matter. Accordingly, as Mrs Humphry Ward put it, the pulpit filled up with men endeavouring 'to fit a not very exacting science to a very grading orthodoxy'.[399]

This did not mean that Darwin was entirely effaced. There was a noticeable veneration of his works (and Wallace's), in contradistinction to what Wilfrid Ward, the Catholic essayist and editor, described as the sweeping generalisations of their second-rate followers.[400] Much of this exploited the sorts of character assessments we have already seen, with religious supplement. So Darwin's painstaking research was held up as an example of the faith of the scientist.[401] (By the same token, the loss of appreciation of music

THE DEATH OF DARWIN AND AFTER   291

and art which became one of the most frequently cited anecdotes of the *Life* could be mobilised to challenge his competence to speak on matters of spirituality.)[402] The prominent Methodist preacher Frank Ballard (1851–1933) even suggested that the young Darwin was a committed theist, and it was only in later life 'when senile atrophy set in' that he fell into agnosticism.[403] The convoluted positioning such considerations could produce in the late-Victorian laity is illustrated by the Guildford architect Edward Langridge Lunn (1856–1922?), who painstakingly positioned himself as 'discarding the dynamic view of Haeckel…and the ultra-development theory of Darwin, and being unable to grasp the philosophic cult of Spencer, yet see[ing] in a modified Theistic theory of evolution as a process,…a *modus vivendi* between liberal Christianity and inductive science'.[404]

Evolutionary disputes inflected theological controversies in unprecedented ways. By the later 1880s and 1890s a new group of mostly late-Victorian scholars, led by Samuel Rolles Driver (1846–1914), Pusey's successor as Professor of Hebrew at Oxford, were actively promoting German biblical criticism, including the repudiation of the Mosaic authorship of the Pentateuch. Driver remained determined to defend his right to judgement, 'provided I take my scientific facts from scientists'.[405] Older critics saw in this the dangers of an evolutionary theology, not just the argument that the religious dispensations of the early Jews were the result of human evolution rather than divine dispensation, but an *a priori* assumption of biological evolution.[406] Driver himself accepted the intrinsic plausibility of the scientific theory of the evolution of man, but, along with late-Victorian apologists generally, continued to deploy the characteristic arguments of the high Victorians, including the safety valve of first causes. Darwinism could not answer the mystery of God and creation, the origin of life, and the beginnings of man on earth. Huxley's verdict was withering: 'If Satan had wished to devise the best means of discrediting "Revelation" he could not have done better'.[407]

The 1880s and early 1890s saw two major late-Victorian attempts to develop arguments for the congruence of Christianity and evolution, Henry Drummond's *Natural Law in the Spiritual World* (1883) and *The Ascent of Man* (1894), and the collection of Anglican essays published as *Lux Mundi* (1889). Drummond was Professor of Natural Science at the Free Church College, Glasgow, a charismatic and occasionally flamboyant figure who travelled widely and made a genuine contribution to contemporary understandings of the geology of Africa.[408] Although his early teaching aimed at enlisting evolution on behalf of natural theology, he managed to maintain his reputation for orthodoxy even within the often fiercely conservative atmosphere of Scottish Protestantism.[409] 'A man's theology', he was fond of saying in a pointedly generational as well as gendered construction, 'could not be of the same special colour as his grandmother's'.[410] Part of his rhetorical strategy was to claim the *Origin* as perhaps the most important contribution to the literature of Apologetics which the nineteenth century had produced, seeking to supplement rather than dismiss Darwinism, while representing himself as rejecting Darwin's 'maimed account' of evolution.[411] *Natural Law in the Spiritual World* challenged the conventional distinction between material and spiritual worlds, arguing for the identity of the Spiritual Law and the Natural Law. As many critics pointed out, it was more a series of reflections on the way the language of evolutionary biology might be applied to spiritual topics than a reasoned piece of metaphysics, but it had enough credibility to draw approval not just from Anglican Darwinists like F.A. Dixey, but also the unreligious A.C. Haddon.[412] *The Ascent of Man*, with its implicit challenge to *Descent*, was a more substantial intervention in Victorian debates over evolutionary science. Here Drummond argued that evolution had been misunderstood since Darwin, that it was God's method in creation, and although natural selection was a cause of evolution, it was not the only cause. Just as the neo-Lamarckians sought to add the dynamic of use-inheritance, so Drummond argued that evolution was as much an ethical as a biological phenomenon, and that the physical struggle for life which brought the survival of the fittest needed to be supplemented by a parallel struggle for the survival of others that was not individualist and selfish, but collective and altruistic.

Drummond's works rivalled Kidd's *Social Evolution* as the publishing sensation of the late nineteenth century. *Natural Law* quickly sold 40,000 copies, reaching 130,000 by the end of the century; into the mid-1890s it was the religious book most frequently issued by the public libraries, and was very widely cited in the sprawling press debates on evolution at the *fin de siècle*.[413] *Ascent* sold fewer copies but was reviewed more respectfully.[414] Responses were probably shaped primarily by religious considerations, and by the support Drummond provided for Christian believers shifting from older literalist views to more modern positions (although *Ascent*'s anti-individualist arguments meant that it was also widely quoted in socialist circles).[415] But generational modulations are still visible. The early and mid Victorians generally refused to be seduced. For Shaftesbury, *Natural Law* was a 'singularly pernicious publication', while James Martineau dismissed it as a 'system of illusory analogies'.[416] Where mid Victorians were more positive it was because they were able to read into *Natural Law* key tenets of their own thinking. J.J. Murphy approved because he saw it as strengthening arguments for design.[417] *Ascent* was attacked widely by early and mid Victorians as a crude rehash of Darwin and Spencer which verged on plagiarism. Neither high-Victorian agnostics like Leslie Stephen nor ministers like Dallinger and Iverach had anything positive to say about what Stephen disparaged as 'evolutionary theories which started from [such] a transcendental base', producing 'mere cobwebs of the brain'.[418] Nevertheless, *Ascent* frequently struck a chord for late Victorians looking to strengthen the accommodation of their faith with contemporary science, and in the 1890s it was being wielded by students at King's College, Aberdeen in resistance to the anti-evolutionary lectures of David Johnston (1836–99), the Professor of Divinity.[419] Hugh Price Hughes enthusiastically recommended *Ascent*, both for its challenge to purely Darwinian theories of evolution, and also for its ability to explain original sin as the taint of animality left by mankind's evolutionary history. Not so the Edwardian George Wyndham (1863–1913): reading *Natural Law* in 1885 he dismissed it as 'sheer rubbish'.[420]

Lux Mundi was a more weighty affair. Traditionally discussed as a landmark in the acceptance of the higher criticism in late-nineteenth-century Britain, it also represented the strengthened accommodation with evolutionary thought of an influential late-Victorian coterie. Evolution, argued one its contributors, J.R. Illingworth (1848–1915), was the inescapable 'category' of the age.[421] Of course, the idea of evolution developed in *Lux Mundi* was not specifically Darwinian. It quite deliberately embraced analogies of evolutionary and divine progress, and it certainly pressed Darwinian motifs into service without compunction, although often with compromises which became increasingly strained as theological liberalism marched forwards.[422] The exception was Aubrey Moore, who was committed to a more rigorous Darwinism, and who had already in the 1880s published several essays arguing for evolution's active assistance to theology.[423] For the theologian Clement Webb (1865–1954), looking back, the contributors 'belonged to a generation which had been profoundly affected by the impulse given to an evolutionary view of things by the discoveries and hypotheses of Darwin'.[424] The twelve essayists were born between 1844 and 1856,[425] and their ideas had been forged in the discussions of the group over the previous decade, not least in regular gatherings at Illingworth's rectory.[426] As he freely confessed, there was a tone of 'fortyness' about the book.[427] Its editor, Charles Gore (1853–1942), had already caused a brief storm of controversy in 1884, when soon after his appointment as librarian of the 'Pusey Memorial' in

294 DARWINISM'S GENERATIONS

Oxford, he had welcomed the compatibility of 'frank acceptance' of evolutionary thought and orthodox Christian belief.[428] *Lux Mundi* took this process further and attempted to reinterpret the incarnation as a culmination of the evolutionary process. Liddon, who had taught many of the essayists, was pained by such boldness; 'outrun', as one memoir put it, 'by his pupils'.[429]

Lux Mundi also contributed to two central characteristics of late-Victorian thought: its renewed organicism, and more especially its move from the ideas of divine transcendence popular with the high Victorians to an incarnationalist version of immanence.[430] For *Lux Mundi*, the natural world should be seen less as a machine on the old watchmaker analysis and more as a living organism, a 'system in which, while the parts contribute to the growth of the whole, the whole also reacts upon the development of the parts' as Illingworth put it.[431] This conception left no room for chance or even contingency. Evolution was the gradual unfolding of a germ, of a seed packed with that potentiality from the beginning of its existence.[432] The function of evolution was not external, rather it was to develop its own perfection. Theology came to share the greater propensity for progressivist understandings of evolution elsewhere visible in late-Victorian thought, albeit frequently with an Hegelian rather than a Spencerian flavour.[433]

At the same time, Moore's contribution sought to repudiate the conception of God as an 'occasional visitor', recognising that as science developed natural explanations for apparently miraculous phenomena, this conception was progressively undermined; in this context, he argued, by enriching the notion of divine immanence, Darwinism, 'under the disguise of a foe, did the work of a friend', rescuing God from being entirely excluded from the world. Gaps in nature of the sort implied by miracles or special creations were unnecessary: for an immanent divinity, acts of nature were acts of God and vice versa.[434] Moore's writings were received with enthusiasm by many high- and late-Victorian scientists.[435] Even Lankester responded warmly, and the tone of their surviving correspondence is dramatically different from Lankester's treatment of earlier opponents.[436]

Here again, the imprecision with which ideas of 'design' were invoked and the eagerness with which controversialists sought to turn the words of opponents against themselves makes absolute distinctions impossible. Teleology for the late Victorians was not the old belief in precedent design, but a new confidence in divine destiny in which a developing theology of the incarnation was critical.[437] With Christ incarnate, the relationship between God and man became a matter more of ethics than of law, more of fatherhood than sovereignty, while the evolution of the human body to prepare for the incarnation of Jesus, 'explain[ed] the evolution and the nature of the world'.[438] Immanence added the conscious agency of a personal creator to instilled capacity or endowment, and provided an escape from mechanical versions of natural law, God-given but absolute and blind, which appeared to push the divine to the margins.[439] The distinction is dramatised by a conversation in Joseph Compton-Rickett (1847–1919)'s *Quickening of Caliban. A Modern Story of Evolution* (1893), in which the minister speaks for this view, in opposition to the scientist who contended that Natural Law, rather than requiring a Supreme Intelligence, was quite able to take care of itself.

Conclusion

There is no convenient dividing line, real or symbolic, between the period when late-Victorian conceptions dominated and the new world of the Edwardians. The 1890s were a strange confusion of lingering Victorianism and *fin de siècle avant garde*. But a number of interventions in 1894 hint at some of the underlying shifts. H.G. Wells' science fiction, the first instalment of which, *The Time Machine*, appeared in the *National Observer* between March and June 1894, and the first signs of Edwardians imposing themselves on the directions of scientific debate, in William Bateson's seminal study of variation, *Materials for the Study of Variation* (1894), both speak to new forces emerging and will be taken up in the next chapter. They coincided with T.H. Huxley's Romanes Lecture on 'Evolution and Ethics', which although delivered and first published in 1893, was followed up with an extended second edition including an important additional 'Prologomena' in 1894. All three, in one way or another, speak of both the persistence and the crisis of a Darwinian world view.

The Romanes Lecture had been founded by G.J. Romanes in 1891. The first, by W.E. Gladstone in 1892, was a rather inconsequential sketch of the history of universities. But the second was an altogether more significant affair.[440] In large part, *Evolution and Ethics* was a renewal of Huxley's longstanding debate with Herbert Spencer over the claims of individualism or collectivism, and with the theologians over their claims of evolution's divine and progressive underpinnings (as extended by the Idealists' attempt to claim the evolutionary perfection of mankind as an end immanent in nature). Huxley rejected all these positions. Social progress required the

296 DARWINISM'S GENERATIONS

checking of the evolutionary process at every step, he argued, and the substitution for it of another, which could be called the ethical process. It was in many respects a typically mid-Victorian response, inspired by Jowett, welcomed by Mivart,[441] a restatement of mid-Victorian liberalism; a rejection of the absolute laissez-faireism of Spencer, but also of the collectivism of late-Victorian social reform.[442] But its apparently un-Huxley-like pessimism about the essential implications of the evolutionary process encouraged appropriations of all kinds, and it became a frequent point of reference in subsequent debates.

In a symptomatic misreading, *Evolution and Ethics* was met with expressions of regret from high Victorians like Stephen, whose response was to reaffirm a progressive ethical evolution against Huxley's agnosticism, and by Thiselton-Dyer, who rejected Huxley's 'remarkable pronouncement' as irreconcilable with Darwinian theory.[443] (As Huxley subsequently pointed out, his point had not been to deny that evolution produced morals, but that the morality evolution produced was necessarily virtuous.) Late-Victorian idealists could not accept Huxley's antithesis of cosmic process and ethics, nor his dualism of nature and human consciousness.[444] In contrast, Huxley's pessimism appealed to many Edwardians not only because it decoupled what *is* from what *ought to be*, but because it pushed the dynamic of natural selection into the background, foregrounding instead the role of purposeful human intervention, and because it repudiated the crude reduction of Darwinism to the equation of might and right which was becoming characteristic of one strand of European thought.[445]

Standing at the lectern in 1893, the ageing Huxley was already a symbol to his younger hearers of a disappearing past, a feeble old man, increasingly deaf, 'pale, spare and fragile', who had to husband his voice, and so could not make himself heard in the galleries of the Sheldonian theatre.[446] Looking back from the middle of the twentieth century, Herbert Samuel (1870–1963) remembered the lecture as part of the time when the great controversy on evolution was drawing to a close; 'the younger generation were altogether with the victorious scientists'.[447] But even at the time, the lecture, Benjamin Kidd remarked, belonged to 'a phase of thought beyond which the present generation feels itself...to have moved'.[448] Huxley confessed feeling 'chill with age', conscious 'that I am regarded by my contemporaries as a relic of the past'.[449] In the following two decades the Darwinian orthodoxy he had championed for so long came under pressure from a new generation.

THE DEATH OF DARWIN AND AFTER   297

Notes

1. *The Field*, 24 April 1886.
2. George Howell to T.H. Huxley, 28 April 1882, f.286, Box 18, Huxley Papers, ICL.
3. Mary Elizabeth Braddon, *The Day Will Come* (1889), III, 25–6.
4. W. Shairp to J.C. Coleridge, 25 November 1882, ff.193–6, Coleridge Papers, Add. MS 86306, BL.
5. '...dogmas have produced sectaries, and we hear of Lamarckians and Neo-Lamarckians, Neo- and Ultra-Darwinians; the apostles of Weismann and the disciples of Spencer', review of Romanes, *Darwin and after Darwin: Volume III, The Zoologist* 4th ser., 1 (1897), 522. For Lankester as a quarrelsome and intemperate loose cannon, see J. Howarth, 'Science Education in Late Victorian Oxford', *English Historical Review* 102 (1987), 347–8.
6. *Times*, 21 April 1882.
7. For E.J.J. Browell (1827/8–1914), Darwin's cautious empiricism was to be contrasted with the 'unphilosophical language' and overeagerness to attack opponents of some of his disciples, 'Presidential Address', *Transactions of Natural History Society of Northumberland, Durham and Newcastle* 8 (1884–9), 173.
8. Havelock Ellis, *The New Spirit* (1890), 5. This was a very common trope from 1880, not least for those otherwise careful not to articulate their specific 'consent'; for example Eliza Brightwen, as in *Glimpses into Plant Life* (1897); North's *Recollections of a Happy Life*, II, 87, 214–15.
9. *Whitby Gazette*, 29 April 1882, Francis Paget to Stephen Paget, 23 April 1882, Stephen Paget and J.M.C. Crum, *Francis Paget* (1913), 71.
10. George Howell to T.H. Huxley, 28 April 1882, f.286, Box 18, Huxley Papers, ICL.
11. Michael R. Watts, *The Dissenters*, Vol. 3: *The Crisis and Conscience of Nonconformity* (2015), 66.
12. Wedgwood to Asa Gray, 4 May 1882, Asa Gray Correspondence, Wa–Wh Harvard.
13. H.D. Rawnsley, *Harvey Goodwin: Bishop of Carlisle* (1896), 223–4; *Westminster Gazette*, 2 July 1918; see J.R. Moore, 'Charles Darwin Lies in Westminster Abbey', *Biological Journal of the Linnean Society* 17 (1982), 97–113. Dale's sermon offered a typically cautious and partial endorsement, suggesting Darwin's evolutionary speculations were 'still waiting judgement from the court of final appeal', while also conceding that his theory 'was invested with a high degree of probability, and henceforth it would largely control the whole movement of scientific inquiry', *Birmingham Daily Post*, 24 April 1882.
14. Barton, 'Royal Society Politics', 81.
15. *Life and Letters of Thomas Henry Huxley*, II, 422. 'To us scientific Socialists whose sociological theories are based upon recorded facts of evolution and development, Darwin's work serves as a sure foundation', Jones, *Social Darwinism*, 70 (quoting *Justice*, 13 June 1885).
16. See recollection of W.H.F.C., *West Middlesex Gazette*, 3 October 1913; in 1888 the *Christian World* was prompted to note that the Congregational Union of that year 'had no cheer to spare for the great name of Darwin', although cheers did greet reference to his loss of literary taste, *Christian World*, 10 May 1888.
17. For example, 'The Home of a Naturalist', *Good Words*, 34 (1893); Leonard Huxley, 'Charles Darwin: A Centenary Sketch', *Cornhill Magazine* ns 26 (1909), 376–89. For a discussion of Romanes' personal memories of Down House in the context of his Darwinian beliefs, see Pleins, *In Praise of Darwin*, 155–206.
18. W.L.M., 'A Ramble near Darwin's Home, 4 June 1910', *South Place Ethical Society Record*, August 1910, 4; lecture of George Beesley Austin (1856–1936?), *Norwood News*, 31 March 1900.
19. Lloegryn (George Newman (1836–1911?)), *Kentish Mercury*, 22 June 1893. In July 1909 the works outing of Francis Campion, Ltd, of East Greenwich, included a visit to Down House, *Kentish Independent*, 2 July 1909. Even while Down House was occupied as a school, it was still open to the occasional visitor, Darwin's study and sand walk carefully preserved; see Bland-Sutton, *Story of a Surgeon*, 103–4.

298   DARWINISM'S GENERATIONS

20. J.T. Cunningham, *Charles Darwin* (1886), G.T. Bettany, *Life of Charles Darwin* (1887), Grant Allen, *Charles Darwin* (1888), Charles Frederick Holder, *Charles Darwin, His Life and Work* (1891), B.O. Flower, *Life of Charles Darwin* (1892), Poulton's *Charles Darwin and the Theory of Natural Selection* (1896), Edward Woodall, *Charles Darwin: A Paper Contributed to the Transactions of the Shropshire Archæological Society* (1884), E.A. Parkyn, *Darwin His Work and Influence: A Lecture Delivered in the Hall of Christ's College Cambridge* (1894); in the United States, C.F. Holder, *Darwin, His Life and Work* (1891). See Bernard Lightman, 'The Many Lives of Charles Darwin: Biographies and the Definitive Evolutionist', *Notes and Records of the Royal Society* 64.4 (2010), 339–58.

21. For example, Lawrence J. Tremayne (*c.*1873–1959), 'Darwin: A Biographical Sketch', for the Grocers' Company School Natural History Society, LMA/4760/C/03/01/011, LMA.

22. *Ardrossan and Saltcoats Herald*, 29 November 1889.

23. William Erasmus Darwin to Murray, 13 June 1882, f.59, MS.40314, Murray Papers, NLS.

24. *Sunderland Daily Echo*, 2 February 1898; the annual report of the Clerkenwell Public Library suggested that 'in recent years' the library's two copies of *Descent* had been issued nearly two hundred times; there was a similar picture from Stroud Library, *Stroud Journal*, 20 July 1894; see also the account of reading of one storesman reader at the Clerkenwell Public Library in 1895, *Bideford Weekly Gazette*, 6 February 1896. For an exception, see Cwmpark Public Library, where in 1899 *Descent* was still on the proscribed list, along with Renan's *Life of Jesus, Glamorgan Free Press*, 20 May 1899.

25. *Dublin Daily Express*, 27 November 1882; see also the books presented to Ernest Groom (*c.*1860–1944), of Wrexham Bicycle Club, including the *Origin, Descent, Expression of the Emotions*, and *Worms*, along with others, *Wrexham and Denbighshire Advertiser*, 7 October 1882; see also the account of the presentation to E. Bennett of the Laverstock Reading Room, *Hants and Berks Gazette*, 29 June 1901.

26. Linton also had Henry Drummond's *Ascent of Man*, though Gould described 'the scorn that gleamed in Mrs. Linton's eyes as she pointed out this book of plausibilities', Gould, *Chats with Pioneers*, 20, 26–7; A.J. Hamilton, 'Henry Arthur Jones', *Munsey's Magazine* 11 (May 1894), 177; it was very much part of Jones' public persona that he was an avid reader of such texts, Helen C. Black, *Pen, Pencil, Baton and Mask: Biographical Sketches* (1896), 82. Interview with R. Blathwayt, *Idler* 4 (1894), 70.

27. Emily Spender (1841–1922), whose *The Missing Link* (serialised 1888) includes a picture of the parlour of a farmer full of books, including the *Origin*, and works of Lubbock and Spencer in an old-fashioned glass-fronted bookcase.

28. As in the case of the London residence of the Aberdeens, *The Gentlewoman*, 6 September 1890.

29. 'Mr Grant Allen at Home', *American Magazine* 6 (October 1887), 721.

30. Raymond Blathwayt, *Interviews* (1893), 71.

31. *The Field*, 8 April 1882.

32. Roy Compton, 'Amongst the Lions: A Chat with Mr J.T. Nettleship', *The Idler* 10 (1897) includes reference to 'a clever pen-and-ink Darwinian sketch by Du Maurier' in his study, 362. Browne's chapter 'Darwin in the Drawing Room' discusses cartoons etc., Browne, *Darwin*, II, 375.

33. W.C. Williamson to Frank Darwin, 3 June 1882, MSS.DAR.198/220, Darwin Papers, CUL; *PMG*, 2 June 1891. See also 'Sir Robert Ball at Home', *The World* [6? September 1901], quoted in *Formby Times*, 14 September 1901.

34. 'Fame and Photography', *Globe*, 18 February 1889.

35. Jane C. Burdon Sanderson to John Burdon Sanderson, 18 August [1883], ff.84–7, MS.20030, J. Burdon Sanderson Papers, NLS. (J.S. Haldane declined a copy, on the basis that 'I never saw Darwin and somehow generally find it difficult to read into a man's portrait the idea of him gathered from his writings', Haldane to Alexander, 9 January, nd, ALEX/A/1/1/110/4, Samuel Alexander Papers, MJRL.)

36. *Sheffield Independent*, 1 October 1888.

37. *Hampstead and Highgate Express*, 5 May 1883. Kelvin's study was decorated with portraits of Darwin, Joule, and Faraday, Arthur Warren, *London Days: A Book of Reminiscences* (1920), 112; the journalist Edmund Yates (1831–94) had portraits of Darwin and Huxley,

THE DEATH OF DARWIN AND AFTER 299

alongside Tennyson, Salisbury, Chamberlain, and Irving, Henry How, 'Illustrated Interviews XXIV: Mr Edmund Yates', *The Strand Magazine* 6 (1893), 84; even Samuel Smiles (1812–1905), for whom there is no evidence of Darwinian engagements, had a portrait of Darwin, along with George Stephenson and Mill, 'Mr Samuel Smiles at Kensington', *The World*, reprinted in *Brisbane Courier*, 10 April 1883.

38. A.R. Wallace, 'The Debt of Science to Darwin', *New Century Magazine* 25 (1882–3), 420–32; collected as part of his Darwiniana by the autograph collector John Crosse Brooks (1812/3–97), in the collection of the Newcastle Society of Antiquaries.

39. Bland-Sutton, *Story of a Surgeon*, 22. See also the collection of John T. Page (1855/6–1919), 'Rambles among my Autographs', *East London Advertiser*, 27 February 1909.

40. Edward R. Pease, *The History of the Fabian Society* (1916), 14–15.

41. Philip John Dear (1854–1934), *Are These Things So?* (1936), 4–5.

42. See W.B. Carpenter to John Lubbock, 5 May 1882, ff.138–9, Avebury Papers, Add. MS 49645, BL. See, for example, David W. Simon (1830–1909), 'Present Day Revelations', in his *Some Bible Problems* (1898), 6; other instances include the verdict of the feminist Barbara Bodichon (1827–91) ('What a wonderful book! Who after perusal can help believing in Evolution, and that everything improves as we go on?', as reported in Matilda Betham-Edwards, *Friendly Faces* (1911), 93); see *The Guardian*, 21 October 1891; Rev. F.S. Ross, sermon reported in *Church Weekly*, 13 August 1897; Drummond Grant (*c.*1836–1909), *Coleraine Chronicle*, 27 August 1898.

43. W.B. Carpenter to WTTD, 7 July 1875, f.132, W.T. Thiselton-Dyer Papers, In-letters I, RBGK.

44. J.D. Hooker to Frank Darwin, 4 May 1882, MSS.DAR.189/94, Darwin Papers, CUL.

45. Eventually published by Macmillan as *Charles Darwin: Memorial Notices* (1882).

46. Feeling 'the work would gain both in justness and dignity by delay', Frank Darwin to Hooker, 6 May 1882, JDH/1/2/6/13, Hooker Papers, RBGK.

47. Frank Darwin to Gray, 10 May [1882], and 28 December 1886, Asa Gray Correspondence, D, Harvard.

48. For this and the wider initial reputation management, see the correspondence in JDH/1/2/6, Hooker Papers, RBGK.

49. Mary St. Leger (Kingsley) Harrison (1852–1931) [Lucas Mallet], letter extracted in Clifford Harrison, *Stray Records, or Personal and Professional Notes* (1892), I, 246.

50. See Henry Sidgwick to F.W. Myers, 18 November 1887, Sidgwick Papers, c100/151, Trinity College, Cambridge: 'Frank D does not want the reviewing to be entirely done from the scientific point of view'; Sidgwick clearly also approached Stephen; see Leslie Stephen to Henry Sidgwick. 19 November 1887, Sidgwick Papers, c100/104, Trinity College, Cambridge.

51. For example, the *Daily News* review, 19 November 1887, focussed almost entirely on volume 1 and the second and third volumes were dismissed in one final paragraph which managed to say more about his politics and his manner of dying than about his science. The extent of this engagement is very clear in the collection of letters Frank Darwin collected from family and others receiving presentation copies in which Darwin's scientific work is almost entirely absent, and focus is securely on his character; even Thiselton-Dyer was swept up in this perspective, telling Frank that it was 'like reading the experiences of one of the old saints', WTTD to F. Darwin, 21 November 1887, MSS.DAR.199.5/30–1, Darwin Papers, CUL. The exceptions came almost exclusively from high Victorians like Judd, who praised the book for presenting the *Origin* as the central pivot, flanked by discussions of preparation and vindication, Judd to F. Darwin, 27 November 1887, MSS.DAR 199.5/50, Darwin Papers, CUL.

52. [Alfred Newton], 'The Life of Darwin', *Times*, 19 November 1887. For Newton's authorship, see correspondence with Mowbray Morris, MS Add. 9839/1M/709–13, Newton Papers, CUL.

53. Dicey to H.E. Litchfield, tps, undated (but dated by Litchfield to F. Darwin, 27 November 1887, MSS.DAR.199.5/56, CUL), MSS.DAR.219.7/5, Darwin Papers, CUL.

54. Bonney, review of the *Further Letters of Darwin*, *Nature*, 9 April 1903; Joseph McCabe, 'The Centenary of Charles Darwin', *South Place Magazine* 14 (April 1909), 97–8; an

300 DARWINISM'S GENERATIONS

approach visible in F.W.H. Myers, 'Charles Darwin and Agnosticism', in his *Science and a Future Life* (1893), 51–75.

55. An excellent example is the fiercely evangelical Primitive Methodist Rev. John Snaith (1836–1923), whose contributions the correspondence columns of the *Nottingham Journal* in 1898 revealed an intimate knowledge of the *Life and Letters*, and some of Darwin's critics, most notably J.H. Stirling, but little sign of engagement with Darwin's main works (and also a facility for quoting Darwin's early doubts in the years immediately after the initial publication of the *Origin*, as evidence of his mature stance), *Nottingham Journal*, 28 March 1898; see also comments of Rev William Guest (1818–91), *Nonconformist*, 8 December 1887. This did not stop others, including Samuel Kinns, from concluding that by Darwin's own confession, the adoption of his theory had driven him into agnosticism; see *West Surrey Times*, 3 December 1887.

56. Text of sermon of T. Dixon Rutherford (1864–1943), *Hackney and Kingsland Gazette*, 2 January 1901.

57. See the extensive correspondence in MS.40314, Murray papers, NLS.

58. Frank Darwin to Poulton, 6 September 1891, Box 3, Poulton Papers, OMNH.

59. Frank Darwin to Bonney, 15 March 1888, Bonney Correspondence, Royal Geographical Society, London.

60. Not that this was always done heavy-handedly, or that the Darwins exercised any automatic influence; in the case of Poulton's volume Frank Darwin tried, but failed, to persuade Poulton to modify his treatment of Huxley's views on natural selection, and also failed to spot the less than emphatic account of the Huxley–Wilberforce encounter Poulton provided; see F. Darwin to Hooker, 7 January 1897, JDH/1/2/6/25, Hooker Papers, RBGK; Poulton to C. Lloyd Morgan, 5 December 1896, #85, Lloyd Morgan Collection, DM128, University of Bristol.

61. As in Patrick Geddes' proposal for a 'Primer of Evolution' in 1882.

62. This can be the only real explanation for Poulton's willingness to engage with the eccentric and perhaps mentally unstable James Philip Mansel Weale (1838–1911); see Box 11, Poulton Papers, OMNH. It was explicit in Poulton's anxiety to avoid publicity distasteful to the Darwin family even in the late 1920s; see Poulton to DWT, 8 March 1928, ms19342, D'Arcy Wentworth Thompson Papers, St Andrews.

63. Henrietta Maria Stanley to John Lubbock, 18 June [1888], ff.56–7, Avebury Papers, Add. MS 49651, BL.

64. *Burnley Express*, 2 August 1884.

65. *Sheffield Independent*, 13 March 1896. It was a view apparently shared by Edward King (1829–1910), Bishop of Lincoln, who after reading the *Life* apparently talked of Darwin as a pattern for the patience needed by the clergy; see recollections of [Frederick William] Puller (1843–1938), *Yorkshire Post*, 12 March 1910. Helen Gladstone (1849–1925) wrote of how the *Life* reinforced the simplicity she had glimpsed in her friendship with Darwin, Helen Gladstone to Nora Sidgwick, 27 April 1906, Sidgwick Papers, c.103/96, Trinity College, Cambridge.

66. A.C. Benson's *The House of Quiet: An Autobiography* (3rd ed., 1907, or. 1904), 121. This account is based on Rev. John Andrewes Reeve (1847–1911), who is quoted in similar terms in A.C. Benson, *The Trefoil, Wellington College, Lincoln and Truro* (1923), 219.

67. Farrar, 'Charles Darwin', in *Social and Present Day Questions* (1893), 306.

68. Thomas Hodgkin to Howard Lloyd, 16 December 1887 in Louise Creighton, *Life and Letters of Thomas Hodgkin* (1917), 149–50.

69. *Royal Cornwall Gazette*, 16 June 1882, which rehearsed a whole series of standard arguments against the *Origin* and *Descent*; for Whiteley, see *Quarterly Journal of the Meteorological Society* 17 (1891). In response, John Mugford Quicke (1829–1914) defended Darwin, arguing that Darwin's work had been a delight to the scientific world since the publication of his *Beagle* narrative, and evidence for his theories grew year by year; see various contributions *Royal Cornwall Gazette*, 12, 26, May, 23 June 1882, 10 December 1886.

70. See *The Churchman*, 3 September 1885.

THE DEATH OF DARWIN AND AFTER 301

71. See J.S. Blackie to Eliza Blackie, 3 May [1889], ff.47–9, MS.2638, J.S. Blackie Papers, NLS; Owen to Argyll, 16 September 1884, 1209/1631, Prof R. Flint to Duke of Argyll, 2 January 1885, 1209/1635, Argyll Papers, Inveraray Castle.

72. *Good Words* 29 (1888), 166–73, 265–70, 330–3; the articles were then of course used by other opponents; see letter of 'Anti-Darwinian', *Newcastle Chronicle*, 19 May 1888.

73. Johnston, *Life and Letters of H.P. Liddon*, 275 (diary 22 April 1882); see the discussion in Nixon, *Gerard Manley Hopkins*, 126–32.

74. Huggins to H.W. Acland, December 1883, ff.234–6, MSS Acland d.63, Bodleian.

75. Lankester's contempt for Rolleston, despite his earlier impeccable credentials as a supporter of Darwin and the *Origin*, is all too clear in his colourful correspondence; see E.R. Lankester to Robert Gunther, 8 November 1911, A Gunther Collection 20, Letters and Printed Matter relating to E.R. Lankester, NHM; Lankester to Burdon Sanderson, nd, ff.41–4, MS.20030, J. Burdon Sanderson Papers, NLS. Relations were so bad that he told Macmillan in 1878 that if they sent Rolleston a copy of Carl Gegenbauer's *Elements of Comparative* Anatomy (1878) which he had translated, this should be clearly marked as from the publisher, and not from him personally, Lankester to Macmillan, 17 November [1878], ff.19–20, Macmillan Papers, Add. MSS 55219, BL. There are suggestions that in general Rolleston was not sympathetic to the young evolutionary biologists by the 1870s; see for example Geddes to Beddard, 10 April 1878, D'Arcy Thompson to Beddard, 14 March 1914, Frank E. Beddard Papers, COLL 623, Edinburgh. Lankester had form: he had little time for H.W. Acland, either, describing him on one occasion as a pompous out of date fool, Lankester to Huxley, 18 December 1872, ff.39–42, Box 21, Huxley Papers, ICL.

76. Sir Henry Newbolt, *My World as in My Time* (1932), 116.

77. He 'pronounces me to be an infidel, and refuses his licence', complained Henry George Day (1831–1900?), Day to George Maw, 4 April 1882, GSM/GX/Mw/5/142, British Geological Survey.

78. Huggins to H.W. Acland, December 1883, ff.234–6, MSS Acland d.63, Bodleian.

79. An accusation faced by Lankester during his campaign for Edinburgh in 1881; see Lankester to J. Burdon Sanderson, 30 October [1881], ff.21–2, MS.20030, J. Burdon Sanderson Papers, NLS, and in file of letters to Isaac Bayley Balfour, Correspondence, Box L, Bayley Balfour Papers, RBGE.

80. Frank Young (*c.*1852–1941) to Geddes, 16 December 1884, f. 326, MS.10523, Geddes Papers, NLS.

81. Murray to Geddes, 30 October 1882, ff.182–3, MS.10522, Geddes Papers, NLS.

82. The committee included John Jackson (1811–85), Lord Arthur Hervey (1808–94), Henry Philpott (1807–92), and Harold Browne (1811–91), along with the younger William Maclagan (1826–1910); Browne was a long-standing anti-evolutionist; see *The Holy Bible...with an Explanatory and Critical Commentary* (1871). Other key supporters including the Earl of Shaftesbury (1801–85), Henry Pelham, Earl of Chichester (1804–86), and William Fremantle (1807–95). A number of older mid Victorians were also prominent, including Jabez Hogg (1817–99), consulting surgeon, who served as the committee's secretary, and was also one of the scientific authorities whom Kinns attempted to deploy to counter the criticism he received; see *Knowledge*, 5 September 1884. A brief account of the affair is given in J.H. Brooke and G.N. Cantor, *Reconstructing Nature: The Engagement of Science and Religion* (1998), 62.

83. *Dorset County Chronicle*, 31 January 1884, *Morning Post*, 25 February 1890; the book was into its twelfth thousand in 1890 and fourteenth thousand in 1895.

84. Kinns to G.G. Stokes, 27 July 1891, MS Add. 7656/1K/468, Stokes Papers, CUL.

85. Joseph Prestwich to Acland, 5, 17 January, 20 February 1884, ff.62–9, MSS Acland d.83, Bodleian.

86. William Huggins to H.W. Acland, 5 May 1884, ff.250–1, MSS Acland d.63, Bodleian. Looking back, Henry Woodward regretted the assiduousness with which Carruthers and his allies pursued Kinns, which allowed him to pose as a Christian martyr; it would have been better simply to ignore him, Woodward to Meldola, 15 November 1884, #1868, Meldola Papers, ICL.

302 DARWINISM'S GENERATIONS

87. C. Pritchard to Acland, 4 January 1884, ff.293–4, MSS Acland d.64, Bodleian; as Pritchard remarked later in the month to Stokes, it was the damage that would be done to religion by Kinns' 'total absolute ignorance' which concerned him, though 'I am no more of an evolutionist than you are', Pritchard to Stokes, 24 January 1884, MS Add. 7656/1P/726, Stokes Papers, CUL. Carruthers' part in the affair, which generated significant correspondence, including with Kinns himself, is collected in a large scrapbook, DF BOT/404/12, William Carruthers Papers, NHM.

88. *London Evening Standard*, 5 January 1884; see various letters in BENSON/15/1–50, Lambeth Palace Library. Bonney was greatly reassured at finding common ground, having begun to feel something of a 'theological Ishmaelite'.

89. Edward Benson to Acland, 9 January 1884, ff.48–9, MSS Acland d.83, Bodleian.

90. Kinns to Benson, 11 January 1884, ff.26–7, BENSON/15/1–50, Lambeth Palace Library; *Derbyshire Advertiser*, 24 October 1884.

91. Kinns to Frederick Temple, 28 March 1892, ff.251–2, FP Temple 21, Lambeth Palace Library.

92. H.W. Acland to W.E. Gladstone, 13, 22 December 1885, ff.141–4, 157–60, Gladstone Papers, Add. MS 44091, BL; Acland to Mrs Gladstone, 12 December 1885, ff.139–40, Gladstone Papers, Add. MS 44091, BL. This was a position widely shared by mid Victorians, such as Brownlow Maitland (c.1817–1902?), *Steps to Faith* (1880). Owen was even blunter: 'Genesis, read by a brain-bearing biped to whom the truth-finding faculty has been vouchsafed, tells him that amount of knowledge of his world possessed by the sacred Writer with the conviction of an Almighty Creative Cause', Owen to Gladstone, 7 December 1885, ff.188–9, Gladstone Papers, Add. MSS 44493, BL.

93. Royle, *Radicals, Republicans and Secularists*, citing the *Freethinker*, 8 November 1885. See Gladstone to Acland, 11 December 1885, 12 March 1895, ff.74–5, 97–8, Acland d.68, Bodleian, Oxford. For the range of the advice sought see Gladstone to Robert Scott (1811–87), 4 December 1885, Scott Papers, 1/11/39, PHL.

94. R.W. Church to E.S. Talbot, 8 December 1883, #2, Letters from R.W. Church to E.S. Talbot, 1880–90, PHL. Rather unkindly, Müller conceded that the religious portions of Gladstone's 'brain are petrified—hard as rock', Müller to Huxley, 6 November 1885, f.120, Box 23, Huxley Papers, quoted in Desmond, *Huxley: Evolution's High Priest*, 162. Gladstone apparently sent an intermediary privately to Huxley to try to deflect his attacks, which only allowed Huxley to extract the fact that Owen was largely behind Gladstone's position, which in turn presented this as merely an echo of Owen's earlier indirect attacks through Wilberforce; see Hooker to Asa Gray, 24 January 1886, ff.82–4, JDH/2/22/1, J.D. Hooker Papers, RBGK.

95. See Pollock, *For My Grandson* (1933), 99; Talbot to Poulton, fragment in Box 11, Poulton Papers, OMNH.

96. Which surely misunderstood Huxley's intent to strike at Gladstone and the religiously orthodox, Gladstone to Argyll, 31 September 1890, Argyll Papers 1209/1611, Inveraray Castle.

97. Conan Doyle, *Memories and Adventures* (1924), 32; *A Full Report of a Lecture on Spiritualism* (1919); May Kendall, 'Nirvana', in her *Dreams to Sell* (1887). This was clearly a memorable episode for undergraduates of the time; see also Frank Russell, *My Life and Adventures* (1923), 99, Snell, *Movements, Men and Myself*, 44–5. There is a similar tone in J.A. Thomson's discussion of the exchanges of Argyll and Huxley in the *TNC* in 1887; see 5 December 1887, T-GED/9/17, Geddes Papers, Strathclyde University. For James Adderley (1861–1942), in his *In Slums and Society: Reminiscences of Old Friends* (1916), Huxley's controversies with figures like Henry Wace (1836–1924) were dismissed as part of the dreary (Victorian) past (an indifference enacted in his portrait of the ineffective wife of the clerical aunt-sally Rev. David Bloose, in *Stephen Remarx* (1897), for whom 'it mattered little whether Evolution could be made to square with Christianity or Darwin').

98. See his letter to Joseph Prestwich, 20 June 1895, A.G. Prestwich, *Life and Letters of Joseph Prestwich* (1899), 387–8. Interestingly, even a mid Victorian like Edward White was still confessing in the 1890s that 'I have always felt an extreme difficulty in giving up its [i.e. Genesis's] historicity', White to Stokes, 30 September 1894, MS Add. 7656/1W/531, Stokes Papers, CUL.

THE DEATH OF DARWIN AND AFTER 303

99. Lewis Campbell (1830–1908) to Huxley, 12 August 1894, f.19, Box 12, Huxley Papers, ICL; Clodd to Richards, 14 January 1897, In-letters, Box 2, Edward Clodd Papers, BC MS 19c Clodd, Brotherton Library. For M.E. Grant Duff (1829–1906), it was 'not so much a battle as a massacre, for Gladstone had nothing but a bundle of antiquated prejudices', M.E. Grant Duff to Leonard Huxley, 4 November 1898, f.171, Box 30, Huxley Papers, ICL.

100. A.C. Hingston Quiggin, *Haddon the Head Hunter: A Short Sketch of the Life of A.C. Haddon* (1942), 79.

101. Adrian Desmond, *Evolution's High Priest* (1997), 190, quoting *Collected Essays* V, 334; Desmond gives a vivid picture of Huxley out of time, 'asserting mid-Victorian, male-governing, family values', ibid., 193.

102. Lankester to Welby, 26 October 1889 (tps), 1970–010/009(05), Welby Papers, York University. Foster was more sympathetic, but by the mid-1880s even he was addressing Huxley as 'your Sixty-ship', Foster to Huxley 4 May 1885, quoted in Desmond, *Evolution's High Priest*, 158. Augustus Desiré Waller (1856–1922) was less complimentary, later recalling what he described as 'the senile vanity of Huxley's threatening opening', Waller to K. Pearson, 15 May 1903, Pearson/11/1/22/10–19/14, Pearson Papers, UCL.

103. Meldola to Poulton, 21 February 1891, Meldola Papers, OMNH.

104. There is some coverage of this in Clark, *Bugs*, 105–31; for one sign of these connections at the very outset of the episode, see the comments of W.L. Distant on the society and the anti-Darwinian work of Pascoe, Distant to Raphael Meldola, 10 April 1891, #1954, Meldola Papers, ICL. See also the reference to disputes over an unidentified 'Anti-Darwinist' in the context of discussions about the presidency in 1887, W.A. Weber, ed., *The Valley of the Second Sons* (2004), 111.

105. Meldola to Poulton, 15 December 1892, and previous letter, 14 December 1892, and also Meldola to Poulton, 17 December 1887, Meldola Papers, OMNH. Opponents did include some younger figures, like David Sharp (1840–1922), who while not apparently adopting a straightforwardly anti-Darwinist position, did complain about the overly speculative cast of 'modern "Biologists"', Sharp to Poulton, 10 January 1893, Box 10, Poulton Papers, OMNH.

106. H.J. Elwes to Poulton, 25 December nd, Box 3, Poulton Papers, OMNH. Elwes had certainly in the 1880s shown an interest in the limits of variation which would not have endeared him to mainstream Darwinists; see *Cheltenham Examiner*, 11 October 1882.

107. J.W. Tutt to Poulton, 7 January 1893, Box 11, Poulton Papers, OMNH; Tutt's list of potential supporters was almost exclusively late Victorians and Edwardians.

108. Poulton to H. Goss, 11 December 1892, Box 14/1, Druce Papers, Royal Entomological Society; 'Election of a President', flyer dated 4 January 1893, Box 14/3, Druce Papers, Royal Entomological Society; the others on the list were Tutt, J. Jenner Weir (1822–94), Philip Crowley (1837–1900), G.A.J. Rothney (1849–1922), F.H. Hill (1852–1933), William Farren (1862–1965), and Francis A. Walker (1841–1905). For the response see the series of letters from Thomas de Grey, 6th Baron Walsingham (1843–1919) to Poulton, including 13 December 1892, in Box 12, Poulton Papers, OMNH; also the minutes of the Entomological Society, 7, 14 December 1892, Royal Entomological Society.

109. W.G. Blaikie, *David Brown, D.D., LL.D.: Professor and Principal of the Free Church College, Aberdeen: A Memoir* (1898), 182–3.

110. C. Elam, 'The Gospel of Evolution', *Contemporary Review* 37 (1880), 713–40.

111. Andrew Common (1815–96?), apparently accepting of Darwinian versions of evolution; see lecture *Sunderland Daily Echo*, 2 March 1888; and William Connor Magee (1821–91)'s acceptance of evolution in a sermon at St Mark's Peterborough in December 1885, Owen Chadwick, *The Victorian Church* (1970), II, 24.

112. R. Strachey, *Lectures on Geography* (1888).

113. Alexander Balloch Grosart (1827–99); see his letter to Samuel Alexander, 8 October 1894, ALEX/A/1/1/106/2, Samuel Alexander Papers, MJRL; William Bonner Hopkins (1823–90), sermon at Great St Mary's, *The University Pulpit: Supplement to the Cambridge Review*, 3 May 1882; Felicia Mary Frances Skene (1821–99) to H.W. Acland, 28 April 1896, ff.86–7, MS Acland, d.76, Bodleian.

114. James Backhouse to Gray, 24 October 1882, 4 January 1883, Asa Gray Correspondence, B–Ba, Harvard.

304   DARWINISM'S GENERATIONS

115. *Northern Whig*, 15 September 1902. This echoed Ruskin's stance, as reported by Archibald Stodart Walker (1869–1934), that 'no-one should read Darwin's *Origin of Species*, for the simple reason that it is man's duty to know what he is, not what he was', Walker, 'Some Celebrities I Have Known', *Chambers's Journal* ser. 6, 12 (1908–9), 47.

116. See Catherine Marshall, Bernard Lightman, Richard England, eds., *The Metaphysical Society (1869–1880): Intellectual Life in Mid-Victorian England* (2019).

117. Martineau to W. Knight, 25 April 1896, ff.198–9, Carpenter MS 6, Harris Manchester, Oxford.

118. Wilfrid Ward to A.J. Balfour, 27 March 1896, ms38347/VI/2/13, Balfour to Ward, 8 April 1896, ms38347/VII/19/10, Ward Papers, University of St Andrews; for subsequent discussions of this theme see Ward, 'Mr Balfour's Gifford Lectures', *Edinburgh Review* 223 (1916), 59–82. There were others, for example, George Tyrrell in 1899, Ward to Henry Sidgwick, 29 May 1899, ms38347/VI/30/33, Ward Papers, University of St Andrews.

119. Talbot was by birth a young high Victorian, but as a member of the *Lux Mundi* party (see below) was strongly aligned intellectually and personally with the late Victorians. The society's chairmen were equally high Victorian in character (Talbot, A.C. Lyall, R.B. Haldane, Oliver Lodge, and Balfour), with Henry Sidgwick as the only exception.

120. A.J. Balfour to Wilfrid Ward, 8 April 1896, ms38347/VII/19/10, Ward Papers, St Andrews; A.C. Lyall to Haldane, 9 September 1899, ff.203–5, Haldane Papers MS.5904, NLS; in like manner James Bryce commented that it was interesting 'observing how many different planes men's minds may move in when trying to discuss the same problem', Bryce to James, 22 April 1897, Additional Correspondence, b MS 1092, William James Papers, Harvard. For the broader history, see W.C. Lubenow, 'Intimacy, Imagination, and the Inner Dialectics of Knowledge Communities: The Synthetic Society, 1896–1908', in M. Daunton, ed., *The Organization of Knowledge in Victorian Britain* (2005), 357–70.

121. Lankester had played truant as a schoolboy to hear Huxley and Tyndall lecture, J. Lester, *E. Ray Lankester and the Making of Modern British Biology* (1995), 13. Huxley always offered 'fatherly advice and assistance', *Dundee Courier*, 8 May 1882.

122. Including also Sydney Vines (1849–1934), George Buckston Browne (1850–1945), Frederick Orpen Bower (1855–1948), and Harry Marshall Ward (1854–1906), D.E. Allen, 'The Biological Societies of London, 1870–1914: Their Inter-Relationships and Responses to Change', *Linnean* 14 (1988), 23–38. 'Conwy Lloyd Morgan', in C.E. Murchison, *History of Psychology in Autobiography* (1930), II, 237–64, recalled that for him Huxley's *Lay Sermons* (1870), led to the *Origin* and *Descent*.

123. See quote from H.R. Haweis, *Current Coin* (1876), 11.

124. *National Secular Society Almanack for 1884*, cited in C.K. Krantz, 'The British Secularist Movement: A Study in Militant Dissent', unpublished PhD, University of Rochester (1964), 67.

125. *Nonconformist*, 1 July 1886.

126. See the debate over the paper of Thomas Karr Callard (1822–89), 'Breaks in the Continuity of Mammalian Life in Certain Geological Periods Fatal to the Darwinian Theory of Evolution', *JTVI* 16 (1882–3), 170–200. For an account of the Institute's anti-Darwinian tone in the 1890s, see *Academy*, 13 April 1895.

127. Recollections of James Trengrove Nance (1852–1942?), *Market Harborough Advertiser*, 10 May 1898.

128. See A.E. Shipley, *Cambridge Cameos* (1924), 162–3; T.G. Bonney to Haddon, 10 December 1895, Envelope 23/1/4, Haddon Papers, CUL, on losing students like Haddon from geology to biology; *Cambridge University Natural Science Club 1872–1903*, copy at ms45513, D'Arcy Wentworth Thompson Papers, St Andrews. The Oxford Club heard from Michael Foster in 1883 urging them 'to follow Darwin's method, even upon Darwin's works', *Oxford Journal*, 9 June 1883. At Edinburgh from 1891 a University Darwinian Society offered papers on evolutionary topics; see the syllabus for 1891–2 in Correspondence, Box Da–Dy, Isaac Bayley Balfour Papers, RBGE.

129. Gadow to Newton, 5 April 1883, 19 September 1884, MS Add. 9839/1G/7,11, Newton Papers, CUL. See Lankester's account of the transformation of natural sciences he and

THE DEATH OF DARWIN AND AFTER   305

Lawson effected in the mid-1870s, in the face of the opposition of Rolleston, Lankester to Robert Gunther, 8 November 1911, A Gunther Collection 20, Letters and Printed Matter relating to E.R. Lankester, NHM.

130. Shipley, *Cambridge Cameos*, 163–4.
131. Peter F. Clarke, *Liberals and Social Democrats* (1978), 12.
132. Cited in Cock and Forsdyke, *Treasure Your Expectations*, 12.
133. Karl Pearson, 'Old Tripos Days at Cambridge', *Mathematical Gazette* 20 (1936), 32.
134. Constance Louisa Maynard, Autobiographical memoir, GCPP Stephen 3, Girton College, Cambridge; Philip E. Smith and Michael S. Helfand, *Oscar Wilde's Oxford Notebooks* (1989). See also G.S. Stephenson (1848/9–1929), *Reminiscences of a Student's Life at Edinburgh in the Seventies* (1918), 41. Lewis R. Farnell, *An Oxonian Looks Back* (1934) notes there was 'scarcely any orthodoxy of belief', but very little militant atheism, except among the 'Natural Scientists', 57–8.
135. See recollections of the botanist George Claridge Druce (1850–1932), *Northampton Mercury*, 22 March 1918, and Druce, 'The Formation of the Northamptonshire Natural History Society', *Journal of the Northamptonshire Natural History Society* 19 (1918), 131–42. The core group included Druce, science and art master Beeby Thompson (1848–1931), S.J.W. Sanders (1846–1915?), headmaster of the Northampton Grammar School, and R.G. Scriven (1845–1938?), land agent, and subsequently drew in figures including Walter Drawbridge Crick (1857–1903), who collaborated with Darwin in studying freshwater mollusca; these late Victorians were joined by various high Victorians with pro-Darwinian views, including Thomas Littleton Powys, Lord Lifford (1833–96), and the Rev. Henry Crosskey. In response to his address 'The religious side of evolution', Canon Robert Bevan Hull (1844–1900) was accused by one local Methodist minister of replacing the Gospel of the Word with the 'Gospel of Worms', and of 'filling his coach with a company of scientists leaving the place of honour to Charles Darwin', rather than sticking to the 'old coach, with its driver Jesus Christ', *Northampton Mercury*, 28 January, 11 April 1888.
136. The prime movers were William Maxwell Ogilvie (1851–1929), first president, James B. Orr (1854/5–1931), science teacher and long-time secretary/curator, William Neish Walker (1849–1927), known as a Darwinian enthusiast (*Dundee Advertiser*, 10 June 1884), Frank Young (c.1852–1941), who became science master at Dundee High School, with high-Victorian allies including James Durham (1840–1905), W.B. Simpson (1842–1910), and Dr James Brebner (1839–1933), teacher at the Dundee Institution, whose alpine activities had brought him into contact with Tyndall. See F.W. Young, *Coming of Age of the Dundee Naturalists' Society* (1895), which makes clear the ardent Darwinism of many of the initial promoters, and for the early Darwinian debates, *Dundee Courier*, 12 March 1930.
137. Examples include Rev. Percy Watkins Fenton Myles (1849–91) to the Ealing Microscopical and Natural History Society, *Middlesex County Times*, 26 May 1888; Henry Worsley Seymour Worsley-Bennison (1845–1918), *Charles Darwin: Highbury Microscopical & Scientific Society: President's Address for the Year 1885* (1885).
138. R.A. Baker, 'In Search of a Naturalist: Natural History in Warrington from 1870 to 1910 and the Records and Collections of Linnaeus Greening (1855–1927)', *Archives of Natural History* 34.2 (2007), 235–43.
139. See, for example, *Bristol Mercury*, 10 February 1888, *Western Daily Press*, 24 April 1885.
140. Havelock Ellis to Clodd, 3 letters, In-letters, Box 2, Edward Clodd Papers, BC MS 19c Clodd, Brotherton Library.
141. These included the 1893 translation of Weismann's *Germ Plasm: A Theory of Heredity*, as well as Lloyd Morgan's *An Introduction to Comparative Psychology* (1894) and Ellis' own *Man and Woman* (1894). See Phyllis Grosskurth, *Havelock Ellis: A Biography* (1981), 114–17.
142. Copy of letter to J.A. Thomson [presumably from Patrick Geddes and Colleagues], 18 August 1896, f.72, MS.10588, Geddes Papers, NLS.
143. *Gardeners' Chronicle*, 21 December 1889; G.T. Bettany to Hooker, 4 March 1891, JDH/1/2/2/268, Hooker Papers, RBGK. Bettany's introduction focused entirely on the

306 DARWINISM'S GENERATIONS

Journals as the foundational text of Darwin's evolutionary theories, directing readers to read it through this lens.

144. F.J. Gould, *The Life Story of a Humanist* (1923), 27, 50, 51; Bland-Sutton, *Story of a Surgeon*, 30; for Bland-Sutton's Darwinism, see his *Evolution and Disease* (1890). For an example of *Descent* as a presentation volume, see *Falkirk Herald*, 3 July 1895.

145. See the works of Henry Neville Hutchinson (1856–1927) and Edward Step (1855–1931). For Hutchinson's shifting position, which involved a gradual dilution of his commitment to natural selection, and increasing incorporation of Lamarckian elements, see Lightman, *Popularizers*, 450–60. Typical examples of Step's work include *Plant-Life: Popular Papers on the Phenomena of Botany* (1891), and *The Romance of Wild Flowers: A Companion to the British Flora* (1899).

146. *South Wales Echo*, 4 January 1898.

147. Described by Baldwin as 'the Changing of Britain's Scientific Guard, 1872–1895', involving, 'the younger generation of men born in the 1840s and 1850s'; Baldwin, *Making Nature*, 48–73, 49.

148. William A. Herdman (1858–1924), 'An Ideal Natural History Museum', *Proceedings of the Liverpool Literary and Philosophical Society* 41 (1886–7), 61–80. Even fifteen years later it was possible for a well-connected museum curator to suggest that although it was by then generally accepted that museum displays should be organised on evolutionary principles, there was not yet a single museum in the country systematically organised along such lines; see Henry Ogg Forbes (1851–1932), Director of Liverpool's Free Public Museums, quoted in Tony Bennett, *Pasts beyond Memory* (2004), 72–3. Forbes was a field naturalist in his own right, author of *A Naturalist's Wanderings in the Easter Archipelago* (1885).

149. *Morning Post*, 25 September 1891. For the history of the NHM and some of the wider practice, see A. Macgregor, 'Exhibiting Evolutionism: Darwinism and Pseudo-Darwinism in Museum Practice after 1859', *Journal of the History of Collections* 21.1 (2009), 77–94, especially 89–90, and W.T. Stearn, *The Natural History Museum at South Kensington* (1981).

150. According to Lord Sudeley (1840–1922), letter to the *Times*, 10 October 1910, quoted in Stearn, *Natural History Museum*, 100–1.

151. The work of Richard Lyddeker (1849–1915), *Evening Mail*, 14 March 1904.

152. See David van Keuren, 'Museums and Ideology: Augustus Pitt-Rivers, Anthropological Museums, and Social Change in Later Victorian Britain', *Victorian Studies* 28 (1984), 171–89, and Sadiah Qureshi, 'Dramas of Development: Exhibitions and Evolution in Victorian Britain', in Lightman and Zon, *Evolution and British Culture*, 261–85. From 1893 the collections were curated by Henry Balfour (1863–1939), who continued to champion the Pitt-Rivers approach until his retirement.

153. See Haddon in *Folklore* 18 (1907), 229–30.

154. *Nature*, 15 October 1885.

155. Including the Dublin Natural History Museum, the Manchester University Museum, the Liverpool Museum, the Dundee Museum, and in 1906 the newly opened Cuming Collection at the Southwark Museums and Libraries. See the discussion in Alexander Scott, 'The "Missing Link between Science and Showbusiness: Exhibiting Gorillas and Chimpanzees in Victorian Liverpool', *Journal of Victorian Culture* 25 (2020), 1–20; J. Adelman, 'Evolution on Display: Promoting Irish Natural History and Darwinism at the Dublin Science and Art Museum', *BJHS* 38 (2005), 411–36; *Morning Post*, 28 November 1906, *PMG*, 28 November 1913.

156. See the penny guide, *Handbook to the Collection Arranged as an Introduction to Animal Life* (1904); in 1908 the Horniman Museum was busy developing exhibits to show the zoological affinities of man and higher apes, Bennett, *Pasts beyond Memory*, 76. In respect of lectures, the Horniman again led the way; see the flyers for a course of free lectures on evolution in 1912 at LCC/PUB/11/01/173, LMA.

157. These included not just Haddon and Ogg (see note 148 above), but also Elijah Howarth (1853–1938), first editor of the *Museum's Journal*, at Sheffield, Henry Maurice Platnauer (1857–1939) (with Howarth the instigator of the Museums Association in 1888) at the Yorkshire Museum (York), Butler Wood (1854–1934) at Bradford, W.E. Hoyle

THE DEATH OF DARWIN AND AFTER 307

(1855–1926) at the Manchester University Museum, and Robert Francis Scharf (1858–1934) and George Herbert Carpenter (1865–1939) at Dublin. Possibly also Seth Lister Mosley (1848–1929) at Huddersfield Technical School and Keighley, an early member of the Huddersfield Naturalists' Society. At Dundee, a leading role was taken by Alexander Meek (1865–1949). Many of these examples are taken from Sam Alberti's ground-breaking study of late-Victorian museum developments in Yorkshire ('Museums and Natural History', in his 'Field, Lab and Museum: The Practice and Place of Life Science in Yorkshire, 1870–1914', unpublished PhD thesis, University of Sheffield (2000), 93–133), and detailed research into other regions would no doubt increase the number. For some reflections on the generational dynamics of this group see Kate Hill, '"A Rather Undefined Social Position and Public Recognition": Professionalisation, Status and Masculinity in Provincial Museums, c.1870–1930', *Gender and History* 33 (2021), 448–69.

158. See F.A. Bather, 'How May Museums Best Retard the Advance of Science?', *Report of the Proceedings of the Museums Association* 7 (1896), 92–105.

159. '"Colin Clout" at Home', *PMG*, 4 November 1889; see also Allen to Thiselton-Dyer, [April 1883], f.54, W.T. Thiselton-Dyer Papers, In-letters I, RBGK. For Allen see his essays 'Spencer and Darwin', *FR* 67 (February 1897), 251–62, Peter Morton, '*The Busiest Man in England': Grant Allen and the Writing Trade in Britain, 1875–1900* (2005), William Greenslade and Terence Rodgers, *Grant Allen: Literature and Cultural Politics at the Fin de Siècle* (2005), especially Heather Atchison's essay 'Grant Allen, Spencer and Darwin', 55–64, and Lightman, *Victorian Popularizers*, 266–89.

160. Allen to Kidd, 26 February [nd], MS Add. 8069/A19, Kidd Papers, CUL.

161. Allen to William James, 18 March [1881], I.K. Skrupkelis and E. Berkeley, eds., *Correspondence of William James*, Vol. 5: *(1878–84)* (1997), 155–6; Allen to Havelock Ellis, 16 October [nd], Havelock Ellis Papers, Yale.

162. Though not so disappointing as to dissuade Chatto and Windus from subsequently publishing a companion volume of Allen's essays to the *PMG*. By 20 June 1881 *The Evolutionist at Large* had sold only 416 copies; see letter from Chatto and Windus to Allen, in the Chatto and Windus Archive, University of Reading, partly transcribed at https://sites.google.com/site/petermortonswebsite/home/grant–allen–homepage/primary-sources [accessed 24 November 2019]. Allen's study *Force and Energy* (1888) also struggled and was eventually pulped, to Allen's great annoyance, Allen to Edward Clodd, 17 March [n.y.], Transcripts Box 1, Edward Clodd Papers, BC MS 19c Clodd, Brotherton Library.

163. See the comprehensive list at Peter Morton's Grant Allen website, https://sites.google.com/site/petermortonswebsite/home/grant-allen-homepage/ [accessed 24 November 2019].

164. Lecture Notes: Evolution of Plants, April–June 1890, T-GED/18/1/77/1, Geddes Papers, Strathclyde University.

165. A description applied to himself by Louis Robinson (1857–1928), 'Evolution and the Amateur Naturalist', *Blackwood's Edinburgh Magazine* 161 (1897), 567.

166. Stephen Paget, *I Have Reason to Believe* (1921), 50. It was said of the Oxford historian Frederick York Powell (1850–1904) that 'above all, the vaguer applications of Darwinism were one source of his creed…he looked more and more on life and history from that point of view', Oliver Elton, *Frederick York Powell: A Life and Selection from His Letters and Occasional Writings* (1906), I, 406; see also the confession of the architect William Carr Crofts (1846–94) that ever since he had been able to form his own opinion, he revered Darwin and had 'enthusiastic faith in the great truths' he made plain, Crofts to F. Darwin, 26 November 1887, MSS.DAR.199.5/12, Darwin Papers, CUL. For Joseph Jacobs (1854–1916), Australian-born folklorist and social scientist, 'there was a confident feeling, among those of us who came to our intellectual majority in those years, that Darwinism was to solve all the problems', Joseph Jacobs, *Literary Studies* (1895), xix.

167. [J.F. Nesbit], 'Our Handbook', *The Referee*, 12 August 1894.

168. Okey, *Basketful of Memories*, 43. 'The fact of evolution I think we took for granted', recalled Sollas, 'at any rate we were abundantly supplied by the evidence on which it rests', Sollas, 'The Master', Supplement, *Nature*, 9 May 1925. Albert Fleming (1846–1923),

308   DARWINISM'S GENERATIONS

lawyer and friend of Ruskin, remarked to Tyndall that 'one might as well be stormfully eloquent against gravitation as against evolution', 5 May 1881, *Letters of John Tyndall*, Vol. 16 (forthcoming).

169. As in the case of William Cecil Marshall (1849–1921); see Frances Partridge, *Love in Bloomsbury: Memories* (1981), 15–16.

170. Quoted in James G. Nelson, *Sir William Watson* (1966), 27, 135–6.

171. Eliza Margaret Humphreys (1850–1938), *Recollections of a Literary Life* (1936), 181–2.

172. Hingston Quiggin, *Head-Hunter*, 12–18.

173. *The Seen and the Unseen, from the Religious Writings of William Robertson Nicoll* (1926), 29. For Wilde, Mill as 'a man who knew nothing of Plato and Darwin gives me very little'. 'But Darwinism has of course shattered many reputations besides his', Oscar Wilde to W.L. Courtney, [? January 1889], published in Merlin Holland and Rupert Hart-Davis, *The Complete Letters of Oscar Wilde* (2000), 387–8.

174. Allen, 'Obituary', *Academy*, 20 April 1882. As Alexander Graham Bell (1847–1922), recently emigrated to Boston via Canada, confessed to his parents in 1873, in the light of geological discoveries, 'I cannot understand the prejudice with which many people view an honest and hard-working investigator like Darwin', Bell to Alexander Melville Bell, Eliza Symonds Bell, Carrie Bell, 27 January 1873, Bell Papers, Library of Congress.

175. Lyttelton, *Memories and Hopes* (1925), 272–3.

176. Constance E. Plumptre, 'On the Neglected Centenary of Harriet Martineau', *Westminster Review* 158 (1902), 669–75.

177. Shaw, 'Socialism and Medicine' (1909), in *The Doctor's Dilemma* (1911), cited in S. Hynes, *Edwardian Frame of Mind* (1975), 132. For Havelock Ellis, being born in the same year as the publication of the *Origin*, 'one of the greatest dates in the whole history of science', was a particular point of pride, Ellis, *My Life* (1939), 41.

178. Although, if Gissing's more downbeat reading of the barrenness of Zillah Denyer's engagement with Darwinism in *The Emancipated* (1890) is anything to go by, there was considerable irony in this description. Compare with his own reflection, 'Spent evening in troubled state of mind, occasionally glancing at Darwin's "Origin of Species"—a queer jumble of thoughts', Paul R. Mathiesen, et al., *The Collected Letters of George Gissing*, Vol. 4:*1889–91* (1993), 139.

179. Another high Victorian from a clerical background, E. Kay Robinson, who read the *Origin* and *Descent* at twenty-one, recalled that although previously a greedy reader, it was only with these books 'that I found any light at all'; see his *The Religion of Nature* (1906), 176, 179.

180. Frank Harris (1855–1931), *My Life and Loves* (1927), 44; Havelock Ellis, *The Dance of Life* (1923), 198–9; W. Leaf, *Walter Leaf, 1852–1927: Some Chapters of Autobiography* (1932), 106–8; Margaret Wynne Nevinson, *Life's Fitful Fever: A Volume of Memories* (1926), 56.

181. William Scott Palmer [i.e. Mary Emily Dowson (1848–1941)], 'An Agnostic's Progress: I', *Contemporary Review* 89 (1906), 21–40.

182. See J.W. Wallace, 'Memoir', in J.B. Glasier, *On the Road to Liberty* (1921), xi.

183. Poulton recalled that the neurologist Sir Charles Sherrington (1857–1922) was in 1873 persuaded by his mother to take the *Origin* with him on his summer holidays, with the words 'it sets the door of the universe ajar!', *Belfast Telegraph*, 2 September 1937.

184. D.H. Scott to A.C. Seward 12 May 1909, #195, A.C. Seward, 'Correspondence re Darwin and Modern Science 1909', MS Add. 7733, CUL; see also D.H. Scott, *Evolution of Plants* (1911), 9.

185. E.R. Pease, 'Autobiography' (tps), Margaret Cole Papers, H/7/1/6, Nuffield College, Oxford.

186. E. R. Pease, *The History of the Fabian Society* (1916), 18.

187. Conan Doyle, *Memories and Adventures*, 32.

188. Kenealy is interesting in that her fictional Darwinism seems to drift over the years towards the sort of post-Darwinian evolutionary sensibility of many Edwardians, summed up by characters like Alma Wenlith in *The Whips of Time* (1909).

THE DEATH OF DARWIN AND AFTER   309

189. The boorish male anti-Darwinian also appears fleetingly in Clifford's *Mrs Keith's Crime* (1885), and with more nuance in the title character of her *Sir George's Objection* (1910), whose half-understanding and untempered commitment to the survival of the fittest and the 'stamping out' of disease is central to the novel's plot.

190. See notice, *Leicester Chronicle*, 9 January 1892.

191. *Glasgow Herald*, 10 January 1883.

192. Francis Ysidro Edgeworth (1845–1926), *Mathematical Physics* (1881), 132; compare William Herdman's opening address to Zoology section BA, *Nature*, 19 September 1895; likewise R.B. Haldane (1850–1928)'s telling rhetoric in his 'Science and Religion' to the Haddington Literary Society (1899), published in his *Education and Empire* (1902): 'Science…has attacked theology with vast armies at every point of the field. The latter has been driven from her outposts, and has abandoned much of what used to be considered essential ground of defence', 164–5.

193. Isaac Bayley Balfour, *Some Botanical Problems: An Address Delivered at the Opening of the Second Session of the Darwinian Society of the University of Edinburgh, on Tuesday 1st November 1892* (1893). See also the case of the working-class botanist Thomas Whitelegge (1850–1927) to Frank Darwin, 22 June 1882, MSS.DAR.198/216, Darwin Papers, CUL: 'it is from reading your father's works that I was induced to take up the study of Science'.

194. For example, the squabbles between the explorer and plant collector Frederick William Thomas Burbidge (1847–1905) and William Thiselton-Dyer, rehearsed at length in Burbidge to William Bateson, 10 June 1895, B26, [14], Bateson Correspondence, Copy Collection, Vol 2, 97–201, JIC.

195. See James Cossar Ewart, inaugural address, Chair in Natural History, University of Edinburgh, *Dundee Courier*, 9 May 1882; also James Alexander Lindsay (1856–1931), 'Darwinism and Medicine, Bradshaw Lecture, 1909', *Lancet*, 6 November 1909.

196. J.W. Judd to J.D. Dana, 20 December 1891, Dana Family Papers, Box 2, MS164, Yale; Alexander Macalister similarly remarked in 1903 that the 'new generation' could have no conception of the hostility of the response to the *Origin* while 'we of the older generation remember the fierce onslaughts', *The Bookman* 16 (1903), 186.

197. H.H. Johnston, *Story of My Life* (1923), 44; Chalmers Mitchell, *My Fill of Days*, 113. Hence the recollection of Walter Garstang (1868–1949) of his undergraduate days at Oxford in the mid-1880s, the 'heyday of speculation as to the evolutionary origins of different groups of animals', 'Theory of Recapitulation', *Journal of Linnean Society* 35 (1922), 81–101.

198. Cock and Forsdyke, *Treasure Your Expectations*, 35.

199. For Balfour see Brian K. Hall, 'Francis Maitland Balfour (1851–1882): A Founder of Evolutionary Embryology', *Journal of Experimental Zoology* 299B(1) (2003), 3–8; Helen J. Blackman, 'The Natural Sciences and the Development of Animal Morphology in Late-Victorian Cambridge', *Journal of the History of Biology* 40.1 (2007), 71–108.

200. Agnes Crane to A. Hyatt, 1 December 1893, Box 2, Folder 60, Hyatt and Mayer Collection, PUL.

201. W.H. Mallock, 'Religion and Science at the Dawn of the Twentieth Century', *FR* 70 (1901), 812–31, 826.

202. Sharp to Newton, 14 November 1892, MS Add. 9839/1S/1082 Newton Papers, CUL. (In the mid-1860s Sharp apparently accepted evolution, but rejected natural selection; see the letter of Wallace in *Athenaeum*, 1 December 1866.)

203. This is clear in the marginalia in F.W. Mott (1853–1926)'s personal copy of *Descent*, via Internet Archive.

204. F.H. Bradley to C. Lloyd Morgan, 16 February 1895, cited in Robert J. Richards, *Darwin and the Emergence of Evolutionary Theories of Mind and Behaviour* (1987), 105. In a similar vein, Wallas recalled during his time at Oxford, 'fresh from reading Darwin' asking T.H. Green whether his ideas applied to dogs, and Green responding that he was not interested in dogs, Clarke, *Liberals and Social Democrats*, 14, citing *New Statesman*, 25 April 1931.

310   DARWINISM'S GENERATIONS

205. *Mental Evolution in Man* (1888); the argument of Romanes was simply that there was enough evidence that in animals there is the germ of what is visible in humans, to enable the question of why only humankind has developed abstract thought and language to be set aside; see his 'Man and Brute', *North American Review* (1884), republished in C.L. Morgan, ed., *Essays* (1897), 59–74, and his 'Origin of Human Faculty', *Essays*, 86–112. For Morgan, see Murchison, *History of Psychology*, 248–9.

206. Radick, *Simian Tongue*, 71.

207. This was certainly the case for Annie Besant; see her 'The Genesis of Conscience', *Our Corner* 9 (1887), 25–7.

208. Ritchie, *Darwinism and Politics*, 2. For Grant Allen, despite Darwin's reluctance, 'there *must* inevitably be a gospel according to Darwin lying perdu somewhere', 'The Gospel According to Darwin—1', *PMG*, 5 January 1888.

209. Her *Why I Am a Socialist* explained it was 'because I am a believer in Evolution'; Besant was one of the teachers in Huxley's summer class in 1882; by 1884 she was in touch with Geddes and lecturing on the Evolution of Society, Annie Besant to Geddes, 13 October 1884, ff.201–2, MS.10523, Geddes Papers, NLS. For the 'high priest' comment see Karl Pearson, *The Ethics of Freethought* (1888), 7.

210. W.R. Hughes, *Constance Naden: A Memoir* (1890), 46–7. Her 'Evolutionary Ethics' in her *Induction and Deduction* (1890) offers a fully Spencerian account of the inheritance of racial mentalities. See Clare Stainthorp, *Constance Naden: Scientist, Philosopher, Poet* (2019). Another instance is Arabella Kenealy, 'The Talent of Motherhood', *National Review* 16 (1890–1), 446–59, and 'The Physical Conscience', *National Review* 17 (1891), 477–93. Also Lady Florence Douglas Dixie (1857–1905), whose epic poem *Isola* (1903, but written in 1877) aligns evolutionary language with general ideas of progress: Isola is identified explicitly as an 'Evolutionist'; Vergli, the illegitimate son of her estranged husband, is the leader of the 'Evolutionist Party'; see Gates, *Kindred Nature*, 157–62.

211. See Anne Ashley, *William James Ashley: A Life* (1932), 39, 36, 109, 108, and *passim*.

212. Robinson, *Religion of Nature*, 180.

213. Cunningham, 'The New Darwinism', *Westminster Review* 136 (1891), 14–28, 14. For Cunningham, see Obituary, *Nature*, 6 July 1935, Bowler, *Eclipse*, 89–91.

214. For example, Charles Lapworth's reference to the 'crude theorizing of young specialists, all aflame with the newest ideas of the latest /Zoological\ schools', Lapworth (1842–1920) to Gunther, 23 January 1882, 6/53, A. Gunther Collection, NHM.

215. E.R. Lankester, *The Advancement of Science* (1890), 381–2, quoted in Glick, *What about Darwin*, 237. E.R. Lankester, *Degeneration: A Chapter in Darwinism* (1880), 10–11. For Lankester's identification with '"pure" Darwinism', see his review of Wallace's *Darwinism*, *Nature*, 10 October 1889 (which references his lectures 'four years ago at the London Institution' in which he was brought 'to discard even that tincture of Lamarckism which Darwin had admitted and to advocate "pure Darwinism"'); or the freethinker William Platt Ball (1844–1917), whose *Are the Effects of Use and Disuse Inherited* (1890) represented his identity as 'a thorough-going Darwinian' who thought 'it unnecessary to retain any remnants of the Lamarckian hypothesis', Ball to T.H. Huxley, 24 October 1890, f.217, Box 10, Huxley Papers, ICL.

216. Lankester to Poulton, 29 November, nd, Box 6, Poulton Papers, OMNH; Lankester to Victoria, Lady Welby, 22 October 1888, 1970–010/009(05), Lady Victoria Welby Papers, York University, Canada.

217. Lankester to Patrick Geddes, 13 November 1882, T-GED/9/10, Geddes Papers, Strathclyde University.

218. For Poulton and Dixey see Richard England, 'Natural Selection, Teleology and the Logos', *Osiris* 16 (2001), 270–87. As such Poulton is to be found locking horns (fairly intemperately) with Romanes; see Romanes letter, *Nature*, 16 August 1888. A good example is his opening address to the Zoology Section, BA 1896, *Nature*, 24 September 1896. For Meldola, see extracts of his address to the Entomological Society, *Nature*, 13 February 1896.

219. For example, James Allen Harker (1847–94), Professor at the Royal Agricultural College, Cirencester; letters to S.S. Buckman, GSM/GX/BM/1/1 British Geological Survey.

220. WTTD to Huxley, 24 December 1894, f.232, Box 27, Huxley Papers, ICL.

THE DEATH OF DARWIN AND AFTER 311

221. *Natural Science* 5 (1894), 223.
222. Lester, *Lankester*, 38.
223. Lankester reviewed the proofs of *Essays*; see Poulton correspondence, OMNH. Weismann's *The Germ Plasm* (English translation 1893) was translated by another of this late-Victorian group, William Newton Parker (1857–1923). Other advocates of Weismann include Harry Croft Hiller (1846–1934), *Against Dogma and Free Will and for Weismannism* (1893); Bertram Coghill Alan Windle (1858–1929), *London Medical Recorder*, 20 August 1889; Albert Wilson (1854–1928), *Unfinished Man* (1910).
224. For a flavour see Lankester in *Nature*, 6, 27 March 1890; F.C.S. Schiller, review of Ritchie's *Darwin and Hegel*, *The Philosophical Review* 2 (1893), 585.
225. Julian Huxley, *Evolution: The Modern Synthesis* (1943), 23, quoted in Glick, *What about Darwin*.
226. William Graham (1839–1911) in his *The Creed of Science* (1881); see B. Lightman, 'The Creed of Science and Its Critics', in M. Hewitt, ed., *The Victorian World* (2012), 449–65. See, for example, A.W. Benn, 'Higher Criticism and the Supernatural' (1895), in Benn, *Revaluations, Historical and Ideal* (1909), 126–50; or J.H. Japp's *Darwin and Darwinism* (1899).
227. Galton to Poulton, 21 September 1889, Box 3, Poulton Papers, OMNH. Thisleton-Dyer recognised its importance, but still tended to steer clear himself and recommend others to do the same; see WTTD to Charles Baron Clarke (1832–1906), 15 November 1894, f.7, A Gunther Collection 16, NHM.
228. See *Aberdeen Journal*, 12 September 1889. (Flower was 'weak', was Meldola's unsurprising verdict, Meldola to Poulton, 17 September 1889, Meldola Papers, OMNH); likewise the opposition of Sir Edward Fry, Fry, *Memoir of Rt Hon Sir Edward Fry*, 96–7.
229. Michael Foster to T.H. Huxley, 20 June 1894, f.379, Box 4, Huxley Papers, ICL; Alfred W. Bennett, 'Mimicry in Plants', *Popular Science Review* 11 (1872), 1–11.
230. Stephen, 'The Ascendancy of the Future', *TNC* 51 (1902), 795–811, 795. One exception was Edward Clodd, who was happy to embrace Weismannism as disproving Lamarck, 'Darwinism up to Date: A Reply', *T.P.'s Weekly*, 15 April 1910. For the mid Victorians it was perhaps more simply an irrelevance; H.B. Tristram in his opening address to the Biology section of the BA in 1893 fell back on the common refuge of incomprehension; see *Nature*, 21 September 1893; likewise John Clavell Mansel-Pleydell, presidential address. *Proceedings of the Dorset Natural History and Antiquarian Field Club* 12 (1891), in which he just throws up his hands at Weismann's work and claims that this will always remain 'among the hidden arena of Nature's mysteries', 15. Even Hooker remained bemused, telling Huxley in 1890 that 'Darwinism is all a dream to me now. Please enlighten me', Hooker to Huxley, 3 November 1890, f.365, Box 3, Huxley Papers, ICL.
231. Wright, 'The New Dogmatism', 192–213.
232. For one study along these lines Karl Jordan (1861–1959) in his 'On Mechanical Selection', *Novitates Zoologicae* (1896), quoted in Johnson, *Karl Jordan*, 96. For one attempt to delineate the various factions of the 1880s see A.G. Tansley on 'Neo-Darwinianism', *Hardwicke's Science Gossip* 27 (1891), 207–10.
233. Romanes reiterated his claim that while the Weismann school might be called 'neo-Darwinian; 'pure Darwinian it certainly is not', *Nature*, 30 August 1888. For Romanes, see J. David Pleins, *In Praise of Darwin* and Pleins, *The Evolving God: Charles Darwin on the Naturalness of Religion* (2013).
234. Pleins, *In Praise of Darwin*, Hale, *Political Descent*.
235. Letter, *Times* 16 September 1889; Romanes to Poulton, 2, 11 November 1889, Box 9, Poulton Papers, OMNH; extract headed 'The Darwinian Cult', printed in many provincial papers, see for example *Rugby Advertiser*, 26 February 1890. See also Romanes' series on 'Post Darwinian Questions', the Rosebery Lectures at Edinburgh in the winter of 1890–1.
236. For Beddard, see his *Animal Coloration* (1892) (see Poulton's long hostile review in *Nature*, 6 October 1982), and his review of the second edition of Haeckel's *History of Creation*, *Saturday Review*, 18 February 1893, Beddard Scrapbook, Box 4/7, Evolution Collection, Mss 28, UC-Santa Barbara.

312 DARWINISM'S GENERATIONS

237. See Scott, *Evolution of Plants*.
238. For Headley see his *Problems of Evolution* (1900), and *Darwinism and Modern Socialism* (1909); for his doubts, see Headley to James Cossar Ewart, 3 October [1903?], COLL 14/9/9/108, James Cossar Ewart Papers, Edinburgh.
239. 'The Darwinian Theory of Instinct' *TNC* 16 (1884), reprinted in Romanes, *Essays* (1897), 25–58. See also his correspondence with Poulton in the later 1880s, Box 9, Poulton Papers, OMNH.
240. Romanes to Meldola, 5 May 1886, #25, Meldola Papers, ICL.
241. Of course 'panmixia' had a much wider history, as discussed in Hale, *Political Dissent*, *passim*.
242. See the discussion in Donald Forsdyke, *The Origin of Species Revisited: A Victorian Who Anticipated Modern Developments in Darwin's Theory* (2001), 47–63. It was an argument which carried weight even with those inclined to dismiss all the implications Romanes sought to draw out (see for example C. Lloyd Morgan to Wallace, 27 October 1897, ff.27–8, Wallace Papers, Add. MSS 46437, BL).
243. Romanes to Meldola, 16 September 1886, #1944, Meldola Papers, ICL.
244. See A.R. Wallace to Morgan, 14 September 1905, #142, Conwy Lloyd Morgan Collection, DM128, University of Bristol. Similarly, in discussing drafts of what becomes *Darwinism* with Wallace, Meldola urged him to strengthen the argument against the attacks of 'our enemies', 'and wipe out physiological selec[tion]!! Your refutation of Romanes is splendid', Meldola to Wallace, 10 April 1888, ff.186–7, Wallace Papers, Add. MS 46436, BL. See Forsdyke, *Origin of Species Revisited*, 208–9, 222–4.
245. Meldola to Poulton, 22 February 1889, Meldola Papers, OMNH.
246. Meldola to Poulton, 17 September 1889, 15 May 1891, Meldola Papers, OMNH.
247. Hale, *Political Descent*.
248. For Bates' attitudes, see E. Clodd to WTTD, 8 November [possibly October] 1888, f.158, W.T. Thiselton-Dyer Papers, In-letters I, RBGK.
249. G.A. Gibson, *Life of Sir William Tennant Gairdner* (1912), 212.
250. Quoted in Baldwin, *Making Nature*, 59–60; also Lankester to Poulton, 19 April 1890, Poulton Papers, OMNH. As Lankester put it in his article 'Darwinism', *Nature*, 7 November 1889, Darwin 'considered his theory of natural selection to be a theory of the origin of species. Mr Romanes says it is not. I say this is an attack on Mr Darwin's theory, and about as simple and direct an attack as possible.' Wallace was especially keen that Romanes' attempts to position himself as the successor of Darwin should be stopped, Wallace to Meldola, 28 August 1886, Meldola Papers, OMNH.
251. Baldwin, *Making Nature*, 59; A. Meadows, *Science and Controversy: A Biography of Sir Norman Lockyer* (2008), 218.
252. F.V. Dickins to Thiselton-Dyer, 5 October 1888, f.198, W.T. Thiselton-Dyer Papers, In-letters I, RBGK. Henry Festing Jones expressed similar suspicion in his correspondence with Samuel Butler, 10 September 1884, ff.136–7, 19 September 1888, f.262, Samuel Butler Papers Add. MS 44030, BL.
253. G.J. Romanes to Lockyer, 30 October 1886, Norman Lockyer Papers, EUL110, Exeter University.
254. Romanes to Ward, 8 October 1889, Subseries E. Autograph letter books, Lester F. Ward Papers, Brown University.
255. Romanes to Gulick, 3 December 1889, in Gulick, *Evolutionist and Missionary*, 417.
256. For the Lamarckism of Spencer's ideas, see Peter J. Bowler, 'Herbert Spencer and Lamarckianism', in Mark Francis et al., *Herbert Spencer: Legacies* (2014), 203–21.
257. Quoted in Forsdyke, *Origin of Species Revisited*, 222.
258. See his 1915 review of Cannan on Samuel Butler, B. Tyson, *Bernard Shaw's Book Reviews*, Vol. 2: *1884–1950* (1996), 300–9.
259. Letter to Charles Rowley reprinted in *Labour Leader*, 8 March 1907.
260. *Fabian News* 14 (1906), 13–14.
261. Although, as Lightman shows, he engaged closely in several, 'Greatest Living Philosopher', in Hesketh, ed., *Imagining the Darwinian Revolution*, 37–57.

THE DEATH OF DARWIN AND AFTER   313

262. As examples of a substantial literature, see B. Lightman, 'Spencer's British Disciples', in M. Francis and M.W. Taylor, eds., *Herbert Spencer's Legacies* (2015), 222–43.

263. A.H. Sayce to Spencer, 7 November 1884, MS791/183, Herbert Spencer Papers, Senate House Library, UL. Other examples might include Thomas Welbank Fowle (1835–1903), cleric, social reformer, author of three articles reconciling Christianity and evolution in *TNC*, July 1878, March 1879, September 1881, who in his undergraduate days, according to the recollection of Rev. A.W.S. Young, 'was full of the thoughts of the day, Darwin and evolution, Huxley, Herbert Spencer', *Life of Samuel Barnett*, I, 23.

264. M. Schabas, *The Natural Origins of Economics* (2006), 144–5; Marshall, 'On a National Memorial to Herbert Spencer', *Daily Chronicle*, 23 November 1904.

265. Lightman, *Popularizers*, 464; Gates, *Kindred Nature*, 61–3; Petrunic, 'Evolutionary Mathematics', 89–110, William H. Brock, 'Chemical Characters: Sir William Crookes (1832–1919)', in G.D. Patterson and S.C. Rasmussen, eds., *Characters in Chemistry: A Celebration of the Humanity of Chemistry* (2013), 73–99.

266. This is very clear from the extensive correspondence regarding the efforts to get a memorial to him in Westminster Abbey after his death in the Meldola Papers, ICL; see Hannah Gay, 'No "Heathen's Corner" Here: The Failed Campaign to Memorialize Herbert Spencer in Westminster Abbey', *BJHS* 31.1 (1998), 41–54.

267. James Collier, 'The Centenary of Herbert Spencer' [1920], James Collier Notebooks, MLMSS, 1693/2, State Library of New South Wales.

268. *Northern Daily Telegraph*, 9 December 1903.

269. Sorley to A. Campbell, 27 May 1904, Box 18, Dep.208, A. Campbell Fraser Papers, NLS.

270. Ryle to Pearson, 5 August 1903, Pearson/11/1/17/102, UCL.

271. For descriptions of Spencer's influence see Hobhouse, *Development and Purpose* (1913), xv; Geoffrey M. Hodgson and Thorbjorn Knudsen, *Darwin's Conjecture: The Search for General Principles of Social and Economic Evolution* (2010), 15, suggests that 'Spencer overshadowed Darwin in the period 1880–1900'; or, as Wilfrid Ward put it, 'In the 'eighties Herbert Spencer enjoyed much of the popularity which had once been Mill's', *Men and Matters* (1914), 195.

272. Collier, 'The Centenary of Herbert Spencer', Collier Notebooks, MLMSS 1693/2/284, State Library of New South Wales.

273. Havelock Ellis, Diary, entries 1 February, 21 July, 24 November 1878, A 6904/1, Havelock Ellis Papers, SLNSW. Ellis' commonplace books from these years show that there was nothing at all biological in his engagement with Darwin, or with evolution more broadly, and that his interest was almost entirely focused on human development.

274. See Bowler, *Eclipse*, 58–106.

275. See also Joseph Compton-Rickett, *Origins and Faith: An Essay of Reconciliation* (1909).

276. See Ruth Livesey, 'Morris, Carpenter and Wilde, and the Political Aesthetics of Labour', *Victorian Literature and Culture* 32.2 (2004), 601–16.

277. Stevenson to Sidney Colvin, December 1887, Sidney Colvin, *Collected Works of R.L. Stevenson*, Vol. 24: *Letters II* (1918), 100; recollection of Donkin, *BMJ*, 13 September 1909, and Mercer's 'The Transmission of Acquired Characteristics', *Contemporary Review* 94 (1908), 705–15. Notably in his contribution to the *British Weekly* series 'Books Which Have Influenced Me', Stevenson paid tribute to Spencer, but did not mention Darwin, *Books Which Have Influenced Me* (1887), 8–9.

278. D'Arcy, *Adventures of a Bishop*, 49; Morgan suggested that 'to none of my intellectual masters do I owe a larger debt of gratitude than to you', quoted in Richards, *Evolutionary Theory*, 245; Gould, *Chats with Pioneers*, 122.

279. J.A. Hobson, 'Herbert Spencer', *South Place Magazine* 10 (January 1904), 49–55, 49.

280. See Hector Macpherson, *A Century of Intellectual Development* (1907).

281. Quoted in *Edinburgh Evening News*, 11 June 1904, and Hector Macpherson, *Books to Read and How to Read Them* (1904), 103.

282. See William [Francis] Barry, 'Samuel Butler of "Erewhon"', *Dublin Review* 145 (1914), 322–44. Olivier recalled being introduced at university to evolutionary controversies, 'Spencer's works...much impressed me', and that Wallas introduced him to Samuel

314 DARWINISM'S GENERATIONS

Butler, *Erewhon* and *Force and Habit*, which 'gave me a frame for a tenable conception of evolution', Francis Lee, *Fabianism and Colonialism: The Life and Political Thought of Lord Sydney Olivier* (1988), 29–31.

283. A shift visible in his correspondence with E.B. Poulton and with Wallace, for example Morgan to Poulton, 25 November 1890, 12 April 1896, Box 7, Poulton Papers, OMNH, Morgan to Wallace, 9 April 1896, ff.8–9, Wallace Papers, Add. MSS 46437, BL; see also Weismann to Morgan, 26 November 1896, #82, Conwy Lloyd Morgan Collection, DM128, University of Bristol; Morgan, 'Dr. Weismann on Heredity and Progress', *The Monist* 4 (1893), 20–30. See Evan Arnet, 'Conwy Lloyd Morgan, Methodology, and the Origins of Comparative Psychology', *Journal of the History of Biology* 52 (2019), 433–61.

284. Y.C. Drazin, ed., *My Other Self: The Letters of Olive Schreiner and Havelock Ellis* (1992), 43.

285. Others included Andrew Wilson (1852–1912), the botanists Sydney H. Vines and Walter Gardiner (1859–1941), the physician Walter Aubrey Kidd (1853–1929), the biologist Walter Heape (1855–1929), and the zoologists Hans Gadow (1855–1928) and Marcus Hartog (1851–1924). For Frazer, see R.A. Ackerman, *J.G. Frazer: His Life and Work* (1987); in a letter of 1885 Frazer described his intellectual debt to Spencer as 'deep and life-long', 22. For Wilson, see his response to Archdall Reid, *ILN*, 24 November 1900; and also his natural history journalism in the 1900s, for example, *Cardiff Times*, 15 November 1902. Gardiner made clear his rejection of the 'too rigid a view of the absolute finality of the Darwinian theory', which he saw in Thiselton-Dyer's 1902 *Edinburgh Review* article, and the need to keep open space for the transmission of acquired characteristics affecting nutrition and reproduction; see Gardiner to WTTD, 23 January 1903, W.T. Thiselton-Dyer Papers, In-letters II, RBGK. See the discussion in Mark E. Borrello, *Evolutionary Restraints: The Contentious History of Group Selection* (2010), 25–6. For a non-scientific example, see John Beattie Crozier (1849–1921)'s *My Inner Life: Being a Chapter in Personal Evolution and Autobiography* (1898), 443–4. Key works include Hartog's *Germ Plasm* (1891), Allen's *The Colour-Sense: Its Origin and Development* (1879), and the later contributions of Kidd, including *Design in Nature's Story* (1900), and *Use-Inheritance Illustrated by the Direction of Hair on the Bodies of Animals* (1901).

286. Edward Clodd, *Grant Allen* (1900), 24. See Grant Allen to Spencer, 10 November 1874, MS791/102 (and various others at 104, 108, 118, 122), Herbert Spencer Papers, Senate House Library, UL. Croom Robertson described him as 'a Spencerian pur sang', postcard to Lester F. Ward, 14 April 1884, Subseries E. Autograph letter books, Lester F. Ward Papers, Brown University; for Cunningham see *Nature*, 9 March 1893.

287. Geddes to Welby 23 March [1891], 1970–010/006(02), Lady Victoria Welby Papers, York University, Canada. For Geddes see also his article on 'Evolution' in *Chambers's Encyclopaedia* (1896). In the 1890s Geddes positioned his work explicitly against 'the Darwinian and Weismannian school'; see Patrick Geddes to Spencer, 22 July 1898, MS791/268, Herbert Spencer Papers, Senate House Library, UL.

288. The immediate response of Wallace was to look for rebuttals: Wallace to W.B. Hemsley, 26 August 1888, HEM/1/2/176, W.B. Hemsley Papers, RBGK.

289. See 'Notes' for an article by Geddes and Thomson, for the *Revue de Morale Sociale*, December 1898, T-GED/18/1/161, Geddes Papers, Strathclyde University.

290. Cunningham remarked in 1891 that they are 'few in number', J.T. Cunningham to 'Dear Sir', 14 November 1891, Box 2, Folder 65, Hyatt and Mayer Collection, PUL.

291. See, for example, John M. Robertson, as reported in Gould, *Chats with Pioneers*, 122.

292. See Anthony S. Travis, 'Raphael Meldola and the Nineteenth-Century Neo-Darwinians', *Journal for General Philosophy of Science* 41 (2010), 89–118, 102–3.

293. See F.B. Churchill, 'The Weismann–Spencer Controversy over the Inheritance of Acquired Characters', *Human Implications of Scientific Advance: The 15th International Congress of the History of Science* [1977], ed. E.G. Forbes (1978), 451–68. Crichton-Browne dodged the BA in Newcastle in 1889 precisely to keep out of the discussions of Weismann's views on heredity, 'which is sure to be the principal biological topic at Newcastle', Crichton-Browne to Victoria, Lady Welby, 5 September 1889, 1970–010/004(04), Victoria, Lady Welby Papers, York University, Canada.

294. Cunningham to Geddes, 11 May 1893, ff.182–3, MS.10525, Geddes Papers, NLS.

THE DEATH OF DARWIN AND AFTER 315

295. Including contributions from its editor, John M. Robertson, and H. Croft Hiller (1846–1934): Robertson, 'Spencer or Weismann', *National Reformer*, 30 April 1893, 'Weismann and Sociology: A Reply', *National Reformer*, 25 June 1893.

296. See letters of George Bowdler Buckton (1818–1905), J. Jenner Weir (1822–94), Hyde Clarke (1815–95), and William Duppa Crotch (1831–1903), responding to Spencer's 'Inadequacy of Natural Selection' essay, sent to Patrick Geddes, MS.10525, Geddes Papers, NLS.

297. See the example of R.J. Ryle and Pearson discussing ideas of the relative variability of males and females; Ryle was exercised by a memory of a reference in Darwin, and searched the *Origin* without success trying to find the passage, Ryle to Pearson, 16 April 1896, Pearson/11/1/17/102, UCL.

298. A. Weismann to Ewart, 7 October 1894, COLL 14/9/2/5, Cossar Ewart Papers, Edinburgh. For the willingness of mainstream biologists to leave space for telegony in their ideas of heredity, see J.D. Biggars, 'Walter Heape, FRS: A Pioneer in Reproductive Biology', *Journal of Reproductive Fertility* 93 (1991), 173–86.

299. Charles Herbert Hurst to S.S. Buckman, 19–20 December 1893, GSM/GX/BM/1/1 British Geological Survey. There is a further long account in his letter to Huxley, 7 September 1894, ff.201–2, Box 18, Huxley Papers, ICL; see also 'Biological Theories' in *Natural Science* 1 (1892), 578–87, ibid., 3 (1893), 195–200.

300. Ellis to Pearson, 25 October 1897, Pearson/11/1/5/19, UCL.

301. By far the best account of this is in Radick, *Disputed Inheritance*, 113–16; see also Nicholas W. Gillham, *A Life of Sir Francis Galton: From African Exploration to the Birth of Eugenics* (2001), 276–7. Pearson was the natural person for R.D. Roberts to approach for a lecture on evolution at the 1900 University Extension Meeting; see Pearson/11/1/17/50, UCL.

302. W.F.R. Weldon to Pearson, 13 October 1899, Pearson/11/1/22/40, UCL.

303. Ryle to Pearson, 24 September 1898, Pearson/11/1/17/102, UCL.

304. W.F.R. Weldon to Pearson, 8 August 1896, 28 September 1899, Pearson/11/1/22/40, UCL. Unsurprisingly, his relations with Poulton were also difficult, and in 1905 there was said to be 'bitter war being raged' between the two, R. Shelford to H.N. Ridley, 27 November 1905, HNR/2/1/5/285, Ridley Papers, RBGK.

305. Havelock Ellis, *The New Spirit* (1890), 12. For discussions of this 'sociological turn', see for example Jose Harris, 'Platonism, Positivism and Progressivism: Aspects of British Sociological Thought in the Early Twentieth Century', in Eugenio F. Biagini, ed., *Citizenship and Community: Liberals, Radicals and Collective Identities in the British Isles, 1865–1931* (1996), 343–60.

306. Mrs Humphrey Ward, *A Writer's Recollections* (1918), I, 222–3.

307. *Bristol Times and Mirror*, 25 February 1895.

308. P. Geddes, 'A Naturalists' Society and Its Work', *Scottish Geographical Magazine* 19 (1903), 89–95, 141–7, 94.

309. As indeed was his later *Reconstruction of Belief* (1905); see also Bernard Bosanquet, 'Socialism and Natural Selection', in his *Aspects of the Social Problem* (1895), 290.

310. Karl Pearson, 'Socialism and Natural Selection', *FR* 62 (1894), 1–21, 1.

311. Gordon Epperson *The Mind of Edmund Gurney* (1997), 54. An evolutionary approach which explicitly modelled itself on the Malthusian mathematics of progression, variation, struggle, and the survival of the fittest was also the hallmark of the late-Victorian leaders of the Edwardian folk song revival, including Cecil Sharp (1859–1924) (see Ross Cole, 'On the Politics of Folk Song Theory in Edwardian England', *Ethnomusicology* 63.1 (2019–20), 22–3) and Lucy Broadwood (1858–1929); see letter, *Morning Post*, 2 February 1904).

312. Epperson, *Edmund Gurney*, 85, 108; Gurney to William James, 23 September 1883, Gurney Folder, William James Papers, MS Am 1092.9/A, Harvard. See also Bennett Zon, 'The "Non-Darwinian" Revolution and the Great Chain of Musical Being', in Lightman and Zon, *Evolution and Victorian Culture*, 208–10, who also quotes poet, clergyman, and armchair ethnomusicologist John Frederick Rowbotham (1859–1925), who 'reads Darwin—selectively and with suspicion', 212.

316 DARWINISM'S GENERATIONS

313. Vernon Lee, 'Vivisection: An Evolutionist to Evolutionists', *Contemporary Review* 41 (1882), 788–811.

314. Lee to Matilda Paget, 3–4 July 1894, printed in Mandy Gagel, 'Selected Letters of Vernon Lee', unpublished PhD dissertation, Boston University (2008).

315. Pearson, *National Life from the Standpoint of Science* (1901). For Henley see Peter D. McDonald, *British Literary Culture and Publishing Practice, 1880–1914* (1997), 30–3; take for example Henley's *The Song of the Sword* (1892), which urged Britons to fulfil their destiny by 'Sifting the nations,/The slag from the metal,/The waste and the weak,/From the fit and the strong', cited ibid., 46.

316. Even within Fabian circles, Darwinian monogenesis could be given an overtly racial gloss: Sydney Olivier invoked 'Darwinian principles' as justification for views that while all humanity shares a common evolutionary tree, the evolution of different branches can still be judged to have produced 'limitations, excrescences or shortcomings of Humanity', *White Capital and Coloured Labour* (1910), 11–12.

317. Stack, *First Darwinian Left*, 98–111.

318. John H. Muirhead, *Reflections of a Journeyman Philosopher* (1942), 134.

319. *St James's Gazette*, 7 May 1904. A theme he returned to in the preface to *Man and Superman* to warn that 'we defeat natural selection under cover of philanthropy', and suggest that what was needed was a 'State Department for Evolution', Daniel J. Kevles, *In the Name of Eugenics: Genetics and the Uses of Human Heredity* (1985), 86.

320. Notes of Geddes' paper 'On the Application of Physics and Biology to Practical Economics' [British Association, 1887], MS119/7B/4, Mavor Papers, Thomas Fisher Library, University of Toronto.

321. See Judith R. Walkowitz, 'Science, Feminism and Romance: The Men and Women's Club, 1885–1889', *History Workshop Journal* 21.1 (1986), 37–59, which notes that most of the female members were slightly older women, 'a transitional generation, straddling two worlds', but most of these were still late Victorians, including Elizabeth Cobb (1846–?), Laetitia Sharpe (1850–?), Maria Sharpe (1853–1928), Constance Parker (1858–?), Olive Schreiner (1855–1920), Henrietta Muller (1845/6–1906), Emma Brooke (1844–1926), Robert Parker (1857–1918), Ralph Thicknesse (1856–1923), and H.B. Donkin (1842–1927). The membership of the Rainbow Circle was Herbert Samuel (1870–1963), Sir Richard Stapley (1842–1920), Graham Wallas (1858–1932), Herbert Burrows (1845–1922), J.A. Hobson (1858–1940), William Clarke (1852–1901), Charles Trevelyan (1870–1958), Ramsay MacDonald (1866–1937), Russell Rea (1846–1916), J.A. Murray Macdonald (1854–1939), Rev. William Douglas Morrison (1852–1943), J.M. Robertson (1856–1933), Sydney Olivier (1859–1943), Henry Salt (1851–1939), and Maurice Adams (1849–1933) (Hobhouse was elected but never attended); see Clarke, *Liberals and Social Democrats*, 54–61, Michael Freeden, ed., *Minutes of the Rainbow Circle, 1894–1924* (1989), and Chris Nottingham, '"The Broader Desire to See Things Fitly Ordered": Some Fabians and New Liberals in the Context of the Generation of the Eighteen Nineties', in A. Blok et al., eds., *Generations in Labour History: Papers Presented to the Sixth British–Dutch Conference on Labour History, Oxford 1988* (1989), 109–28.

322. Grant Allen, 'Character Sketch: Professor Tyndall', *Review of Reviews* 9 (1894), 21–6.

323. J.M. Robertson, *The Economics of Progress* (1918), 239. Similarly he observed that 'Darwin entirely failed to relate his biology to a scientific sociology, uttering in that direction mere empiricism', Robertson, *Buckle and His Critics* (1895), 20. For a similar position see Henry (Harry) Quelch (1858–1913), *Malthusianism vs Socialism* (1899).

324. Hence Andrew Seth (Pringle-Pattison) (1856–1931), while accepting of the role of natural selection in the development of society, was also concerned to warn that it could not on its own explain the evolution of morality, see 'Professor Huxley on Nature and Man', *Blackwoods* 154 (1893), 823–34.

325. S. Olivier, 'Morals', in Shaw, ed., *Fabian Essays* (1889), 108; Olivier, *White Capital and Colonial Labour*, 11. Olivier's view was to accept the current analysis, but to reject its permanence, Olivier, 'Long Views and Short on White and Black', *Contemporary Review* 90 (1906), 491–504.

THE DEATH OF DARWIN AND AFTER 317

326. Crook, *Darwin's Coat-Tails*, citing Wiener, *Wallas*, chapter 7; Wallas, 'Darwinism and Social Motive' (1906), in his *Men and Ideas* (1940), 93. The Edwardian philosopher G.E. Moore found the whole approach, with its emphasis on utility and progress, 'wretched' and 'beastly'; see Paul Levy, *Moore: G.E. Moore and the Cambridge Apostles* (1981), 181.

327. See Webb to Wallas, 23 July 1908, WALLAS/1/38/58–60, BLPES. See Beatrice Webb, Diary entry, October 1884, Webb, *My Apprenticeship* (1938), 164.

328. Jones, *Social Darwinism*, 70 (quoting *Justice*, 13 June 1885).

329. Stack, *First Darwinian Left*.

330. Keir Hardie, *From Serfdom to Socialism* (1907), 92, 88–94. Similarly other writings, such as 'The International Socialist Congress', *TNC* 56 (1904), 562; Hardie's *Labour Leader* continued to emphasise the importance of Darwin over Marx, to the dismay of the less Darwinian *Justice*; see *Justice*, 20 February 1904; for Mann see John Laurent, 'Tom Mann on Science, Technology, and Society', *Science & Society* 53.1 (1989), 84–93. Interestingly Mann's *Memoirs* (1923) have no direct reference to Darwin, and only one passage invoking evolutionary conceptions.

331. See B. Tillett, *An Address on Character and Environment* (nd [1896]), and *Memories and Reflections, an Autobiography* (1931).

332. For Tochatti see *Freedom*, March 1894, where he traced his anarchism to the influence of Darwin, Huxley, Morris, Kropotkin, and Malatesta.

333. See J. Trevor, *God and My Neighbour* (1904). Blatchford's Darwinism was largely, if not entirely, second-hand; 'Pure science has no charms for me. Darwin was a great man; but I cannot read his books', he confessed in *My Favourite Books: Essays* (1901), 133. He also seems to have come to Darwin relatively late; Albert Neil Lyons remarked that 'Darwin, who is one of "R.B.'s" great literary heroes, came to him at forty', *Robert Blatchford: The Sketch of a Personality* (1910), 157, possibly via an encounter with a nursing gorilla at the Belle Vue pleasure grounds in Manchester, before he had read any Darwin; see *Westminster Gazette*, 31 March 1911. This didn't stop him placing Darwin first in his list of those thinkers who underpinned *The New Religion* (1892), 3, or indeed in bemoaning the fact that the working classes had not read important books like the *Origin*, 'In the Library', *Clarion*, 12 December 1902.

334. For a careful account of Kropotkin's Lamarckian influences, see Hale, *Political Descent*, 221–50, and also Crook, *Darwin's Coat-Tails*, 71–8; Livesey, 'Morris, Carpenter and Wilde', 610.

335. For Kidd's place in the intellectual culture of late-Victorian Britain, see Collini, *Public Moralists*, 199–250.

336. Macmillan to Kidd, 27 November 1907, MS Add. 8069/M199, Kidd Papers, CUL.

337. For sales figures drawn from *The Bookman* see Dixon, *The Invention of Altruism*, 278. The 11th thousand was issued in May 1895; Frederick Macmillan to Kidd, 27 November 1907, MS Add. 8069/M199, Kidd Papers, CUL. For the currency *Social Evolution* continued to have, see the comments on its regular circulation from Bury public library, in A. Holt, 'The Lancashire Artisan: A Protest and an Appeal', *The Independent Review* 2 (1903–4), 622–6.

338. See correspondence with the society, including Henry Richardson, MS Add. 8069/L35–37, Kidd Papers, CUL.

339. Charles Booth to Kidd, 17 June 1894, MS Add. 8069/B165, Kidd Papers, CUL.

340. Kidd to Wallace, draft, 2 April 1902, MS Add. 8069/K72, Kidd Papers, CUL.

341. R.C. Bannister, *Social Darwinism: Science and Myth in Anglo-American Social Thought* (1979), 158.

342. Hughes to George Macmillan, 30 March 1894, MS Add. 8069/H86, Kidd Papers, CUL. For evidence of his appeal to older mid Victorians who embraced *Social Evolution* as a way of reclaiming evolution from scientific materialism, see Emelia Russell Gurney (1823–96) to Kidd, March nd, 'a most welcome fact, a sort of scientific buttress of our faith', Kidd Papers, MS Add. 8069/G77, CUL; Henry Toynbee (dates not known, but he noted that he had retired at seventy, and so was a mid Victorian) to Kidd, 20 April 1894,

318   DARWINISM'S GENERATIONS

MS Add. 8069/T47: 'Your book is quite a comfort to my mind'; H.J. Berry (describes himself as in his later eighties) to Kidd, 5 July 95, MS Add. 8069/B141, Kidd Papers, CUL.

343. *The Bookman* 6 (April 1894), 23.

344. J.H. Bridges, 'The Darwinist Utopia', in his *Illustrations of Positivism* (2nd ed., 1915), 378; see also the comment that 'the evolutionist must study the actual world, and not pretend to base his speculations upon data outside of all possible experience', Leslie Stephen, 'The Ascendancy of the Future', *TNC* 51 (1902), 795–810, 802; the hostility of conventional economists is clear from Marshall to Kidd, 27 May 1902, MS Add. 8069/M256, Kidd Papers, CUL. For others prepared to accept the 'close affinity' of their ideas with Kidd's see, for example, Alfred Milner (1854–1925) to Kidd, 4 March 1894, MS Add. 8069/M333, Kidd Papers, CUL.

345. Mackenzie to Kidd, 14 April 1894, MS Add. 8069/M24, Kidd Papers, CUL. See also Beatrice Chamberlain to Neville Chamberlain, 17 April 1894, NC1/13/2/26, Neville Chamberlain Papers, CLB.

346. J.S. Mills, 'The Limits of Darwinism', *Observer*, 17 February 1918.

347. As in the account of the observations of 'Judge Grove', presumably the physicist Sir William Robert Grove (1811–96), at a dinner party in 1881, K.E. Farrer to T.C. Farrer, 14 March [1881], 2572/1/10/81, Letterbooks of T.C. Farrer, 1880-1, Farrer Papers, Surrey History Centre.

348. W.T. Stead, *The Last Will and Testament of Cecil Rhodes* (1902), 86. Rhodes' intellectual history is not clear, but it can be suggested that he lost his early religious beliefs under the influence of Winwood Reade, Darwin, and Gibbon; see Godfrey Elton, *General Gordon* (1954), 309. He was inspired by the possibilities opened out by the doctrine of evolution: what mankind had achieved and might yet achieve; see the account of Thomas E. Fuller, *Cecil Rhodes* (1910), 246-7. For Selous, see his *Sunshine and Storm in Rhodesia* (1896), 67. Selous noted that Darwin's 'far-reaching theories—logical conclusions based upon an enormous mass of incontrovertible facts—have revolutionised modern thought, and destroyed for ever many old beliefs that had held men's minds in thrall for centuries', *Travel and Adventure in South East Africa* (1893), 286. After this death, his brother Edmund described Selous as a 'Darwinian evolutionist' with no belief in any creed or revelation; see *Richmond Herald*, 24 February 1917. Im Thurn regarded Darwin as 'a magnificent man'; see Everard F. im Thurn to WTTD, 9 May 1888, DC/204/340, RBGK. (We might compare im Thurn with an earlier generation of colonial governor like Sir Richard Temple (1826–1902), apparently unaffected by Darwinism.)

349. See, for example, Sir John Alexander Cockburn (1850–1929), a long-term emigrant, and Premier of South Australia briefly in 1889–90, whose address 'The Evolution of Empire', in *Compatriots Club Lectures 1st Series* (1906), 175-195, was a grab-bag of evolutionary prejudices and clichés.

350. See Wilfrid Scawen Blunt, *My Diaries: Being a Personal Narrative of Events*, Vol. 1: *1888–1900* (1919), entry for 27 September 1897.

351. For Pearson see B. Semmel, *Imperialism and Social Reform* (1960). See also Pearson's letter to the *Times* of 1910, cited in Francisco Louca, 'Emancipation through Interaction: How Eugenics and Statistics Converged and Diverged', *Journal of the History of Biology* 42.4 (2009), 649–84, 662. The initial impact of *Hereditary Genius* was limited; Louisa Galton's diary on its reception commented that 'Frank's book not well received, but liked by Darwin and men of note', K. Pearson *Life of Francis Galton* (1924), 88.

352. See the discussion in D.B. Paul, 'Wallace, Women and Eugenics', in C.H. Smith and G. Beccaloni, eds., *Natural Selection and Beyond* (2008), 263–78. For one example of late-Victorian Weismannite eugenics, see the works of Harry Campbell (1860–1938), including *Evolution Past and Future* (1923).

353. See James Crichton-Browne to Buckman, 18 October 1906, GSM/GX/BM/1/1 British Geological Survey.

354. See Lang to Welby, 5 May 1904, 1970–010/009(04), Lady Victoria Welby Papers, York University, Canada.

355. Alexander Graham Bell, 'How to Improve the Race', *Journal of Heredity* 5.1 (1914), 1–7; 'A Few Thoughts on Eugenics', *National Geographic* 19 (1908), 119–23.

THE DEATH OF DARWIN AND AFTER 319

356. Higgs was educated at Girton, and was the first woman to study for the Natural Sciences degree; see her *Glimpses into the Abyss* (1906), which employed notions of psychological evolution and reversion to explain the emergence of a hereditary underclass.

357. M. Cruikshank, 'Mary Dendy 1855–1933', *Journal of Educational and Administrative History* 8 (1976), 26–9.

358. Poulton to C. Lloyd Morgan, 26 December 1909, #229, Lloyd Morgan Collection, DM128, Bristol.

359. See his *Darwinism and Race Progress* (1895), 20 etc. For other examples of the readiness with which late Victorians adopted eugenicist approaches, see comments in W.L. Lindsay, 'Darwinism and Medicine', *BMJ*, 6 November 1909, 1325–31, especially 1331, Ernest Bruce Iwan-Müller (1853–1910), 'The Cult of the Unfit', *FR* 86 (1909), 207–22.

360. J. Macdonell, 'The Question of the Native Races in South Africa', *TNC* 99 (1901), 367–76, 375.

361. Annotation of Saleeby's 'The Dawn of Natural Eugenics: Maternity Benefit in Being', from the *PMG*, T-GED/8/3/474, Geddes Papers, Strathclyde University.

362. Including D.G. Ritchie, J.H. Muirhead, Charles Gore, R.B. Haldane, Arnold Toynbee (1852–89), Sir Henry Jones (1852–1923), James Bonar (1852–1941), and also Charles Loch (1849–1923), Arnold Toynbee (1852–89), and W. R. Sorley; see Andrew Vincent and Raymond Plant, *Philosophy, Politics and Citizenship: The Life and Thought of the British Idealists* (1984).

363. Muirhead, *Reflections*, 97–8, including volumes by Ritchie, Bosanquet, James Bonar.

364. As W.J. Maunder points out, Caird's account of the development of the idea of evolution in his *Evolution of Religion* managed only one passing reference to Darwin, while T.H. Green was equally unsympathetic, W.J. Maunder, *British Idealism: A History* (2011), 261. For this shift between what he sees as the first- and second-generation Idealists see Chris Renwick, *British Sociology's Lost Biological Roots: A History of Futures Past* (2012), 102.

365. A. Seth [Pringle-Pattison] and R.B. Haldane, eds., *Essays in Philosophical Criticism* (1883). For the significance to Seth, see G.F. Barbour, ed., *Balfour Lectures on Realism* (1933), 17; this generational character was clearly widely recognised; see the review in *Dundee Advertiser*, 31 January 1883. The authors, apart from Caird who wrote a preface, were Seth, Haldane, Bosanquet, Sorley, Ritchie, Jones, Bonar, with T.B. Kilpatrick (1857–1930), who moved to Toronto, and W.P. Ker (1855–1923), literary scholar. For two studies which flesh out the philosophical background and implications of the argument here, see E. Neill, 'Evolutionary Theory and British Idealism: The Case of David George Ritchie', *History of European Ideas* 29.3 (2003), 313–38, and David Boucher, 'British Idealism and Evolution', in W.J. Mander, ed., *The Oxford Handbook of British Philosophy in the Nineteenth Century* (2014), 306–23.

366. James Bonar's 'The Struggle for Existence', in *Essays in Philosophical Criticism*, 214–45, 226; Bonar is interesting as a very early systematic thinker about the application of evolutionary ideas to societies; though Hodgson and Knudsen, *Darwin's Conjecture*, 9, stresses that he repeatedly warned of the need to be cautious in application. At this stage Ritchie, 'hadn't then paid very much attention' to evolution theory; see Ritchie to Alexander, [end of 1888 or early 1889], ALEX/A/1/1/236/4, Samuel Alexander Papers, MJRL.

367. Hastings Rashdall, 'The General Functions of the State', in J.E. Hand, ed., *Good Citizenship* (1899), 10. For Pringle-Pattison's suspicion of the materialism of the German Darwinism he found, see Barbour, *Balfour Lectures*, 17–19. As Haldane remarked, physiology and biology have 'inflicted loss on themselves and distorted these very facts, by mere dogmatic assumption that the relations of life are reducible and must be so to those of mechanism', Haldane to Mary Ward, 27 May 1890, ff.167–71, Haldane Papers, MS.5903, NLS.

368. *Darwinism and Politics* (1889); the 2nd edition (1891) was much more obviously social Darwinian, with two additional chapters; see Hawkins, *Social Darwinism*, 159–62. At the same time, note Ritchie's later questioning of the role which Alexander was prepared to give to natural selection in the progress of human understanding; natural selection, he pointed out, had only produced 'a very imperfect adaptation' of likes and dislikes in taste

320   DARWINISM'S GENERATIONS

and smell to what is life-furthering or life-hindering, Ritchie to Alexander, 24 January 1897, ALEX/A/1/1/236/4/10, Samuel Alexander Papers, MJRL.

369. D.G. Ritchie, 'Evolution and Democracy', in Stanton Coit, ed., *Ethical Democracy: Essays in Social Dynamics* (1900), especially 14–15, and see Hale, *Political Descent*, 197–202, and Matt Carter, *T.H. Green and the Development of Ethical Socialism* (2016), 58–61.

370. F.H. Bradley, 'Some Remarks on Punishment', *International Journal of Ethics* 4 (1894), 269–84, 279. Bradley's rather abstract ideas of evolution are visible in the tortured formulations of 'My Station and Its Duties', in his *Ethical Studies* (1876), 145–92.

371. Bosanquet to F.H. Peters, 16 September 1877, in J.H. Muirhead, ed., *Bernard Bosanquet and His Friends* (1935), 42.

372. Bosanquet, 'Socialism and Natural Selection', in his *Aspects of the Social Problem* (1895).

373. Muirhead, *Bosanquet and His Friends*, 107–8, and W.R. Sorley, *Recent Tendencies in Ethics* (1904). For a discussion of Bosanquet's marriage of evolutionary forces with mind and will, see L. Goodlad, *Victorian Literature and the Victorian State* (2004), 201–5.

374. See his strongly evolutionary (and implicitly Darwinian) *The Elements of Ethics: An Introduction to Moral Philosophy* (1892), which went as far as accepting the possibility of general laws of society, while pointing out that these would not trench on a study of ethics as judgement of conduct. For a more trenchant scepticism that evolutionary theory could inform the study of ethics, see W.R. Sorley, *The Ethics of Naturalism: A Criticism* (1885).

375. Bradley, 'Some Remarks on Punishment', *passim*.

376. For example, Henry Jones, *Is the Order of Nature Opposed to the Moral Life?* (1894).

377. Michael A. Weinstein, *Unity and Variety in the Philosophy of Samuel Alexander* (1984), 3: 'The privilege of Alexander's generation [born in the late 1850s and in the 1860s] was to make the transition between the minds of the nineteenth and the twentieth centuries.' For Alexander's distance from pragmatism see his comment that 'I have got a better understanding of late of what you pragmatists are at', Alexander to Schiller, 23 February 1912, Folder A–B, Box 1, F.C.S. Schiller Papers, UCLA.

378. Ritchie to Alexander, 23 July 1889, ALEX/A/1/1/236/6, Samuel Alexander Papers, MJRL; 'Natural Selection in Morals', *International Journal of Ethics* 2 (1892), 409–39.

379. Alexander to Welby, 20 November 1910 tps, 1970–010/001(06), Lady Victoria Welby Papers, York University, Toronto.

380. Macpherson, *Books to Read*, where the debt to Haldane and Pringle-Pattison is explicit.

381. For Robinson, see Holmes, *Darwin's Bards*, 37–40; the founders of the London Ethical Society (established in 1886) were Bonar, Muirhead, Chubb, with Bosanquet, and John Stuart Mackenzie, along with the American Stanton Coit (1857–1944), I.D. MacKillop, *The British Ethical Societies* (1986). See also Trevor, *My Quest for God*. As Mark Bevir has recognised, Trevor's immanentism envisaged the Labour Church as the most advanced expression of the evolutionary process.

382. Maynard, Autobiographical memoir, GCPP Stephen 3, Girton College, Cambridge.

383. Gibson, *Life of Sir William Tennant Gairdner*, 271.

384. M'Kinney's volume accused Darwin of muddle-headedness, trawled the writings of other evolutionists for linguistic chinks, reverted to the limitations of the fossil evidence, ridiculed the anthropological evidence of man's antiquity and development and paraded the arguments of mid Victorians like Müller on language.

385. Harry Marshall Ward (1854–1906) to Selina Mary Kingdon ('Lina'), 17 October 1877, 28 September 1879, HMW/1/3 [unfoliated bundle], H.M. Ward Papers, RBGK.

386. As in the case of G.B. Longstaffe (1849–1921), for whom they were 'the turning of the scale', Longstaffe to F.A. Dixey, 7 April 1906, ff.19–20, MSS. Eng. Lett e.163, Bodleian; 'My scepticism is so profound', Longstaffe commented, 'that Darwinism is almost as difficult to me as Christianity.'

387. See the account of a lecture to the YMCA, *Carlisle Patriot*, 15 March 1889. See also Rev. George Martius Macdermott (1863/4–1939), *Evolution and Revelation* (1897).

388. *North Wales Chronicle*, 17 January 1885, *Wrexham and Denbighshire Advertiser*, 14 February 1885.

THE DEATH OF DARWIN AND AFTER   321

389. William Henry Muncaster (c.1850–1921), as reported in *Supplement to the Nonconformist*, 28 October 1897; A. Balmain Bruce (1831–99), 'Modern Apologetic Problems in the Sphere of Natural Theology', *Scotsman*, 17 March 1899.

390. Hinchliff (citing Owen Chadwick), *God and History*, 22.

391. Dorothea Price Hughes, *Life of Hugh Price Hughes* (2nd ed., 1904), 117; other instances include Rev. Samuel John Woodhouse (1846–1917), Sermon on evolution, *Northampton Mercury*, 3 March 1888; J. Armitage Robinson (1858–1933), *Some Thoughts on the Incarnation* (1913), viii, and Arthur Foley Winnington-Ingram (1858–1946), *Reasons for Faith and Other Contributions to Christian Evidence* (1907).

392. *Blackburn Standard*, 20 December 1890.

393. *Cornish Telegraph*, 10 April 1884. Hughes' fellow Methodist minister, W.L. Watkinson offers a very similar balance of general acceptance of evolution, emphasis on the potential debates within Darwinism, a penchant for Darwinian reference, but also frequent sceptical digs which at times can persuade hearers that his attitude is fundamentally hostile, *Wakefield Free Press*, 5 May 1883, *Cornish Guardian*, 13 October 1905.

394. Ellen Wilkinson (1891–1947), in H.H. Asquith, *Myself When Young* (1934), 405.

395. One common strategy was to invoke the 'the evolutionist' as an 'other', as in Thomas Gunn Selby (1846–1910), *The Unheeding God* (1899).

396. Momerie, 'Evolution and Design', in Frederick Hastings and A.F. Muir, *Christianity and Evolution: Modern Problems of the Faith* (1887), 53–73; *Islington Gazette*, 8 January 1901. For Momerie, see Black, *Pen, Pencil, Baton and Mask*, 170–8. For Dennis Hird (1850–1920), socialist clergyman and later principal of Ruskin College, 'the teaching of evolution has done more to clear up the mysteries of life than has any previous view of the origin of things', John Beatson-Hird, *Dennis Hird: Socialist Educator and Propagandist, First Principal of Ruskin College* (1999).

397. One anecdote of Henry Drummond was of a conversation on a Royal Institution lecture on the inner ear; and Drummond remarking 'Yes…the Almighty knew what he was about—the ear is a marvellous organ', Hunter Boyd, 'Recollections of the Late Professor Henry Drummond', Acc.5890/3 [not foliated], Drummond Papers, NLS.

398. H. Scott Holland, 'Faith in Jesus Christ', in W.E. Bowen, ed., *The Faith of Centuries* (1897), 81.

399. Ward, *A Writer's Recollections*, I, 187. For one example, see the comments of Robinson, in his *Some Thoughts on the Incarnation*, which offers a strikingly blasé reconciliation of Genesis and evolutionary science: 'the general idea is wholly in harmony with our latest scientific thought. The details may be slightly out of order here and there: but…This is merely the literary setting of the thought, and need not detain us', 10–11.

400. W. Ward, 'New Wine in Old Bottles', *TNC* (1890), reprinted in *Witness to the Unseen* (1893), 94.

401. As in Rev. Arthur Chandler (1859?–1939) 'Faith in God', in Bowen, *The Faith of Centuries*, 3–4.

402. See Charles Robert Lloyd Engström (1842?–1922) lecture, *Reading Mercury*, 9 February 1889; see entry for 6 November 1892, J.C. Gibson, ed., *Diary of Sir Michael Connal, 1835–1893* (1895), 341.

403. F. Ballard, *Clarion Fallacies* (1908), 7. Ballard was another who drew heavily on Darwin's *Life* to point to his inclination to design and sympathy with Christianity; see also Ballard, *The Miracles of Unbelief* (1900), *Methodist Recorder*, 29 February, 7 March 1884.

404. See letter to *West Surrey Times*, 10 December 1887.

405. S.R. Driver to Acland, nd, ff.161–2, MSS Acland d.68, Bodleian; John Rogerson, 'What Difference Did Darwin Make? The Interpretation of Genesis in the Nineteenth Century', in S.C. Barton and D. Wilkinson, eds., *Reading Genesis after Darwin* (2009), 75–92. See Driver, 'Evolution Compatible with Faith' (1883), in his *Sermons on Subjects Connected with the Old Testament* (1892), 1–27.

406. Hinchliff, *God and History*, 105–6. See the essays of Richard Valpy French (1839–1907) and Alexander Stewart (1835–1909) in French, ed., *Lex Mosaica, or The Law of Moses*

and the Higher Criticism (1894), and F.E. Spencer, *Did Moses Write the Pentateuch after all?* (1892), 3.

407. T.H. Huxley to E.B. Poulton, 19 February 1886, Box 5, Poulton Papers, OMNH.

408. David S. Cairns, 'Recollections of Henry Drummond', *North American Student* 2.2 (1913), 54–62.

409. George Adam Smith, *Life of Henry Drummond* (1901), 47; *The Speaker*, 20 March 1897.

410. 'Dr John Hunter', *T.P.'s Weekly*, 4 March 1910.

411. 'The first step in the reconstruction of Sociology will be to escape from the shadow of Darwinism—or rather to complement the Darwinian formula of the Struggle for Life by a second factor which will turn its darkness into light', Drummond, *Ascent of Man* (1894), 27.

412. A.C. Haddon, 9 September 1883, ff.307–8, MS.10522, Geddes Papers, NLS; F.A. Dixey, thought it 'a book…containing many true thoughts strikingly expressed', even if much open to criticism, Dixey to Acland, 30 March 1895 copy ff.162–3, MSS Acland d.68, Bodleian.

413. *Edinburgh Evening News*, 12 August 1893; James Young Simpson, *Henry Drummond* (1901), 122.

414. Drummond, *Ascent of Man* (1894), 333, cited in Bernard Lightman, 'Darwin and the Popularization of Evolution', *Notes and Records of the Royal Society* 64.1 (2010), 5–24. Dixon sees *Ascent of Man* as the zenith of a tradition of 'Theistic, moralistic and altruistic Darwinism', *Altruism*, 158.

415. See, for example, letter of E. Tasker, in *Warwick and Warwickshire Advertiser*, 28 May 1887; Jones, *Social Darwinism*, 70.

416. James R. Moore, 'Evangelicals and Evolution: Henry Drummond, Herbert Spencer and the Naturalisation of the Spiritual World', *Scottish Journal of Theology* 38 (1985), 383–417, 399; Martineau to J.H. Allen, 28 March 1890, Box 3, Joseph Henry Allen Correspondence, bMS416, Andover Theological College, Harvard.

417. Joseph John Murphy, [*Natural Law in the Spiritual World*], *British Quarterly Review* 80 (1884), 114–29, and Murphy, *Natural Selection and Spiritual Freedom* (1893).

418. Leslie Stephen, 'The Ascendancy of the Future', *TNC*, 51 (1902), 795–810; 'An Interview with the Rev. W.H. Dallinger', *The Young Man* 8 (1894), 363–70; *Nonconformist*, 24 May 1894. (Dallinger had been equally unimpressed with *Natural Law*; see 'Notes of a conversation', 1970–010/004(07), Lady Victoria Welby Papers, York University, Canada.)

419. *Aberdeen Press and Journal*, 19 December 1895; see Johnston's willingness to accept that there is 'much which is true' in the theory of evolution combined with an absolute refusal to accept the evolution of man, *Orkney Herald*, 1 January 1896. For Robert William Barbour (1854–91), it was 'known and loved as a "true evolution" by men of science', Barbour to Drummond, 13 September 1883, Acc.5890/2 [not foliated], Drummond Papers, NLS. Drummond also appeared in a number of the accounts of influential books provided by Labour MPs to the *Review of Reviews* 33 (1906), including James O'Grady (1866–1934), G.H. Roberts (1868–1928), Arthur Richardson (1860–1936), and G.J. Wardle (1865–1947).

420. G. Wyndham to G.K. Chesterton, 7 November 1909, f.94, Add. MS 73241, Chesterton Papers, BL.

421. Hinchliff, *God and History*, 111. For a recognition of the generational gulfs of this evolutionary stance, see Harold Anson, *T.B. Strong: Bishop, Musician, Dean, Vice–Chancellor* (1949), 14.

422. For Gore's continued evolutionary sensibility, see Charles Gore, 'The Social Doctrine of the Sermon on the Mount', *Economic Review* 2 (1892), 154. For his scepticism about Darwinism without some prior tendency in nature towards certain forms see Gore to Ethel Romanes, 11 August 1893, ff.51–2, Romanes Letters, MS Eng d.3826, Bodleian. Likewise, Gore told Oliver Lodge in 1897 that his conception did not absorb Christ into nature: he remained supernatural, Gore to Lodge, 25 February 1897, [not foliated], Mansbridge Papers, Add. MS 65362, BL.

423. His address at the Reading Church Congress asserted that 'there is in the doctrine of evolution much which ought to render it especially attractive to those whose first thought

THE DEATH OF DARWIN AND AFTER   323

is to hold and to guard every jot and tittle of the Catholic Faith', *Literary Churchman*, 4 January 1884; Gouldstone, *Anglican Idealism*, 109–33.

424. C.C.J. Webb, *A Century of Anglican Theology and Other Essays* (1923), 32. In his essay on 'Bishop Gore and the Church of England' (1908), reprinted in *Outspoken Essays: First Series* (1919), 106–36, W.R. Inge argued that *Lux Mundi* meant that 'High Church clergy have been able without fear to avow their belief in the scientific theories associated with Darwin's name'.

425. The group were Gore, Walter Lock (1846–1933), William James Heathcote Campion (1851–92), J.R. Illingworth (1848–1915), Arthur Lyttelton (1852–1903), Aubrey Lackington Moore (1848–90), Henry Scott Holland (1847–1918), Robert Campbell Moberley (1845–1903), Francis Paget (1851–1911), and Robert Lawrence Ottley (1856–1903).

426. For the close personal bonds of the group, see G. Rowell, 'Historical Retrospect: *Lux Mundi* 1889', in R. Morgan, ed., *The Religion of the Incarnation: Anglican Essays in Commemoration of 'Lux Mundi'* (1989), 205–17, Schlossberg, *Conflict and Crisis*, 155–63. The group continued to hold annual reunions at Longworth until 1915; see Stephenson, *Talbot*, 63–4.

427. A.L. Illingworth, ed., *The Life and Work of John Richardson Illingworth Edited by His Wife* (1917), 159. The younger T.B. Strong (1861–1944) apparently suggested it might have been called 'There's life in the old dogma yet', Anson, *T.B. Strong*, 21.

428. *Literary Churchman*, 4 January 1884.

429. J.S.B., *Canon Liddon: A Memoir* (1890), 11.

430. Supported by Brooke, 'Christian Darwinians', 54–5, and David Knight, 'Response: Further Varieties of Christian Darwinian', in A. Robinson, ed., *Darwinism and Natural Theology* (2012), 68–76.

431. Quoted in Hinchliff, *God and History*, 111.

432. J.R. Illingworth, *The Doctrine of the Trinity, Apologetically Considered* (1907), 9–10. Compare with the comment of the Liberal MP Theodore Cooke Taylor (1850–1952), rejecting the old [high-Victorian] view of God as 'a shadowy artificer, who constructed the world, wound it up, and set it going according to the laws of evolution', *Batley Reporter and Guardian*, 8 March 1901.

433. For example, Josiah Nicholson Shearman (1846–1915), *The Natural Evolution of Theology* (1915).

434. See Moore's essay in *Lux Mundi*, and the discussion in England, 'Natural Selection'. Significantly, this denial of the need for miracles increasingly became a problem for Gore himself, as revealed in his correspondence in the later 1890s as represented by the Mansbridge Papers, Add. MS 65362, BL. In 1909 W.T. Stead (1849–1912) was deploying this view of Darwin as strengthening faith by converting creation from a distant event to a continuous ongoing marvel, *Dundee Evening Telegraph* 30 July 1909.

435. Moore's articles in *The Guardian* were praised by Flower as 'thoroughly satisfactory—so clear, simple and concise, just containing all that is necessary to show how the difficulties that many people feel on the subject may be removed without any redundant details or recondite or tedious discussions', Flower to Moore, 23 February 1888, Moore Autograph Collection, B M781, APS. Moore continued to operate as a yardstick for late-Victorian Anglicans, like Poulton and Dixey; see Dixey's comment in 1908 that 'What Aubrey Moore called "the new teleology" wants driving into people's heads', Dixey to C.W. Formby, 2 January 1908, quoted in Richard England, 'Natural Selection, Teleology and the Logos', *Osiris* 16 (2001), 285; see also the discussion in John Hedley Brooke, 'Genesis and the Scientists: Dissonance among the Harmonizers', in Barton and Wilkinson, *Reading Genesis*, 93–109.

436. Lankester praised Moore's articles as 'excellent', Lankester to Moore, 16 February [1889?], Moore Autograph Collection, B M781, APS; and later described Moore as 'one of the not very many persons whom I warmly esteemed', Lankester to Victoria, Lady Welby, 8 February [1889], 1970–010/009(05), Lady Victoria Welby Papers, York University, Canada. Older readers were less impressed. Lionel Beale fretted that this continual adjustment of religion to science would soon leave no faith left to modify, Beale,

324   DARWINISM'S GENERATIONS

'Introductory Lecture on Progress at King's College and the Nature of Life', *The Lancet*, 7 October 1893.

437. Frank Ballard's arguments in favour of 'design', in which Darwin, Huxley, and Asa Gray were invoked against Haeckel and Romanes, are a vivid example of this; see his *The Miracles of Unbelief* (1900).

438. J.S. Lidgett, *The Victorian Transformation of Theology* (1934), 59.

439. F.A. Dixey, lecture to the Christian Evidence Society, *Guardian*, 13 December 1905.

440. There are several detailed discussions, including Hinchliff, *Jowett*, 204–8.

441. For the role of Jowett, see Hinchliff, *God and History*, 70.

442. Some of these implications are discussed in Michael S. Helfand, 'T. H. Huxley's "Evolution and Ethics": The *Politics of Evolution* and the Evolution of Politics', *Victorian Studies* 20.2 (1976–7), 159–77.

443. [WTTD], 'The Rise and Influence of Darwinism', *Edinburgh Review* 196 (1902), 366–407, 406. Poulton observed that it was perhaps not surprising, in that Huxley had hardly been engaged in science in his later years, focusing entirely on religious and moral questions, E.B. Poulton to WTTD, 18 July 1908, f.134, Thiselton-Dyer Papers, In-Letters III, RBGK.

444. See, for example, Bosanquet reviewing *Evolution and Ethics and Other Essays* (1894) in the *International Journal of Ethics* 5 (1895), 390–2, and E.S. Talbot to Welby, 3 June 1893, 1970–010/017(02), Lady Victoria Welby Papers, York University, Canada. For a general discussion see Michael Freeden, 'Biological and Evolutionary Roots of the New Liberalism in England', *Political Theory* 4 (1976), 471–90.

445. Waggett, *Religion and Science*, 44. There is a sense that this was a constant reference point for a wide variety of positions. For John Neville Figgis (1866–1911), 'Huxley...showed in his famous Romanes Lecture that the best things in human life had not come from natural evolution, but from the human will set upon good and resisting cosmic development', *Hopes for English Religion* (1919), 18. In the same way, F.R. Tennant assimilated much of Huxley's lectures into explorations which culminated in the 1901–2 Hulsean Lecture entitled *The Origin and Propagation of Sin*, where he integrated evolutionary ideas into a Christian synthesis.

446. Michael Sadler, 1314/PA/225, Michael Sadler Papers, Brotherton Library. In the same way, James Mavor's reaction to Huxley at the British Association the following year was that 'the fire had burned out of him. He was emaciated and obviously ill', Mavor, *Windows*, 329.

447. H. Samuel, *Memoirs* (1945), 11, 18. This was very much the sense of D'Arcy Thompson's account of the lecture; see Ruth D'Arcy Thompson, *D'Arcy Wentworth Thompson: The Scholar Naturalist, 1860–1948* (1958), 58.

448. Kidd, *Social Evolution*, 17.

449. Huxley to Foster, 2 November 1894, in W.F. Bynum and C. Overy, eds., *Michael Foster and Thomas Henry Huxley, Correspondence, 1865–1895* (2009); Huxley to N.P. Clegg, 2 January 1895, ff.234–5, Box 12, Huxley Papers, ICL.

The Reception of Darwinian Evolution in Britain, 1859–1909: Darwinism's Generations. Martin Hewitt,
Oxford University Press. © Martin Hewitt 2024. DOI: 10.1093/9780191982941.003.0005

5

Darwinian Debates at
the *Fin de Siècle*

The Edwardians

The spirit of the age does not move all in a piece. What is old in one quarter is in another quarter just arriving. Controversial weather is a matter of latitude. It is a matter of altitude....In provincial towns, in training colleges, among the highly intelligent but scantily informed scholars of our state schools, among men who have turned late in life from the pasturing of souls to the construction of books, in all ranks of the great nation which has lately learned to love reading, there is widespread unrest on the old Darwinian grounds. The very possibility of buying, the possibility, say rather, of selling Darwin for a shilling and Laing for sixpence, marks the truth of this statement. You have a whole new generation experiencing the thrill which Darwin's own generation experienced, the thrill of thinking, and that in a new language, and about a new set of facts....That old blunderbuss which we thought had for ever missed fire, fetched down in antic mood from over the hearth, goes off at last in a tragic manner; goes off in the hands of a generation which did not see it loaded. (P.N. Waggett, *Religion and Science: Some Suggestions for the Study of the Relations between Them* (1904), 20–1)

In 1892 T.H. Huxley had looked around complacently at the rout of Darwin's opponents, observing that 'I may consider it unnecessary to discuss any more controverted Questions'.[1] The glowing tributes which followed his death in 1895 would have confirmed him in his belief. The *Pall Mall Gazette* remarked that the *Origin* and Huxley's *Man's Place in Nature*, 'which were anathema to the generation passing away, have become the standards of scientific thought today, blessed by bishops and quoted by rural deans'.[2]

326   DARWINISM'S GENERATIONS

Everywhere one looks as the century concludes there is evidence of the permeation of evolutionary thought, and continuing fascination with Darwin as one of the great men of Victorian civilisation. Reviews of the nineteenth century consistently emphasised the fundamental importance of his thought. The *Origin* not only routinely featured in the lists of '100 Best Books' which became a fashion at the *fin de siècle*, but was frequently identified as the most important single work of the century.

While most early-Victorian thinkers seemed largely to have passed out of public view, Darwin was still an author urged on readers. Sales of the *Origin* and *Descent* in cheaper popular editions in the 1890s were kept buoyant by the Darwin family's strategic programme of reissues. The standard editions continued to sell steadily, supplemented first by half crown versions, and then by shilling editions.[3] The inclusion of the *Origin* in Routledge's 'Hundred Books' series created a whole new audience, and *More Letters of Charles Darwin* (1903) provided further insights into Darwin's thought. Murray brought out a 1/- paperback edition in 1901, and the expiry of copyright on the first edition of the *Origin* in 1902 prompted a rush of cheap reprints, even though by this time second-hand copies of early issues were being bought for as little as 4½d, even by the generally non-book-buying artisans of London's East End.[4] These editions placed the *Origin* in the holiday reading class for 'the modern, eager, restless intelligence'.[5] The *Origin* was available in a cloth edition in the Hutchinson classics series for only 9d by 1906, and 8d in the general section of Cassell's People's Library by 1909, and the 'flood' of sixpenny paperbacks which characterized the early 1900s included several Darwinian texts, including Grant Allen's *Evolution of the Idea of God*, 30,000 of which were printed by the Rationalist Press Association, whose titles circulated widely in the working men's clubs of the north.[6] Darwin was being made available to new audiences, but also to previous readers who now had the chance to own and carefully study their own copy, many of which became the repository of newspaper cuttings and personal annotations, by which they were transformed almost into personal relics.[7]

Darwinism had seeped deep into the cultural fabric. Take the opening of John Galsworthy's archetypical Edwardian rumination on family and change, *Man of Property* (1906), with its sustained reflection on the meanings of a family gathering in which biological analogies jostle with anthropological associations and generational relationships, heavily underscored by reference to hereditary fitness—the prevailing of the Forsytes against a 'hundred other plants, less fibrous, sappy and persistent'.[8] Students in the universities, trainee ministers at theological colleges, ambitious professionals in their

DARWINIAN DEBATES AT THE *FIN DE SIÈCLE*: THE EDWARDIANS   327

literary societies, young men on the make in the colonial civil service, brothers and sisters in kitchens and drawing rooms, all were reading their Darwins.[9] By this point, in the picture presented by Richard Le Gallienne's *The Romance of Zion Chapel* (1898), for a new generation the old Lit and Phils were the preserve of mild old men offering little more than treasurers' reports 'with an occasional paper on fossils', being confronted by younger groups interested in Morris wallpaper, Ibsen, Wagner, the nude in art, and 'The Darwinian Theory'.

Not that Darwinism had been entirely domesticated. Indeed, what is striking about the *fin de siècle* is the durability of the controversies and the controversialists of the 1860s. The correspondence from his audience which Karl Pearson's public lectures to Gresham College prompted in the early 1890s shows that Darwinian topics, from the origins of language to the authority of Weismann, continued to trouble many.[10] When Henry Vivian, Baron Swansea (1821–94), told a meeting of the Welsh National Eisteddfod in 1893 that 'I don't believe in evolution', the reaction of the audience apparently suggested general agreement.[11] The responses to Lord Salisbury's presidential address to the British Association in 1894 reinforce this impression. Salisbury's broadside against evolution, skilfully framed as an exploration of the dangers of dogmatising on topics where scientific proof remained elusive, caused a considerable stir. Despite an effective riposte from Huxley, who pointed out that Salisbury had in effect conceded the crucial proposition of the evolutionary history of the earth, editorial writers in the conservative press were emboldened to restate previously submerged doubts, not just about the Darwinists' denial of design, but even about the fact of evolution itself. The private correspondence of the scientific community buzzed with outrage at Salisbury's effrontery, and young Edwardian undergraduates looking on ridiculed him as a reactionary old chemist.[12] But as a thoughtful editorial in *Nature* observed, the wider response showed how far the general public, rather than being convinced of evolution, merely seemed in the face of a mass of technical argument and the endorsement of leading men of science to have 'relapsed into sullen acquiescence', ready, even eager, to overthrow their adherence if they felt they realistically could.[13] Salisbury's clever attack had shown that a skilled debater might make a considerable breach in what had been taken to be an impregnable stronghold, and many followed with enthusiasm.

Newspapers continued to ring periodically with clashes of pro- and anti-evolutionists. The reissue of books like Max Müller's *Chips From a German Workshop* (1895), and new interventions like the Duke of Argyll's

328   DARWINISM'S GENERATIONS

Organic Evolution Cross-Examined (1898), or *Doubts on Darwin by a Semi-Darwinian* (1903) by the wealthy property owner Charles Morrison (1816/7–1909), periodically recharged discussion.[14] Debating societies still found Darwin a fruitful source of disagreement, even if opponents were now more clearly in the minority. When the John O'Groat's Debating Society considered 'Creation vs Evolution' in 1900, not one of the members came forward willingly to advocate evolution, and creation was not surprisingly carried by a large majority.[15] Undergraduates still heard echoes of the earlier controversies, faint but recognisable as the remnants of past thunder.[16] Passions were undimmed in the autodidact culture of the north. In her mid-teens the older family of Lancashire millgirl Annie Kenney (1879–1953) were, as she recalled, 'reading Haeckel, Spencer, Darwin. Mother would be as interested as we were until the arguments got so heated that she felt it was wise to close the discussion because of the younger children'.[17] In South Wales, as a young clerk, Thomas Jones (1870–1955) was driven to write to W.H. Dallinger for advice about how to sustain the evolutionary argument in his furious debates with a co-worker.[18] Evolutionary frames saturated theological discourse, and yet in many congregational circles bounds to the acceptance of Darwinism continued to be vigorously policed.[19] Take the Paisley Presbyterian minister Andrew Henderson (1825?–?), lecturing in 1900. Henderson pronounced his willingness (albeit reluctantly) to accept the fact of evolution, but not anything labelled as 'Darwinism' or accidental evolution. Instead he presented himself as aligning with Charles Kingsley's view of an 'ordered universe in which rational purpose was being wrought and in which an intelligent mind was working out its design'.[20] These continuities were reinforced by the presence of a figure like William Thomson (now Lord Kelvin), digging in at the Victoria Institute in 1897, and also eliciting rebuttals from William Thiselton-Dyer, Karl Pearson, and W.H. Mallock, when in 1903 he asserted that in the light of science people 'were absolutely forced…to admit and to believe…in a directive Power'.[21]

Perhaps in retrospect these were increasingly irrelevant debates. St George Mivart found himself socially ostracised and intellectually isolated as the century closed; 'a sort of English, scientific "Dreyfus"', he lamented.[22] Max Müller was criticised for merely re-enunciating old solutions when the conditions of the problem had altogether changed, and the Duke of Argyll's *Organic Evolution Cross-Examined* (1898) was dismissed as 'a kind of evis-cerated Bridgewater Treatise with an aggressive binding'.[23] But all help to remind us that the 'eclipse of Darwin' around 1900, which has been traced by historians including Peter Bowler, was a complex phenomenon, part the

DARWINIAN DEBATES AT THE *FIN DE SIÈCLE*: THE EDWARDIANS 329

last gasps of those who had always been unconvinced, part a growing anxiety of the original Darwinians at the ways the theory was being applied, and part new doubts about its scientific basis, all stirred in with the continued energetic promotion of Darwinism by high and late Victorians. Argyll was all too aware that he was arguing in terms which were becoming ever more obviously outmoded.[24] Surviving mid Victorians continued to provide a sounding board against which some Edwardians shaped their own responses to Darwin. These currents were not insulated from each other, but rather intersected in mutually reinforcing ways. The journalist H.W. Massingham (1860–1924), having abandoned the Puritan religion of his upbringing, seems to have been deterred from any dalliance with Darwinian ideas by the hostility of the mid-Victorian Swedenborgian who edited the newspaper where he was first employed in the 1880s.[25]

Spectating high Victorians railed, in Edward Clodd's words, against the 'indifferentism' born of people knowing nothing of the long fight for the acceptance of evolution.[26] If anything, their Darwinism became more assertive. Meanwhile, late Victorians continued their skirmishing across the frontier between neo-Darwinism and Lamarckism.[27] Twenty years after the first volleys of the Weismann controversies, the discussion continued to simmer in the intellectual periodicals.[28] Edward Poulton was in America in 1894, where he was alarmed at the prevalence of Lamarckism.[29] Along with Lankester, still hypersensitive to the slightest suggestion that Darwin's conception was being undermined by more recent scientific discoveries, Poulton was becoming the *de facto* organiser of the Darwinist cause in Britain, more Darwinist even than the Darwins. In 1912 Poulton even went as far as launching a new quarterly, *The Bedrock*, designed to sustain the purist version of the Darwinian faith.[30] He continued to fulfil this role into the interwar period, regaling Oxford undergraduates with accounts of his conversations with Darwin, and campaigning for the reinstatement of Darwin's statue to its original spot in the Natural History Museum—discreetly of course, to save the Darwin family's public blushes.[31]

The Edwardians were largely unmoved by such doctrinal disputes. They wanted light, not warmth.[32] Argyll was an irrelevance, the positions he had struggled for so long to maintain having been entirely swept away by the passage of time.[33] Huxley, still fighting the old fight, 'penetrated to the marrow of his bone' by his struggles with theology, struck them as from a past era. To those of the new generation, 'the generation [he] "made possible"', Peter Chalmers Mitchell conceded, Huxley's later essays seemed unnecessarily combative.[34] Lankester's posturing as 'the "brutal scientist"' also smacked of a throwback to 'those days when Darwinism served as a brick to be hurled at the head of bishops'.[35] Attention was shifting to the new physico-chemical discoveries, subatomic particles and the electrical basis of matter, and the Edwardians were more likely to develop an adolescent passion for electricity than for natural history.

Perhaps the distinctiveness of generational engagements with evolution was lessening. The point in adolescence or early adulthood at which individuals encountered Darwinian thought was becoming more and more variable. Darwin's ideas had obtained such wide and often indirect currency through paraphrase and quotation that they were frequently unconsciously adopted, even by opponents. And the juvenile experience of the Edwardians was often of Herbert Spencer rather than Darwin as the most important evolutionary thinker, even if it was uncommon for him to remain so. The result of the greater attention to Spencer was often commitment to a diffuse cosmic evolution, and a generalised Darwinism in which the significance of the *Origin* was less in its particular arguments, and more that, as one Edwardian put it, it 'gave force and authority to the older doctrine of the continuous development and progression of life'.[36] This was a gradual eclipse, more clearly articulated after 1918, when invocation of the foundational nature of the *Origin* or *Descent* came to be often entirely gestural, and Darwinism for many became irreparably tainted with the international struggle for supremacy which had caused the carnage of World War 1; but the drift in this direction was clearly visible before 1909.

The Edwardian Frame of Mind

Poulton's object in *The Bedrock* was as much to challenge the new heresies of mutation theory and Mendelism as it was to resist old heresies like Lamarckism. By the time of its launch it was clearly around the new ideas of a younger generation that the crucial debates were clustering, the generation described by Neville Talbot (1879–1943), son of E.S. Talbot, as those who 'never knew—they were not themselves moulded by—the times before the "sixties". They were not born, as their parents were, into the atmosphere of pre-"critical" and pre-Darwinian religion'.[37] This was a generation educated in the context of the general hagiography of the years after Darwin's death, and who were emerging into cultural and then political leadership in the years after 1901. A significant part of the contribution of this Edwardian generation came in the radically altered context of post-1914 Britain, but

DARWINIAN DEBATES AT THE *FIN DE SIÈCLE*: THE EDWARDIANS   331

their presence in the journals and periodical press of the 1890s and 1900s helped once again to add a new layer to Darwinian debate.

For most Edwardians Darwinism was what they were brought up with: the position espoused by their parents and inculcated by their teachers, the current orthodoxy. Darwinian texts were a normal presence in Edwardian family life, and a broad acquaintance with Darwin's works was characteristic of educated Edwardians of all classes. Chalmers Mitchell remembered having grown up in an intellectual atmosphere 'impregnated by the Doctrine of Descent', and both Hilaire Belloc (1870–1953) and the traveller Mary Kingsley (1862–1900) traced their Darwinian beliefs directly to their family upbringing; not without intimations of unease at what Kingsley described as 'space and atoms and Darwinism and all that sort of Ju-ju'.[38] For the novelist Arthur Quiller-Couch (1863–1944), Darwin was one of the 'household gods' of his youthful memory.[39] If late Victorians felt that Darwinism had transformed their vocabularies, some Edwardians went further, recognising that it had come to structure the way they thought. 'We think in Darwinian terms', remarked the publisher Laurie Magnus (1872–1933), 'We are all intellectual Darwinians, whether consciously or unconsciously'.[40]

Of course, it was still possible to grow up without exposure to Darwinian ideas, and for the reading of the *Origin* to deliver a profound intellectual trauma. The astronomer Richard Gregory (1864–1952), the Irish playwright John Millington Synge (1871–1909), and the Georgian poet Alfred Noyes (1880–1958) all provide autobiographical accounts of delayed and difficult encounters with Darwin akin to those discussed earlier for the late Victorians.[41] May Sinclair's (pen name of Mary Amelia St Clair (1863–1946)) overwhelmingly autobiographical *Mary Olivier* (1919) also bears testimony to the continued potency of the act of reading the *Origin* (borrowed in Sinclair's case along with Spencer and Haeckel from the London Library through a friend) as a mechanism for the destruction of Victorian theological and also parental authority. Given that for those entering adolescence in the 1890s even the culture of provincial nonconformity offered young men the chance for largely unconstrained discussion of Darwin's works, such intellectual crises were likely to come earlier, like G.M. Trevelyan's pre-Harrow revelation, and be less troubled.[42] Hence the easy acceptance of W. Somerset Maugham (1874–1965), rendered in the response of 'Philip Carew', the hero of *Of Human Bondage* (1915).[43]

E.S. Talbot told Poulton in 1897 that he found Darwin's *Life* in every house that he had been in.[44] It was suggested that copies of the *Origin*

332   DARWINISM'S GENERATIONS

multiplied on drawing room tables as a deliberate attempt to demonstrate that if the religious opinions of the house were heterodox, the shift had at least come from an impeccably orthodox source.[45] For the Antarctic explorer Edward Wilson (1872–1912), reading of the *Voyage of the Beagle* as a seventeen-year-old in 1889 prompted a lifelong habit of keeping a detailed journal, which in turn recorded how he, along with Scott and Shackleton, took a copy of the *Origin* with them as their only book on their sledging expedition across the Antarctic in 1902, reading it aloud (on one occasion reading it 'for lunch' because they were short of food) and then discussing it as they hunkered down at night or waited for bad weather to pass.[46]

Overwhelmingly, the educational experience of these Edwardians was a Darwinian one. At public schools, where once libraries contained only refutations of Darwinism, Darwin's books were now widely available.[47] Teaching was increasingly expected to encompass evolutionary perspectives.[48] It was said of the industrialist Alfred Mond (1868–1930) that as a twelve-year-old at Cheltenham he lost respect for a teacher who confessed to not having read Darwin.[49] Darwinism helped to shape underlying educational approaches. For M.E. Braddon, this was the time when 'schoolgirls with flowing hair and short petticoats pronounce themselves Darwinians, and ask if anyone can really believe the old-fashioned creeds that sustained their grandmothers'.[50] When in the later 1880s the future Labour MP James O'Grady (1866–1934) began exploring the implications of the organicism of Spencer, he was not only able to read through Darwin, Weismann, Haeckel, and others, but also immerse himself in lectures on evolution and biology.[51]

At university the Edwardians learned from the Darwinians of previous generations. Arthur Keith (1866–1955), who became one of the leading evolutionary writers of the first half of the twentieth century, recalled that at the University of Aberdeen while studying medicine, his enthusiasm for Owen, Huxley, and Darwin became more important than the arts of healing.[52] Darwin's works were routinely distributed as prizes. Chalmers Mitchell was awarded the *Origin* at Aberdeen in the early 1880s: 'Darwin amazed me' he later recalled, 'full of exciting facts... [and] conclusions that seemed inevitable'.[53] The teaching of Ruskin Hall, Oxford, founded in 1899, was said to be entirely 'based on Evolution, and the student is enabled to view all facts in the perspective of Evolution'.[54]

Evolution was the new orthodoxy. And perhaps in consequence, in many respects the early 1880s marked the high water of Darwinian influences in student culture. University continued to be a site of religious questionings

DARWINIAN DEBATES AT THE *FIN DE SIÈCLE*: THE EDWARDIANS   333

and the active embrace of evolutionary ideas, but the excitement of breaking new ground had long gone, and the frisson of intellectual rebellion was now to be found not in the passage from Christianity to agnosticism, but from evolutionary enthusiasm to Darwinian disillusion.[55] While the heroes of undergraduates in the 1860s and 1870s might have been Darwin, Huxley, and the authors of *Essays and Reviews*, the heroes of the young men of 1890s Cambridge were more often 'Strindberg, Nietzsche and (for a time) Wilde'; 'The young admired their passion', Bertrand Russell recalled, 'and found in it an outlet for their own feelings of revolt against parental authority'.[56]

Most Edwardians lacked even the dimmest memories of the intellectual universe before Darwin's intervention. The very oldest Edwardians, like the zoologist Sir Arthur Shipley (1861–1927), who went up to Cambridge in 1880, were just about able to remember being contemporaneous with Darwin's later works, and the feeling of being not so far off from the *Origin*.[57] But more typical was the perspective of the slightly younger civil servant Sir Lionel Earle (1866–1948), for whom Darwin was merely one of the great Victorians whose funerals he remembered from adolescence.[58] The explorer Francis Younghusband (1863–1942) recalled his twenties as a time when although Darwin was dead, Wallace, Huxley, and Spencer were contending furiously with the opponents of evolution; and he could read the *Origin* and luxuriate in the spaciousness of the theory and the sense of interrelationship of all beings it offered.[59] During his student days at the Normal School of Science in the mid-1880s, H.G. Wells listened to Huxley, read his books, and clubbed together to buy the *Nineteenth Century* out of his weekly guineas whenever Huxley 'rattled Gladstone or pounded the Duke of Argyle [sic]'.[60] But the controversies of the 1860s seemed almost prehistoric, located back beyond the reaches of personal memory, and their contemporary echoes more ritual sport than vital debate.[61] The positions of the early and mid Victorians appeared simply indefensible. As the anthropologist Robert Ranulph Marett (1866–1943) put it, 'when it came to a stand up fight between Max Müller's words and Tylor's institutions as championed by Lang, I sided at once with the party using what was plainly the heavier artillery'.[62] 'We have almost forgotten the terror of a godless universe, which was inspired in the first generation who read Darwin and Huxley', the clerical essayist W.R. Inge (1860–1964) reflected in 1906.[63]

It was impossible for the Edwardians not to recognise that it was the Darwinists who now represented the mainstream. For the actress Mrs Patrick Campbell (1865–1940), Darwin impinged first and foremost as establishing the lines of mental development which distinguished her

334 DARWINISM'S GENERATIONS

parents.[64] The conventional loss of faith plot so common in the novels of the 1880s lost much of its resonances. Both the fact of evolution and the transmutation of species by natural selection had been fully naturalised. There was a matter of factness about Edwardian perspectives. Darwin might have invalidated the Mosaic account of creation, but the Edwardians experienced this as only incidentally threatening.[65] Some sort of accommodation with Darwin was established by all but the most recalcitrant religious writers, even if it was a reconciliation that often had little engagement with Darwin's particular theories, and made no attempt to address the challenges they presented to central Christian doctrines.[66] In the 1890s G.K. Chesterton, who was later to develop a strongly anti-Darwinian rhetoric, shared an educated Catholicism that had accommodated itself to evolutionary thought, although he felt his sensibility was a world away from Huxley and his contemporaries, and the assumptions 'at the back of the minds of most men of that generation I ever knew'.[67]

The result was a somewhat schizophrenic stance towards Darwin. As the symbol of evolutionary thought, he remained the cornerstone of the Edwardians' world view. Even the ardent Ruskinian Bellerby Lowerison (1863–1935) could warn that 'He who knows not Darwin and his master-key, evolution, enters nowhere today'.[68] And yet at the same time, once Darwinism was installed as the conventional wisdom it was always likely to become the target of young inquirers seeking revelation and rebellion. Parental enthusiasm for Darwin became an object of disdain.[69] Elders were called to account for their acceptance of Darwinism.[70] Repudiating cultural orthodoxy was for the first time more likely to involve anti-Darwinism.

From the 1890s new Edwardian voices began to mould a self-consciously post-Darwinian world. As his standing waxed, so Darwin's grip on contemporary thought waned. Darwin slipped below the horizon of popular natural history.[71] Shrill late-Victorian assertions of Darwin's significance left the Edwardians unmoved. Darwin, although still revered, was no longer required; his achievements were detaching from the pressing issues of contemporary debate; they were terminus of the battle between scholasticism and science, not a point of departure for modern thought.[72] Commissioned to write one of the clutch of biographical studies of Darwin that appeared in the decade after Darwin's death, the young biologist D'Arcy Thompson (1860–1948) found that he just couldn't connect with the subject, and abandoned the attempt.[73]Attention was shifting not just from the fact of evolution to its dynamics, but from dynamics to their implications. The first night at Somerville of one new student in the mid-1890s was taken up with a keen argument over cocoa, not about the *Origin* or *Descent*, but about Benjamin Kidd's *Social Evolution*.[74]

The Edwardians refused to accept Darwinism as a settled body of dogma; generally, they were not just uninterested in, they were often actively fearful of, fixing an evolutionary orthodoxy. They sought to redefine Darwinism not as the '*ipsissima verbe*' of Darwin himself, but as 'the living doctrine' that had developed from his writings.[75] The biologist Arthur Willey (1867–1942) accepted Darwin's work as a guide; but he was anxious to avoid a recrudescence of oppressive orthodoxy.[76] The war correspondent William Beach Thomas (1868–1957) was able to combine an abstract commitment to the evolutionary process with a running counterpoint of sly digs at 'the gospel according to Darwin'.[77] For Ford Madox Ford (1873–1939) and his contemporaries, 'the Distinguished Unorthodox, the followers of Darwin, Huxley, and Ingersoll', were as unpalatable as cultural guides as the 'Deans and Archdeacons' had been to the late Victorians.[78] Evolutionary understanding was seen as itself in evolution.[79] As D'Arcy Thompson put it in his inaugural lecture in 1885, it was a mistake to suppose that evolution had yet settled down into a consolidated scheme, or that its supporters were pledged 'to the stereotyped clauses of an evolutionary creed'.[80] Nor were the Edwardians particularly interested in advocating a rival confession. Instead, attention was focused on moving the debate beyond this impasse.

Revealing in this respect was the Edwardians' rehabilitation of Samuel Butler. A rather forlorn and frustrated figure for much of his life, after death Butler was recuperated as a poster boy of anti-Darwinism. The popularity of Butler amongst the younger Bloomsbury set is well known, but the correspondence of Butler's literary executor Henry Festing Jones (1851–1928) reveals a surprisingly broad sympathy for Butler in the Edwardian years. When the journalist William Purvis (1869–?) wrote an appreciation of Butler in the later 1900s, he received several hundred letters seeking further information.[81] Not unsurprisingly, some older readers pounced on Butler's republished works as an opportunity to reheat ancient arguments, converting Butler into a freedom fighter 'against the Sanhedrin over which Professor Huxley ruled with a rod of iron'.[82] For Oscar Wilde's champion Robert Baldwin Ross (1869–1918), without understanding Butler's science, the story of his dispute with Darwin illuminated the 'preposterous attitude which belonged to the whole school'.[83] But what was especially striking was the warmth of the younger generations. The Mendelian A.D. Darbishire (1879–1915) was reportedly a tremendous admirer, as indeed was the

336 DARWINISM'S GENERATIONS

architect of the welfare state, William Beveridge (1879–1963).[84] Butler's brand of vitalism was also useful for New Liberals as a way of combatting the individualist version of social Darwinism.[85] This was in part the appeal of Butler the subversive Victorian (despite his personal dislike, Ford Madox Ford came to see Butler's trenchant novel of filial anti-Victorian revolt, *The Way of All Flesh* (1903), as of 'vastly more use to us today than is *The Origin of Species*'),[86] but it was also a fascination with Butler's criticism of Darwinism, and sympathy with his hostility to the Darwinian establishment.

There is no doubt that as a result of several unfavourable forces, not least professionalisation, the spread of laboratory science, and the emergence of new and potentially rebarbative genetic approaches to evolutionary questions, the Darwinian scientist became less heroic, and often a great deal more sinister. This shift was presaged by the rise of the anti-vivisection movement in the 1880s, but it was greatly extended by the Edwardians. Its *locus classicus* is of course H.G. Wells' *The Island of Doctor Moreau* (1896), and the horror which it expressed at the implications of eugenic experimentation.[87]

Edwardians who had imbibed Darwinism as adolescents were, by their twenties and thirties, beginning to ask searching questions about its explanatory capacities. The short story the ILP activist Margaret McMillan (1860–1931) published in the *Clarion* in 1895 presented the debates over Darwinism as too detached from the realities of working-class lives, and offered a vividly jaundiced version of the increasingly common Darwinian conversion narrative. The story opens with 'Young Hope', the socialist activist, reading the *Origin* by candlelight. 'He had read it years before, and found it dull. Tonight every word came to him as a revelation. He forgot that he was cold and hungry'. He jumps to his feet. '"I see it!", he cried, "it is life that struggles; it is life that conquers; it is life that transforms"'. But at that moment he is visited by Death, gloating over the fatal toll of diseases multiplying in the unventilated mills and workshops, but also in Labour Churches and trade union rooms. Hope tries to drive off Death, 'groping for his Darwin, which had fallen to the floor', but he succumbs, and within three days he too is dead.[88]

The Shift to Heredity and Variation

The science of evolution was moving into a fresh phase. In the later 1880s, while public attention was engrossed in the Spencer–Weismann debate,

DARWINIAN DEBATES AT THE *FIN DE SIÈCLE*: THE EDWARDIANS   337

younger researchers, frustrated at the speculative nature of Darwinian understandings of variation and heredity, exasperated with their over-reliance on 'abstract theories', and impatient with techniques which seemed to have reached the limits of their usefulness, were turning to experimental approaches in search of a way beyond the sectarian divisions of the late Victorians.[89] It was clear, H.G. Wells commented in 1895, that '[t]he study of variation is the newly favoured method of attacking the problem of species'.[90] Too much of evolutionary theory struck Edwardians as assumptions made to explain observed phenomena, relying too much on guesswork, especially in respect of questions of inheritance.[91] 'Horace knew probably as much about heredity as most of us do', remarked the social reformer Clara Collet (1860–1948) in 1900.[92] The Edwardians became convinced that mor-phological studies of species pedigrees had gone as far as they could, and that cytology had more to offer than endlessly extended phylogeny.[93] Young scholars found themselves negotiating a narrow line between their desires to open up new avenues of exploration, the pressure of supervisors and mentors still wedded to Darwinian approaches, and the icy disapproval of the high-Victorian Darwinian old guard.[94] The anatomist turned anthro-pologist Grafton Elliot Smith described W.H. Flower's 'particularly freezing' response to his work as 'exhibit[ing] in its highest development the South Kensington contempt for morphologists who recognise something more in an animal than a mere factor in a phyletic tree'.[95] While books like the late Victorian F.O. Bower's *The Origin of a Land Flora* (1908) attempted to reinstate the study of development 'in its proper position in phyletic argu-ment', the younger physiologist Horace Middleton Vernon (1870–1951) was working on *Variation in Animals and Plants* (1903), which from the outset emphasised Vernon's desire to address what he saw as one of the most sig-nificant lacunae in Darwin's researches.[96]

The hard exclusions of Weismann's germ plasm theory remained influen-tial, but across not just natural science but also philosophy and theology, natural selection was increasingly interpreted as a purely negative factor, which could do nothing without variation.[97] When Darwinists spoke of the omnipotence of natural selection, the philosopher F.C.S. Schiller (1864–1937) noted, they failed to observe that their opponents had really turned their flank, solving problems which, on the basis of Darwinism alone, could not be discussed, and leaving the Darwinians without 'any logical locus standi'.[98] In works such as Marion Newbigin (1869–1934)'s *Colour in Nature: A Study in Biology* (1898), natural selection was not renounced, but it was relegated to the ranks of secondary explanation.[99]

338 DARWINISM'S GENERATIONS

Characteristically, the Scottish biologist J. Arthur Thomson (1861–1933) described himself as believing in natural selection 'but in more besides.'[100] By 1915 it was possible for one leading Edwardian biologist to go as far as to suggest that it was ludicrous to assert that natural selection and the struggle for existence had any claims to be regarded as scientific law.[101] This dethronement was particularly pronounced in botany, leaving contemporaries amazed at the way the field had become almost entirely Lamarckian by the end of the Edwardian period.[102] The popular botanical works of G.F. Scott-Elliot (1862–1934) or George A.B. Dewar (1862–1934) offered a sort of diluted Darwinism; Scott-Elliot's with Lamarckian concessions, in which natural selection could not be rightly described as a cause, Dewar's with a sort of Richard Jefferies inspired romanticism that the utilitarian laws of adaptation could not explain all the beauty and wonder of nature.[103]

A.G. Tansley (1871–1955), later to become the leading figure in the development of British ecology, provides a characteristic instance of the trajectory of Edwardian biological thought. Taught by Lankester, Tansley's early work was conventionally Darwinian, examining the evolution of the vascular system of Filicinean ferns.[104] In the 1890s he wrote popular essays defending the neo-Darwinians, and helped Herbert Spencer with his revised edition of the *Principles of Biology*.[105] But Tansley's 1894 scholarship essay on natural selection clearly shows the crowding in of new questions, including variation, which traditional Darwinian approaches were ill-equipped to address, and the *New Phytologist*, which Tansley founded in 1902, became a vehicle for his advocacy of a shift from questions of origin and descent to the activities of living organisms and the problems of mechanism and process.[106] Tansley's stance in the years before 1914 in respect of evolutionary questions is captured in a lecture on 'Modern Views of Heredity and Evolution', which he gave to the London Working Men's College in March 1909, a broad-ranging survey which stressed the inadequacy of both neo-Darwinian and neo-Lamarckian positions. Instead, he told his audience, through the new science of heredity, 'a flood of light has been thrown upon these questions, and the whole subject has been immensely advanced, so that we have definitely escaped from the period of stale speculation and fruitless polemic of 15–20 years ago.'[107]

During the 1890s, the Cambridge biologist William Bateson (1861–1926) took up the question of heredity with fresh determination. Bateson was one of the oldest of the Edwardian generation, imbibing, his biographers suggest, art, culture, and Darwinism with his mother's milk, and very much in the vanguard of the first truly post-*Origin* cohort, whose early professional

DARWINIAN DEBATES AT THE *FIN DE SIÈCLE*: THE EDWARDIANS 339

careers were taking shape in the second half of the 1880s and early 1890s.[108] His preoccupation from the outset was the impasse in evolutionary theory. 'My brain boils with Evolution. It is becoming a perfect nightmare to me', he confessed in 1888.[109] Bateson was quickly convinced of the need for a new approach. 'Five years hence no one will think anything of that kind of work, which will be very properly despised', he wrote privately of the phylogenetic work which had helped gain him his Cambridge fellowship in 1885.[110] He was soon locking horns with late-Victorian Darwinists, even those only a few years older than he was. One Cambridge colleague, Adam Sedgwick, bluntly told Bateson in 1890 that while his work was not unimportant, his 'views on zoology—on the morphology side—[were] stupid and narrow'.[111]

Bateson's first major publication, *Materials for the Study of Variation, Treated with Especial Regard to Discontinuity in the Origin of Species* (1894), could not have been clearer in announcing its revolt against orthodox Darwinism. The 'forced pleading' of natural selection, established morphological approaches, arguments from recapitulation, and the tendency of Darwinists to assert rather than demonstrate the adaptive purposes of structures, were all questioned.[112] *Materials* offered nearly nine hundred examples of discontinuous variation in the evolution of species. Francis Galton was enthusiastic, but Wallace and the ultra-Darwinians were unsurprisingly sceptical.[113] Bateson's questioning of some of the foundational ideas of Batesian mimicry had already brought him into conflict with Poulton, who he virtually accused in one letter to *Nature* in 1892 of selecting his facts to fit 'the sustenance of a facile hypothesis', and now he added William Thiselton-Dyer to his list of public antagonists, via another ill-tempered exchange in *Nature*.[114] Even the more sympathetic late Victorians baulked at Bateson's ideas. F.E. Beddard recognised *Materials* as epoch-making, but remained unconvinced that it provided a basis for new theories, while the neurologist Charles Scott Sherrington (1857–1952) praised an early draft as 'very valuable and intensely interesting', but confessed he still believed that infinitely gradated continuous action could produce discontinuous effect.[115] By contrast, Bateson's distinction between continuous and discontinuous variation had an immediate impact on the work of younger biologists.[116] Initially developing their research careers under the supervision of late Victorians, the Edwardians gradually repudiated the established methods and the old questions. Zoology moved rapidly from the study of transitional species to the study of heredity, and from the field to the farm and the laboratory.[117]

340 DARWINISM'S GENERATIONS

Younger observers displayed a more general tendency to be suspicious of theory and to rely exclusively on experimental deduction. The high and late Victorians had not of course been opposed to laboratory science, or to experimentation. Many of Darwin's insights had been informed by his experiments at Down, and Huxley had championed comparative anatomy as an alternative laboratory-based science, much to the annoyance of many amateur naturalists. The 'New Botany', somewhat nebulous, but encapsulating an interest in plant functionality and process, and an approach to research based on experiment as well as observation, and the sort of breeding experiments undertaken in the 1890s by James Cossar Ewart and C. Lloyd Morgan were both indications that the shift to laboratory or experimental science was not the responsibility of any one of the Victorian generations.[118]

Nevertheless, in attitudes to experimentation also, there was a generational divide. At its heart, Darwinism had been a field science, and many of the leading Darwinians were reluctant to accept the criticism implicit in the calls for systematic laboratory investigation. Wallace attacked the modern school of laboratory naturalists in 1889, and to the end remained suspicious of the new work of the Edwardians, arguing for example that the inability to replicate experimental results in the wild was fatal to new mutational theories, and suggesting that the artificial circumstances of the laboratory 'introduces some provocative condition that is lacking (or latent) in nature'.[119] T.G. Bonney, in his 1890 Boyle lectures, and Thiselton-Dyer in his address to the British Association in 1895 both challenged the move towards the laboratory, as something which had not contributed much to the Darwinian theory and which would have left Darwin 'exhausted by the infinity of detail'.[120] Late Victorians were generally unsympathetic to this sort of rhetoric, but Lankester too criticised the increasing number of younger students, 'who have little or no interest in natural history beyond what is derived from contemplation of ribbons of sections dyed like Joseph's coat'.[121]

Experimentation and the Evolution Committee

As the century drew to a close, interest in greater experimental rigour resolved in particular into questions of how it might be possible to organise sustained explorations of heredity, and the course of these debates effectively illustrates the lines of generational division. Initially, efforts to establish

DARWINIAN DEBATES AT THE *FIN DE SIÈCLE*: THE EDWARDIANS 341

a sustained programme of laboratory investigation appear to have foundered on the lack of enthusiasm of the leading Darwinians. One base might have been the Marine Biological Association laboratory established in Plymouth in 1888; but with Lankester as a dominant presence, innovative experimental investigation seems to have been largely baulked.[122] In 1890 Wallace did propose to Galton some sort of field station for the investigation of evolutionary questions, but Herbert Spencer and Thiselton-Dyer threw cold water on the suggestion.[123] Darwinians instinctively fell back on the lone scholar model: 'the ideal', Alfred Newton told Bateson, 'would be to work out the business alone—as did Darwin'.[124] But the question was not going to go away while younger, Edwardian, investigators were increasingly invested in it. 'Of course Darwin had to invent the subject itself ab initio', Bateson retorted, 'but it has often struck me as a noticeable fact that he did not succeed in getting much systematised /or statistical\ information on these points'.[125] Bateson persisted in pressing the 'quite exceptional importance' of an experimental farm which would allow for continuity of observation.[126] By the turn of the century there were conversations in Scotland about the creation of an Inter-University Institute of Experimental Evolution, enthusiastically supported by J.A. Thomson.[127]

For a while these debates aired differences between high and late Victorians. In 1894 J.T. Cunningham circulated a proposal for the formation of a Society for the Investigation of the Problems of Evolution (or 'The Darwinian Society', he suggested).[128] What was needed, he argued, was a much more experimental approach in both zoology and botany, and his proposed society was to provide the resources needed to sustain the long-term investigations required. The proposal met fierce hostility from the older Darwinians. Huxley and Foster were quick to ridicule Cunningham's pretensions and question his motives.[129] Lankester relapsed into personalities, warning Huxley that the proposed society would simply be the means for 'a few creatures like Mr Cunningham to climb on Darwin's shoulders'.[130] In dismissing Cunningham as narrow-minded, competent in descriptive work, but incapable of rising to effective speculation, Thiselton-Dyer denigrated the whole detailed experimental approach he was advocating.[131] In contrast, for Patrick Geddes, it was the Darwinian name which caused him to pause. It rather begged, he thought, one of the key questions.[132] Instead, in 1894 some of the most prominent late-Victorian Darwinians, including Poulton and Meldola, with Galton as chair, established the 'Committee on the Measurable Characteristics of Plants and Animals' under the auspices of the Royal Society, and eventually in 1896 Bateson's persistent pressure from

342 DARWINISM'S GENERATIONS

the sidelines persuaded them to convert this into an 'Evolution Committee', with an expanded membership including Bateson himself.[133]

The Evolution Committee was an ill-fated enterprise, not least because, riven by generational divisions, it failed to establish any sort of consensus as to priorities. For a while the committee became entangled in discussions over the possibility that Down House might be turned into some sort of memorial to Darwin.[134] Its acquisition as a permanent biological research station using the house for residential accommodation and the surrounding land for experimental work could well have been a solution to the problems of cost and remoteness which the various proposals faced, but despite intermittent efforts to raise funds no concrete scheme materialised.[135] (Eventually, in the late 1920s, after a chequered career which included time as a private girls' school, Down was bought by Sir George Buckston Browne in recognition of Darwin's achievements, and control was vested in the British Association.)[136]

Meanwhile, Galton wanted the committee to focus on the theories Darwin had developed in the *Variation of Animals and Plants under Domestication*.[137] Bateson, while prepared for the sake of expediency to situate his interests in the context of Darwin's work, was much less willing to allow the Darwinian canon to determine entirely the lines of investigation.[138] By the later 1890s he was attempting systematic breeding experiments with poultry to ascertain how far characteristic features of varieties were capable of blending when they were interbred.[139] The scientific importance of this class of work could not be overestimated, he told George Darwin in 1899, adding with characteristic disregard for the *amour propre* of the preceding generations, 'It is for want of continued observations on inheritance in animals and plants that so little progress has been made in the science of Evolution since Darwin's work'.[140] By 1899 Raphael Weldon was warning that the committee was 'in rather a mess', and things came to a head in 1900 when he and Galton resigned, along with Pearson, Thiselton-Dyer, and Meldola.[141] The committee did organise some experimental work from 1897, initially funding Poulton and Bateson, and presented a series of reports in the years after 1901 prepared by Bateson and the geneticist Edith Saunders (1865–1945), but by this time attention was already shifting towards the new questions thrown up by Mendelism. Ultimately, in 1908 Arthur Balfour anonymously funded a professorship in Biology at Cambridge, with a specific remit for the study of heredity and variation, to which Bateson was appointed, and in 1910 the Evolution Committee was finally allowed to lapse. Balfour had hoped that the Cambridge

DARWINIAN DEBATES AT THE *FIN DE SIÈCLE*: THE EDWARDIANS   343

professorship might have been called the 'Darwin Chair in Biology', but the University Senate, probably to Bateson's relief, decided that it would be inappropriate to attach Darwin's name to an unendowed chair.

Into the 1900s: Mutationism and Mendelism

What the history of the Evolution Committee reinforced was that the shift to questions of variation and the dynamics of heredity was a shift into areas where Darwin was not an effective guide, or even necessarily a useful starting point.[142] After the rediscovery of the pioneering genetic work of Gregor Mendel, first introduced into Britain in a lecture by Bateson to the Royal Horticultural Society in May 1900, the distance widened further.[143] Bateson's investigations in the later 1890s led him naturally to embrace Mendel's original insights into the patterns of inheritance visible in pea plants, and his challenge to the previously dominant theories of blended inheritance. At Cambridge, Mendelism took hold, primarily through Bateson's advocacy and leadership, but also through the researchers, in particular the group of pioneering women biologists, most prominently Edith Saunders, he gathered around him.[144] As the decade progressed, Bateson's denunciations of the limits of Darwinism became more forthright. As president of the Zoology section at the 1904 British Association he warned that while Darwin had opened up fruitful lines of investigation, his conception of species was barren and unnatural, and his 'indiscriminate confounding of all divergences from type into one heterogeneous heap under the name "Variation"' had been positively harmful. He continued his campaign at the third international conference on genetics in 1906, describing the *Origin* as a distraction from the process of evolution itself, and the cause of biologists' diversion into arid channels of palaeontology, classification, comparative anatomy, and distribution.[145] Bateson and Mendelism loom large over accounts of the Edwardian debates on evolution; sometimes too large. There are plenty of Edwardians, both outside the natural sciences, but also within, who shared in the generational contrasts which Bateson epitomises, while showing little direct influence of Mendelism itself. In this sense Mendelism was part of a broader shift which included the closely linked but never identical mutation theory of the Dutch botanist Hugo de Vries, and potentially also some of the work of the biometricians and of the emerging ecologists.

The dislocation from Darwin was not absolute. Chalmers Mitchell told Bateson in 1913 that he 'had been saying much that differed only by a hair's

344 DARWINISM'S GENERATIONS

breadth from much that you say, and that nonetheless I thought myself to be expanding Darwin's theory'.[146] Nevertheless, Bateson's personality both brings generational conflicts into stronger relief and overwrites them with personal animus. Wells described him in the 1920s as having a schoolboy pleasure in making trouble and a Samuel Butler-like hatred for Darwin.[147] Even by his own confession Bateson was hot-headed. The language of his private correspondence was scurrilous: his *bete noire* Karl Pearson was a 'rascal', Pearson's journal *Biometrika* was 'disgraceful', and so it went on.[148] It was his personality as much as the underlying science which meant that Mendelism emerged in opposition to, rather than as part of a development of, Darwinism, and not all Mendelians adopted his extreme anti-Darwinian tone.[149] Weldon's response to the rediscovery of Mendel's work was briefly enthusiastic, and although his efforts to incorporate some elements of Mendelian analysis into his work proved short-lived, this was largely a result of the anti-Darwinian uses to which Bateson was determined to put Mendel's methods.[150] As a result, the biological debate after 1900 largely resolved itself into a feud between Mendelians and mutationists on the one side, and the biometricians and older Darwinists on the other. Pearson and the biometricians were happy to play their part in sustaining the bad blood; even to one of Bateson's known allies, Pearson described Bateson's position as 'unintelligible', his work full of 'blunders', his tone 'vulgar', and dismissed Mendelian work more generally as 'superficial'.[151] The confrontation in the Zoology section at the 1904 British Association had echoes of Wilberforce and Huxley forty-four years earlier; Weldon attacking Mendelism, with beads of sweat dripping down his face, was met with Bateson brandishing *Biometrika* as incontrovertible evidence of the folly of the biometricians.[152] At times the surviving sources read like the records of gangland rivalries, as for example Weldon's widow, endorsing the establishment of a Weldon medal shortly after his early death in 1906, on the understanding that 'no elector must ever be a Mendelian'.[153]

The generational dimensions of these conflicts were more than usually ragged. There was no consistent late-Victorian resistance to Mendelian ideas; rather a grudging and limited acceptance of some of the Mendelians' findings, albeit with scepticism about their wider implications.[154] Lamarckians could see in Mendel further ammunition in the fight to cut down the scope for natural selection.[155] Weldon and Bateson were born scarcely more than a year apart, at the margins of the schematic dividing line between late Victorians and Edwardians; although Weldon briefly taught Bateson, they were to all intents and purposes contemporaries at Cambridge in the

DARWINIAN DEBATES AT THE *FIN DE SIÈCLE*: THE EDWARDIANS   345

1880s.[156] The work of the biometricians, seeking as it did to elucidate the dynamics of variation and heritable change, has more conceptually in common with Bateson and his circle than with that of Poulton and the late-Victorian neo-Darwinians, and did attract its own followers from the generation born after 1860 (including the statistician G. Udney Yule (1871–1951)), just as Bateson's work in the 1890s had drawn heavily on the same statistical inspiration, and shared much with the breeding investigations of younger late Victorians like Cossar Ewart.[157] But the older generations engaged with the debates primarily via investigation into Lamarckian acquired-character inheritance, and were inclined to reject mutation theory as a sort of reformulated special creation.[158] For a high Victorian like Thiselton-Dyer, who confessed that even a full grasp of Weismann's theory had been beyond him, Mendelism was a leap too far; for Poulton it merely came down to whether it was Darwin's earlier or later views which were correct.[159] He was happier concentrating his ire on 'the veiled Lamarckism which is all the fashion', as he did in an extended essay in the *Edinburgh Review* in 1902.[160] Meanwhile, many Darwinians probably shared Romanes' belief that mathematics in biology was 'like a scalpel in a carpenter's shop'.[161]

At times, the methodological disagreements between the biometricians and orthodox Darwinians brought their own tensions amongst Bateson's opponents. Riled at Lankester's observation in his 1905 Romanes lecture that nothing was being done to collect knowledge on the inheritance of mental and physical qualities, Weldon launched into an intemperate attack, describing Lankester as 'palaeolithic in [his] anthropomorphic childishness'.[162] Yet ultimately, the biometricians found themselves increasingly aligned with the late-Victorian Darwinians. It was not just that when the chips were down the Darwinians sided with the biometricians, but also that biometrics remained happy to work within the wider presumptions and protocols of late-Victorian Darwinian thought.[163] In contrast, it was Mendelism, along with theories of mutation, which despite their restricted popular appeal resonated with the Edwardian frame of mind. Before 1914 there were plenty of Edwardians still operating within an orthodox Darwinian world view, but on the other hand, virtually all the British scientists who might be identified as Mendelians were Edwardians or younger, and Bateson's often intemperate interventions were not enough to dissuade even protégés of Pearson like Udney Yule from acknowledging the significance of the new science.[164]

An interesting example of the ambivalences involved and the way in which they mapped onto generational as well as ideological structurings is

346 DARWINISM'S GENERATIONS

offered by the Southsea doctor George Archdall Reid (1860–1929). Reid is another liminal figure, with a confused response to questions of evolution and its dynamics. In the 1890s he appeared to offer characteristic Edwardian frustration at the lack of progress in heredity science; but although initially attracted to Mendelism, he was never quite able to accept it, and he eventually repudiated experimentation in the Mendelian mode.[165] By the end of 1909 he was rejoicing that his *Laws of Heredity* (1910) had quite destroyed the Mendelian and mutation hypotheses.[166] 'The more you study man's present evolution', he wrote in 1911, 'the clearer grows the proof that evolution is due to Natural Selection, and to <u>nothing</u> else.'[167] What Mendelians and mutationists might prove experimentally was too narrow for anything more than 'wild speculation'.[168] Rather than conduct artificial experiments, Reid argued, Darwin 'appealed to the great mass of experiments which nature is constantly conducting—to the facts of ordinary experience'.[169]

The manoeuvrings of Poulton and Lankester flag up the extent to which responses to Mendelism echoed the initial resistance of the scientific establishment to Darwin, only this time with the Darwinians as the upholders of orthodoxy. There were the same complaints about what Wallace described as 'ludicrously exaggerated claims and utterly inconclusive reasoning'.[170] There were the same aspersions about speculation. And there was the same willingness to condemn on at best partial engagement with the new literature.[171] There was a familiar concerted effort to mobilise institutional authority to hold the new ideas in check. In 1911 Bateson was outraged at what he saw as the reluctance of the Royal Society to elect his young protégés to fellowships.[172] (It is also probable that some of the Darwinian leadership blocked Bateson's membership of the Savile Club, which had become the recognised centre of science in London's clubland; whatever the reality, Bateson was convinced of dirty tricks by 'Ray Lankester and Co'.)[173]

The late-Victorian biologists certainly kept up a vigorous opposition. In the early 1900s the biometricians took the lead. Pearson and Thiselton-Dyer continued to assert stridently that there was no evidential basis for Mendelism.[174] Lankester used his presidential address to the British Association in 1906 to proclaim that the proofs of the key principles of Darwinism had never been stronger.[175] Throughout the 1900s Poulton vigorously resisted Mendelism at Oxford, encouraging younger entomologists to continue developing the evidence base for natural selection.[176] In 1907 he was circulating fellow late Victorians and those of the previous generation, seeking approval for a wide-ranging protest against Mendelism, which eventually appeared as *Essays on Evolution* (1908), which trained its guns

DARWINIAN DEBATES AT THE *FIN DE SIÈCLE*: THE EDWARDIANS   347

on Bateson and de Vries in at times rather hysterical ways.[177] Contemporaries rallied round; Bayley Balfour confessed to being appalled by the disastrous influence of the Bateson school. 'No more sterilising hypothesis than mutation as now advanced has appeared since we got rid of "constancy of species"', he expostulated, while F.O. Bower was so pleased with one of Thiselton-Dyer's anti-Mendelian articles in 1910 that he issued an instruction that all his staff at Glasgow must read it.[178] The same was true of late-Victorian Lamarckians like J.T. Cunningham, who worked from his *Sexual Dimorphism in the Animal Kingdom* (1900) to his attempt in *Hormones and Heredity* (1921) to argue that hormones were the mechanism for the inheritance of acquired characteristics, and to subordinate the new science to his Lamarckism.[179]

Indeed, even allowing for the sorts of generational lags and overlays that are central to my overall argument, there is little sign of general enthusiasm for Mendelian or mutationist ideas in Britain before 1909. Persuading public opinion of the potential insights to be gained from plant breeding was an uphill struggle. The study of heredity and variation had none of the easy access offered by industrious earthworms, elegantly contrived orchids, or mimetic butterflies. Mendelism's focus was narrowly defined, its methodology complex and its language rebarbative. Darwin wrote for a general audience, Bateson a specialist one. Within a few years of publication, his *Variation* was being sold off at heavily discounted prices, apparently 'utterly unsaleable'.[180] De Vries' early writings were equally demanding, with the added complication that for a while they circulated only in the original German, and were thereafter only available as expensive academic editions.[181] Exiled in Southsea, even an ardent controversialist like Archdall Reid found it hard to get hold of the literature.[182]

We should not, however, exaggerate the barriers to popular engagement. By the turn of the century enthusiastic amateurs like Alice Balfour (1850–1936) were designing their own heredity experiments alongside traditional practices of nature study and collecting.[183] Within a year or two of Bateson's initial lecture, scientifically informed clerics like the popular Anglican preacher Philip Waggett (1862–1939) were citing Mendel.[184] There was an obvious appeal to breeders, even if older breeders were more likely to be attracted to the methods of Cossar Ewart than Bateson, and continued to pour the new ideas into old Darwinian bottles.[185] In the provinces Mendelism was often merely fresh ammunition for those dissenting from Darwinism, but by the later Edwardian years experimental Mendelism was spreading amongst provincial horticultural societies.[186] New flower variants

348 DARWINISM'S GENERATIONS

brought its advantages to a wide audience.[187] By 1908 it was claimed that 'Mendelism is the daily talk now in the sporting papers!', and the *Gardeners' Chronicle* was regularly promoting Mendelian investigations, approving Bateson's championing of genetics 'heedless of the talking Darwinians'.[188] In 1909 the cultivation of the first strain of hard wheat capable of growing in English conditions by R.H. Biffen (1874–1939) created something of a stir in agricultural circles. Younger pigeon-fanciers were swept up in the possibilities, and in 1909 London's most prestigious club, the City Columbarian Society, created two annual trophies, the Mendel Memorial and the Darwin Centenary (the second something of an afterthought to balance the first).[189]

Conscious of the preliminary status of their knowledge, the leading Mendelians shied away from any attempt at a comprehensive reworking of evolutionary understandings. The one or two introductory accounts which were published during the 1900s were not especially helpful. Bateson's polemical *Mendel's Theory of Heredity: A Defence* (1902) was directed more at challenging Weldon than explicating Mendel.[190] R.C. Punnett's more explicitly popularising *Mendelism* (1905), might have filled the gap, but Weldon thought it, too, merely 'a very good mimicry of Bateson's semi-biblical rhetoric'.[191] Supporters made occasional efforts to summarise the new ideas on the lecture platform, and Bateson's long-time ally, the Leicester nurseryman, Charles Chamberlain Hurst (1870–1947), fed Mendelian materials to the country gentlemen who made up the readership of *The Field*, but there was relatively little wider engagement.[192] Bateson himself was generally uninterested in the work of popularisation, and even his less specialised performances made few concessions to popular understanding.[193] 'Why, why is it that men like Bateson cannot write intelligibly?', grumbled even the sympathetic physician and zoologist Gerald Rowley Leighton (1868–1953) in 1904. 'Spread of knowledge is utterly impossible when it is dressed in language of this sort.'[194]

This left the field open to a number of dubious champions, most notably G.P. Mudge, a lecturer at the London Medical School, and later an outspoken eugenicist and supporter of Nazism.[195] From around 1906 Mudge became perhaps the country's most ardent Mendelian, peppering Bateson and other Mendelian scientists with gossip and encouragements to take the Mendelian gospel into the wider world, and becoming one of its most visible advocates.[196] Bateson was driven to fresh paroxysms of contempt.[197] But Mudge did represent a strand of generational response, and was in 1907 able to organise a Mendelian Society, although, notwithstanding its publication (briefly) of the *Mendel Journal* (1909–1912), this was really little more

DARWINIAN DEBATES AT THE *FIN DE SIÈCLE*: THE EDWARDIANS   349

than an informal and occasional dinner party group, centred around the salon of his wealthy patron Rose Haig Thomas (*c*.1854–1939).[198] Mudge also floated the idea of a London Biological Society, for the discussion of variation, heredity, and evolution amongst the educated classes.[199] If it were possible, he was even more confrontational than Bateson, remarking in 1908 that Pearson had 'been /so\ beastly contemptuous [sic] and dogmatic about Mendelism, that I am simply dying to give him the best "knocking about" that I can.'[200]

Mudge was also a leading figure in the emerging eugenics movement. Eugenicist ideas garnered wider support in the context of anxieties about national efficiency widespread after the Boer wars. The Eugenics Education Society was established in 1907, and followed by the *Eugenics Review* in 1909. But it was really only after 1909 (and indeed after the first British Eugenics Conference in 1912) that eugenics metamorphosed from fad into fashion.[201] As a polysemic identity, eugenics attracted supporters from across the generations; but there is a sense that as an organised movement it was primarily an Edwardian phenomenon. Chesterton characterised the eugenicist par excellence as part of a generational parasitic elite: 'born about 1860, and...a member of parliament since 1890'.[202] Some Edwardians, like Wells and Thomson, were only peripherally involved.[203] Others, like W.R. Inge, Schiller, the birth-control advocate Charles Vickery Drysdale (1874–1961), and the social reformer Violet Rosa Markham (1872–1959), gave more comprehensive support.[204] Institutionalised eugenics tended to adopt extreme neo-Malthusian propositions wrapped up in Darwinian clothes, although generally supporters offered little evidence of any meaningful engagement with Darwin's work. Despite the fact that eugenicist ideas were anathema to those Edwardians more inclined to Mendelism,[205] most Edwardian supporters tended to conceive of eugenics primarily in hereditary terms, as the fear of degeneration encouraged by the dilution of natural selection gradually gave way to a more general desire for the suspension of the struggle for existence for humans.[206] It was perhaps in this context that someone like Inge could feel in the later 1900s that the younger generation were no longer moved by eugenic ideas.[207]

Renewed Saltationism

The wider ramifications of all this quarrelling were not immediately apparent. It is noticeable, for example, how little impact Mendelism had within

350 DARWINISM'S GENERATIONS

official eugenic circles before 1909; perhaps in part as a result of the influence of Leonard Darwin. However, there is plenty of evidence that the application of de Vries and Mendel to the problems of heredity brought into question the gradualism which had dominated not just Victorian science, but Victorian thought. Relatively suddenly, appropriately, saltationism was once more on the agenda. In another of those repurposings of old anti-Darwinian ideas which litter the unfolding of the Darwinian debates, one of Bateson's central questions had been if species were the product of innumerable small variations, and the environments in which these species lived varied continuously, why should the intermediate degrees become extinct, leaving chasms between the extremes? Perhaps, he suggested, the intermediate forms never existed, and the variations were more abrupt and discontinuous than Darwin allowed for.[208] Even the history of Mendelism itself after 1900 suggested to Bateson how progress in the history of ideas came 'like those of evolution, not by imperceptible mass improvement, but by the sporadic birth of penetrative genius'.[209] Inevitably, Huxley's own doubts about Darwin's wholesale rejection of saltationism were widely cited, as an impeccable Darwinian imprimatur for these lines of thought.[210]

The impact was transformative. It was still possible in 1909 for T.H. Huxley's son Leonard Huxley to pointedly assert that all reasoned thought must rest on the idea of continuity in change, and for Bertrand Russell to note that even if biology was coming to see development as discontinuous, then neither philosophers nor the general public had been influenced by the change.[211] But in reality Edwardian thinkers *were* absorbing the wider implications of the Mendelian demonstration that whatever evolution might be, it was not continuous.[212] As J.A. Thomson put it in 1914, 'It used to be a dogma, "*Natura non facit saltus*" (Nature makes no leaps) but evidences of "*Natura saltatrix*" (Nature is continuously leaping) are rapidly accumulating'.[213] 'Twenty years ago I was a pure Darwinian, believing that natural selection turned small chance variations into new species', remarked Graham Wallas in 1921. 'Now I learn from the biologists that mere "variations"...don't count. What matters is, they say, "mutations", big, sudden hereditary changes.'[214]

Here again, the experience of the First World War greatly sharpened awareness of the potential for historical caesura. But the war was not itself the cause. There were signs of a reaction against gradualism even before 1900. Younger zoologists had been quick to incorporate saltational dynamics into

DARWINIAN DEBATES AT THE *FIN DE SIÈCLE*: THE EDWARDIANS 351

their evolutionary accounts.[215] The young turks of the modernist literary magazine *The New Age* championed de Vries against Darwinist gradualism; just as the new generation of socialists in the SDF saw in saltational ideas a vindication of revolutionary change over the frustrating gradualism of Ramsay MacDonald and Keir Hardie and the late-Victorian Labour leadership.[216] Amongst the academic disciplines, the popularity of saltationist approaches was most immediately apparent in anthropology, perhaps because human history was demonstrably less gradualist than biological evolution proper, perhaps just because of the much shorter time spans under consideration. Edwardians repudiated a supposed evolutionary view of history which downplayed the importance of happenstance and incident.[217] Sudden transitions were at the core of diffusionist social anthropologies of the sort espoused by Grafton Elliot Smith, the geographer and anthropologist William James Perry (1868–1949), and the anthropologist and neurologist W.H.R. Rivers (1864–1922), who questioned adaptation to environment as a mechanism of cultural development, argued for the importance of material factors such as inventions in producing human progress, and posited an episodic past, with periods of stasis punctuated by moments of dramatic change.[218] Similarly in his influential *Herd Instinct* (1916), based on ideas published in 1909, Wilfred Trotter (1872–1939), neurosurgeon and social psychologist, suggested there had been periods of rapid change potentially perceptible to direct observation.[219] Even the avowedly scientific approach of J.B. Bury (1861–1927), as outlined in his essay 'Darwinism and History' (1909), stressed the chance concatenations of events, and 'the decisive actions of individuals, which cannot be reduced under generalisations and which deflect the course of events'.[220]

Ultimately, there are signs of the younger late Victorians shifting in saltational directions. One example is the concept of 'emergent evolution' advanced by Conwy Lloyd Morgan's 1922 Gifford Lectures, according to which evolution was not a continuous gradual process, but was periodically interrupted by something new which could not be deduced from what had gone before. But the late Victorians who acknowledged the newly introduced element tended to fall back immediately to the overriding significance of continuity.[221] As the botanist Walter Gardiner (1859–1941) told Thiselton-Dyer in 1903, in nature, discontinuous variation, so-called, was merely the sum of continuous variation; the dam might suddenly break, bringing dramatic change, but the essential cause was gradual.[222]

Orthogenesis or Vitalism

The questioning of the sufficiency of natural selection, the renewed acceptance of Lamarckian dynamics, and especially the attention to hereditary transmission, encouraged a recrudescence of directional and programmatic versions of evolution. After all, the principle of random variation was no more, as F.C.S. Schiller observed, than assumption.[223] In substance, there was nothing new in this; it was a feature of earlier German transcendentalism and equally of Herbert Spencer, and had been an element of religiously inflected understandings in every generation.[224] There was relatively little to differentiate the arguments that Argyll was making along these lines in the 1890s (as in the 1860s) from those of many Edwardians.[225] The Anglican clergyman, Harold Anson (1867–1954) was able to claim that it was Darwin's own writings which had changed his mind about the purposefulness of nature.[226] But for most of the younger generation, the rhetoric of anti-Darwinism increasingly arose from a repudiation of the apparently random and purposeless frame of Darwinian evolution, and its replacement by an explicitly directional, if not actually directed, evolution.

What distinguishes the Edwardians is the relaxed eclecticism of their use of vitalism as a counter to the survivals of Victorian monism, and their willingness to elevate ideas of a specific directional impulse to evolution into the crucial *biological* dynamic, in contradistinction to what they saw as the 'purely negative' sieve of natural selection. By the later 1880s under the influence of Patrick Geddes, J.A. Thomson was already challenging Darwinian assumptions of indefinite spontaneous variation.[227] By 1900 Thomson suggested that there was a distinct school (which he called neo-Lamarckian or neo-Nagelian, but to avoid confusion is perhaps better thought of as 'Orthogenetic'), committed to the idea of variation in specific directions, sometimes progressive but also often degenerative.[228] There was a Lamarckian version, promulgated by George Henslow, that variation occurred only or predominantly in the direction of adaptation to the environment, as part of what he called the 'responsive power of protoplasm', taken up by figures like the pathologist J.G. Adami (1862–1926).[229] In his lectures in the late 1890s, Bateson was also already espousing the idea of non-random, non-equally distributed variation.[230] Mutations were interpreted as some progressive principle of life inherent in organisms. Chalmers Mitchell, studying the osteological modifications of kingfishers, demonstrated that they were 'marching in a definite direction', which suggested that 'the direction of variation is one of the characters that define organic

DARWINIAN DEBATES AT THE *FIN DE SIÈCLE*: THE EDWARDIANS   353

groups'.[231] For many this was associated with what Schiller described as 'an intelligent force to which we must ascribe the progression and direction of the process of Evolution'.[232]

Late Victorians like Lloyd Morgan were prepared to concede the presence of a specific characteristic of living matter, although not as the motor of evolution;[233] but at least some of the younger late Victorians went further, including Geddes and the Catholic anatomist Bertram Windle (1858–1929), who by 1912 was drawing widely on the writings of Bateson and of the American Vernon L. Kellogg to advocate orthogenesis, dismissing 'the old Darwinism'.[234] To this more formalised version the high and late Victorians generally resisted, although by the early twentieth century even ultra-orthodox Darwinians were not inured from toying (at least in private) with ideas that life might have 'certain tendencies' or 'ancestral bias' which could assert themselves when not over-mastered by natural selection.[235] But generally they regretted the fashion for internal and external guiding agencies, and the unpopularity of what Meldola described as the 'much-abused Naturalistic Philosophy'.[236]

This progressivism was associated with a renewal of vitalist ideas. Vitalism's strongest current centred on the French philosopher Henri Bergson and his idea of an 'élan vital', a vital force or impulse that permeated the universe and everything in it and which drove evolutionary change (even though Bergson's key evolutionary text *L'Evolution Creatrice*, published in 1907, was not translated into English until 1911). But because vitalism had so much in common with more long-standing critiques of the materialism and anti-teleology of Darwinism, and was sufficiently loose and capacious to have a broad appeal, it was influential before Bergson. And as Emily Herring has shown, Bergson's assimilation of vitalism with evolution offered an attractive and productive development of earlier theories.[237] In any case, whatever the limits of his direct readership before 1909, Bergson was reviewed and discussed in the specialist philosophy and psychology journals and correspondence networks, and informed the non-Darwinian evolutionary neo-vitalist theories of figures including Thomson and the Scottish physician and poet R.C. Macfie (1867–1931).[238] In the general enthusiasm which ensued, even Darwin could be claimed as 'a thorough-going vitalist'.[239]

In 1903 Kelvin was able to reignite the old debates about the relative provinces of science and religion by claiming biology was turning once again to the principle of a vital creative force. Despite the outrage of Lankester, such ideas were certainly a temptation to younger late Victorians,

354 DARWINISM'S GENERATIONS

like the physicist Joseph Larmor (1857–1942), who was prepared to announce himself a vitalist (in 1915).[240] In his deliberately idiosyncratic and provocative way, George Bernard Shaw illustrates how vitalist elements of late-Victorian Lamarckism could flow after 1900 into a more thorough-going anti-Darwinism. Ever attuned to the cultural temper, Shaw placed Jack Tanner, the expositor of his ideas of evolved humanity in his *Man and Superman* (1903), 'in the grip of the Life Force'.[241]

By the 1920s, as he made vividly clear by the preface to *Back to Methuselah* (1921), Shaw had become progressively more Bergsonian, repelled by what he saw as Darwinism's mechanistic assumptions.[242] His championing of creative evolution was, he later claimed, his most significant achievement—it was 'my religion. It is the religion of the twentieth century'.[243] The speed with which Bergson's ideas were taken up after 1911, even by Edwardians who had previously shown only minimal interest in evolution, including the novelist Arnold Bennett (1867–1931) and the theologian John Neville Figgis (1866–1919), is testimony to the extent to which ideas of an *élan vital* resonated.[244] Often for high and late Victorians, admiration of Bergson was tempered with unease at his opacity, his mysticism, and his penchant for extending the operation of evolution to all phenomena. H.B. Donkin was not alone in complaining at the way 'Bergson's rot has so imposed upon so many people who ought to know better', and by 1911 J.W. Judd was regretting the numbers of even good Darwinians going back to the 'existence of innate impulses'.[245] But the Edwardians turned more readily to approaches which, 'accepting all that science can affirm, have taught us to look deeper still and have revealed to us the truly real which lies at the back of all natural phenomena'.[246]

Interest in Bergson was part of a reaction against the physio-chemical monism of the 1870s and 1880s which had sought to reduce all life to material processes, a reaction against the popularity of treatments like Haeckel's *Riddle of the Universe* and of what Edwardians saw as the crude unbelief encouraged by it.[247] The elevation of vitalism reinforced claims that science could not get at the secret of life. As Bertrand Russell remarked in 1903, 'the scientific Darwin-Huxley attitude' was 'too external, too coldly critical, too remote from the emotions'.[248] J.A. Hobson and Hugh Black (1868–1953), the Scottish-American theologian, welcomed the implication that while science could determine more or less exactly the formal mechanics of evolution, the determinant nature of the energy which played through these processes eluded it.[249] When science had done its best or its worst, Black remarked, the poet, the prophet, and the seer were still needed to interpret nature, not

DARWINIAN DEBATES AT THE *FIN DE SIÈCLE*: THE EDWARDIANS   355

by analysis but by constructive imagination.[250] Le Gallienne's *The Religion of a Literary Man* (1893) suggested that the fundamental division was coming to be not between churchmen and scientists, but between the natural materialists, who could not conceive of anything not visible through a microscope, and the natural spiritualists, who saw these as merely symbols of a deeper mystery.[251] Even the short fiction of the most obviously Darwinian of Edwardian novelists, H.G. Wells, not just the *Island of Doctor Moreau*, but also the *Time Machine*, and *War of the Worlds*, displays not just a greater suspicion of the claims of science and scientists, but a belief that the spiritual rather than biological future of man was the central evolutionary problem.[252]

Edwardian Theological Responses

Bergson's profile in Britain was given a significant boost by his appearance in the spring of 1914 as Gifford Lecturer at the University of Edinburgh, where the doors of the lecture theatre had to be closed an hour before commencement because of the crowds wanting to attend. The Gifford lectures, established to promote the discussion of natural theology, and first delivered in 1888, offer a microcosm of the enduring evolutionary preoccupations of theological and philosophical debate in these years. Concurrently established in four Scottish universities, the Giffords offered a temporary platform for over forty lecturers in the years up to 1914, and inevitably many trenched on evolutionary questions. The first series included J.H. Stirling offering his strongly unDarwinian Hegelianism to the delight of the aged J.S. Blackie, and Max Müller, who explicitly distanced himself from a single evolutionary descent, but who was still too evolutionary for many.[253] Subsequent significant series included James Ward's *Agnosticism and Naturalism* (1899), which offered an evolutionary perspective strongly inflected with his late-Victorian idealism. At the end of the decade the lectures of the American pragmatist William James (1842–1910) delivered a further influential counterblast to scientific naturalism. In turn these lectures were the catalyst for repeated bouts of newspaper debates about the status of Darwinism, which once again demonstrated not only that if outright anti-evolutionary attitudes had largely disappeared, they had in many quarters been replaced by versions of evolutionism erected in explicit opposition to the neo-Darwinism of the late Victorians.[254] By the time Bergson arrived to promote his brand of vitalism, Gifford audiences had

356   DARWINISM'S GENERATIONS

already heard the German biologist Hans Driesch (1867–1941) tell them that while evolution was an incontrovertible fact, Darwinism had been shown to fail 'all along the line'.

Edwardian theology remained overwhelmingly evolutionist. In part in that it was entirely reconciled to the geological overthrow of Genesis, and had no time for the early horror at Darwin, which the Anglican Herbert Hensley Henson (1863–1947) compared to the persecution of St Paul by the 'Thessalonian bigots'.[255] Darwin became subsumed into a broader 'biologism' on which was superimposed an evolution of personality or of spirit, or into a longer philosophical tradition stretching back to Kant if not further.[256] Attacks on Darwinism were increasingly likely to surrender the science and seek to engage with the 'speculative philosophy' derived from it.[257] In part this reflected an increasingly derivatist approach, considering theology itself as a branch of natural history.[258] Theological writers reinterpreted the doctrine of original sin not so much as hereditary taint but rather as a state of only partly developed perfection.[259] Biblical criticism applied evolutionary dynamics to the assembling of the New and Old Testaments. Abandoning any residual desires to insulate humankind from its full implications, Edwardian Nonconformists argued that evolution could not be confined to the physical sphere, but must illuminate the moral and spiritual life of humanity. Only in that way, it was suggested, was it possible to avoid the error of regarding the incarnation as 'a break in the continuity of the Universe,…an afterthought of the divine love'.[260]

The Edwardians showed greater confidence in asserting the separate provinces of the scientist and the theologian. This had always been a resource of religious apologetics, but one which had for earlier generations been defensive, even desperate. The fragmentation of evolutionary science, and the emergence of formidable challenges encouraged a more confident assertion of the greater scope for the incarnate Christ to express his personality. Cosmo Gordon Lang (1864–1945), eventually Archbishop of Canterbury, recalled the way his introduction to Hegelian thought at Glasgow University at the end of the 1870s liberated him from the 'bondage' of 'mere physical science' which seemed to lay so heavily on older generations, and by 1905 he was welcoming science's greater modesty in comparison with 'the pride of its first great conquests, and the early challenges of biblical criticism,…forty or fifty years ago'.[261] The critiques of de Vries and the Mendelians were deployed to deny any sense that biology demonstrated that evolution was merely a mechanical product of natural forces.[262] Vitalism helped to bolster established traditions of immanence. Nonconformist

DARWINIAN DEBATES AT THE *FIN DE SIÈCLE*: THE EDWARDIANS   357

adherents of the 'New Theology', like the Congregationalist T. Rhondda Williams (1860–1915), seized on vitalist ideas as representing a retreat by science generally from purely mechanistic views of natural laws.[263] And in turn pulpit preaching was more confident in its rejection of a narrow empiricism. 'Faith is in the science of religion what experiment is in the science of matter', proclaimed the Congregationalist John Henry Jowett (1864–1923).[264]

As the Edwardian period progressed, the quondam radicals of *Lux Mundi* were increasingly ranged on the side of caution in the face of the greater daring of theological modernism. By 1907, J.R. Illingworth had shifted from advocacy of evolutionary approaches to warnings that they were being taken too far, and Charles Gore's increasing tetchiness with the new theology of R.J. Campbell was more and more redolent of exactly the treatment his mentor Liddon had given to *Lux Mundi*.[265] For their part, Edwardians like Inge had little time for the hesitancies, and the cautious evolutionary consciousness of *Lux Mundi*.[266] Inge was a leading contributor to the collection *Contentio Veritatis* (1902), perhaps the closest there was to an Edwardian equivalent to *Lux Mundi*, produced by a similarly generationally coherent group which met occasionally at the house of the clerical historian and Christian socialist A.J. Carlyle (1861–1943) in Oxford during the later 1890s, and which was greeted as chiming with the mindset of the younger generation.[267] *Contentio Veritatis* was less distinguished, less controversial and certainly less influential, and it was also less directly informed by Darwinian science, or its Edwardian supersessions. But it not only treated evolution as a given, but also saw evolutionary theory as constructive rather than destructive, especially in its assertion that not just morality but also doctrine was constantly evolving. It was even more explicit in its rejection of interruptions to the operation of natural laws as a basis of faith, and although it appeared too early to be greatly affected by the new biologies of the 1900s, it did also hint at fresh saltational sensibilities, presenting, for example, the incarnation not just as an episode in the evolution of morality, but as inaugurating a new era, just as Darwinian methods brought advance 'by leaps and bounds'.[268]

Vernon Storr (1869–1940) and F.R. Tennant (1866–1957) provide examples of the ways Edwardian theological thinking sought not so much to push back against Darwinian thought as to move on beyond it. Storr's study of evolutionary biology brought him an invitation to review in the area for the (Church) *Guardian* newspaper in his early twenties. His *Development and Divine Purpose* (1906) argued that theology had to accept

358   DARWINISM'S GENERATIONS

the widest scope for evolution, not just in the evolution of humanity, morals and consciousness, but even in the creation of life itself, and his address to the pan-Anglican congress at Winchester in 1908 argued that truth was not fixed, but vital and growing and the Bible must be treated critically as a means of God's gradual and continuous revelation.[269] At the same time, he also used the writings of Bateson, Schiller, and Henslow to argue for the limits of Darwinian mechanisms and the importance of Divine will as an evolutionary force. For Tennant, who had studied science and mathematics at Caius College, Cambridge (1885–89) and been a science teacher before preparing for ordination, Darwin had nothing to say about 'the real "origin" of the species', and theories of an immanent God were a metaphorical position lacking 'any definite meaning' and prone to exaggeration.[270] Both were convinced that each generation needed to remake God's truth from their own standpoint. The apologetics, Tennant observed, 'of one generation becomes antiquated and irrelevant to the next'.[271]

These generational shifts were most clearly inscribed in the tensions over the emergence of varieties of religious modernism from the 1890s onwards, although the sheer range of approaches that could be subsumed under this rubric greatly complicated the picture. Chesterton commented that he had been in revolt against new theology, which was nothing more than 'the admirable religion of one's aunts and uncles...the vague Victorian optimism in which we all grew up', all his life.[272] Certainly the 'New Theology' as articulated by R.J. Campbell (1867–1956)'s *The New Theology* (1907) and promoted by the League of Progressive Thought and Social Service which Campbell launched in 1908, was as much as anything an attempt to embed evolutionary logic even deeper into religious belief, arguing that truth itself must evolve, and proposing a divinity whose manifestation in the world involved its own evolutionary development, a sort of hyper-immanentism in which God and humanity became fused, and which advocated the need to abandon many of the traditional tenets of Christian belief, including the doctrine of the fall and the virgin birth. Older opponents were confronted with the need to define the limits of their accommodations with evolution, and it is striking how much of the debate which briefly raged fiercely sought, both positively and negatively, to position Campbell's ideas in the context of the Darwinian legacy. The high-Victorian religious thinkers, while affirming the truth of evolution, denied it was a power in itself rather than a manifestation of divine action leading to a definite end, and so retained the possibility of occasional miraculous interventions.[273] Late-Victorian responses

DARWINIAN DEBATES AT THE *FIN DE SIÈCLE*: THE EDWARDIANS   359

conceded that in the light of Darwinian science and biblical criticism theology required a new language, that the verbal authority of the Bible had
been shattered, but worried that immanence was being taken so far as to
call into question the incarnation of God in Christ.[274] Of course there were
still plenty of Edwardians who were suspicious of the new theology as '[a]n
evolutionary philosophy, masquerading as spiritual religion,'[275] especially in
the intense initial controversy which the appearance of Campbell's book
prompted. But it was clearly amongst this younger generation that his ideas
garnered the bulk of their support.[276]

Many were unwilling to accept the logic of an all-embracing evolutionary approach, that the latest phases of the doctrinal tradition were the
truest.[277] The lines of division were most naked in the Catholic Church,
where liberal theology quickly butted up against an assertively conservative
leadership.[278] Mivart's older brand of biological determinism, compounded
by his deteriorating mental stability, was an inevitable casualty, leading to
excommunication and a lonely death, but more illuminating of Edwardian
developments was the complicated history of one of the leading figures of
the small band of Catholic modernist thinkers, George Tyrrell (1861–1909).
Although Papal proscription ultimately forced Tyrrell to try to construct a
distinction between his evolutionism and 'Darwinism' (a task made easier
by the vagaries of his understanding of Darwin's writings), what marks his
interventions in the Edwardian period is his ready employment of biological language.[279] Tyrrell attempted to situate himself within the tradition
of developmental thought of John Henry Newman's 'Essay on Development';
but he rejected the efforts of late Victorians like Wilfrid Ward to construct a
via media between evolutionary science and scholastic theology on the
basis that the Church's role was to interpret an existing 'deposit of faith'. For
Tyrrell revelation was not a deposit, but a living thing.[280] Instead of seeing
(like Ward or Illingworth) the evolution of doctrine only as 'unfolding', he
sought to use the concept of 'epigenesis' to present it as the growth of new
parts, of transformation in growth which needed to be continually adapting
to the advance of humanity, an adaptation which required the testing of
variations through competition, and a largely unteleological process, in no
way predetermined inherently to any fixed goal.[281] His formulation that
'a heresy is only a rejected variation, but the principle of heresy is a principle of progress and life' was always likely to bring him into conflict with
the Church authorities.[282] But it also brought him a wide approval and
influence.[283]

360   DARWINISM'S GENERATIONS

Applied Darwinism

Tyrrell's theology exemplifies the extent to which Edwardian thought retained strong biological underpinnings. In some senses these impulses may even have intensified. The Edwardians certainly continued to resort to biology not just for analogies but for explanations. Take Grafton Elliot Smith, labouring through much of the 1900s to bring a new biological rigour to Egyptian archaeology.[284] Or William Henry Winch (1864–?), whose educational theory wrestled with the implications of the apparent triumph of Weismannism and recapitulation theories to try to develop an effective pedagogy.[285] In Ramsay MacDonald, the Edwardians furnish a figure in whom evolutionary science and socialism remained inextricably intertwined.[286] But the Edwardians generally found common cause with the high Victorians' suspicion at the 'Darwinisation' of intellectual life. Patrick Geddes continued through force of charisma to gather followers, many of whom were born after 1860, but D.G. Ritchie was unable to replicate this success in Oxford.[287] John Morley, still giving voice to his long-standing unease with the importation of biological terms into politics, was echoed by the Oxford biologist Gilbert Bourne (1861–1933), who warned the listeners to his Spencer lecture in 1909 that it 'was not the business of biologists to offer solutions to sociological questions.'[288]

In many fields of knowledge the dominant evolutionary approaches of the nineteenth century were being marginalised or abandoned. The 'new geography' as promoted by Halford Mackinder (1861–1947) illustrates this discomfort. Although Mackinder studied zoology under the committed Darwinian, H.N. Moseley, and then read for a History degree with the intention of exploring 'how the theory of evolution would appear in human development', apart from his social Darwinian imperialism, he showed little interest in Darwinian ideas in his printed works before 1914, and later in his *Democratic Ideals* (1942) explicitly distanced himself from the fatalist Darwinism of the previous century.[289] To the extent that early twentieth-century geography was shaped by evolutionary ideas this was more through the sort of nature-mysticism incorporating a vision of spiritual as well as scientific progress which characterised Francis Younghusband and Vaughan Cornish (1862–1948).[290]

Edwardians increasingly dwelt on what F.C.S. Schiller called 'the precariousness of progression'.[291] There was a rejection of the smugness which saw evolution as axiomatic evidence that contemporary man was superior to all beasts and to all previous humanity. This mentality was acidly lampooned

DARWINIAN DEBATES AT THE *FIN DE SIÈCLE*: THE EDWARDIANS   361

by May Kendall (1861–1943)'s poem 'The Lay of the Trilobite' (1885) ('How wonderful it seemed and right,/The providential plan,/That he should be a Trilobite,/And I should be a Man!').[292] Darwinian thought was considered tainted by Manchester School economics and individualist politics.[293] The idea that society was an organism and should be approached biologically was treated as an increasingly threadbare and even dangerous cliché.[294] The philosopher Henry Cecil Sturt (1863–1946) provides one example of the continued acceptance of the evolutionary context, but also the placing of humanity on a plane above biology, where development was primarily due not to the struggle for existence, but to the striving of living things to maintain and extend their life.[295]

Equally, there was a revulsion in many quarters at Darwinism's apparent environmental determinism, and the contemporary tendency to the 'listlessness' induced by Darwinian beliefs.[296] In the years before and after World War I, George Bernard Shaw, who in his Fabian Society days had dismissed Darwin as a 'pigeon-fancier',[297] became particularly outspoken in this version of anti-Darwinism. He confessed to having 'the most unspeakable contempt' for natural selection, remarking later that nobody born after the mid-1850s still believed that life could have been produced by Darwin's natural selection.[298] Edwardians were unwilling to acquiesce in the slow gradualism implied by evolutionary dynamics, and attacked the high-Victorian version of social Darwinism which resisted social reform on the basis that it attempted to interfere in the necessary discipline of the survival of the fittest.[299] Whereas the Idealists had largely focussed on the natural selection of ideas and social institutions, Edwardian thinkers were more interested in the ability of the self-reflecting mind to transcend the effects of natural selection. Mankind was passing out of the control of its old schoolmaster, natural selection, proclaimed one Mendelian in 1912, and entering into adult life, when the old evolutionary landmarks would be lost sight of, and when the power of controlling its environment would allow humanity to take a large share in the shaping of its own destiny.[300] As the sociologist Leonard Hobhouse (1864–1929) put it in 1886, he began by investigating human psychology, working on the evolution theory, but believing that if he could show mind acting on itself in further improvement he could prove that 'the growth will not cease, but—no longer the sport of surrounding forces—the mind of man will set its own improvement or development before itself as its great object'.[301]

While the high Victorians defended the ethical consequences of the survival of the fittest, the Edwardians increasingly interpreted Darwinian evolution as amoral. As one Edwardian critic put it, 'What does evolution care about conscience? What does the law of natural selection say to the soul of man?...the very foundation of the law of natural selection is the fact that it acts in defiance of the individual.'[302] Huxley's *Evolution and Ethics* (1893) was significant in this regard because it was amenable to readings which challenged confidence in the purposefulness of evolution, setting it up in opposition to moral progress. The Edwardians, and especially Edwardian Liberals, wanted none of the high Victorians' pieties about the mercies of the brutal realities of the struggle for survival. They revolted against the preying of species on species, against parasites destroying loveliness for their own mean and meaningless ends.[303] Darwinism was a false superstition which was corrupting national morals and celebrating the triumph of strong over weak.[304] The only use to which Darwinian arguments had hitherto been put, suggested Bertrand Russell in 1907, was to justify the spoliation of so-called 'inferior races', a use which had discredited such arguments with 'honest people'.[305] There was particular discomfort with the implications that this was a random process driven entirely by chance. For many, perhaps most, Edwardians this was an utterly inadequate account of the ordered beauty and harmony of nature.

The Boer War quickened Edwardian suspicion of the way that Darwinism had been exploited to fuel jingoism and imperialism. The irony that as the conflict with the Boers gathered momentum 'Pulpits which shrieked against Darwin are now twittering with pseudo-Darwinisms' was not lost on the Liberal journalist J.L. Hammond (1872–1949).[306] Late Victorians were quite likely to conclude that in respect of South Africa, the question would only be solved by the operation of the law of evolution, the struggle for existence, and the survival of the fittest.[307] In contrast, Edwardians vigorously rejected any attempt to equate the ethics of force and of progress, describing war as an excrescence on modern civilisation, and a relic of barbarism.[308] As international tensions mounted in the years after, these tendencies were expressed more directly. The members of the Rainbow Circle were by the mid-1900s arguing for a thorough revision of the principles of evolutionary thought and its applications.[309] Probably the most influential proponent of this position was Norman Angell (1872–1967), whose *The Great Illusion* (1909) became one of the foundational texts of twentieth century internationalism. Angell's direct engagement with Darwin was slight, and *The Great Illusion* essentially addressed only the diffuse 'social Darwinist' appropriation of the survival of the fittest in support of militarism. But it intervened explicitly in the politics of reading Darwin, arguing that an

DARWINIAN DEBATES AT THE *FIN DE SIÈCLE*: THE EDWARDIANS   363

earlier generation which had read about Darwin instead of reading him, had 'fixed upon an interpretation of Darwinism in applying to social phenomena which Darwin feared they would give it, against which he expressly warned them.'[310] Generational divergences become more apparent as we get closer to World War 1. A good example of the Edwardian trajectory is the prolific journalist Harold Begbie (1871–1929), who represents his generation's hostility towards Darwin and Darwinism from an evangelical Christian perspective, but also a particular version of post-war anti-Victorianism, in which Darwinism and the failure of late-Victorian society to resist it became the fundamental cause of the catastrophe of the war.[311]

This was a central preoccupation of Alfred Orage (1873–1934), and the writers and readers of the literary magazine *The New Age*, which he edited. Notwithstanding their paradoxical belief in the need to go back in order to go forwards, Orage and his circle professed a firm belief in evolution.[312] At the same time, this was a diffuse Hegelian version which championed Bergson and Nietzsche, and sustained a fiercely anti-Darwinian rhetoric, especially in reference to what it called 'the paleo-Darwinian school of Manchesterism', associated with laissez-faire individualism. The *New Age* gave a platform to various strands of scepticism, including the revived Ruskinianism of the guild socialist Arthur Joseph Penty (1875–1937), with its suspicion of Darwin and Marx.[313] It was entirely typical of the journal that it that it should publish a poem like 'Darwinism' by 'Kennington Cross', which serves as the epigraph to this book, and that it should accuse the Egyptologist Flinders Petrie (1853–1942) of belonging 'to an obsolete school of sociology, the school of Spencer and Darwin', and take him to task for conceiving 'human society as a species exactly comparable to a species of vegetable or animal in the jungle.'[314]

The *New Age* cannot be offered as straightforwardly representative of Edwardian literary culture, but its lack of sympathy with Darwinian thought was widely replicated. Certainly, Edwardian literary critics were much less interested in the specifically Darwinian drivers of Victorian writing, broadening any evolutionary analysis out to embrace Newman, Emerson, and Whitman, and diluting it into an acceptance of physical derivation and universal brotherhood operating alongside principles of individual emancipation.[315] Edmund Gosse might sneer at Edwardian popular literature as offering sleeping draughts '[f]or the sons of men who used to sit up half the night discussing the *Origin of Species*', but Darwin's influence on Victorian literature was increasingly seen as a source of weakness.[316] For Arthur Compton-Rickett (1869–1927), the *Origin* had infected poetry and

364   DARWINISM'S GENERATIONS

fiction with a destructive determinism, leaving the 'bacteriologist [as] our doctor of divinity, and the engineer our evangelist'.[317] The pre-war studies of Browning and Tennyson of Arthur Waugh (1866–1943), which sidestepped entirely questions of evolution, reflect this dislocation from Darwinian debates. Yet, in Waugh's later *Tradition and Change* (1919), 'the trail of Darwinism' and the doubts it engendered became not only central to his analysis, but part of the canker at the heart of Victorianism, the taint of science which could say nothing about men's souls.[318]

In philosophy, evolutionary approaches were a powerful component of the Edwardian reaction against Idealism. Again, Tyrrell is revealing in this respect not least in his 'Adaptability as a Proof of Religion' (1899) essay, which was especially concerned to problematise the extension of the evolutionary frame to philosophy, and the 'insupportable burdens' which this produced.[319] Many of these philosophical tensions were visible in the debates of the Aristotelian Society during the 1890s and 1900s. In 1893 the Society's long-time secretary George Hicks (1862–1941) was challenging the Darwinism of Samuel Alexander's 'Natural Selection in Morals', on the basis that it was unlikely to be the final form of evolutionary science, and invoking the importance of purposive acts of will in the process of organic adaptation.[320] From Oxford, Schiller championed the move of Edwardian philosophy under the influence of William James' *Varieties of Religious Experience* towards Pragmatism.[321] Schiller was particularly scathing about late-Victorian neo-Darwinism, which he described as the narrowest, most self-righteous and unphilosophical of biological sects, accusing Idealists like Ritchie of defending 'the infantile metaphysics of the biologist', and maintaining a long-running and often ill-tempered squabble with F.H. Bradley.[322] James was never comfortable with this fiercely controversial style, but Schiller saw no reason why he should not 'play the Huxley to your Darwin'.[323] Pragmatism, as Schiller championed it, involved an abandonment of the inductive epistemology of Darwinian science, and the evolutionary ethics of Darwinian philosophy. Theory was to be judged on consequences not antecedents; truth became a matter of usefulness—as soon as something ceased to be useful it ceased to be true.[324] Where pragmatism retained a congruence with Darwinism was in its rejection of psychological individualism, and its understanding of consciousness as the result of social forces. In this sense it was G.E. Moore and Bertrand Russell rather than the pragmatists who took the revolt of the Edwardians to its ultimate end. Moore's 'The Refutation of Idealism', and his *Principia Ethica*, and Russell's *Principles of Mathematics*, all published in 1903, rejected evolutionary

DARWINIAN DEBATES AT THE *FIN DE SIÈCLE*: THE EDWARDIANS   365

ethics in their entirety, abandoning biological approaches for logic.[325] Moore led the reaction against what he saw as the tendency to collapse evolution into progress, which he described as 'the naturalistic fallacy'; science showed development, he asserted, but only philosophy could judge progress.[326] Philosophy became (at least as far as its late-Victorian critics were concerned) reduced to the formalist analysis of the meaning of propositions.[327]

These trends are even more clearly visible in the emergence and institutionalisation of sociology in Britain. Before 1900 the central figure in British sociology was the late-Victorian Patrick Geddes. Geddes was a transitional figure, commencing as a research microbiologist, immersed in evolution, but sceptical about all the main variants of Darwinism. He was wont to say that Darwinian evolution 'accounted for our survival by explaining the deaths of our uncles and aunts, and that it was consequently rather a theory in necrology than biology.'[328] Without entirely shedding his initial organicist instincts, Geddes was brought via French sociology to an idiosyncratic, local, intimate, version of social progress, in which the central biological insight was the need for a holistic nurturing approach. He attracted a circle of disciples who found him forthright in his opposition especially to orthodox Darwinians.[329] In many respects Geddes was perfectly placed to assume the intellectual leadership of the new discipline of sociology as it coalesced and institutionalised in the early 1900s. Why that failed to come about, leaving Geddes marginalised, is revealing of the generational dynamics at work.

The ultimate choice of Hobhouse as editor of the *Sociological Review* and the first Professor of Sociology at University College, London, encapsulates the shift from Geddes' late-Victorian biologism to Hobhouse's conviction that the creation of an effective discipline was dependent on freeing sociology from biology.[330] '[T]he uncritical application of biological principles to social progress', Hobhouse argued in his *Social Evolution and Political Theory* (1911), 'results in an insuperable contradiction.'[331] The choice of Hobhouse was an explicit rejection of the organicist and historicist vision for sociology Geddes outlined in two programmatic presentations in 1904 and 1905, which his listeners heard as an 'application of the view of a biologist to Sociology.'[332] Geddes had his support, not least from other late Victorians like Archdall Reid, whose paper 'The Biological Foundations of Sociology' sought to vindicate the organic approach, and from Victoria Welby, who assiduously promoted the importance of the 'witness of biological science' to the study of social questions.[333] But ultimately, Victor Branford (1863–1930), who ended up acting as honest broker behind the

366 DARWINISM'S GENERATIONS

scenes, recognised that Geddes' vision did not attract general support, and Hobhouse was installed. His inaugural lecture accepted the necessity of taking into account all the biologist might say about physical conditions of life, heredity, and natural selection, but warned that this still left the need to study social institutions, not deduce their nature from biological analogies.[334] It was a stance shared by Edward Johns Urwick (1867–1945), who later became director of the London School of Sociology, forerunner of the L.S.E., and by his successor William Beveridge.[335]

Anthropology, which was emerging as a distinct discipline in this period, offers striking parallels. Edwardian anthropologists, while similarly steeped in biological thinking, were also increasingly inclined to place narrow limits on its explanatory capacity.[336] It is easy to overlook this tendency. After all, a leading figure such as R.R. Marett was frequently fulsome in his Darwinian professions, and argued explicitly in 1912 that anthropology was 'the child of Darwin. Darwinism makes it possible. Reject the Darwinian point of view, and you must reject anthropology also'. At the same time Marett exemplifies the growing suspicion of 'biological' thinking. His calls to contemporaries to 'Darwinize actively' need to be recognised as an embrace of a broad historical approach and the recognition of the continuity of all things, rather than the application of any particular biological framework.[337] As his reviews for the *Economic Review* in the years before 1909 make clear, he had no time for Spencerean biological analogies, and he was especially severe on the approaches of late Victorians like Benjamin Kidd, condemning Kidd's attempts to refurbish social-biology by focussing on society as an organism as merely combatting one heresy with another 'equally gross'.[338] Attitudes of this sort encouraged the shift to functionalist approaches, from general comparative syntheses to the intensive study of particular societies and cultures, of the sort undertaken by the seminal Torres Straits expedition in 1898 (which was very much Edwardian in personnel, apart from its leader A.C. Haddon), and especially the work of W.H.R. Rivers, which anticipated, for example, the anti-evolutionary assumptions of the functionalist anthropology of A.R. Radcliffe-Brown (1881–1955) and Bronislaw Malinowski (1884–1942).[339]

Where evolutionary approaches continued to attract interest, as in the question of atavism in Edwardian social psychology,[340] they came to be associated, in the writings of Wilfred Trotter and Rivers, with new theories of instinct, in which the conscious mind was understood to suppresses unconscious evolutionary survivals. For Rivers, mental disorders were mainly dependent on the surfacing of older attitudes which had been

DARWINIAN DEBATES AT THE *FIN DE SIÈCLE*: THE EDWARDIANS 367

previously controlled or suppressed by the overlays of evolution.[341] Ultimately, this led to Freud and the new understandings of the unconscious. Looking back, for some Edwardians this seemed to be one of the fundamental divisions of their world view and the Darwinian or nineteenth century one. Victorian versions of Darwinism had largely presented it as humanity's escape from its animal origins; under the influence of psychological and especially psychoanalytical theories, the Edwardians were more conscious that nothing had been left behind, that the whole evolutionary backstory was carried in the psyche; that, as the novelist M.P. Willcocks (1869–1952) put it after World War 1, 'The stock-broker has somewhere in him that ancestor of his who built an altar to earth, or slaved in the sun under the whip-lash'.[342] The Victorian age believed in law and reason, observed Hobhouse, 'Its sons have come in large measure to believe in violence, and in impulse, emotion, or instinct'.[343]

A New Anti-Darwinism?

The Edwardian retreat from applied evolutionary perspectives did not necessarily involve a rejection of evolution as a process, or even of Darwin's specific evolutionary theories. At the same time, explicit repudiations did become both more common, and more scientifically respectable. In the years leading up to the fiftieth anniversary of the *Origin* there was a noticeable shift in the tone of debate.[344] By 1895 Frederic Harrison perceived the age of Darwin as passing and confessed that he himself was not prepared to adopt as proven, 'all the wonderful hypotheses about man and the organic world which loosely pass current as Darwinism'.[345] It was an increasingly common stance. The Oxford historian, George C. Brodrick (1831–1903), trembled in confessing to Huxley that he was 'not quite sound on the doctrine of Evolution, if it be heresy to doubt whether it is established in the same sense as the law of Gravitation'.[346] The Irish poet and critic George William Russell (1867–1935) believed Darwin 'almost obsolete'.[347] C. Lloyd Morgan, never any sort of Darwinian ultra, confessed in 1909 to being fed up with hearing people concluding that the bubble of Darwin's reputation had been pricked and had collapsed.[348] Jaundiced pictures of the implications of Darwinism leached into popular literature, for example in the duplicitous industrialist Baxter in Guy Thorne's *When It Was Dark* (1903), or the brutally ambitious John Barnard of W.B. Maxwell's *The Rest Cure* (1910), who despite his dislike of reading, 'once read sixty pages of the

368   DARWINISM'S GENERATIONS

Origin of Species' and found there the secret of business success.[349] By 1908 Arabella Kenealy was offering a much more downbeat account of Darwinian reading; Alma Wenlith, the young heroine of *The Whips of Time*, finding her much-anticipated reading of the *Origin* 'a horrible disappointment...a story with the point left out, like a road which led to nothing'. As we have seen, Galsworthy's Forsyte novels were littered with explicitly Darwinian language; but the most direct reference to Darwinism in his fiction comes in his essay 'Distant Relative' (1911), in which the highly unsympathetically portrayed relative remarks that 'They say Darwin's getting old-fashioned; all I know is he's good enough for me. Competition is the only thing.'[350] Inevitably, Lytton Strachey, who was to make his name by his biographical debunking of *Eminent Victorians* in 1918, eyed up Darwin as a great subject for a biography, and Darwin was one of the authors singled out to be blasted in Wyndham Lewis' modernist manifesto, 'Blast years 1837–1900', which also included 'Blast...fraternizing with monkeys'.[351]

This reaction was the confluence of several not always compatible currents. Surviving mid Victorians doubled down on their doubts, welcoming the decay of the hegemony that natural science had seemingly acquired around 1870, and the recuperation of older truths.[352] For high Victorians it was more a matter of fatigue and a loss of earlier enthusiasm. So for the royal chaplain and literary historian Stopford Brooke (1832–1916), the excitement and liberation of the Darwin of the 1860s gave way gradually to a more jaundiced view of the narrowness of scientific knowledge, its lack of interest in beauty or love, its dreary quarrels with theology, encouraging him to turn back to poetry and to William Morris.[353] The increasingly ubiquitous reference to Darwin's loss of aesthetic sensitivity, which was beginning to really irritate Darwinian loyalists like Poulton by 1909, was symptomatic of this disillusionment.[354]

Darwin ceased to be the point of reference and authority that he had been within the high- and late-Victorian debates, and the attacks of the geneticists and growing scepticism as to the implications of Darwinian mechanisms did open up space for the sort of wholesale rejection of evolutionary thought which had otherwise survived for nearly half a century largely as a relic of the early Victorians. The surprisingly general interest in Bergson suggests perhaps the cultural obsolescence of Darwin. Edwardian intellectuals like the classicist Gilbert Murray (1866–1957) might speak of the 'vast scheme of biological evolution', but it was a scheme which had lost any moorings in biology, never mind in Darwin.[355] There was a tendency to succumb to the temptation provided by the 'eclipse' of Darwinism to take

DARWINIAN DEBATES AT THE *FIN DE SIÈCLE*: THE EDWARDIANS   369

up more disputatious positions. The attempt to oppose scientific discovery with theological orthodoxy might have been a disastrous mistake; but it was no longer necessary when one scientific theory could be ranged against another. For the evangelist James Gilchrist Lawson (1874–1946), in their debates the Lamarckian and Darwinian schools overthrew each other.[356] The Edwardians did not shy away from Darwin, trumpeted their possession of his standard works, and their intimate knowledge of them, but this was now as likely to justify denial as assent.

Public rejections of Darwinism became more frequent and more prominent. As a young banker in Rangoon in the early 1890s, Charles Stewart Addis (1861–1945) could summarily reject the broad outlines of evolutionary naturalism, denying especially that humans were processes of evolution.[357] This was the context of the outspoken hostility of a figure such as Harold Christopherson Morton (1870–1936), whose *The Bankruptcy of Evolution* (1924) became one of the first of the twentieth century's outright denials of evolution in Britain. Unlike much interwar anti-Darwinism, Morton's antagonism was established well before the disillusioning effects of the 1914–18 conflict. As early as the mid-1890s he was refusing to accept the pitiless struggle to the death of evolution, the fact that in order for the race to flourish, 'the many less fit must be sacrificed to the few more fit'.[358] The later creationist Douglas Dewar (1875–1957) offers a more typical trajectory; his *The Making of Species* (1909), co-authored with Frank Finn (1868–1923), sought to offer a balanced interpretation which rejected the dogmatic squabbles between ultra-Darwinism and Lamarckism.[359] It was only after 1918 that he moved to out-and-out creationism.

The precise nature of the positions being represented is often as opaque here as earlier: repudiation of 'Darwinism' did not necessarily imply a denial of evolution, human or general. The Edinburgh-educated emigrant D. Stewart Maccoll (*c.*1863–1938), who seems to have been sufficiently interested to acquire a copy of the *Origin* for his library soon after arriving in Australia in the later 1880s, was thirty years later able to proclaim that 'not a single tinct of Darwinism is held today by any modern scholar or evolutionist'.[360] Often it was less an abandonment of evolution as disillusionment with its implications. Many came to associate Darwinism with an arid materialism. Edwardians spoke of the 'dark days of Darwinism'.[361] As we have seen, this was certainly the case for George Bernard Shaw, whose interventions started with suspicion that the *Origin*, by encouraging a materialistic and soul-destroying conception of the universe, had become an obstacle to progress, but quickly, as his long-running debate in the

370 DARWINISM'S GENERATIONS

Clarion with Robert Blatchford demonstrated, became a uncompromising assault on 'Darwinism' itself. 'I don't believe', he observed in 1907, that 'Charles Darwin knew anything about evolution'.[362]

For others, though, the repudiation was more fundamental. John Ross (1842–1915), Scottish Presbyterian Missionary to Manchuria, told an audience in 1912 that he had been a believer in evolution until he had read Darwin's works and realised how slim the basis for it was.[363] The Japanologist Sir Ernest Satow (1843–1929), who was also a keen amateur botanist, seems for reasons that are not clear to have lost faith in the central articles of Darwinian evolution.[364] The naval surgeon turned university administrator F.V. Dickins (1838–1915) admitted in 1906 to the 'calamity' that 'malgre moi and [to my] grand regret. I am losing my faith in Darwinism, more and more it seems to me a mistake argumentatively based upon confusions and inaccuracies of thought'.[365] Others who might previously have expressed their doubts tentatively were emboldened to more definite opposition. Take the case of Arthur Balfour, whose *A Defence of Philosophic Doubt* (1879) had presented a 'sceptical unease' about the implications of evolution, but who by 1903 was baldly confessing, at least in private, that 'I don't believe one bit in evolution, or in the "origin of species", nor "the descent of man"'.[366] The Cambridge-educated Congregational minister, W.J. Cunningham Pike (1871–1925) told his congregation in 1913 that evolution, whether Darwinian or otherwise, was 'an exploded fallacy', quoting various biologists to support his view.[367]

For the Edwardians in particular, conversion experiences were now as likely to involve a turn away from Darwinism as a turn towards. The 1900s offer numerous examples. The Fabian psychoanalyst David Eder (1865–1936) remembered being 'brought up in the narrowest of Darwinian teachings', and Sunday walks with the novelist Israel Zangwill (1864–1926), 'literature, history, philosophy...Darwin, and love and friendship', before Samuel Butler's *Unconscious Memory*, 'bought 18 years ago at a second-hand bookshop, opened a new life to me', although 'even then I had to dwell many years in the forests, remote from books and men, ere I could free myself of all Darwinian fetters'.[368] The Irish emigrant to Australia, Timothy M. Donovan (c.1863–1950), read 'nearly all the works of the Big Four—Darwin, Spencer, Huxley and Tyndall', and adopted their materialism before later returning to Catholicism, 'their rank materialism...shattered to pieces by the discoveries of present-day science'.[369] The Bermondsey socialist, Alfred Salter (1873–1945), whose early Methodism had been

DARWINIAN DEBATES AT THE *FIN DE SIÈCLE*: THE EDWARDIANS   371

converted into agnosticism in large part by his reading of Darwin as a medical student, was reconverted to Christianity and a more Spencerian view of evolution as a young doctor because he could not accept what he called its 'accidental' version of the world.[370]

But by far the most widely read Darwinian apostate of this generation was the journalist, minor novelist, and man of middlebrow letters, G.K. Chesterton. 'I was brought up to believe in Darwinism, and now I disbelieve in Darwinism', he confessed in 1931.[371] His was another childhood 'in a world that almost deified Darwin'.[372] His early observations on evolution suggest accommodation tempered with an underlying Catholic suspicion. But by the early 1900s hostility to Darwinism, first in passing gibes and then sweeping dismissals, became a staple of his journalism, and in 1905 he engaged in a public debate with Robert Blatchford in the *Clarion*, countering Blatchford's advocacy with the blunt message 'For God's sake *don't* read Darwin'.[373] Although Darwin himself, his 'dullness', his 'crass materialism', did not escape Chesterton's censure (the *Origin* was 'a back number'), at times he sought to distinguish between the fact of evolution as an innocent scientific description of how certain earthly things came about and Darwin's specific scientific theories, and he often took issue more with 'the hush around the holy name of Darwin', and the dogmatism of the Darwinians and the 'materialist sociologists', especially Haeckel, whose 'bogus heraldry' he found especially abhorrent.[374] At other times he railed against progressivist versions of evolution: it was the universal error of the evolutionists to suggest that evolution knew what it was doing.[375] This was not in any sense a scientific recantation: Chesterton disclaimed any knowledge of biology, and had little to say about natural selection or heredity; 'I have not written a book called "The Origin of Species"', he remarked at one point.[376] Rather, it was driven by revulsion at Darwinism the political philosophy which said the weakest must go to the wall: a poison, in Chesterton's eyes, a half-remembered principle with which the ignorant could dispel any scruples they might have about crushing a rival; which led to eugenics and to the reactionary implication that humanity must adapt to circumstance and not change it.[377] 'Under the shock of Darwinism', he proclaimed, 'all that was good in Victorian rationalism shook and dissolved like dust'.[378]

What is particularly telling about Chesterton's stance is that he saw it in clearly generational terms, as a contrast between the generation of his father—the generation that had lived by the religion 'men had when

372 DARWINISM'S GENERATIONS

rationalism was rational', and which 'never understood the new need for spiritual authority'—and his own generation.[379] As he remarked in 1907 in a verse entitled 'I was not born in 1856', prompted by an editorial mistake about his date of birth, and perhaps directed obliquely at Shaw, who of course was born then:

> I am not fond of anthropoids as such,
> I never went to Mr Darwin's school.
> Old Tyndall's Aether, that he liked so much,
> Leaves me I fear, comparatively cool.
> I cannot say my heart with hope is full
> Because a donkey by continual kicks,
> Turns slowly into something like a mule.
> I was not born in 1856.[380]

Conclusion

In the winter of 1906–7 Blatchford's *Clarion* newspaper asked its readers for their choice of Britain's Greatest Benefactor. Darwin emerged as the clear winner, his 136 nominations being more than 100 ahead of all but two other figures (William Caxton (106) and Oliver Cromwell (52)); Ruskin, Marx, and Herbert Spencer managed only six nominations apiece. The justifications that Blatchford printed suggested a primarily theological motivation. Darwin, thought John Wynn (1854–1909), a commercial traveller and insurance agent from Swindon, had 'clarified their intellectual vision concerning the superstitious vagaries of theology'. Alfred Garland, who we can provisionally identify as a forty-one-year-old tinplate worker from Fulham, celebrated the *Origin* as 'the death blow to the old world of thought and theology', while for the young artillery officer Maurice B. Talbot-Crosbie (1881–?), as well as 'destroying current theology', Darwin showed that great changes had to come from within, and could only be brought about by gradual development.[381] Blatchford's editorial reflections confirm the extent to which it was the broad implications of evolutionary thought rather than the specifics of Darwinian natural selection that were being celebrated here.

In reality, as the fiftieth anniversary of the publication of the *Origin* approached, the status of its primary arguments seemed almost as much in the melting pot as had been the case in the months after its first publication, and perhaps for the first time it is possible to argue that Darwinian ideas

DARWINIAN DEBATES AT THE *FIN DE SIÈCLE*: THE EDWARDIANS 373

were if not in full flight, then at least on the defensive. This was perhaps not so much a retreat from Darwinism, as a dilution of its cultural significance, its loss of the status of dogma. Increasingly, it was being treated as a contribution to a wider knowledge, a partial rather than a complete explanation.[382] A great European war was still only a menacing threat. Yet the erosion of the underlying confidence in progress which had sustained so much of the interpretation and implication of Darwinian thought over previous decades was already undercutting the foundations of much of the generational conversations of this half century. To return to Neville Talbot, the Edwardian generation was 'modern in the sense that it never knew the world "before the flood." While it has been growing up the assumptions of Mid-Victorian liberalism have been going bankrupt. Their capital has been running out. Even their last survivor, Progress, has been at grips with a doubt deeper than itself as to man's place in the universe. For the infection of a kind of cosmic nervousness has become widespread.'[383]

The annual meeting of the British Association for 1908 in Dublin offered signs of these times, despite the election of Frank Darwin as president. Several section chairs played their part in anticipating the Darwin anniversaries to come. Sir William Ridgeway (1853–1926) gave the Anthropological Section a strident warning about the dangers of neglecting evolutionary principles in social legislation, and Sidney Harmer (1862–1950) congratulated the zoologists on their role in the development of evolutionary science; but otherwise there was little direct attention to Darwinism in the sections. And Frank Darwin's presidential address was a huge disappointment—if not worse—to many. Bad enough that he opted for a highly technical discussion of the movement of plants which was beyond the understanding of most of the mixed audience. Worse that he offered at best a perfunctory endorsement of the continued relevance of his father's work. But worst of all that he proceeded to defend the very process of the inheritance of acquired characteristics, which for many it had been the fundamental achievement of Charles Darwin to relegate to the margins, if not to dismiss altogether, and then compounded his offence, in the eyes of the neo-Darwinians, by turning on Weismann.[384] Overall, *Science Progress* suggested, although Charles Darwin's name had been in the air, and audiences wanted to hear about him, 'one of the mistakes of the week was that they were allowed to hear so little.'[385] It was at best an ambivalent culmination to fifty years of discussion and debate, as the imminent celebrations of the half-centenary of the publication of the *Origin* were about to confirm.

374 DARWINISM'S GENERATIONS

Notes

1. T.H. Huxley to Sir Henry Thompson 27 April 1892, Huxley Papers, APS Library, Philadelphia.
2. *PMG*, 1 July 1895, quoted in Desmond, *Evolution's High Priest*, 236.
3. See the correspondence in MS.40314, Murray Papers, NLS. It was suggested by the *Field* that it was only with the half crown edition of the *Life* in 1902 that public generally became acquainted with Darwin, *Field*, 21 August 1909.
4. [C.W. Saleeby] *PMG*, 27 September 1907; 'Bookselling in the East End: An Interview with Mr George of Whitechapel', *The Bookman*, February 1894, 150–1.
5. W.L. Courtney, *Daily Telegraph*, 30 May 1903.
6. *South Place Magazine*, 11 (November 1906), 26. For one recollection of the importance of the Rationalist Press Association reprints, including the *Origin*, see 'Books That Have Helped Me, by A Working Miner', reprinted in *Daily Herald* (Adelaide), 17 May 1921.
7. See the recollection of 'A.S.', *Dundee Courier*, 12 December 1953.
8. Interestingly there is virtually no direct reference to Darwin in Galsworthy's writings; but questions of heredity were certainly on his mind; see Galsworthy to E. Garnett, 17 February 1907, H.V. Marrot, ed., *Life and Letters of John Galsworthy* (1935), 666–7.
9. See the recollections of James Leatham (1865–1945), 'Why Not Have Classes in Public Service?', *Aberdeen Press and Journal*, 30 October 1934; Leatham, 'Less Class Feeling and More Goodwill Then', *Aberdeen Press and Journal*, 26 January 1933.
10. See the folder of letters from the audience of lectures in March 1891, Pearson/2/1/8/4, Pearson Papers, UCL.
11. *South Wales Echo*, 4 August 1893. Vivian's rejection seems primarily to have been human evolution, commenting that he didn't believe humans were ever monkeys. The Eistedfodd for the following year named 'Darwin and his Theories' as one of the themes for its essay competitions, obtaining a rather meagre six entries, *Western Mail*, 5 June 1894.
12. See Lewis Campbell to Huxley, 12 August 1894, f.19, Box 12, Huxley Papers, ICL; Clifford Allbutt to Bateson, 23 September [1894], G3q-28, Bateson Correspondence, Copy Collection, JIC. Edmund Clerihew Bentley (1875–1956) to G.K. Chesterton, August 1894, ff.51–4, Chesterton Papers, Add. MS 73191, BL. Bentley became one of the 'Speaker group' of Edwardian New Liberals, along with J.L. Hammond, F.W. Hirst, and others. We can contrast his stance with the measured appreciation both of Salisbury ('an admirable example of popularizing scientific problems'), and Huxley ('a man possessed by an idea'), in Samuel Barnett to Francis Barnett, 11 August 1894, F/BAR/113, Samuel Barnett Papers, LMA.
13. *Nature*, 16 August 1894. Edward Frankland remarked that it was 'useful to be sometimes reminded how much still remains to be done before the darkness even of the <u>intellectual</u> non-scientific mind will be enlightened', Frankland to Huxley, 2 September 1894, ff.276–7, Box 16, Huxley Papers, ICL.
14. These works included the ultimate contributions of the very youngest early Victorians, such as William Willmer Pocock (1813–99)'s *Darwinism a Fallacy* (1891), and of mid Victorians like Newton Crosland (1819–96)'s *Rambles Round My Life* (1898), with its urgent intrusion of a crude attack on human evolution, Thomas Gordon Hake's *Memoirs of Eighty Years* (1892), and James Tait (1829–99), whose crudely anti-Darwinian *Mind in Matter* (1884) reached a third edition in 1892: proof, the *Westminster Review* commented, of 'the popularity of works in which science is made the servant of orthodoxy', 'Philosophy and Religion', *Westminster Review* 139 (1893), 75.
15. *John O'Groats Journal*, 7 December 1900.
16. Recollection of John Grant McKenzie (1882–1963), *Nottingham Journal*, 15 November 1933.
17. Annie Kenney, *Memories of a Militant* (1924), 21. For context, see Erin McLaughlin-Jenkins, 'Annie Kenney on Evolution, Freedom and Fellowship', *Victorian Review* 41.2 (2015), 39–44.
18. Thomas Jones, *Rhymney Memories* (1938), 103–4.

DARWINIAN DEBATES AT THE *FIN DE SIÈCLE*: THE EDWARDIANS   375

19. William Kent (1886–1963), *The Testament of a Victorian Youth: An Autobiography* (1938). E.E. Kellett recalls that he 'knew many confirmed church-goers,…who owned in private that they were Darwinians, anti-Trinitarians, or even agnostics', but who would still feel need to be in their Sunday best on Sunday', Kellett, *As I Remember* (1936), 106.

20. *Paisley & Renfrewshire Gazette*, 27 October 1900. Methodism remained a culture which could sustain anti-Darwinism; see the account of Rev. William Gooderidge (*c.*1839–?), who 'did not give a shadow of countenance to the Darwinian theory of species', *Bexhill on Sea Observer*, 3 November 1900.

21. *Westminster Gazette*, 2 May 1909.

22. St G. Mivart to Meldola, 20 April 1897, #278, Meldola Papers, ICL.

23. Review in *Athenaeum*, 4 April 1896; *Nature*, 5 January 1899; Argyll's 1897 *TNC* article, suggested W.E. Darwin, amounted to little more than a plea for special creation for germs, W.E. Darwin to Hooker, 25 March 1897, JDH/1/2/6/131, Hooker Papers, RBGK.

24. Argyll to John Murray, 22 February 1893, Murray papers, NLS, Online at http://www.nls. uk/collections/john-murray/authors/duke-of-argyll/letter-to-murray/ [accessed 3 March 2023].

25. Massingham, 'The Religion of a Journalist', *Spectator*, 27 September, 4 October 1924. The editor was James Spilling (1823–97).

26. Clodd to Clement Shorter, 18 March 1902, Shorter Correspondence, Box 1/115, Edward Clodd Papers, BC MS 19c Clodd, Brotherton Library.

27. See Dixey, still taking the fight to the neo-Lamarckians in 'Entomology and Evolution' to the Entomological Society in 1911. *Transactions of the Royal Entomological Society of London for 1910* (1911), xcciii–cvi.

28. G. Archdall Reid, 'The Alleged Transmission of Acquired Characteristics', *Contemporary Review* 94 (1908), 399–412, Charles Arthur Mercier, 'The Transmission of Acquired Characteristics', *Contemporary Review* 94 (1908), 705–15.

29. Poulton to Marsh, 6 March 1894, Box 26, Folder 1092, Othniel C. Marsh Correspondence, Yale.

30. See *Westminster Gazette*, 12 December 1908, and Poulton to J.D. Hooker, 27, 31 October 1908, JDH/2/1/17/39,42, Hooker Papers, RBGK. The first issue included Poulton's own 'Darwin and Bergson as Interpreters of Evolution'.

31. See E.B. Ford, 'Some Recollections Pertaining to the Evolutionary Synthesis', in E. Mayr and W.B. Provine, eds., *The Evolutionary Synthesis* (1980), 227; Poulton to DWT, 12 April 1929, ms19343, D'Arcy Wentworth Thompson Papers, St Andrews.

32. Sichel to Elinor Paul, August 1890, in Edith Helen Sichel, *Letters, Verses, and Other Writings* (1918), 52.

33. See C.F.G. Masterman (1874–1927)'s review of Argyll's *Life and Letters*, *Daily News*, 11 July 1906.

34. P.C. Mitchell, 'Huxley', *The New Review* 13 (August 1895), 149. We can see a similar temper in W.B. Yeats' dismissal of Huxley as an example of the sort of obvious, forceful writer who might have appealed to a certain brand of ill-educated self-made men, but which Yeats despised, Yeats to Lady Augusta Gregory [23 January 1901], in John Kelly and Ronald Schuchard, eds., *The Collected Letters of W.B. Yeats*, Vol. 3: *1901–1904* (1994), 19.

35. R.R. Marett, *A Jerseyman at Oxford* (1941), 113.

36. See William Bate Hardy (1864–1934), 'The Physical Basis of Life', *Proceedings of the Royal Institution* 18 (1905–7), 393; Margaret McMillan, *Early Childhood* (1900).

37. Neville S. Talbot, 'The Modern Situation', in Talbot, ed., *Foundations: A Statement of Christian Belief in Terms of Modern Thought* (1913), 4.

38. P.C. Mitchell, 'John Tyndall', *The New Review* 10 (1894), 80; Mary Kingsley to Clodd, 31 January 1898, in Clodd, *Memories*, 79; Joseph Pearce, *Old Thunder: A Life of Hilaire Belloc* (2002); for Chalmers Mitchell, see Crook, *Darwin's Coat-Tails*; Bertrand Russell, too, speaks of a childhood in which 'Darwinism was accepted as a matter of course', and quotes from his notebooks as a fifteen-year-old where he discusses his acceptance of the natural origins of life. Similarly Kathryn Glasier (1867–1950); see her 'How I Became a Socialist', *Labour Leader*, 18 July 1912.

39. Quiller-Couch, *Dickens and Other Victorians* (1925), 5.

376 DARWINISM'S GENERATIONS

40. Magnus, *English Literature in the Nineteenth Century* (1909), 222–3.
41. Gregory to Wells, 9 November 1934, D.C. Smith, ed.,*The Correspondence of H.G. Wells*, Vol. 3: *1919–34* (1996); See B. Cliffe and Nicholas Grene, eds., *Synge and Edwardian Ireland* (2012), 29; A. Noyes, *The Unknown God* (1934), 16.
42. G.M. Trevelyan, *An Autobiography and Other Essays* (1949), 23; recollections of James H. Ashworth (1874–1936), later of Edinburgh University, of the Salem Congregational Chapel's young men's class in Burnley, *Burnley Express*, 22 April 1931.
43. W. Somerset Maugham, *Of Human Bondage* (1915), 394–5. Carew is transparently Maugham himself, not least in being explicitly given the same age; Maugham later wrote of *Of Human Bondage* that 'the emotions are my own', *The Summing Up* (1938), 242. Maugham's reading and thinking in his twenties, in the later 1890s, seem to have led him to an acceptance of natural selection and a rejection that it could be said to have any meaning or purpose; see Maugham, *A Writer's Notebook* (1946), 50–1. Significantly, although Maugham's uncle Henry Macdonald Maugham (1828/9–?) had a copy of the *Origin* on the shelves in his vicarage at Whitstable, it seems he had little or no exposure to Darwinian thought before his medical training in his twenties, *The Summing Up*, 71; see also Anthony Curtis, *Somerset Maugham* (1977), 27.
44. Talbot to Poulton, 29 December 1897 (perhaps 1907, but this would be an odd context for the remark, which almost certainly relates to Poulton's biography of Darwin which had been published in 1896), Box 11, Poulton Papers, NHM, Oxford.
45. Gilbert Cannan (1884–1955), *Samuel Butler: A Critical Study* (1915), 59.
46. G. Seaver, *Edward Wilson of the Antarctic: Naturalist and Friend* (1933), 10; H.G.R. King, ed., *South Pole Odyssey: Selections from the Antarctic Diaries of Edward Wilson* (1982), 32, 34–5, 41; H.R. Mills, *The Life of Sir Ernest Shackleton* (1923), 76.
47. Arthur Granville Bradley, et al., *A History of Marlborough College during Fifty Years, from Its Foundation to the Present Day* (1893), 317. While at Winchester in 1884, Lionel Pigot Johnson (1867–1902) was musing on the purpose of life, invoking Darwin as a part of gospel of development; see *Some Winchester Letters of Lionel Johnson* (1919), 68, 187–8. In contrast, the recollection of the war correspondent Sir William Beach Thomas (1868–1957), who attended Shrewsbury School in the late 1880s was that the school's connection to Darwin went entirely unnoticed, Thomas, *The Way of a Countryman* (1944), 34.
48. Not always without controversy, as the affair of Winifred Mary Gould at Fishlake in South Yorkshire in 1907 demonstrates. Her inclusion of references to evolution in her primary school lessons at the village school brought complaints from at least one parent, censure from the local vicar, Eliezer Flecker (c.1839–1923), dismissal from her post, and eventually a successful prosecution on her part of the vicar for slander. See *Yorkshire Post*, 1 March 1907, *Sheffield Independent*, 14 March 1908.
49. Hector Bolitho, *Sir Alfred Mond, First Lord Melchett* (1933), 44–5; in the same way Norman Angell immersed himself as a twelve-year-old in Mill, Voltaire, Kingsley, Morris, Spencer, Huxley, and Bradlaugh; see Crook, *Darwinism and War*, 103–6.
50. Braddon, *The Conflict*, serialised in *The People*, 1902.
51. See O'Grady, 'Out of the Ranks: From Bottling Factory to Parliament', *Observer* (Adelaide), 2 June 1906; compare with similar accounts of the intellectual formation of Ramsay MacDonald, working as a warehouse clerk in London, taking natural science classes at South Kensington, and spending his lunchtimes in the Guildhall Library, reading Darwin and Spencer, S.V. Brasher, 'Ramsay MacDonald's Story', *The Register* (Adelaide), 8 May 1924.
52. Keith, *Anatomy in Scotland*, 28; Arthur Keith, *An Autobiography* (1950), 74.
53. Mitchell, *My Fill of Days*, 44.
54. Dennis Hird in *Saint George* 4 (1901), 243.
55. David S. Cairns provides a good example; see Cairns, *An Autobiography* (1950), 166–7.
56. Bertrand Russell, 'My Religious Reminiscences', *The Rationalist Annual* (1938), quoted in Dixon, *Altruism*, 355. That said, Russell nevertheless conceded that 'Evolutionism' was 'the prevailing creed of our time', dominating politics, literature and philosophy, Bertrand Russell, *Our Knowledge of the External World as a Field for Scientific Method in Philosophy* (London, 1914), 11. In the same way Benchara Branford (1868–1944), disciple of Patrick Geddes, saw the 'Darwin, Tait and Kelvin school' being superseded by 'Stallo, Pearson

DARWINIAN DEBATES AT THE *FIN DE SIÈCLE*: THE EDWARDIANS 377

and Mach'. See similar confession of the Congregationalist turned Anglican Reginald John Campbell (1867–1956), *A Spiritual Pilgrimage* (1916), 85–6. The intellectual trajectory of Arthur Quiller-Couch (1863–1944) was the abandoning of Spencer and Mill for Balzac, Sir Arthur Quiller-Couch, 'Books and Other Friends', in *Fifty Years: Memories and Contrasts: A Composite Picture of the Period 1882–1932* (1936, or. 1932).

57. Wollaston, *Life of Alfred Newton*, 102–3.
58. Lionel Earle, *Turn over the Page* (1935), 13.
59. F.E. Younghusband, *A Venture of Faith* (1924), 13.
60. During his student days he and his classmates 'read his speeches, we borrowed the books he wrote', Wells, 'Huxley', *Royal College of Science Magazine* 13 (1901), 209–11 (quoted by John Partington, 'H.G. Wells' Eugenic Thinking, 1892–1944', https://www.academia.edu/1693185/H_G_Wells_s_Eugenic_Thinking_1892_1944), and in his autobiography Wells states that 'that year I spent in Huxley's class was, beyond all question, the most educational year of my life' (Wells, *An Experiment in Autobiography* (1934), 161). (For a detailed account of Wells' Darwinism, see Hale, *Political Descent*, 252–300.) In contrast, Arthur Keith noted in his autobiography that he was glad that he was 'not a pupil of the great Huxley…I am still sorry for the modern students who are reared on a mere knowledge of the Huxleyan types', Keith, *An Autobiography*, 73.
61. Alan Campbell Swinton (1863–1930), 'Early Distrust of Darwinism', *Manchester Guardian*, 13 March 1924.
62. Marett, *Jerseyman at Oxford*, 84; see the very similar verdict in Lewis R. Farnell (1856–1934), *An Oxonian Looks Back* (1934), 114–15.
63. Inge, 'Truth and Falsehood in Religion' (1906), in his *Lay Thoughts of a Dean* (1926), 153–4.
64. Mrs Patrick Campbell, *My Life and Some Letters* (1922), 14, where her father John Tanner (1829–95) is described as 'a cheerful believer in the Darwinian theory'. Similarly, for the prison reformer Margery Fry (1874–1958), her father, Sir Edward Fry, perhaps even more Darwinian than he himself realised, typified his generational position; see her contribution to James Marchant, ed., *What Life Has Taught Me* (1948), 53.
65. E.F. Benson, *As We Are: A Modern Revue* (1932), 63; A.N. Whitehead, *Essays in Science and Philosophy* (1948), 26–33.
66. J. Kent, *From Darwin to Blatchford* (1966), 37; James Moffatt, ed., *Letters of James Denney to W. Robertson Nicoll* (1920), 28 November 1903, 39.
67. *Daily News*, 14 March 1903; address to the Christian Social Union, May 1904, quoted in William Oddie, *Chesterton and the Romance of Orthodoxy: The Making of GKC, 1874–1908* (2008), 237. In this sense his recollection that in this period he was a socialist because not to be one necessitated being 'some hoary horrible old Darwinian who said that the weakest must go to the wall' (Chesterton, *Autobiography* (1936), 111) was a telescoping of his later anti-Darwinian reaction.
68. H. [Bellerby] Lowerison, *The Clarion*, 21 July 1911; viz V.F. Storr's description of evolution as 'the keyword which unlocks the mind of the present age', quoted in M. Wellings, *Evangelicals Embattled, 1890–1930* (2003), 191.
69. For a flavour see the constant invocation of Darwin by the father John Hazleden in Hugh Westbury (i.e. Hugh Farrier (1857–1918)), *Frederick Hazleden* (1887).
70. George Herbert Perris (1866–1920), review of Kidd's *Social Evolution, Inquirer*, 7 July 1894; R. Le Gallienne, 'The Passing of Mrs Grundy' (1912), in his *Vanishing Roads and Other Essays* (1915), 42–54; Le Gallienne, 'Grant Allen', *FR* 66 (1899), 1024.
71. Examples would include Oswald Hawkins Latter (1864–1948) and Richard Kearton (1862–1928). Kearton's *The Fairy Land of Living Things* (1907) offers perfunctory praise and operates within a set of Darwinian assumptions, but is largely indifferent to, or at least uninterested in, evolutionary questions.
72. H.N. Brailsford (1873–1958), *Daily News*, 12 February 1909.
73. William Sharp to Ernest Dowden, Friday [1886], f.808, Dowden Papers, TCD; DWT to Mary Lily Walker, 22 December 1886, ms44434, D'Arcy Wentworth Thompson Papers, St Andrews. In the 1900s Walter Garstang (1868–1949) was commissioned to write a study of Darwin for the 'People's Books' series of T.C. and E.C. Jack, but did not deliver the volume.

378 DARWINISM'S GENERATIONS

74. Hilda Oakeley (1867–1950), *Adventures in Education* (1939), 71–2.
75. J. Arthur Thomson, *The Secrets of Animal Life* (1919), 226.
76. Arthur Willey (1867–1942), tps on Organic Evolution, MS Add. 9914/19, Arthur Willey Papers, CUL.
77. See his *Village England* (1935); see also his refusal, even conceding Darwinism, to allow it to explain everything: 'we are not worshipping an idol of the forum in trying to track everything down to a single cause', W.B. Thomas, 'The Song of Birds', *Macmillan's Magazine* 90 (1904), 460, and his early championing of Mendelian ideas, 'Heredity in Plant Life', *Daily Mail*, 9 December 1907. In like manner, for Lawrence Pearsall Jacks (1860–1955), *The Alchemy of Thought* (1910), 233, it was the 'Pontiffs of Evolution' who now threatened intellectual tyranny.
78. Ford, *Mightier Than the Sword* (1938), 124. The same stance is visible in Arthur Machen (i.e. Arthur Llewellyn Jones (1863–1947))'s suggestive 'The Apostolic Ideal', published in the *Academy* in October 1907, with its faux prayer to Darwin, Tyndall, Spencer, Clifford, and Huxley, *The Academy*, 5 October 1907.
79. J. Arthur Thomson, *The Science of Life* (1900), 226–8.
80. R. D'Arcy, *D'Arcy Wentworth Thompson: The Scholar Naturalist* (1958), 71; by the BA of 1894 Thompson was explicitly 'stating his doubts and difficulties in accepting "Darwinism"', *Natural Science* 5 (1894), 221. In similar vein Bateson talked about 'Faith' having 'given place to agnosticism', 'Evolutionary Faith and Modern Doubts', in B. Bateson, *William Bateson, Naturalist: His Essays and Addresses* (1928), 391.
81. William Purvis, 'A Philosopher at Ease: The Private Life of Samuel Butler' [*Sunday Chronicle*, 19 October 1910?], VIII/23/1, Butler Papers, St John's College, Cambridge.
82. Barry, 'Samuel Butler of "Erewhon"', *Dublin Review* 145 (1914), 339. Edward Lyttelton (1855–1942) was another late Victorian whose valuation of Darwin was challenged by a much later reading of Butler; see his *Memories and Hopes* (1925), 277.
83. Robert B. Ross to R.A. Streatfeild, 10 December 1911, VIII/18/1, Butler Papers, St John's, Cambridge.
84. E.T. Cook to Henry Festing Jones, 20 November 1911, VIII/18/1, Arabella B. Fisher [i.e. Arabella Buckley] to Festing Jones, 18 February 1910, VIII/12/3, Butler Papers, St John's, Cambridge. Forster to Beveridge, 13 May 1908, BEVERIDGE/II/B/7/22, Beveridge Papers, BLPES.
85. Gal Gerson, *Margins of Disorder: New Liberalism and the Crisis of European Consciousness* (2004), 111. For Butler and vitalism see Jonathan Rose, *The Edwardian Temperament* (1986), 76: 'Sydney Olivier and Graham Wallas praised him for saving their souls from Charles Darwin.' For Edwardians drawing parallels between Butler and Bergson see James Houston to H. Festing Jones, 22 November 1911, VIII/18/1, Butler Papers, St John's College, Cambridge; plus others in this collection.
86. Ford Madox Ford, *Return to Yesterday* (1931), 182.
87. The link to Darwinian science was made by the fact that Prendick, one of the castaways on the Island, had apparently studied with Huxley. See Hale, *Political Descent* and J. Glendening, *The Evolutionary Imagination in Late Victorian Novels: An Entangled Bank* (2016). This antithesis is echoed in the undergraduate correspondence of Lowes Dickinson in GLD/AJG II/5, Goldsworthy Lowes Dickinson to [ALG], nd [*c.* October 1884], Lowes Dickinson Papers, King's College, Cambridge.
88. Margaret McMillan, 'The Two Conquerors', *Clarion*, 6 April 1895.
89. G.E. Allen, 'Mechanism, Vitalism and Organicism in Late Nineteenth and Twentieth Century Biology', *Studies in the History and Philosophy of Biological and Biomedical Science* 36 (2005), 261–83; M. Richmond, 'The 1909 Darwin Celebration: Re-Examining Evolution in the Light of Mendel, Mutation, and Meiosis', *Isis* 97 (2006), 474–8. See, for example, P.N. Waggett (1862–1939), *The Scientific Temper in Religion* (1905), 4, or James Johnstone's dismissive comments about 'speculators', in James Johnstone (1870–1932) to DWT, 4 July 1917, ms19953, D'Arcy Wentworth Thompson Papers, St Andrews.
90. [H.G. Wells], 'Discoveries in Variation', *Saturday Review*, 9 March 1895. J. Lionel Tayler, 'The Relation of Acquired Modifications to Heredity', *Natural Science* 11 (1897), 247–50. For Wells' anonymous science journalism in the *Saturday Review* and *PMG*, see

DARWINIAN DEBATES AT THE *FIN DE SIÈCLE*: THE EDWARDIANS 379

Robert M. Philmus and David Y. Hughes, eds., *H.G. Wells: Early Writings in Science and Science Fiction* (1975).

91. G.C. Bourne, 'Epigenesis or Evolution', *Science Progress* 1 (1894), 105–26.

92. Continuing 'and the average person's principal debt to Darwin is his emancipation from the bondage of Hebrew mythology', Clara E. Collet, 'Mrs Stetson's Economic Ideal', *Charity Organisation Review*, March 1900, reprinted in Collet, *Educated Working Women* (1902), 120–1.

93. Richmond, '1909 Darwin Celebration', 476; Woodward, *Outlines of Vertebrate Palaeontology* (1898), xxiii; Arthur Keith, 'Pithecanthropus Erectus', *Science Progress* 3 (1895), 348–69. For other late-Victorian embryologist-evolutionists, see A.C. Haddon, *The Study of Embryology* (1887) (discussed in Stocking, *After Tylor*, 98–115).

94. See for example, James Johnstone, who was initially fascinated by the phylogeny founded on comparative anatomy and embryology being practised by the zoologist G.B. Howes (1853–1905), James Johnstone to DWT, 8 September 1917, ms19955, D'Arcy Wentworth Thompson Papers, St Andrews.

95. Elliot Smith to J.T. Wilson, 1 October 1896, ff.532–3, Dawson Mss, Elliot Smith Correspondence, Add. MS 56303, BL. (The letter appears to refer to 'H. Flower' but the full context makes it clear that Sir William is being described.)

96. Bower to WTTD, 12 February 1908, f.94, W.T. Thiselton-Dyer Papers, In-letters I, RBGK; Bower had also recognised four years earlier that his work was out of kilter with the age, and that his collected studies might well appear as 'a curious exhibition of the prejudices and theorisings of a past age', Bower to I. Bayley Balfour, 8 March 1904, Correspondence, Box B, Isaac Bayley Balfour Papers, RBGE. See Vernon to J.B. Sanderson, 5 November, 10 December 1894, ff.207–8, 209–13, J. Burdon Sanderson Papers, MS.20501, NLS.

97. See comments of Goldsworthy Lowes Dickinson (1862–1932) to Graham Wallas, 7 January [1908?], WALLAS/1/39/6, Wallas Papers, BLPES, also the correspondence of Archibald Edward Garrod (1857–1936) with Bateson in 1902, AD Mss 8634/H12, Bateson Papers, CUL.

98. F.C.S. Schiller (1864–1937), 'Darwinism and Design', *Contemporary Review* (1897), reprinted in *Humanism: Philosophical Essays* (2nd ed., 1912), 136.

99. Newbigin, *Colour in Nature*, 4.

100. Thomson, 'The Endeavour after Well-Being', *Natural Science* 8 (1896), 21.

101. Mitchell, *Evolution and War*, quoted in P. Crook, *Darwinism: War and History* (1994), 74–5. At the same date see Arthur Dendy, arguing that evolution would happen 'even if there were no such thing as natural selection', 'Progressive Evolution and the Origin of Species', *The American Naturalist* 49 (1915), 156.

102. Kropotkin to William Wray Skilbeck [acting editor, *TNC*], 19 November 1909, 14 April 1910, 0716/5/84,90, Knowles Correspondence, Westminster Archives.

103. F.G. Elliot Smith, *Botany of Today* (1910). For Dewar see his 'Country Notes' journalism of the 1900s, for example *Birmingham Daily Gazette*, 6 August 1906, as well as volumes like *The Faery Year* (1906) and *This Realm, This England* (1913). Note that this was not associated with any anti-Darwinism; both Elliot Smith and Dewar were fulsome in their praise of Darwin's works, although after 1918 Dewar expressed sympathy with the current 'great revolt' against the bald materialism which followed Darwin, 'My Books and I', *Sheffield Independent*, 24 September 1919.

104. See 'Lectures on the Evolution of the Filicinean Vascular System', *The New Phytologist* 6.2 (1907), 25–35. For Tansley see Peter Ayres, *Shaping Ecology: The Life of Arthur Tansley* (2012) and John Forrester and Laura Cameron, *Freud in Cambridge* (2017), 7–56.

105. Tansley, 'Neo-Darwinianism', *Hardwicke's Science Gossip* 27 (1891), 207–10.

106. 'Natural Selection Considered as a Special Example of the General Principle of Evolution', Essay for the Arnold Gerstenberg Studentship, 1896, MS Tansley/A17, CUL. Essays submitted by students at St Andrews in 1887 for the Gray Essay prize on the topic of evolution were noticeably sceptical about evolution as a wider philosophy, but also about the Darwinian hypothesis in particular; see John Gray, 'The Doctrine of Evolution Philosophically and Historically Considered', M4329/A, Gray Essay Prize Collection, St Andrews. In the same way, the 'age and area' hypothesis of John Christopher Willis

380 DARWINISM'S GENERATIONS

(1868–1958), while acknowledging the importance of Darwin's work in establishing organic evolution, rejected natural selection.

107. Tansley, 'Modern Views of Heredity and Evolution', ms annotated 'WMC 13 March 1909', MS Tansley/E10, CUL.

108. Their birthdates, as Cock and Forsdyke, *Treasure Your Exceptions*, note of Bateson, 'allowed time for those who might influence [their] education, to digest their Darwin and loosen the ties of religious orthodoxy'.

109. Bateson, *Memoir of William Bateson*, 39. For a discussion of the context see Radick, *Disputed Inheritance*, 85–9.

110. Quoted in Cock and Forsdyke, *Treasure Your Exceptions*, 20. See correspondence with Poulton, especially 18 September 1891, in Box 1, Poulton Papers, OMNH; also Gillham, *Francis Galton*, 286–8; and the first chapter of Bateson's *Materials for the Study of Variation*.

111. Sedgwick to Bateson [October 1890], MS Add. 8364/H36, Bateson Papers, CUL.

112. *Memoir of William Bateson*, 217, quoting his essay from *Darwin and Modern Science* (1909); for an account of Bateson's various anti-Darwinian interventions in these years, see Radick, *Disputed Inheritance*, 100–8. Bateson later gained passing notoriety for his address to the American Association of Science in Toronto in 1922 when he announced that although a believer in evolution, he was agnostic as to causes, and asserted that Darwinism itself could no longer sustained; Bateson, 'Evolutionary Faith and Modern Doubts', *Science*, 20 January 1922.

113. Discussion in Gillham, *Francis Galton*, 292–6. Lankester told Bateson that while variation itself was important, he was 'not convinced' of the importance of 'discontinuous variation', Lankester to Bateson, 13 February 1894, G3q-8, Bateson Correspondence, Copy Collection, G3p-G3q, JIC. (Huxley, by contrast, seems to have written to support the potential importance of discontinuous variation; see Bateson to Huxley 26 February 1894 [draft], G3q-14, ibid.; compare the forthright rejection of Wallace, in 'The Method of Organic Evolution', *FR* 57 (1895), 211–24, 435–45).

114. *Nature*, 20 October 1892; Radick, *Disputed Inheritance*, 111–12. The contours of this disagreement are played in out in the letters the two exchanged in the 1890s extant in the Bateson Correspondence, Copy Collection, G3h-G3n, JIC; Romanes told Poulton that the correspondence was 'quite absurdly aggressive, even supposing that he proves to be right', Romanes to Poulton, 2 June [?] 1892, Box 9, Poulton Papers, OMNH. See also evidence that Bateson fell out with Lankester, whose early scarcely veiled antagonism comes across in E.R. Lankester to Bateson, 17 February [1892 from postmark], tps (5) C18, Bateson Correspondence, Copy collection Vol 1, 1–96, JIC.

115. See 'The Secret of Evolution', review of Bateson's *Materials*, Scrapbook f.92, Box 4/7, Evolution Collection, Mss 28, UC-Santa Barbara. Beddard remained unconvinced about mutational approaches; Sherrington to Bateson, 14 August 1892, Bateson Papers, AD Mss 8634/H65, CUL.

116. See Thomson's notes on J.W. Barclay's *Nature's Method in the Evolution of Life* (1894) and covering letter to Patrick Geddes, 28 February 1895, T-GED/18/2/11, Geddes Papers, Strathclyde University; H.M. Vernon to John Burdon Sanderson, 10 December 1894, ff.209-13, J. Burdon Sanderson Papers, MS.20501, NLS.

117. R. Macleod, 'Embryology and Empire: The Balfour Students and the Quest for Intermediate Forms in the Laboratory of the Pacific', in R. Macleod and P.F. Rehbock, eds., *Darwin's Laboratory: Evolutionary Theory and Natural History in the Pacific* (1994), 154; Meldola to Wallace, 28 June 1904, f.146, Wallace Papers, Add. MS 46437, BL.

118. Michael B. Jackson, 'One Hundred and Twenty-Five Years of the Annals of Botany: Part 1: The First 50 Years (1887–1936)', *Annals of Botany* 115 (2015), 1–18.

119. Wallace, *Darwinism*, also Wallace, 'The Present Position of Darwinism', *Contemporary Review* 94 (1908), 129–41, quoting a Thiselton-Dyer article in *Nature* in 1907.

120. T.G. Bonney, *Old Truths in Modern Lights: The Boyle Lectures for 1890 with Other Sermons* (1891), 11; Thiselton-Dyer, 'The Botanical Work of the British Association', *Nature*, 26 September 1895.

DARWINIAN DEBATES AT THE *FIN DE SIÈCLE*: THE EDWARDIANS   381

121. *Nature*, 10 October 1889. For a discussion of some of the issues at play here see Graham Gooday, '"Nature" in the Laboratory: Domestication and Discipline with the Microscope in Victorian Life Sciences', *BJHS* 24 (1991), 307–41.

122. Although Weldon did use it to conduct systematic experiments into species variation; see Gillham, *Francis Galton*, 280–3.

123. Galton to Wallace, 12 February 1891, ff.248–50, Wallace Papers, Add. MS 46436, BL.

124. Newton to Bateson [January 1890], AD Mss. 8634/H65, Bateson Papers, CUL. The Birmingham writer and indexer F.H. Collins (1857–1910) accepted the need for funding, but not a predetermined plan of work, urging that money be given without strings. 'What were Darwin's own epoch-making experiments, but brains & make shifts?', he asked in 1897, Collins to Galton, 27 January 1897, Galton/2/5/4/2/5, Galton Papers, UCL.

125. See his earlier approach to Newton in January 1890, Draft of Bateson's response to Newton, 22 January 1890, AD Mss. 8634/H65, Bateson Papers, CUL.

126. Bateson to Galton, 3 December 1896, Galton/2/5/4/2/1, Galton Papers, UCL.

127. J.A. Thomson to J. Cossar Ewart, 20 July 1901, COLL 14/9/5/42, James Cossar Ewart Papers, Edinburgh.

128. For his ideas see copy of his manuscript circular, T-GED/18/2/10/1, Geddes Papers, Strathclyde University, and his letter to Galton, 26 December 1894, Galton/3/3/3/36, Galton Papers, UCL.

129. Huxley to Foster, 23 December 1894, in Bynum and Overy, *Michael Foster and Thomas Henry Huxley, Correspondence*, Foster to Huxley, 29 December 1894, f.388, Box 4, Huxley Papers, ICL.

130. E.R. Lankester to Huxley, 26 December 1894, f.162, Box 21, Huxley Papers, ICL. For Cunningham's own sense of the antagonism he faced, especially within the Royal Society, see J.T. Cunningham to Bateson, 16 February 1894, c162 [66], Bateson Correspondence, copy collection Vol 1, 1–96, JIC.

131. Thiselton-Dyer to Huxley, 24 December 1894, f.232, Box 27, Huxley Papers, ICL. It would only, thought Thiselton-Dyer, be 'a platform for drivel like the Victoria Society'. (This was rather to avoid the point that Cunningham was protesting at precisely this sort of evolutionary speculation, calling for detailed experimental investigation; see his 'The Evolution of Flat Fishes', *Natural Science* 1 (1892), 191–9.)

132. Geddes note, T-GED/18/2/10/2, Geddes Papers, Strathclyde University.

133. For a useful summary of the history see Gillham, *Francis Galton*, 299–301; Radick offers an account less sympathetic to Bateson in *Disputed Inheritance*, 118–21. Members of the initial committee were Frank Darwin, A. Macalister, Meldola, Poulton, and Weldon, to which were added S.H. Burbury and Galton in the first year; Adam Sedgwick (Jnr) was also closely involved. The Committee was expanded in December 1897 to include Bateson, Sir Edward Clarke (1841–1931), Godman, Walter Heape (1855–1929), Lankester, Edward Joseph Lowe (1825–1900), Maxwell T. Masters (1833–1907), Karl Pearson, Osbert Salvin (1835–98), and Thiselton-Dyer.

134. Bourne to Galton, 1 December 1896, Galton/2/5/4/2/2, Galton Papers, UCL. See the account of a visit from a group of thirty from Toynbee Hall, 'A Visit to Charles Darwin's Home', *Westminster Budget*, 8 September 1893.

135. See scattered correspondence in JDH/1/2/6, Hooker Papers, RBGK.

136. Alexander Keith, 'Darwin Remembered', *Rationalist Annual* (1930), 19–25.

137. Galton to Bateson, 1 January 1897, [258], Bateson Correspondence, Copy Collection, Vol 3, 202–83, JIC.

138. Bateson to Galton, 1 February 1897 [draft] C199 [240], Bateson Correspondence, Copy Collection, Vol 3, 202–83, JIC.

139. William Bateson to Arthur Cecil, 13 November 1898, COLL 14/9/4/24, James Cossar Ewart Papers, Edinburgh.

140. Typescript draft of memo to George Darwin, 21 July 1899, AD Mss 8634/H63, Bateson Papers, CUL.

141. Weldon to Bateson, 5 June 1899, [282], Bateson Correspondence, Copy Collection, Vol 3 202–83, JIC; 'Evolution Committee Minutes', CMB/65, Royal Society, which details the

382 DARWINISM'S GENERATIONS

vagaries of the committee's active membership; Radick, *Disputed Inheritance*, 120-1, makes it clear that Weldon's departure was a direct result of his refusal to accept Bateson's attempt to distinguish his statistical approaches from 'Natural History' (i.e. Darwinian) ones.

142. As one of many possible examples see J. Arthur Thomson, 'On the Facts of Inheritance', *Proceedings of the Royal Institution* 14 (1899–1900), 346–59. For this dismissal of Darwin as unable to explain variation, see F.W. Bain, *On the Realisation of the Possible* (1899). For general accounts of the confrontation of Darwinism and Mendelism see Radick, *Disputed Inheritance*, and Gayon, *Darwinism's Struggle*, 253–317.

143. Bateson, 'Problems of Heredity as a Subject for Horticultural Investigation', *Journal of the Royal Horticultural Society* 25 (1900), 54–61.

144. Including not just Edith Saunders, but also Agnes Isabella Mary Elliot (1863–1946), Helena Maude Caulfield (née Martyn) (1869–1951), Florence Margaret Durham (1869–1948), Mary Hart-Davis (1875–1934), and Hilda Blanche Killby (1877–1962); see Marsha L. Richmond, 'Women in the Early History of Genetics: William Bateson and the Newnham College Mendelians, 1900–1910', *Isis* 92.1 (2001), 55–90; and also others born after 1875, the generation of many of Bateson's male followers, including Robert Heath Lock (1879–1915), Reginald Philip Gregory (1879–1918), Thomas Barlow Wood (1887–1929), William Lawrence Balls, and Leonard Doncaster (1877–1920).

145. *Nature*, 25 August 1904, 22 August 1907.

146. P.C. Mitchell to Bateson, 8 November 1913 [copy], Bateson Collection WB/K.65/3263, JIC.

147. Wells to Julian Huxley, 23 February [1928], *Letters of H.G. Wells*, III, 259.

148. See the letters in MS Add. 8364/H16, Bateson Papers, CUL. As Radick's *Disputed Inheritance* shows, the opposing camp were equally outspoken.

149. See, for example, the botanist John Bretland Farmer (1865–1944). For Farmer's commitment to Mendelism, but in a much less aggressively anti-Darwinistic form, see Radick, *Disputed Inheritance*, 160, and his letter to WTTD, 25 November (nd, but *c*.1907), f.17, W.T. Thiselton-Dyer Papers, In-letters II, RBGK; Poulton's verdict was that Farmer had 'a tendency to shout with the crowd', Poulton to WTTD, 1 December 1907, f.133, In-Letters III, Thiselton-Dyer Papers, RBGK.

150. Doncaster to Bateson, 21 June, Naples [1902?], WB/K.29 C1428, Bateson Collection, JIC. By 1903 relations were breaking down entirely. 'I am resolved to do as little with them as I can', Bateson to Hurst, 24 March 1903, MS Add. 7955/3/12, Hurst Papers, CUL.

151. Pearson to Hurst, 9 January 1904, AD Mss 7955/4/69, Hurst Papers, CUL.

152. D. Lipset, *Gregory Bateson* (1982), 39; Gillham, *Francis Galton*, 321–3. In December Hurst was exulting in 'the smashing [Bateson] gave them ['the opposition'] at Cambridge', after which 'they have lost heart. W[eldon] appears to be silenced, and K. P[earson] seems to be giving off a few dying kicks before adopting Mendel to his purpose!', Hurst to Bateson, 9 December 1904, MS Add. 8364/H18, Bateson Papers, CUL. For a less dramatic account which emphasises the continued resistance of Weldon and Pearson, see Radick, *Disputed Inheritance*, 232–59.

153. F.J. Weldon to Pearson, 26 June 1906, Pearson/11/1/22/39, Pearson Papers, UCL.

154. See, for example, F.W. Mott, *Nature and Nurture in Mental Development* (1914), 34–5. The orchid specialist Robert Allen Rolfe (1855–1921) was also cautious about Mendelism, noting that Weldon's criticisms 'voice[d] several of my difficulties', Rolfe to Hurst, 6 April 1902, MS Add. 7955/2/33, Hurst Papers, CUL. Reading 'Bateson's recent paper, addenda and postscripts' in June 1902 he noted that Bateson 'has a full crop of difficulties' [Rolfe to Hurst, 10 June 1902, AD Mss. 7955/2/34], and that his paper 'contained much that was nebulous and inconclusive', Rolfe to Hurst, 10 July 1902, AD Mss. 7955/2/66, Hurst Papers, CUL. See also Henry Knipe, *Evolution in the Past* (1912).

155. A persistent attitude; see William Barry to Sir James Marchant, 15 February 1915, ff.26–7, MS. Eng. Lett. d.314, Bodleian, where Mendel quickly slides into Butler's *Luck or Cunning*.

156. Radick gives an illuminating analysis of how even a year's difference became significant because of the timing of the death of F.M. Balfour, much more influential for Weldon than Bateson, Radick, *Disputed Inheritance*, 73–5.

DARWINIAN DEBATES AT THE *FIN DE SIÈCLE*: THE EDWARDIANS 383

157. Even if there was a marked contrast between the rather old-fashioned approach of Cossar Ewart, with his apparent indifference to wider evolutionary questions, his links to horse racing, and his network of maverick correspondents like Frank Challice Constable (1846–1937), practical breeders and racing enthusiasts, and the systematic, theory-driven and aggressively academic work of Bateson and his collaborators. For Constable's engagement with Darwinian ideas, see his *Curse of the Intellect* (1895), a weird tale of an atheist and the monkey he trained to speak and which eventually killed him, also his contributions to the *Socialist Review*, 1908 and 1910, noted in Stack, *Darwinian Left*, 144. The *Curse of the Intellect* is discussed in V. Richter, *Literature after Darwin* (2010), 107–10.

158. See the account of a conversation between the entomologist John Jenner Weir and Darwin, given in T.D.A. Cockerell to Cossar Ewart, 25 April 1902, COLL 14/9/8/31, James Cossar Ewart Papers, Edinburgh; James Collier in *Lyttleton Times*, 5 July 1913.

159. See WTTD to Charles Baron Clarke, 17 November 1894, f.8, A Gunther Collection 16, NHM; Poulton, review of Vernon's *Variation in Animals and Plants*, *Nature*, 12 January 1905. Lankester, too, was apparently in no hurry to read Mendel in the early 1900s; see Radick, *Disputed Inheritance*, 175.

160. Thiselton-Dyer to Wallace, 22 December 1903, ff.142–3, Wallace Papers, Add. MS 46437, BL; WTTD, 'Rise and Influence of Darwinism', 366–407.

161. Romanes to Poulton, 2 November 1889, Box 9, Poulton Papers, OMNH.

162. Weldon to Pearson, 4 June 1905, Pearson/11/1/22/40, Pearson Papers, UCL.

163. R.H. Lock to Bateson, 5 June 1903, AD Mss 8634/H27, Bateson Papers, CUL.

164. Including James Wilson (1861–1941) (professor of Agriculture at the Royal College of Science, Dublin, and the author of a number of Mendelian works, including *The Evolution of British Cattle and the Fashioning of Breeds* (1909), and *The Manual of Mendelism* (1916), and various articles in the *Proceedings of the Royal Dublin Society*), Gertrude Clarke Nuttall (1868–1929) (see her 'Mendel and His Theory of Heredity', *FR* 83 (1908), 528–38), C.C. Hurst, the seedsman William Cuthbertson (*c*.1859–1934); see the interview published in *Jersey Independent*, 13 February 1909, Edgar Chamberlain (1868–1931?) (author of *The Homing Pigeon* (1907), who consulted with Hurst). For Yule, see James G. Tabery, 'The "Evolutionary Synthesis" of George Udny Yule', *Journal of the History of Biology* 37.1 (2004), 73–101.

165. G. Archdall Reid to WTTD, 17 December 1907, f.163, W.T. Thiselton-Dyer Papers, In-letters III, RBGK; Reid to Meldola, 4 November 1907, #2348, Meldola Papers, ICL.

166. Reid to Wallace, 27 December 1909, ff.111–12, Reid to Wallace, 5 January 1910, ff.113–18, Wallace Papers Add. MS 46438, BL.

167. G. Archdall Reid, 'Methods of Research', *Eugenics Review* 3 (1911–12), 241–64, 254–5.

168. Reid to Meldola, 4 November 1907, #2348, Meldola Papers, ICL.

169. *St James Gazette*, 9 November 1900. It should be observed that personal as well as intellectual dynamics were in operation here. As a provincial GP, Reid was acutely aware of his lack of academic status, and inclined to seek more institutionally secure patrons, and although he initially quarrelled with Lankester over his book *The Present Evolution of Man*, Lankester eventually saw the advantages of encouraging Reid's hostility to Bateson and the Mendelians, persuading, for example, Lockyer to publish his attacks on Mendelism in *Nature*; and Reid gratefully accepted this patronage. See Reid to WTTD, 9 December 1907, f.161, W.T. Thiselton-Dyer Papers, In-letters III, RBGK, and his correspondence with William James, especially Reid to James, 19 June 1901, 12 September 1895, Reid Folder, William James Papers, MS Am 1092.9/A, Harvard.

170. M. Shermer, *In Darwin's Shadow* (2002), 214–15.

171. The suggestion of Poulton, *Essays on Evolution*, xiii, is that he did not pay attention to the Mendelism literature until well into the 1900s. Meldola confessed hadn't taken the time to read Bergson, relying as late as 1911 on the abstracts of his lectures published in the *Times*, Meldola to Poulton, 5 November 1911, Meldola Papers, OMNH.

172. Bateson to Hurst, 3 March 1911, MS Add. 8364/H17, Bateson Papers, CUL.

173. Typescript note, MS Add. 8364/H64, Bateson Papers, CUL.

174. See Mudge to Hurst, 21 November 1908, MS Add. 7955/8/111, Hurst Papers, CUL; Report of a letter from Thiselton-Dyer, in Weldon to Pearson, 16 February 1902, H2.a.49, Bateson Correspondence, Copy Collection, JIC.

384   DARWINISM'S GENERATIONS

175. See *Report of the British Association for the Advancement of Science, 1906* (1907), 3–42.
176. Gilbert C. Bourne to Bateson, 12 May 1909, AD Mss 8634/H62, Bateson Papers, CUL. See, as one example, his correspondence with G.D.H. Carpenter (1882–1953), Poulton Papers Box 3, OMNH.
177. Bateson was the 'arch-sneerer', the 'Cambridge clique' 'narrow-minded and ignorant', Poulton to F.A. Dixey, 29 July 1907, Box 5, Dixey Papers, OMNH.
178. I. Bayley Balfour to Poulton, 29 December 1907 [copy], Correspondence, Box P, Isaac Bayley Balfour Papers, RBGE; Bower to WTTD, 13 July 1910, f.97, W.T. Thiselton-Dyer Papers, In-letters I, RBGK. For a further glimpse of his opposition to Bateson, see A.D. Boney, 'The Summer of 1914: Diary of a Botanist', *Notes and Records of the Royal Society of London* 52 (1998), 323–38, 328–32.
179. See his 'The Evolution of Man', *Mendel Journal* 1 (1909–11), 29–44.
180. R.J. Ryle to Karl Pearson, 21 July 1898, Pearson/11/1/17/102, Pearson Papers, UCL.
181. The first substantial English version of de Vries' work in English was D.T. McDougal, ed., *Species and Varieties: Their Origin by Mutation: Being Lectures Delivered at the University of California* (1905). Even someone like Gerald Leighton confessed in 1905 that, not being able to afford de Vries' work, he was left relying on tapping the brains of his correspondents, Leighton to Hurst, 13 September 1905, MS Add. 7955/5/92, CUL.
182. Reid to Hurst, 11 August 1905, Hurst Papers, MS Add. 7955/5/102, CUL; at least before 1904 Clifford Allbutt remained non-committal; see Allbutt to Victoria Welby, 24 May 1904, tps, 1970–010/001(07), Welby Papers, York University, Canada.
183. Alice Balfour to Cossar Ewart, 4 June 1898, COLL 14/9/4/13, James Cossar Ewart Papers, Edinburgh. See Donald L. Opitz, '"Behind Folding Shutters in Whittingehame House": Alice Blanche Balfour (1850–1936) and Amateur Natural History', *Archives of Natural History* 31.2 (2004), 330–48. For another example, see the Edinburgh educationalist and mathematician Arthur John Pressland (1865–1934), who in the late 1890s under the influence of the early works of Bateson and Galton was investigating the inheritance of colour in cattle; see Pressland to Cossar Ewart, 27 January 1899, COLL 14/9/5/6, James Cossar Ewart Papers, Edinburgh.
184. Waggett, *Religion and Science*, 112–13.
185. It was Ewart's writings in the *Livestock Journal* which initially attracted the interest of established horse breeders like Captain Matthew Horace Hayes (1842–1904). See Hayes to James Cossar Ewart, 26 March 1902, COLL 14/9/8/25, Cossar Ewart Papers, Edinburgh. Perhaps not surprisingly at least some late Victorians were quite able to absorb him as an opponent of natural selection into their Lamarckism; see Henry Neville Hutchinson (1856–1927) to Clodd, 28 November 1911, In-letters, Box 4, Edward Clodd Papers, BC MS 19c Clodd, Brotherton Library. In the same way, for John Sidney Turner (1843–1920?), surgeon and for twenty-two years chairman of the Kennel Club, breeding experiments were contextualised against the ideas of the *Origin* and Darwin's panmixia theory, Turner to Cossar Ewart, 22 November 1897, COLL 14/9/3/11/15, James Cossar Ewart Papers, Edinburgh.
186. Bunyard notes 'some lively opposition' to previous attempts at introducing Mendelian ideas at his local horticultural club, Bunyard to Hurst, 13 December 1907, MS Add. 7955/7/75, Hurst Papers, CUL. Where there was greater interest, it tended to come from Edwardians like Redcliffe Nathan Salaman (1874–1955), botanist and potato breeder.
187. For example, Charles John Bond (1857–1939), Leicester surgeon, correspondence with Hurst in 1904, including 28 March 1904 (after Hurst had given a talk on Mendelism to the Leicester Lit & Phil in 1903), MS Add. 7955/4/25–33, CUL. For flowers see the new sweet pea variety developed by the amateur horticulturalist Alexander Malcolm (1863/4–?), *Berwickshire News*, 8 June 1909.
188. Hurst to Bateson, 1 March 1908, and 30 March 1908 ('All agree with Mendel's Law and will not hear of an exception'), MS Add. 8364/H19, Bateson Papers, CUL; *Gardeners' Chronicle*, 6 March 1909.
189. It was noted that 'Mendelism does not appeal so much to the "City Fathers"…It is the younger members who are…disposed to breed on Mendelian lines', Thomas R[ichard] Beaufort (1857–?) to Hurst, nd [1908?], MS Add. 7955/8/36, Hurst Papers, CUL; Beaufort to Hurst, nd [1910], MS Add. 7955/10/20, Hurst Papers, CUL.

DARWINIAN DEBATES AT THE *FIN DE SIÈCLE*: THE EDWARDIANS 385

190. See Bunyard to Hurst, 18 July 1903, MS Add. 7955/3/42, Hurst Papers, CUL.
191. Weldon to Pearson, 4 June 1905, Pearson/11/1/22/40, Pearson Papers, UCL.
192. Report of Thomson's lecture to the Scottish Natural History Society in 1902, *Aberdeen Journal*, 12 November 1902; for Hurst at Leicester see *Transactions of the Leicester Philosophical Society* 12 (1908); see also the account of his lecture to Monks Kirby Farmers' Club, *Rugby Advertiser*, 23 November 1907. Radick explores Hurst's early conversion to Bateson's views in *Disputed Inheritance*, 195–7. Maxwell T. Masters, long-time editor of the *Gardeners' Chronicle*, confessed to finding it difficult to follow the Mendelian hypothesis and the arguments of Bateson and others, remaining at best a sympathetic sceptic; see his correspondence with Hurst, 31 December 1901, MS Add. 7955/2/23, 30 December 1904/4/68, Hurst Papers, CUL. Bunyard noted that 'Mendel [did] not seem very enthusiastically welcomed by many hybridisers' (including George Herbert Engleheart (1851–1936), the daffodil breeder), Bunyard to Hurst, 11 June 1903, MS Add. 7955/3/38, Hurst Papers, CUL.
193. Sir William Martin Conway (1856–1937) to Karl Pearson, 23 October 1914, Pearson/11/1/3/101, Pearson Papers, UCL.
194. Gerald Leighton to Hurst, 21 October 1904, MS Add. 7955/4/67, Hurst Papers, CUL. Leighton was an open-minded enquirer rather than an opponent; his correspondence with Hurst continued to explore the claims of Mendelian theory, and of the mutational ideas of de Vries. See also the observation of Henry John Reynolds-Moreton, 3rd Earl of Ducie (1827–1921), 'Mendelism frightens me', Reynolds-Moreton to WTTD, 6 November 1910, f.242, W.T. Thiselton-Dyer Papers, In-letters I, RBGK. R.J. Ryle could not get to grips with Bateson's *Reports to the Evolution Committee of the Royal Society*, describing himself as too old and stupid and too unfamiliar with the constantly shifting terminology of modern heredity theory to grasp it, Ryle to Pearson, 6 June 1902, Pearson/11/1/17/102, Pearson Papers, UCL.
195. See G. Lebzelter, *Political Anti-Semitism in England 1918–1939* (1978).
196. As in his contribution to the Royal Society of Medicine discussion in December 1908, *Proceedings Royal Society of Medicine* 2.1 (1909), 108–9.
197. Bateson to Hurst, 30 November 1908, MS Add. 8364/H17, Bateson Papers, CUL.
198. See Mudge to Hurst, 23 October 1907, which describes this as the 'Mendelian Society meeting', MS Add. 7955/7/110–11, 117, Hurst Papers, CUL. P. Mazumdar, *Eugenics, Human Genetics and Human Failings: The Eugenics Society, Its Sources and Its Critics in Britain* (1991) discusses Mudge's *Mendel Journal* editorials, but without identifying Mudge's marginal status. Haig Thomas was the sister of the suffragette Sybil Thomas, Viscountess Thomas, and wrote a number of nature books.
199. Mudge to Hurst, 3 November 1907, MS Add. 7955/7/112–13, Hurst Papers, CUL.
200. Mudge to Hurst, 27 November 1908, MS Add. 7955/8/113–15, Hurst Papers, CUL.
201. A point made by G.K. Chesterton, *Eugenics and Other Evils* (1922), 191.
202. Chesterton, *Eugenics and Other Evils*, 115.
203. Wells' restless exploration of arguments of all kinds makes him especially difficult to place. Some of his earlier work rehearsed eugenic ideas of the extinction of the unfit in particularly brutal ways; but by *Mankind in the Making* (1903) his focus had shifted to birth control; see the discussion in Hale, *Political Descent*, which shows the pull of eugenic ideas even for Edwardians who were prepared to downplay the role of biological selection in the advance of humanity.
204. C.V. Drysdale, *Neo-Malthusianism and Eugenics* (1922). For Inge, see P.J. Bowler, 'Conflict Avoidance? Anglican Modernism and Evolution in Interwar Britain', *Endeavour* 22.2 (1998), 65–7. Schiller gets significant coverage in G.R. Searle, *Eugenics and Politics in Modern Britain, 1900–1914* (1976). Other Edwardian eugenicists include the writer and photographer Ethel Brilliana Tweedie (1862–1940) and the detective story writer Richard Austin Freeman (1862–1943).
205. See L.T. Hobhouse, 'The Value and Limitations of Eugenics', *Sociological Review* 4 (1911), 281–302.
206. See W.C.D. Whetham and Catherine Durning Whetham, *An Introduction to Eugenics* (1912).
207. W.R. Inge to Galton, nd (1907?), Galton/3/3/9/3, Galton Papers, UCL.

386 DARWINISM'S GENERATIONS

208. Similarly, Punnett's *Mimicry in Butterflies* (1916) rejected natural selection and argued for mutational leaps. For the Manchester science lecturer Solomon Herbert (1874–?), mutation theory helped overcome the 'one weak point of Darwinism', how small initial steps in direction of an adaptive change could be selected, *First Principles of Evolution* (1913), 223–4.

209. Bateson, 'Address by the President', *Report of the 84th Meeting of the British Association for the Advancement of Science, 1914* (1915), 3–40.

210. See F.W. Mott's Huxley Lecture 'On Hereditary Aspects of Nervous and Mental Diseases', *BMJ*, 8 October 1910; James R.A. Davis (1861–1934), *Thomas H. Huxley* (1907).

211. Huxley, 'Charles Darwin: A Centenary Sketch', 387; Bertrand Russell, 'Pragmatism', *Philosophical Essays* (1910), 120.

212. Thomas John Gerrard (1871–1916), *Bergson: An Exposition and Criticism from the Point of View of St. Thomas Aquinas* (1913).

213. Thomson, *The Wonder of Life* (1914), 584. For another example, see Charles Samuel Myers (1873–1946) to C. Lloyd Morgan, c.8 July 1914, #304b, DM128, Lloyd Morgan Papers, Bristol.

214. Quoted in Jones, *Social Darwinism*, 93. For another view which might reflect the shifts of the Edwardian period, see J.H. Balfour Browne (1845–1921), *Recollections, Literary and Political* (1917), e.g. 169–70.

215. See Walter Garstang in his 1894 paper on salps; see Nicholas D. Holland, 'Walter Garstang: A Retrospective', *Theory in Biosciences* 130 (2011), 247–58.

216. David Eder, 'Good Breeding or Eugenics', *The New Age*, 16 May 1908. In the years around 1909 de Vries was also taken up by members of the succeeding generation, like T.E. Hulme (1883–1917) and Thomas Quelch (1886–1954) (the son of Harry Quelch). Coming across de Vries in November 1911, Hulme seized upon his ideas as preferable to Darwin's 'antiquated theory of evolution' with its implications of progress, taking de Vries to argue that once a new species had been created by mutation, it was thereafter 'absolutely constant, so that there would then be no hope of progress for man'; see Louise Blakeney Williams, *Modernism and the Ideology of History: Literature, Politics and the Past* (2002), 95. Quelch reviewed Punnett as a sign of the growing awareness that nature does not move gradually and incrementally, something 'discreetly disregarded by the Mid-Victorian bourgeois intellectuals', and that this casts light in the 'dark places of Darwinism', *Justice*, 26 June 1909.

217. See the comment of A.S.T. Griffith-Boscawen (1865–1946), in his *Fourteen Years in Parliament* (1907), 7. For the wider context see Ian Hesketh, *The Science of History in Victorian Britain: Making the Past Speak* (2016).

218. Kuklick, *The Savage Within*, 76.

219. Trotter, *Herd Instinct*, 101; Bain, *On the Realisation of the Possible*, 198.

220. Bury, 'Darwinism and History' (1909), contributed to *Evolution in Modern Thought* [1917] and reprinted in his *Selected Essays* (1930). Bury's Darwinism was not sufficient for him to include Darwin in his list of the six books with most real influence on the mental evolution of the previous three hundred years; see Harold Begbie, 'Master Workers: Right Honourable John Morley, MP', *Pall Mall Magazine* 31 (1903), 503–8.

221. For example, Oliver Lodge, *Continuity* (1913), 29.

222. Walter Gardiner to WTTD, 23 January 1903, ff.39–42, W.T. Thiselton-Dyer Papers, In-letters II, RBGK.

223. See Slater, Review of Kidd's *Social Evolution* in *Seed-Time*, April 1895, 9.

224. See the discussion in M.W. Taylor, 'Herbert Spencer and the Metaphysical Roots of Evolutionary Naturalism', in Lightman and Reidy, eds., *Age of Scientific Naturalism*, 75.

225. Argyll's contributions included 'Professor Huxley on the Warpath', *TNC* 29 (1891), 1–33, *The Philosophy of Belief, or Law in Christian Theology* (1896), and especially *Organic Evolution Cross-Examined* (1898). For more general discussions of vitalism in Victorian and Edwardian Britain see R. Rylance, *Victorian Psychology and British Culture* (2000), Rose, *The Edwardians*, and Bowler, *Science for All*.

226. Harold Anson, *A Practical Faith* (1925), 41. Anson's autobiography reflected the gap he felt between his conceptions and those of his parents (born 1811 and 1826) who 'would not

DARWINIAN DEBATES AT THE *FIN DE SIÈCLE*: THE EDWARDIANS 387

have read any of [Darwin's] works, and would have felt him a dangerous person', *Looking Forward* (1938), 85.

227. Thomson to [James] Oliphant, 14 December 1888, T-GED/18/2/3/1, Geddes Papers, Strathclyde University; made clear in his 'Wallace's Darwinism', *Theological Review and Free Church College Quarterly* 4.1 (1889), 15–24.

228. In this I follow E. Asprem, *The Problem of Disenchantment: Scientific Naturalism and Esoteric Discourse, 1900–1939* (2014), and Bowler, *Eclipse of Darwinism*; see also Thomson, *Science of Life*, especially 226–7. It was certainly clear to Herbert Wildon Carr (1857–1931), who became one of the most vocal of the English Bergsonians, that orthogenesis had become one of the three schools of evolutionary thought (along with neo-Darwinian and neo-Lamarckian); see *Henri Bergson: The Philosophy of Change* (1911), 77–8.

229. J.G. Adami, *Medical Contributions to the Study of Evolution* (1918), and his Croonian Lectures on 'Adaptation and Disease' delivered in 1917. For another example, see the botanist, Philip Sewell (1865–1928), 'The Colouring Matters of Leaves and Flowers', *Transactions of the Botanical Society of Edinburgh* 17 (1888), 276–308.

230. George Henslow to [Alpheus Hyatt], 4 March 1889, Box 4, Folder 26, Hyatt and Mayer Collection, PUL.

231. P.C. Mitchell, 'On the Anatomy of the Kingfishers', *Ibis* 8th ser., 1 (1901), 121. We can see this in G.C. Bourne, *Herbert Spencer and Animal Evolution* (1909); it was also very much the tone of A. Dendy, *Outlines of Evolutionary Biology* (1912). In the same way, it was said of the pioneering experimental embryologist J.W. Jenkinson (1871–1915) that he rejected the view of the organic world as a 'welter of aimlessly competing forms, chance-begun and chance-ended. Somehow there must be determination towards an end, a movement and growth responsive to a central principle of order', R.R. Marett, 'Biographical Note', in J.W. Jenkinson, *Three Lectures on Experimental Embryology* (1917), xii–xiii.

232. F.C.S. Schiller, review of Ritchie's *Darwin and Hegel*, *The Philosophical Review* 2.5 (1893), 584–90, 587. See also Schiller, 'Darwinism and Design', 155; Schiller, *Riddles of the Sphinx* (1894), 202–6. Another example is Sydney Herbert Mellone (1869–1956), who explicitly uses Bateson and de Vries as a counterweight to Darwin's reference to random change, *God and the World* (1919), 96.

233. See C. Lloyd Morgan, 'Vitalism', *The Monist* 9 (1899), 177–96.

234. As early as his 1888 *Encyclopaedia Britannica* article with Thomson, Geddes had committed to non-promiscuous variation along specific lines; see C. Renwick, *British Sociology's Lost Biological Roots* (2012), 86; Windle, 'Darwin and the Theory of Natural Selection', *Dublin Review* 150 (1912), especially 313–16; Jones, 'Darwinism in Ireland' notes Windle's rejection of natural selection as the motor of evolutionary change, in favour of immanent evolution and/or vitalism. Windle displays an interesting trajectory from late-Victorian positions to something more Edwardian, traced in his articles in the *Dublin Review* in the mid-1900s.

235. Quoted in Lester, *Lankester*, 89; E.B. Poulton to A.G. Mayer, 23 December 1902, Box 6, Folder 34, Hyatt and Mayer Collection, PUL. In contrast, D.G. Ritchie ridiculed the whole idea as 'lame science and...unphilosophical theology', 'Evolution and Democracy', in Stanton Coit, ed., *Ethical Democracy: Essays in Social Dynamics* (1900), 6.

236. Meldola to Poulton, 5 November 1911, Meldola Papers, OMNH.

237. Emily Herring, 'The Vital Impulse and Early 20th-Century Biology', in M. Sinclair and Yaron Wolf, eds., *The Bergsonian Mind* (2021), 318–31.

238. J. Arthur Thomson, *Darwinism and Human Life* (1910), Macfie, *Heredity, Evolution and Vitalism* (1912). For one response to Bergson in 1909 see the tempered enthusiasm of A.S. Pringle-Pattison to William James, 29 July 1909, Pringle-Pattison Folder, William James Papers, MS Am 1092.9/A, Harvard.

239. Victor Branford and Patrick Geddes, *The Coming Polity: A Study in Reconstruction* (1917), 9.

240. Larmor to DWT, 26 July 1915, ms41223, D'Arcy Wentworth Thompson Papers, St Andrews; see J. Arthur Hill's review of *Unconscious Memory* in *Hibbert Journal* 9 (1909–10), 226.

388 DARWINISM'S GENERATIONS

241. A complex of parody and self-parody continued in the appendix, 'A Revolutionist's Handbook'.
242. Discussed in Bowler, *Eclipse of Darwinism*, 103–5; Piers J. Hale, 'The Search for Purpose in a Post-Darwinian Universe: George Bernard Shaw, "Creative Evolution", and Shavian Eugenics', *History and Philosophy of the Life Sciences* 28.2 (2006), 191–213.
243. See 'Shaw Looks at Life at 70', *London Magazine* 59 (1927), 616.
244. Arnold Bennett, *Mental Efficiency, and Other Hints to Men and Women* (1911). For the impact of Bergson see K. McConkey, *Memory and Desire: Painting in Britain and Ireland at the Turn of the Twentieth Century* (2002).
245. Donkin to WTTD, 2 May 1912, f.220, W.T. Thiselton-Dyer Papers, In-letters I, RBGK; Judd to WTTD, 10 November 1911, f.205, W.T. Thiselton-Dyer Papers, In-letters II, RBGK. Lankester did not mince his words, dismissing Bergson as 'worthless and unprofitable'; see his 'Introduction' to Hugh Elliot, *Modern Science and the Illusions of Professor Bergson* (1912), vii; and in private describing him as 'a mere windbag', Lankester to Clodd, 19 March 1912, In-letters, Box 5, Edward Clodd Papers, BC MS 19c Clodd, Brotherton Library. Henry Armstrong was similarly sensitive; see Armstrong to Meldola, 20 January 1912, #1232, Meldola Papers, ICL.
246. F.E. Younghusband, *Mutual Influence* (1915), ix–x. His *Within: Thoughts during Convalescence* (1912) argued for evolution driven by an 'inherent impulse within', 53–4.
247. See John Scott Haldane, 'Vitalism', *TNC* 94 (1898), 400–13; for Haeckel, see the material on the Leamington and Warwick Clerical Society, ff.130–3, Wallace Papers, Add. MS 46442, BL; John Arthur Hill (1872–1951), 'The Fallacies of Materialism', *The Independent Review* 12 (1907), 77–88. It will be seen that in the following I part company with interpretations such as G.E. Allen, 'Mechanism, Vitalism and Organicism in Late Nineteenth and Twentieth Century Biology', *Studies in the History and Philosophy of Biological and Biomedical Science* 36 (2005), 261–83, which argues that the mechanistic view became a rallying point for a 'younger generation of biologists'; I do so partly on the basis of the distinction that Allen makes between an epistemological approach and a wider ontological approach (his argument focusses on the former, mine on the latter), partly on potential differences between America (Allen's focus) and the UK, and partly on the attempt to look at a much broader engagement than academic biology.
248. Russell to Goldsworthy Lowes Dickinson, 16 July 1903, quoted in Richard Sams, 'Bertrand Russell's Spiritual Development and the Victorian Crisis of Faith, 1888–1914', unpublished MA thesis, McMaster University, (1980), 87.
249. J.A. Hobson, 'Evolution as Creator', *South Place Magazine* 13 (May 1908), 143–7, 145; likewise Thomas William Hazen Rolleston (1857–1920), *Parallel Paths* (1908), which argued that evolution in biology, ethics, and art moved along parallel paths, all impelled by a [hidden] X-force, the 'life impulse'.
250. Hugh Black, *Edinburgh Sermons* (1906), 7, 78–9.
251. Le Gallienne, *The Religion of a Literary Man* (1893), 112. The notorious 'red Dean', Hewlett Johnson (1874–1966), was encouraged in his reconciliation of evolution and religion by what he saw as the evidence provided by research into telepathy and extra sensory perception of realms beyond the narrow science of the evolutionists. See R. Hughes, *The Red Dean: The Life and Riddle of Dr. Hewlett Johnson* (1987), 16.
252. This sensibility is fully articulated in the reminiscences of the poet and fantasy writer (George) Allen Upward (1863–1926), who recalled 'the vain attempts of biologists to create life in their laboratories, and the equally vain search of others for a magical substance to be named "living matter" '. Life, he concluded, 'ought to be studied as a phenomenon of motion, rather than as a form of matter', Allen Upward, *Some Personalities* (1921), 17. In a short story, 'The Spectre of the Laird', see *Shields Daily Gazette*, 28 June 1900, Upward placed his belief in spirits/spectres in opposition to the 'great wave of materialism set in motion by the discoveries of Darwin'.
253. For Blackie's response, see *John O'Groats Journal*, 25 March 1890. For another example of the mobilisation of Stirling's ideas to underpin anti-Darwinian positions, see the letter of the Methodist minister John Snaith (1836–1923), *Nottingham Journal*, 18 April 1898.

DARWINIAN DEBATES AT THE *FIN DE SIÈCLE*: THE EDWARDIANS   389

254. See, for example, the extended exchanges of letters in the *Glasgow Herald*, February and March 1895.

255. Hensley Henson, *The Liberty of Prophesying* (1909), 266; Henson, 'Man's Thirst for God', in *The Value of the Bible and other Sermons* (1904), 116. We see this is in the sort of condescending allusion to 'the ordinary curate who argues against evolution' made by Margaret Benson, A.C. Benson, *Life and Letters of Maggie Benson* (1917), 323.

256. See the Bampton lectures of Frederick William Bussell (1862–1944), *Christian Theology and Social Progress* (1907), which had no less than five separate references to the fact that Darwin was unable entirely to abandon religion.

257. For example, Herbert Gresford Jones (1870–1958), letter *Sheffield Daily Telegraph*, 28 November 1918.

258. Francis Crawford Burkitt (1864–1935), *Two Lectures on the Gospels* (1901), 5. For others see Frederick Ernest Weiss, *Bearings of the Darwinian Theory of Evolution on Moral and Religious Progress* (1909).

259. F.R. Tennant, *The Origin and Propagation of Sin* (1902).

260. William Boothy Selbie (1862–1944), *The Servant of God and Other Sermons* (1911), 117. See also the comments of William A. Manning (1859–1924) in 'What Is Modernism', *The Modern Churchman* 2 (1912–13), 574–83, and James Bethune-Baker (1861–1951), *Faith of the Apostle's Creed* (1918).

261. J.G. Lockhart, *Cosmo Gordon Lang* (1949), 13; Lang, *The Church of England* (1905), 5, 9; echoed by F.R. Tennant, 'The Theological Significance of Tendencies in Natural Philosophy', *Journal of Theological Studies* 1 (1900), 347–69.

262. Waggett, *Religion and Science*, and Vernon F. Storr, *Development and Divine Purpose* (1906), in which Storr acknowledged his indebtedness to Schiller.

263. Williams, *The Working Faith of a Liberal Theologian* (1914). See Charles Addis Journal entry, 2 November 1893: 'Everyone, scientist and philosopher alike, is agreed that the ultimate of human knowledge is a power (force) underlying all phenomena. But I cannot conceive the exertion of power without will', PP MS 14/002/067/115, Addis Papers, SOAS.

264. J.H. Jowett, *The Transfigured Church* (1910), 42.

265. J.R. Illingworth, *The Doctrine of the Trinity, Apologetically Considered* (1907), Charles Gore, *The New Theology and the Old Religion* (1907).

266. Adam Fox, *Dean Inge* (1960).

267. [W.R. Inge, ed.] *Contentio Veritatis: Essays in Constructive Theology by Six Oxford Tutors* (1902) had contributions from William Cartwright Allen (1867–?), A.J. Carlyle (1861–1943), Herbert Louis Wild (1865–1940), Charles Fox Burney (1868–1925), Inge (1860–1954), and the late-Victorian Hastings Rashdall (1858–1924); for comparisons with *Lux Mundi* see *Westminster Gazette*, 9 April 1902; letter in *Spectator*, 13 September 1902, and the explicitly generational discussion in 'Contentio Veritas', *Church Quarterly Review* 55 (1903), 363–84.

268. C.F. Burney, 'The Permanent Religious Value of the Old Testament', *Contentio Veritatis*, 178–9; William Paterson Paterson (1860–1939), *In the Day of the Ordeal* (2nd ed., 1917).

269. See the detailed account in *Hampshire Chronicle*, 12 March 1908.

270. Tennant, 'The Influence of Darwinism upon Theology', *Quarterly Review* 211 (1909), 418–440, 436; Tennant, 'The Being of God in the light of Physical Science', in his *Essays on Some Theological Questions of the Day* (1905), 55–100, 75.

271. Tennant, 'Theological Significance', 347; a sentiment replicated in Storr's Pan-Anglican congress address, *Hampshire Chronicle*, 12 March 1908.

272. 'The Rejected Oracle', *Daily News*, 10 September 1910.

273. See W.L. Walker, *Christian Theism and a Spiritual Monism* (1907), and Walker, *What about a New Theology* (1907).

274. A.P.F. Sell, *Theology in Turmoil* (1986), 35, citing Gore; Stewart Headlam (1847–1924), *History, Authority and Theology* (1909); Muirhead, 'The Ethical Aspect of the New Theology', *International Journal of Ethics* 20 (1910), 128–40.

275. J.N. Figgis, *The Gospel and Human Needs: The Hulsean Lectures of 1908–9* (1911), 162–3.

276. Including Albert Dawson (1866–1930), editor of the *Christian Commonwealth*, and three congregationalist ministers, T. Gilbert Sadler (1871–1939?), Arthur Pringle (*c.*1867–?), and

390   DARWINISM'S GENERATIONS

A.W. Anderson (*c*.1872–?), who published a letter in support in various papers; see *Daily News*, 23 January 1907.

277. Henson, *The Liberty of Prophesying*, 316.

278. Explored in Josephine Mary Hope-Scott Ward (1864–1932)'s novel *Out of Due Time* (1906).

279. Tyrrell to Alfred Leslie Lilley, 23 June 1903, ms30765, A.L. Lilley Papers, St Andrews.

280. Tyrrell to Edward Thomas, 22 July 1904, Maude Petre, *Letters of George Tyrrell* (1920), 20–1.

281. G. Tyrrell, *Christianity at the Cross-Roads* (1910), 18; Maude Petre, *George Tyrrell: Essays on Immortality and Faith* (1914), 248. It is not clear how widespread a position this was, and it was likely to be met by a rejection of what Henson calls 'the perplexing conclusion [of the "Evolutionary infalliblists"] that the latest phase of the doctrinal tradition are the truest and better than the New Testament gospels', *The Liberty of Prophesying* (1909), 316.

282. Tyrrell to Wilfrid Ward, 1 August 1901, Petre, *Letters of George Tyrrell*, 74; most explicitly brought out in the discussion in Tyrrell, 'Semper Eadem. I', in his *Through Scylla and Charybdis, or The Old Theology and the New* (1907).

283. See the reading of Harry Snell around 1900, Snell, *Movements*, 76–7.

284. Elliot Smith to Arthur Keith, 3 December 1907, ff.217–22, Dawson Mss, Elliot Smith Correspondence, Add. MS 56303, BL.

285. W.H. Winch, 'The Psychology and Philosophy of Play', *Mind* 15 (1906), 177–90, and 'A Modern Basis for Educational Theory', *Mind* 18 (1909), 83–104.

286. Stack, *Darwinian Left*, 53–63, including the illuminating discussion of MacDonald's notebooks in his papers in the National Archives. Stack makes clear that this was Darwin invoked without any real engagement with the specific mechanisms of evolutionary change.

287. '[R]ather mild and uninspiring' was Cosmo Lang's verdict, Lockhart, *Cosmo Gordon Lang*, 36.

288. *Spectator*, 6 January 1912; *Evening Mail*, 3 December 1909.

289. Brian W. Blouet, *Halford Mackinder: A Biography* (1987), 26. In contrast to his limited Darwinian engagements, it is clear that Mackinder was influenced by a broad evolutionary understanding, although not so as to extend to the sorts of biological frames of the late Victorians; see his 'On the Scope and Methods of Geography', *Proceedings of the Royal Geographical Society* 9 (1887), 141–60. See also G. Kearns, *Geopolitics and Empire: The Legacy of Halford Mackinder* (2009), 69–78; Alfred Zimmern described him as 'a Bismarckian Darwinist of the purest Milnerian water', Zimmern to Graham Wallas, 12 May 1908, WALLAS/1/36, BLPES. This post-Darwinian stance is true also of the work of A.J. Herbertson (1865–1915); see H. Meller, *Patrick Geddes: Social Evolutionist and City Planner* (1990), 122–34, and his *Man and His Work: An Introduction to Human Geography* (1899).

290. David Matless, 'Nature, the Modern and the Mystic: Tales from Early Twentieth Century Geography', *Transactions of the Institute of British Geographers* 16.3 (1991), 272–86.

291. F.C.S. Schiller, review of Hobhouse, *Development and Purpose*, proof copy in MS Reviews, Box 10, Authors H–J, Schiller Papers, UCLA.

292. First published in *Punch*, January 1887, republished in Kendall's *Dreams to Sell* (1887). See also Kendall's 'Cold Comfort (The Hope of the Evolutionist)', *Longman's Magazine* 5 (1884), 604–5.

293. See the comments in A.R. Orage, 'Politics for Craftsmen', *Contemporary Review* 91 (1907), 783; James Frederick 'Fred' Henderson (1867–1957), *The Case for Socialism* (1911), 126–7; see also Cecil Desch's rebuttal of the exaggerated Darwinism of Edwardian individualism, 'Peace and Progress', *Positivist Review* 14 (1906), 68.

294. Muirhead, *Reflections*, 87. See, for example, the ambivalence expressed in Joseph Morgan Lloyd Thomas (1868–1955), *A Free Catholic Church* (1907), 67–8. In the same way, Wilfred Trotter's *Instincts of the Herd in Peace and War* (1916) includes a forthright attack on the prevailing efforts to deploy Darwinian precepts in the interpretation of history. These sorts of attitudes, which ignored the extent to which man was equally subject to biological laws as the rest of nature, led to blunders of the gravest sort in contemporary social

DARWINIAN DEBATES AT THE *FIN DE SIÈCLE*: THE EDWARDIANS   391

legislation, the (late-Victorian) William Ridgeway told the Anthropological Section of the British Association in 1908.

295. Sturt, 'The Line of Advance in Philosophy', *Proceedings of the Aristotelian Society* ns 5 (1904–5), 82.

296. T. Rhondda Williams, 'Syndicalism in France and Its Relation to the Philosophy of Bergson', *Hibbert Journal* 12 (1913–14), 402. See also F.W. Bussell (1862–1944), 'The Rural Exodus', *Economic Review* 14 (1904), 390. Likewise William Caldwell (1863–1942), 'Philosophy and the New Sociology', *Contemporary Review* 74 (1898), 411–25, which offered analysis (and endorsement) of the shift to more psychological versions of sociology which vindicate the ability of man to intervene in the process of natural selection.

297. According to the testimony of Richard Whiteing, *My Harvest* (1915), 322.

298. Quoted in Kirsten E. Shepherd-Barr, *Theatre and Evolution from Ibsen to Beckett* (2015), 134; Shaw, 'Where Darwin Is Taboo: The Bible in America', *New Leader*, 10 July 1925, reprinted in *The Shavian* 2.2 (1960), 3–9. Of course, it's always necessary to take Shaw's rhetoric with a pinch of salt; as Kevles, *Eugenics*, 86, observes, we need to balance these sorts of comments with the fact that he took Pearson's work on evolution seriously enough to subscribe to *Biometrika*.

299. For Dickinson, the problem was 'a crude misapplication of Darwinism, combined with invincible ignorance of the true bearings of science upon life', *The European Anarchy* (1916), 51–2; Snell later recalled struggling to develop his political views in a context in which 'the doctrine of evolution' was 'commandeered to the service of the employing classes and the political reactionaries', Snell, *Movements*, 54.

300. Charles John Bond (1856–1939), 'Health and Healing in the Great State', in H.G. Wells, F.E.M. Greville, and G.R.S. Taylor, eds., *The Great State: Essays in Construction* (1912), 178.

301. Hobhouse to Rosalind Frances Howard, 18 October 1886, quoted in Radick, *Simian Tongue*, 213.

302. James Douglas (1867–1940), review of Kidd's *Western Civilization*, *The Star*, 14 February 1902.

303. P. Redfern, *Journey to Understanding* (1946), 53. See also Upward's comment that 'Darwinism and Christianity are saying the same thing. Evolution is the sacrifice of the unfit on behalf of the fit. The scapegoat bears away the sins of the righteous', *Lord Alistair's Rebellion* (1910), 392.

304. Frederic Carrel (1869–1928), 'The Moral Crisis', *Monthly Review* 26 (1906), 31–46. See, for example, Marett's attack on Spencer's *Ethics of Industrialism* as grossly Darwinistic, 'The Ethics of Industrialism', *Economic Review* 2 (1892), 342–50; compare Willie Herbert Utley (1866?–1918), 'The Scientific Aspect of Socialism', *Our Corner* 9 (1887), 82–7.

305. Bertrand Russell, 'The Politics of a Biologist', *Albany Review* 2 (1907–8), 89–98, 89. F.C.S. Schiller spoke of 'the ghastly law of struggle for existence, the cruel necessity which engages every living thing in almost unceasing warfare', 'Darwinism and Design'.

306. 'Foreign and Colonial Policy', in F.W. Hirst, Gilbert Murray, and J.L. Hammond, *Liberalism and the Empire* (1900), 170.

307. William Frederick Bailey (1857–1917), 'The Native and the White in South Africa', *TNC* 59 (1906), 328.

308. G.P. Gooch (1873–1968), 'The Reign of Force', *Westminster Review* 154 (1900), 609–18; also H. Samuel, *Liberalism* (1902), 359; Trotter, *Instincts of the Herd*, 99–100.

309. See the journalist G.H. Perris (1866–1920) on 'International Relationship(s)', in April 1906, which echoed his 1896 paper 'The Evolutionary Standpoint in Social Theory', Freeden, *Minutes of the Rainbow Circle*, 148–9.

310. Norman Angell, 'Introduction' to George Nasmyth, *Social Progress and the Darwinian Theory* (1916), vi. In a similar way, John William Graham (1859–1932), Quaker opponent of German militarism in the run up to the war, directly challenged the Darwinian narrative of national conflict; see Joanna Dales, 'John William Graham and the Evolution of Peace: A Quaker View of Conflict before and during the First World War', *Quaker Studies* 21.2 (2016), 169–92.

392 DARWINISM'S GENERATIONS

311. For his pre-war hostility see his *Tables of Stone* (1908), in which reassurance of the alignment of evolution and religion was folded into a narrative in which Darwinism was idly used as little more than a source of intellectual superiority by London society; yet his poem 'George Frederick Watts', from his collection, *Fighting Lines* (1914), 52, has 'Darwin's toil' as one of a paean to the greats of English civilisation, the 'nobler sounds' that moved Watts. In contrast, his post-war *Life of William Booth* (1920), 435, suggests that Darwin 'taught men a philosophy which could do nothing but ensure them destruction'.

312. T. Gibbons, *Rooms in the Darwin Hotel* (1973), 6.

313. A.J. Penty, 'Aestheticism and History', *The New Age*, 2 April 1914, 683–4. and Penty, *Post-Industrialism* (1922), which had an introduction from Chesterton.

314. *The New Age*, 1 August 1907, 219.

315. For Waugh, see his *Reticence in Literature* (1915), especially 41; there was no mention in his study of Browning (which doesn't even notice 'Caliban upon Setebos'), or in *Tennyson* (1902).

316. For Bennett, see the comment in 'The Human Machine', *T.P.'s Weekly*, 29 May 1908; E. Gosse, 'Ten Years of English Literature', *North American Review* 165 (1897), 139–48, 147. For Bennett, see McDonald, *British Literary Culture*, especially 101–17. Hope's own patrician cynicism is visible in passing references like Valentine Hare in 'What Was Expected of Miss Constantine', in his *Tales of Two People* (1907), whose evolutionary anthropology, published by 'Mr Murray', is an insincere exercise in claiming intellectual credibility.

317. Arthur Compton-Rickett, *London Life of Yesterday* (1909), 383, Compton-Rickett, *Prophets of the Century* (1898), 112. Edward Garnett (1868–1937), reader for Fisher Unwin and intimate of Galsworthy, rejected the scientific method as incapable of offering understanding except as the auxiliary tool of instinctive perceptions, Garnett, *Literary Criticisms and Appreciations* (1922), 17–18.

318. Waugh, *Tradition and Change* (1919), 218.

319. Tyrrell, 'Adaptability as a Proof of Religion', *The Faith of the Millions* (2nd ser., 1904), 264.

320. Hicks, in 'Symposium: The Nature and Range of Evolution', *Proceedings of the Aristotelian Society* 2.3 (1892–4), 137–50.

321. For an expression of this Edwardian reaction against the Idealism of the previous generation, see the collection *Personal Idealism* (1902), edited by Sturt, and with contributions from Schiller, Stout, Marett, Bussell, W.R. Boyce Gibson (1869–1935), G.E. Underhill (1860–1924), along with Rashdall Hastings as the sole representative of the late Victorians.

322. Schiller, review of Ritchie's *Darwin and Hegel*, 584; Bradley to Alexander, 20 July 1904, Samuel Alexander Papers, ALEX/A/1/1/33/3, MJRL.

323. Schiller to James, 4 September 1904, Schiller Folder, William James Papers, MS Am 1092.9/A, Harvard.

324. Some Edwardian pragmatists continued to argue that Darwinism retained a significance as providing an avenue to the 'interest' as the corollary of utility which was the necessary basis of truth; hence Howard Vincent Knox (1868–1960) told Schiller that 'if Darwinism did not exist, pragmatism would have to invent it', Knox to Schiller, 2 March 1904, Folder H.V. Knox #1, Box 1, F.C.S. Schiller Papers, UCLA.

325. Renwick, *Biological Roots*, 102. See Bertrand Russell, 'The Development of Morals' [review of Hobhouse, *Evolution of Ethics*], *The Independent Review* 12 (1907), 210, and 'The Elements of Ethics', in his *Philosophical Essays* (1910), 1–58. The direct nature of Moore's attack on 'the Evolutionists', and the 'wholly irrelevant' nature of arguments based on developmental origin, is visible in Keynes' note of Moore's Cambridge lectures in 1903; see John Laurent, 'Keynes and Darwinism', in John Laurent and John Nightingale, eds., *Darwinism and Evolutionary Economics* (2001), 73. Keynes' heaviest scoring in his copy of *Principia* was reserved for the passage where Moore stressed that the assumption that more involved meant 'higher' 'forms no part of Darwin's scientific theory'.

326. Clarke, *Liberals and Social Democrats*, 148, quoting *Social Evolution and Political Theory*, 8.

DARWINIAN DEBATES AT THE *FIN DE SIÈCLE*: THE EDWARDIANS 393

327. Muirhead, *Reflections*, and Muirhead, *In Defence of Idealism* (1917), 109.
328. Mavor, *Windows*, 213–14. 'I am delighted to find that you are not following orthodox lines in "Evolution"', A.C. Haddon told him in 1887, noting that he had long felt he had not known enough 'to discuss such questions with profit, but I have long felt that there are higher aids to evolution than mere struggle', A.C. Haddon to Geddes, 25 December 1887, T-GED/18/2/1A, Geddes Papers, Strathclyde University.
329. D'Arcy W. Thompson, 'Memories of Pat Geddes', tps, ms14641, D'Arcy Wentworth Thompson Papers, St Andrews. 'He had the magnetic power that attracted young men of unfettered mind to him as students and kept them as disciples.'
330. Renwick, *Biological Roots*, 12.
331. Leonard Hobhouse, *Social Evolution and Political Theory* (1911), 28; see also Hobhouse, *Democracy and Reform* (1905), which challenged the indiscriminate use of the laws of biology; Hobhouse to Lady Welby, 19 October 1904, 1970–010/006(29), Lady Victoria Welby Papers, York University, Canada; Welby protested that the drive to reform came from 'these very "biological conditions" which you would have us ignore as irrelevant', Welby to Hobhouse, 13 May 1905. See also the opinions of G.C. Bourne in his 1909 Spencer lecture, *Evening Mail*, 3 December 1909.
332 Renwick, *Biological Roots*, 143, quoting 'Discussion—Civics: As Applied Sociology', (1904), 122.
333. Sir Charles Lewis Tupper (1848–1910), who urged the establishment of the new discipline on fundamentally evolutionary lines, 'Sociology and Comparative Politics', *Sociological Review* 1 (1908), 209–26, especially 210–11; Welby to Lankester, 18 April 1904, 1970–010/ 009(05), Lady Victoria Welby Papers, York University, Canada.
334. J. Owen, *L.T. Hobhouse: Sociologist* (1974), 72.
335. See his *A Philosophy of Social Progress* (1912).
336. We can see this in Radick's account of the split of physical and cultural anthropology, 'with evolutionism becoming the near exclusive possession of the former', Radick, *Simian Tongue*, 161; supporters included Sir William Ridgeway (1858–1926), Henry Balfour (1863–1939), John Linton Myres (1869–1954); see also Stocking, *After Tylor*.
337. Marett, *Anthropology* (1912), 8–10 and *passim*; and Marett to Lady Welby, 2 March 1908, 1970–010/010(22), Lady Victoria Welby Papers, York University, Canada.
338. Marett, review of Kidd's *Principles of Western Civilization*, in *Man* 3 (1903), 75–7.
339. See I. Langham, *The Building of British Social Anthropology: W.H.R. Rivers and His Cambridge Disciples in the Development of Kinship Studies* (1981). The expedition membership bar Haddon was Rivers (1864–1922), C.S Myers (1873–1946), W. McDougall (1872–1938), and C.S. Seligman (1873–1940); plus S.H. Ray (1858–1939) as linguist (with a younger photographer, Anthony Wilkin); Haddon, *The Study of Man*, quoted by Radick, *Simian Tongue*, 165.
340. We can see this in G. Archdall Reid, *The Present Evolution of Man*; also in Havelock Ellis, *The Criminal* (1890).
341. Quoted in Charles S. Myers, 'The Influences of W.H.R. Rivers', in Myers, ed., *Psychology and Politics* (2011), 171.
342. M.P. Willcocks, *Between the Old World and the New* (1967, or. 1926), 22.
343. Hobhouse, *The World in Conflict* (1915), 29.
344. Wallace, 'Present Position of Darwinism', 129–30, citing *The Inquirer*, 6 April 1907.
345. Frederic Harrison, 'The Reaction and Its Lessons', *FR* 58 (October 1895), 485–96, 490.
346. Brodrick to Huxley, 3 September 1894, f.91, Box 11, Huxley Papers, ICL.
347. Hans Driesch, *The Science and Philosophy of the Organism* (1908); R.B. Davis, *George William Russell* (1977), 136.
348. C. Lloyd Morgan to Meldola, nd (postmarked 28 June 1909), #966, Meldola Papers, ICL.
349. Thorne, pen name of Cyril Arthur Edward Ranger Gull (1875–1923), quoted in P. Waller, *Writers, Readers and Reputations* (2006), 1014; W.B. Maxwell (1866–1938) was one of the sons of M.E. Braddon. See also Arthur Quiller-Couch's *Shining Ferry* (1905) and *Brother Copas* (1911). In *Shining Ferry* Quiller-Couch had one of his characters describe the *Origin* as 'the biggest book of this century, and a new gospel for the next to think out. The conclusion is that the spoils go to the strongest. You may help a man for the use you can

394   DARWINISM'S GENERATIONS

make of him, but in the end every man's your natural enemy'; also see the way Mary (Mrs H.H.) Penrose (1860–?) associated *Descent of Man* with patriarchal views of women's subordination, in her short story 'The Experience of John Pomeroy', *Ardrossan and Saltcoats Herald*, 31 August 1900. (We might compare with the portrait of Robert Burnikel in Walter Besant's *The Master Craftsman* (1896), whose library of well-thumbed Victorian classics, including Darwin, Huxley, Wallace, and Tyndall, symbolised his ambition and drive for self-improvement in a much more sympathetic way.)

350. See, for example, the chapter on 'The Unwritten Suburb' in Lynne Hapgood and Nancy L. Paton, eds., *Outside Modernism: In Pursuit of the English Novel, 1900–1930* (2000), 22–42. And Galsworthy can be balanced with Arnold Bennett, for whom although evolution was a fact, its mechanism in his *Clayhanger* trilogy (1910–15) was the efforts of individuals upon themselves.

351. Hesketh Pearson, *Modern Men and Mummers* (1921), 138.

352. Address of 1897, quoted in G.W.E. Russell, *Edward King (1829–1910), Sixtieth Bishop of Lincoln: A Memoir* (1912), 315–16.

353. S. Brooke, *Religion in Literature and Religion in Life* (1901), 22–3.

354. Including Chesterton's characteristic dig that Darwin 'loses his human love of music for an inhuman love of information', 'Our Note Book', *ILN*, 27 January 1906; J. St Loe Strachey (1860–1922), *The Adventure of Living* (1922); Dorothy Beale, 'Modern Education', *Great Thoughts*, 2 December 1905.

355. See, for example, his essay on 'Stoic Philosophy' in *Tradition and Progress* (1922), 185, with its indifference to whether evolution is Bergsonian or Darwinian.

356. *Greenock Telegraph and Clyde Shipping Gazette*, 2 December 1902. See also the trenchant judgement that Darwinism has been 'knocked out of court' of the colliery manager, Thomas Luther Davies (1861–1936); see *Aberdare Leader*, 25 January 1908.

357. Addis to Mills, 18 July 1893, PP MS 14/002/067/073, Addis Papers, SOAS.

358. *Methodist Recorder*, 7 February 1895. For other Edwardian anti-Darwinians see William Hall Calvert (1861–1917), *The Further Evolution of Man* (1913).

359. Meldola dismissed it as 'pure bombast from cover to cover', Meldola to WTTD, 21 July 1909, In-Letters III, Thiselton-Dyer Papers, RBGK.

360. Letter, *Cairns Post*, 18 August 1923.

361. May Sinclair, *In Defence of Idealism* (1917), 26.

362. Shaw, 'The New Theology', *Christian Commonwealth*, 23, 30 May 1907; see his earlier lecture to the Guild of St Matthew, *Staffordshire Sentinel*, 1 December 1906. For another example, see George Dewar, 'My Books and I', *Sheffield Independent*, 24 September 1919. Dewar was clearly not in any sense denying evolution, or indeed the fundamentals of Darwin's theory, although his pre-war writings had also displayed scepticism about the complete efficacy of the 'hard and relentless method' of natural selection, which drew in part on Bateson; see his *This Realm: This England* (1913), 220–5, 293.

363. *The Scotsman*, 24 May 1912.

364. I.D. Ruxton, ed., *Sir Ernest Satow's Private Letters to W.G. Aston and F.V. Dickins* (2008), 287. For another example of acceptance of a rationalist/Darwinian position and then conversion back to a theist position of some sort, see J.F.C. Fuller, *Memoirs of an Unconventional Soldier* (1936), 458–60.

365. F.V. Dickins to WTTD, 22 May 1906, f.203, W.T. Thiselton-Dyer Papers, In-letters I, RBGK.

366. A.J. Balfour, *A Defence of Philosophic Doubt* (1879), 261; Balfour to Ward, 7 December 1903, VII/226/1/31(a–b), Ward Papers, St Andrews. (Significantly, in public pronouncements Balfour continued to offer carefully ambiguous endorsements of Darwin, as in his speech to the first International Congress on Eugenics in 1912; see Gilham, *Francis Galton*, 345–53.) Even within Unitarian circles it is possible to find descriptions of 'the catastrophic disasters of physical and social evolution', William Lawrence Schroeder (1872–1950), 'Intellectual Difficulties of Our Elder Scholars', *Sunday School Quarterly* 1 (1909–10), 39–42, 41.

367. *Sunday Times*, 9 November 1913.

DARWINIAN DEBATES AT THE *FIN DE SIÈCLE*: THE EDWARDIANS 395

368. Eder, 'Good Breeding or Eugenics', 67; David Eder, *Memoirs of a Modern Pioneer* (1945), 36.
369. Timothy M. Donovan in *The Catholic Advocate* (Brisbane), 13 May 1926.
370. A.F. Brockway, *Bermondsey Story: The Life of Alfred Salter* (1949), 6, 17–18.
371. *ILN*, 12 December 1931.
372. By this he meant he rejected natural selection and its implication of randomness; evolutionary change was from simple to complex, and it 'not only permits of design, it demands design', 'Darwin and the Demagogues', *Nash's Magazine* (May 1927), 28–9, 79–81; Chesterton, *Autobiography*, 51–2, 86.
373. *Daily News*, 23 February 1903, *Clarion*, 1 December 1905.
374. Chesterton, *Autobiography*, 250–1; *ILN*, 21 August 1920, 8 December 1923; see also his various swipes at 'evolutionists' and 'the clotted folly of Eugenics', in *The Uses of Diversity* (1920); by this stage he is prepared to stress that the evolutionary sequence of man from animals is entirely theoretical, that there is no actual evidence of the evolutionary 'missing link'; and ready to repudiate theories of ontological recapitulation; Chesterton, *Everlasting Man*, 47, Chesterton, *Orthodoxy*, 46, 91; Chesterton, *The Victorian Age* (1913), 206; Chesterton, 'The Evolutionists' Error', in *Utopia of Userers* (1917), 48.
375. Compare with Hobhouse's warning about the '"eulogistic" suggestion' the word evolution carried, Hobhouse, 'Sociology and Ethics', *The Independent Review* 12 (1907), 322–31.
376. 'Our Notebook', *ILN*, 21 August 1920.
377. Chesterton, *Utopia of Userers*, 185; Chesterton, *Heretics* (1905), 92; 'My Notebook', *ILN*, 9 January 1909. See the reference to an article in *G.K.'s Weekly*, in Masie Ward, *Gilbert Keith Chesterton*, 374, and especially Chesterton, *What's Wrong with the World* (1910), 126, 324–5.
378. *The Victorian Age*, 209.
379. Chesterton to Ronald Knox, quoted in Dudley Baxter, *G.K. Chesterton* (1973), 21.
380. *The Tribune*, *c*.1906–7, quoted in Cecil Chesterton, *G.K. Chesterton: A Criticism* (1909), 244. Perhaps tellingly, in 1923 Chesterton found himself in a brief controversy over the status of Darwinism with one of those whose birth in 1856 did seem to have encouraged a different engagement with Darwinism, the theatre critic William Archer (1856–1924), *ILN*, 8 December 1923.
381. *Clarion*, 11 January 1907.
382. See, for example, Thomas Bailey Saunders (1860–1928), *Quest of Faith* (1899), or Davis, *Thomas H. Huxley*.
383. Neville Talbot, 'The Modern Situation', in Talbot, ed., *Foundations: A Statement of Christian Belief in Terms of Modern Thought* (1913), 4.
384. It did not help that Frank too was wrestling with exactly how much Lamarckism his father had accepted; see Darwin to Meldola, nd [1908], #976, Meldola Papers, ICL; and in fact Darwin's ideas on botanical evolution, on which he had collaborated most closely with his father, had always been more inclined to Lamarckian mechanisms. (I am indebted to Ian Hesketh for pointing this out to me.) Characteristically, Poulton had a more cynical interpretation, having suggested in 1907 that Frank was a 'great mystery' but seemed to have become 'absolutely hypnotised' by Bateson, Poulton to WTTD, 1 December 1907, f.133, In-Letters III, Thiselton-Dyer Papers, RBGK.
385. *Science Progress in the Twentieth Century* 3 (1908), 177.

The Reception of Darwinian Evolution in Britain, 1859–1909: Darwinism's Generations. Martin Hewitt, Oxford University Press. © Martin Hewitt 2024. DOI: 10.1093/9780191982941.003.0006

Conclusion

Continuity, Conversion, and Counter-Example

It is extremely difficult to put one's self into a proper relation with the age in which one lives—especially as one really lives in a brisk succession of ages, all shipped into a box, and hard to distinguish. (Edmund Gosse to Alfred Noyes, 22 December 1908, Noyes Papers, Boston College.)

1908/9 Celebrations

The Darwin celebrations of 1908/9, marking the centenary of his birth, and the fiftieth anniversaries first of his paper to the Linnean Society in 1858, and then of the publication of the first edition of *On the Origin of Species*, offer an appropriate place to draw this discussion to a close. The anniversaries inevitably offered contemporaries a frame for reflecting on the intellectual changes of the previous half-century, and also of their contemporary resonances. The events of the two years, by requiring a stocktake of the position and progress of Darwinism, shed one final spotlight on the continued valency of generational distinctions. On the surface things proceeded with due decorum, but behind the scenes generational frictions sparked.

The centenaries helped bring Darwin back into the public gaze. Newspapers mused on the fecundity of a year which also saw the birth of Gladstone, Tennyson, and Abraham Lincoln. The provincial press saw a fresh flurry of controversial correspondence.[1] Local scientific societies listened to panegyrics of Darwin's importance. Darwinian themes enjoyed a revival in the satirical magazines (Figure 8). At the Royal College of Art one student obtained the travelling scholarship with a modelled design for a commemorative monument to Darwin, an allegorical group of 'Time and Science Unveiling the Truth'.[2] For many, Darwin still symbolised the intellectual freedoms wrought out of the generational clashes which Gosse's *Father and Son* (1907) and Butler's *The Way of All Flesh* (1903) had so

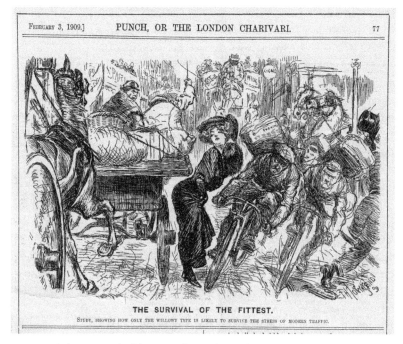

Figure 8 'The Survival of the Fittest', *Punch*, 3 February 1909.

vividly explored.[3] Darwin and Darwinism could still energise enquiring minds. On an exploratory trip to D.H. Lawrence's Nottingham home just before World War 1, Ford Madox Ford was astounded to find young people talking about Nietzsche, Wagner, Karl Marx, and Darwin, displaying a knowledge of things 'that my generation in the great English schools hardly ever chattered about'.[4]

At the same time, the eclipse of Darwin was visibly gathering pace. In the years before 1914 public librarians reported a waning of interest amongst their working class readers.[5] When the educationalist Arnold Freeman (1886–1972) investigated the knowledge of Birmingham teenagers just before World War 1, he found that most boys had not heard of Darwin.[6] There was an open field, even for those whose underlying position was fundamentally sympathetic, to take aim at Darwinian orthodoxy. Typical of the sort of interventions which were becoming more common was Andrew Lang's 'Me and My Governor', which appeared in the *Morning Post* at the start of February 1908, offered as an exploration of 'those contrarious variations between parent and progeny which are among the most notorious

398   DARWINISM'S GENERATIONS

facts of heredity'. Lang portrayed a familial crisis of Darwinian faith in which the father, a Darwinian, albeit only from newspaper articles and the popular science current in conversation, but who was the kind of person who 'explains everything by…[the] "struggle for existence", or "Evolution" mythically regarded as a kind of personage', when confronted with the actual writings of Darwin at the instigation of his son (a nervous, emotional aesthete, constantly in tears over dying trout and dead rabbits), rejects them in horror. Although Lang subsequently clarified his position, claiming his own stance was a middle point between his two protagonists, it was hardly surprising that even a fairly conscientious reading of the press could give the impression that there were scarcely any fully committed Darwinians left.[7]

The actual anniversary of Darwin's birth, 12 February 1909, passed off with little fanfare. A statue was unveiled outside Shrewsbury School, the newspapers generally offered substantial appreciations of the father of evolution, complete with illustrations or commemorative portraits, and there was a celebration in the Examination Schools at Oxford, which allowed E.B. Poulton to add to his engagement list by giving another address, on 'Fifty Years of Darwinism'; but otherwise acknowledgement was muted. For the majority, it would seem, the transformation which Darwin had wrought from pre-*Origin* modes of thinking was already too distant, the established positions he had overthrown already too alien, and his own status too uncertain for any great enthusiasm to be generated. Although tributes flowed in from across the world, proposals in America for an edition de luxe of the *Origin* or a research endowment had eventually to be shelved for lack of support, in favour of the inevitable volume of essays, and a Darwin medal.[8] Johns Hopkins University held a 'Darwin Day' celebration at the start of 1909, at which Poulton spoke, but the formal celebrations at the American Association for the Advancement of Science were postponed, not least because of difficulties in lining up speakers. Frank Darwin refused to give a keynote lecture, leaving the ubiquitous Poulton to be drafted in instead. William James was pressed to contribute, but refused twice; and then it seems after having compromised on a brief after-dinner speech, withdrew belatedly from that as well, telling the organisers he could not find a single word to say about Darwin.[9]

Commemorations are not of course the place to look for balanced evaluation, but discussions in the press, often anonymous and unplaceable generationally, continued to encompass the full range from implacable hostility to uncritical hagiography. Where generational positions can be ascertained, they replicated the patterns traced in the previous chapters. Mid-Victorian voices had all but died away, although of course Alfred Russel Wallace was still very much alive and encouraging forthright defence of Darwinian orthodoxies, and in private it is still possible to find octogenarians seeking solace in the belief that even if 'the militant section of evolutionists adopted all the arts of the political agitator and cowed the multitude', still 'our great men of science were not converted'.[10] High and late Victorians continued to reverence the revolutionary import of the *Origin*: for the geologist Charles Callaway (1838–1916), it had been 'the explosion of a bomb-shell under the very foundations of medieval and Biblical beliefs', and evolution 'the most important thought that man has ever conceived'; while for the physician and anthropologist Daniel John Cunningham (1850–1909), the *Origin* was still 'the crowning event of this epoch'.[11] Late Victorians extended their turf-wars under the guise of tributes, Lankester using his newspaper column in June 1909 to contrast Darwin's decades of painstaking research to Lamarck's 'ingenious guess'.[12] At the end of November the British Academy's celebration of the fiftieth anniversary gave the opportunity for W.R. Sorley and a number of other high Victorians to emphasise the diversity of supplementary theories which had augmented Darwin's own work.[13] Even Patrick Geddes braved the sceptical readership of the *New Age* to reaffirm the Darwinian equation of naturalist and sociologist.[14] Edwardians, increasingly to the fore, generally offered more cautious assessments which gave due weight to subsequent scientific advances, but still lauded Darwin personally, and the *Origin* as 'a veritable lamp of Aladdin'.[15]

Vigorous efforts were made to use the anniversaries to defend Darwinian orthodoxy from its upstart challengers, politely acknowledging the subsequent development of Darwin's ideas, but firmly dismissing any fundamental divergences. An exhibition of Darwinian relics in the central hall of the Natural History Museum offered what the *Graphic* described as 'an illustrated reading' of Darwin's theory; apparently largely ignored by the regular museum sightseers.[16] One of the highlights was itself a vestige of the early fall-out from the publication of the *Origin*, some of the proof sheets of Richard Owen's slashing *Edinburgh Review* article. In effect, given Owen's repeated denial of authorship during his life, he was being pointedly outed as a representative of the losing side in the long Darwinian debate. Leading Darwinists were encouraged to publish supportive articles. Poulton was particularly energetic in campaigning for what he still thought of as 'the cause', delivering a number of addresses and papers, gathered with the addition of a witheringly dismissive dig at Bateson and the Mendelians, as

Charles Darwin and the Origin of Species (1909).[17] His arm twisted, William Thiselton-Dyer offered an uncompromising rebuttal of the mutationists for the *Gardeners' Chronicle*: unable to live with the thought of the tribute being allocated to some 'young Cambridge whipper-snapper'.[18] Raphael Meldola took every opportunity to restate the neo-Darwinian case, astonished at the hold upon Cambridge obtained by Mendelism. 'The lesson', he warned, 'conveyed by the revolution in scientific thought effected in a comparatively short period by one book is in danger of being overlooked by the present generation'.[19]

Poulton pronounced the Linnean Society commemoration in 1908 the most inspiring and interesting event he had ever attended.[20] The eighty-five-year-old Wallace and the ninety-one-year-old Hooker were in attendance to receive two of seven Darwin–Wallace medals to be awarded in honour of the jubilee, and share the stage with Galton and a typically combative Lankester, who dismissed all putative amendments to Darwin's work as of little consequence. Bateson, seated next to Pearson, apparently turned his back and refused even to make eye contact, but could not escape a lecture from Lankester on the persistence of true Darwinism, even if its current detailed work did not appeal to the public mind as easily as theories 'which pretended to deal with wider issues'.[21] In private at least, Lankester was at his most vituperative, describing Bateson as 'the really impudent and foolishly conceited amongst these snarling dogs'.[22]

The Cambridge commemorations were carefully stage-managed. Minute attention was given to the planning of the *Darwinism and Modern Science* volume which the University Press published to coincide with the celebrations. Direction was entrusted to a weighty committee, and ultimately to the safe editorship of the Professor of Botany, A.C. Seward (1863–1941), who was also given charge of arrangements for the commemoration event. Thiselton-Dyer pressed the need to ensure that the volume kept its focus on Darwin and did not become a diffuse celebration of evolution. His ally on the committee, Adam Sedgwick, urged even closer focus on Darwin's writings themselves: arguing that the essential point to be brought out was how attitudes to nature had changed 'because of Darwin's work /(not only his theories)\'.[23] Sedgwick represented the persistence of late-Victorian Darwinism, arguing that the underlying message of the volume should be that 'Darwin's great work was the immense number of observations which he made, collected, and brought to bear to show the importance for the question of organic evolution of three factors, (1) natural selection, (2) heredity,

CONCLUSION   401

(3) variation' and the volume 'ought to show how later work has been solely directed to examine more closely the mode of action of these factors, and how the realisation of their importance has influenced and developed other branches of knowledge'. The slate of contributors was keenly discussed. Bateson wanted H.G. Wells to write on Darwinism in the future, but there was little enthusiasm for this, and when the committee's preferred choice of Lankester said no, the whole topic was abandoned. L.T. Hobhouse was another Edwardian who was mooted as a contributor, but eventually dropped. The essayists wrote under the beady gaze of the Darwins; J.W. Judd's contribution, one of the more personal of the essays, was given particular scrutiny.[24] The gossip amongst the biometricians was that at one point Bateson threatened to withdraw if his nemesis Pearson was invited to contribute, only for Sedgwick, to general laughter, to ask him 'Are you the Pope?'.[25] For all this, Wallace was outraged at the treatment of Darwin's ideas by several of the younger contributors.[26]

Similar generational discrimination guided the selection of speakers at the celebratory events. A suggestion from the newly appointed Vice-Chancellor that Bateson be invited to deliver the Rede Lecture which was being incorporated in the Cambridge celebrations was quietly shelved in favour of the much more 'orthodox' high-Victorian Archibald Geike's safe discussion of Darwin's Geology.[27] This did not prevent Thiselton-Dyer from deciding not to attend, for fear that he would lose his temper at 'being told that Darwin's mantle had fallen upon Bateson—that the Origin of Species had still to be discovered, and that specific differences have no reality'.[28] And it did not prevent observers noticing hints of the vehement controversy occasionally breaking through the stately carapace of the occasion.[29] Lankester on behalf of the Royal Society gave a typically bullish defence of Darwin's theses. At the banquet Arthur Balfour offered careful praise of Darwin's works as landmarks in the intellectual history of mankind. If anyone enjoyed it, it was the high Victorians. But their triumph was tempered: for Thiselton-Dyer, there was a touch of charade about the whole celebration, as the 'younger Cambridge School' did not begin to understand the theory; 'the younger generation don't know the facts and don't read the classical authorities', he lamented.[30] His suspicion is borne out by the apparent lack of enthusiasm amongst younger scientists. The American mathematician Robert Simpson Powell (1849–1924) was awestruck by being able to talk with the last survivors of Darwinian circles, Joseph Hooker, Henrietta Huxley, and Eleanor Sidgwick, and was amazed during his tour of the

402 DARWINISM'S GENERATIONS

Darwinian exhibition at Christ's College to catch sight of the monkey effigy which had been suspended over Darwin as he received his honorary degree in 1877.[31] But there was, for example, little sense of engagement within the extensive correspondence networks of D'Arcy Thompson, although he dutifully travelled down for both the celebrations; and despite their inclusion along with the other Edwardians, J.B. Bury, P.N. Waggett, and William Cecil Dampier Whetham (1867–1952) in *Darwinism and Modern Science*, Bateson and the Mendelians felt deliberately marginalised.[32]

For the first time the Edwardians were themselves experiencing a creeping consciousness of eclipse. By 1909 the oldest, like D'Arcy Thompson, were already sensing that they were taking their place 'among the elderly generation'.[33] A couple of years later, Bateson at fifty felt himself facing a younger generation of intellectual challengers. 'We are settling steadily down to a position which is almost hopelessly agnostic', he told Thiselton-Dyer in 1911, 'and the next generation have got so far into agnosticism that they don't even seem to care to talk about the question at all.'[34] With hindsight it is already possible to see the vanguard of yet another generation, including those Mary Bartley has described as the 'young Darwinians', R.A. Fisher (1890–1962) and Julian Huxley (1887–1975), beginning to react against Bateson's Mendelism, and take the Darwinian debate to the next stage.[35] Other thinkers were coming to the fore. Einstein, and Freud, who was, according to the painter C.R.W. Nevinson (1889–1946), son of H.W. Nevinson, 'to my generation what Darwin was to my father's'.[36] Younger novelists like Gilbert Cannan (1884–1955) became increasingly dismissive of the whole cult of natural history within which Darwin had been enshrined.[37] To the extent that generational continuities allow of such a thing, 1914–8 was a genuine historical caesura. Nothing was quite the same again; much—including popular Darwinism—was transformed out of all recognition. If the slaughter of the trenches taught anything, it was not the survival of the fittest: those who lived through two great wars saw with their own eyes that it was often the bravest, the noblest, and the best who perished.[38] More and more the *Origin* was one those books the post-Edwardians had never read, part of the circumscribed intellectual life of their Victorian parents and grandparents.[39] First editions of the *Origin* continued to command a premium in the collectors market, but fell steadily away from the prices commanded by the Victorian literary greats.[40] Ultimately, we know that the twentieth century evolutionary synthesis reintegrated natural selection into the mainstream of evolutionary science. But that was the story of another time and another generation.

Darwinism

While the generational merry-go-round continued to turn, the events of 1908/9 provide an effective standpoint from which to review the arguments of this study. For all their differences of opinion, no-one involved in the 1909 commemorations doubted the enormous significance of *On the Origin of Species*, and for the avoidance of any misunderstanding, at no point has the intention here been to downplay this, or of Darwin's contribution more broadly, either within the natural sciences, or across the intellectual history and culture of later Victorian Britain. In the half-century under review, Darwinism had a transformative effect. In the wake of the *Origin* it was impossible to address any of the domains of knowledge in which evolutionary understandings were brought to bear without establishing some sort of a position on Darwin. And across these domains the transformations were paradigmatic. The final rebuttal of the authority of religion over science. The dismantling of biblical literalism. The undermining of anthropocentrism. The spread of scientific naturalism. The recognition of the web of life, the interconnectedness of all things. The move from an ontology of fixity to fluidity. The search not for single causes of events as for a plurality of conditions. In all of these Darwinism played a fundamental role.

The extraordinary diversity of material that can be brought to bear on the question of the reception of Darwinian thought demonstrates if nothing else that the interventions of Darwin, his supporters and opponents, led to sustained, wide-ranging, and often intensive engagement not just with evolution broadly conceived, or heredity, or even 'the survival of the fittest', but with the specifics of Darwin's contribution, the man, his works, his disciples, and his particular arguments. It is easy when examining a particular topic in history, especially in the era of keyword searches on disjunct databases, to fall into a tunnel vision which magnifies the actual presence of the focus of the research; and we must recognise that despite their ability to engender controversy in most places and periods, Darwinian debates occupied only a tiny fraction of contemporary attention. But at the same time, the breadth, intensity, and longevity of these debates was remarkable, especially in a culture as ambivalent about science as Britain in the second half of the nineteenth century. It may be that most Victorians did not read Darwin, but few would have been without a rudimentary understanding of his ideas and their implications; and many did read, carefully and conscientiously, in bothies and back kitchens as well as in drawing rooms and studies. Publishers' lists, public libraries, local platforms, and press

404 DARWINISM'S GENERATIONS

correspondence columns were constantly replenished with fresh interventions, even if the positions expressed often seemed to be endlessly recycled. As the rich scholarship of the last forty years has shown, we have much to gain from being attentive to the ways in which Victorian and Edwardian thought was influenced by Darwinian debates. But this impact was often indirect, mediated in complex ways, and as much a matter of resistance and accommodation as of acceptance. Although it is often taken as evidence of the spread of Darwinian beliefs, it was not necessarily so. For this reason, as the foregoing discussion has hopefully demonstrated, we still have much to gain by balancing exploration of indirect influence with attention to direct and explicit engagements.

Of course the meanings of 'Darwinism' remained malleable, but it is precisely in contemporary efforts to situate themselves within the complex of interrelated positions and agendas that its real influence resided, and through which its reception can most effectively be explored, and for this reason the question of whether there was a 'Darwinian revolution' remains relevant. Accepting the danger of constructing an unrealistic threshold for Darwinian belief—a sort of Darwinian thirty nine articles to which not even Huxley would have subscribed—there is value in observing the distinctions between (i) a consciousness of evolutionary ideas as a component of intellectual life, (ii) a broad intellectual condition which accommodated evolutionary approaches, (iii) commitment to evolution as a historical *fact* whatever its nature and cause, and then (iv) the spread of Darwinian *beliefs*, the subscription to specific ideas or propositions about the natural history of the world and its fundamental dynamics, drawn primarily from or situated confidently within Darwin's own writings. This latter distinction is particularly important because it was Darwin's presentation of a credible scientific explanation of the cause of evolution which is usually identified as the basis of his significance in changing minds over the fact of evolution itself, even though for many, including Huxley, the crucial evidence continued to be palaeontological and anatomical rather than biological.[41]

The importance of policing these distinctions may well differ according to context and the particular matters under consideration. The question of for or against Darwin might make less sense for those working within professional science, where the crucial question was increasingly the form of accommodation that individuals were prepared to make. But if we are interested in age-effects, even in this narrow professional frame, the period in which Darwin and his supporters were able to decisively shift the terms of the debate, and the dates at which evolutionary science became widely accepted, still are significant. For the wider culture the terms of engagement were different: here the costs of swimming against the tide, if not entirely absent for those not hankering after recognition by the Royal Society, were generally much less. Contemporaries could, and overwhelmingly did, address Darwinism as a matter of belief which was not primarily to be determined by considerations of personal advancement or intellectual convenience. As a result, the chronology of reception in the wider culture was not the same.

If Darwinian influence spread quickly, Darwinian belief progressed less rapidly. Without doubt, the rallying of the X-Clubbers in the 1860s, and the dominating position they acquired within the British scientific establishment created, as Robert Chambers observed, a 'remarkable' and 'rapid dissolution of institutionalised opposition'.[42] But the scale and longevity of public debate registers not merely the significance of Darwin's ideas, but also the extent to which they remained a matter of dispute. The constant contemporary repetition of assertions of the acceptance of Darwinism, or at least of evolution, by 'all rational men' carried a double indication that acceptance was in fact far from universal. In the months and indeed years after November 1859, Darwin was left despondent at the lack of support within the established scientific community and the wider culture. His small band of supporters included a handful of naturalists, often with their own experience of tropical flora and fauna, and there was a scattering of other adherents who had already committed themselves to existing versions of evolution, including many of John Chapman's radical *Westminster Review* circle who had been exposed to Herbert Spencer's early evolutionary ideas, veterans of the Lamarckian controversies in London medical schools of the 1840s, and other religious radicals who had already seized on evolutionary ideas to strengthen their various anti-religious sentiments. But the number of new adherents outside these specific groups was small, and even the weight of this group dissolves or at least becomes sketchier on closer examination. Many of the Darwinians' supporters aligned wholeheartedly with their stand against religious intolerance, but remained ambivalent about both the precise nature and causes of evolution. Otherwise, support for Darwin's ideas in the 1860s came from younger adults yet to enter their thirties when the *Origin* appeared, many of the most visible of whom were students or young fellows at Oxford and Cambridge, or part of the metropolitan circles of young professional and literary men, including John Lubbock, Leslie Stephen, Walter Pater, Edward Clodd, as well as a new cadre of professional naturalists such as Alfred Newton, William Boyd Dawkins, and W.H. Flower.

406 DARWINISM'S GENERATIONS

The role *Nature* played after 1871 in promulgating Darwinism and scientific orthodoxy needs to be balanced against its own increasingly narrow readership, the failure of earlier attempts to sustain pro-Darwinian periodicals, and the longevity of titles which retained a much more sceptical tone, including the newspaper *Land and Water* and the *Zoologist* magazine. The apparent progress of Darwinism in the Royal Society and the British Association is set against the history of many of the other scientific associations, including the Linnean Society, the Geological Society, the British Ornithological Union, and most determinedly the Entomological Society, which retained strong minorities if not majorities of non-Darwinians into the 1880s and 1890s.

David Hull has suggested that the theological and metaphysical objections to evolutionary theory gradually waned 'not because their authors had been converted, but because no-one whose opinion mattered was listening any more'.[43] But the objections were there for those who care to look, and they continued to be meaningful to many readers, and we need to be careful we are not constructing another condescension of posterity in remaining blind to them. Ignoring anti-Darwinian prejudices condemned by the inexorable march of scientific progress was not a luxury available to George Murray, struggling to develop a career at the Natural History Museum under the disapproving supervision of William Carruthers, or Henry George Day, denied a licence to take a clerical living by Christopher Wordsworth, his bishop. Although this sort of conflict lens is out of scholarly favour, attention to what contemporaries said demonstrates that whatever sort of Darwinian transformation occurred in late-Victorian Britain, it emerged out of an enduring dissent. Taking all this into account, the evidence is clear: outside the narrow bounds of the biological sciences (and even here only with significant caveats) there was no Darwinian revolution in Britain, if by this we mean a rapid and significant subscription to the evolutionary propositions advanced in *On the Origin of Species*. To write as if there was, or even to leave this interpretation open, is to help perpetuate a crude distortion of the historical record. If taken together the changes Darwinism effected were revolutionary in import, and if at particular nodes and in specific moments this change might even be described as revolutionary in its speed as well as its implications, overall they did not constitute a revolution in the temporal sense, if we understand 'revolution' in the perspective of a normal human lifespan.

Once we disentangle the attributable contributions from the enormous mass of anonymous commentary it becomes abundantly clear that age played a vital part in shaping and fuelling these divergences. We can neither explain the degree of conservatism which the Darwinian debates reveal nor map the history of Darwinism and evolutionary beliefs without attending to the question of age. As Darwin prophesied, and contemporaries consistently recognised, age powerfully inflected responses to the *Origin*. This is inscribed vividly in the contrast between the response to the *Origin* of older and younger readers.

It is true that for younger readers in the 1860s and 1870s the embrace of Darwin's ideas involved a process of abandoning previous understandings and beliefs, and that for some, and here one might cite the cases of W.H. Hudson or Robert Buchanan, this was a troubling and sometimes traumatic process involving strongly held beliefs. But for the most part, new friendship groups, active exploration of ideas, and the gradual loss of deference to elders and orthodoxies all exerted powerfully solvent influences, and juvenile opinions inherited from parents were relatively easily abandoned for new Darwinian beliefs. As we move through the rest of the century the encounter with Darwin was likely to come earlier in adolescence or early adulthood, so that although the trauma and sense of loss (as the cases of J.B. Glasier or Margaret Wynn Nevinson exemplify) was not entirely removed, acceptance was easier and often hardly involved the abandonment of any settled conceptions to the contrary. This was less a matter of conversion than of the first formulation of mature convictions.

Otherwise, evidence of individual acceptance, and especially of rapid acceptance, of Darwin's ideas is sketchy at best, often relying, as in the case of Alfred Newton, on retrospective constructions for which there is remarkably little contemporary documentation. There are figures like William Huggins or Francis Galton, whose initial hostility seems fairly rapidly to have been replaced by a firm conviction of the validity of Darwin's positions. There were many more for whom the passage of time brought degrees of retreat from outright hostility to the fact of evolution. But these shifts were reluctant, slow, and partial, and their precise nature is often elusive.

This is not an assertion of absolute fixity of view. That mental faculties became less flexible with age was, as Thomas Huxley commented to one correspondent, a matter of experience; but that does not mean, Huxley emphasised, that there is ever a settled formation of the mind.[44] The willingness of the leading figures in the Darwinian debates to republish controversial essays sometimes decades after their initial appearance, and to affirm explicitly that their stance had not changed, is striking. Preparing her 'Darwinism in Morals' for reissue just before her death, Frances Power

408 DARWINISM'S GENERATIONS

Cobbe commented that she was 'glad to be able to re-circulate it,…It was written just 50 years ago, and I am able to say with truth that I have not seen reason to abandon the position I then took, although the "cocksureness" of 30 can never be maintained to 80!".'[45]

Despite such instances, the argument here is rather a contention about the pace, scope and scale of change. Of those confronted by the *Origin* in middle age, those whose beliefs remained almost entirely unaffected were almost certainly a tiny minority, even allowing for the vehemence with which many continued to protest their uncompromised convictions. Particularly in old age, abandoning strongly held and fiercely espoused views required a rare degree of engagement, energy, and self-honesty. During the 1860s Adam Sedgwick, despite his vigorous initial response to the *Origin*, had reluctantly abandoned the biblical chronology; but he continued to robustly reject Darwinian evolution.[46] By 1872, a year before his death, he was a frail old man, able to read only briefly in strong daylight, just capable of shuffling the hundred yards from his house in the cathedral close to the cathedral itself, and reliant on young relatives to act as amanuenses.[47] It was not a state which was conducive to the abandonment of long-harboured commitments. Often opponents of Darwinism clung to the hollowed-out kernel of their antagonism as much as anything out of a lack of intellectual resolve to concede, while losing the will to continue public resistance against a tide which seemed increasingly to be unstoppable. If individual opposition to Darwinism waned in the years after 1859, this was often a matter of discretion as of conversion.

Attitudes generally evolved. Inevitably, over time, the emotional shock that older readers had felt on first exposure to Darwin dulled. But recent studies have not always avoided the temptation to imply a greater accommodation than is warranted, as in James Moore's tendentious contention that even E.B. Pusey 'eventually allowed for the truth of evolution', which then becomes in the hands of Peter Hinchliff the entirely unjustifiable suggestion that by the end 'Darwin's theory does not appear to have greatly shocked' Pusey.[48] Evolutionary literature is not without its own death bed repentances, but they are generally no more convincing nor more conclusive than contemporary accounts of recanting atheists. Thiselton-Dyer suggested of W.C. Williamson that before he died he had reconsidered and abandoned all his heresies, which although not impossible given the concessions that he did reluctantly offer, is unlikely given his unwillingness in the face of the enduring and concerted pressure from Huxley to publicly avow his complete adherence.[49] Nor is Tyndall's claim that the notoriously

CONCLUSION 409

outspoken Carlyle had come to accept Darwinian ideas towards the end any more credible, in the face of the overwhelming testimony that his hostility strengthened with the passing of years.[50] 'The older I grow', he was reported as saying in 1877, '—and I now stand upon the brink of the eternity—the more comes back to me the sentence in the catechism, which I learned when a child, and the fuller and the deeper its meaning becomes:—…No gospel of dirt, teaching that men have descended from frogs through monkeys can ever set that aside'.[51]

Of course, the process of conversion is not one of absolutes. It was a question of both position and perspective. To the sceptical G.G. Stokes, W.K. Parker appeared to have become an evolutionist; to the committed Darwinian Arabella Buckley he was at best a temporiser. Even so, Carlyle, Pusey, and even Williamson are much more convincing as examples of the remarkable persistence of resistance on the part of those born before 1830. The evidence of publication, of anecdote, and of reminiscence, is overwhelmingly of the viscosity of intellectual positions. It was unfair of Spencer to remark that the positions of 'Max Müller and Co' were 'beyond the reach of reason', but it did reflect an essential truth.[52] The opposition was not just a matter of religious enthusiasts who had taken up science out of Paleyite commitments to natural theology, or cultural conservatives anxious at the broader claims that were being made. Into the 1880s and 1890s the Cambridge botanist Charles Cardale Babington (1808–95), and the Anglican divine J.W. Burgon (1813–88) were still preaching biblical literalism, urging anti-evolutionary tracts on correspondents, or warning students that the *Origin* was a book which undergraduates ought not to read except under the most careful guidance.[53] It was recalled that at one point Burgon 'had protested against the Darwinian heresy…crying with clasped hands "O ye men of science! leave me my ancestors in Paradise, and I will willingly leave you yours in the Zoological Gardens"', while it was said of Westwood that he not only attempted to secure the endowment of an anti-Darwinism readership for Oxford in 1873, but gave the impression, 'that there was some failure of duty or, at any rate, some want of caution in my being allowed to have the [*Origin of Species*] at all!'.[54] George Rawlinson (1812–1902), who contributed to the response to *Essays and Reviews, Aids to Faith,* continued to his final years to fling intemperately at the 'the gross absurdities of Infidel Astronomical, Geological, and Biological theories'.[55] J.F.W. Herschel, the Duke of Argyll, Gladstone, Ruskin, and Frances Power Cobbe all continued to the end to bear witness to their distance from Darwin. Indeed, we should not underestimate the extent to which for many older figures, just as for Kingsley and Tristram, divergences from Darwin hardened over time.[56] Even a secular progressive like Hewett C. Watson, whom Darwin had identified in October 1860 as a transmutationist, confessed to Wallace in 1881 that 'with some of the most fundamental of [Wallace's evolutionary] views I still find myself unable to concur, while unavoidably admitting them to be the views now generally accepted and nearly proved by yourself'.[57] And ultimately, the Darwinians in their turn lived to offer their own witness to the inflexibilities of belief, generally refusing to abandon older theories of heredity in the wake of the Mendelian revolution after 1900.[58]

There was no doubt in part an evidence effect here, the result of individual reluctance to confess, even to oneself, a fundamental change in position. Where the documentation is fuller, in the case of Lyell's partial acceptance of Darwinism, for example, or later for Conwy Lloyd Morgan's shift from Lamarckian to Weismannian positions in the 1890s, we can trace as we would expect shifts, even significant shifts, in position. It is clear that as the balance of opinion tilted in favour of evolutionary thinking, those with doubts tended to disguise their precise position, offering credit where possible, but remaining otherwise obscure in ways in which it is hard to distinguish suppression of belief from suppression of disbelief. Richard Owen offers a characteristic case of this sort of obfuscation, despite the richness of the surviving material. Owen's withdrawal from public debate and his careful management of his private correspondence effectively masks, as it was no doubt intended to do, his opinions for the last twenty years of his life. It is pretty clear that he remained sceptical about human evolution.[59] But as Rupke has argued, there are just enough scattered indications to suggest that that he did otherwise essentially accept the broad outlines of evolution while attempting to maintain his opposition to its Darwinian version.[60]

But due allowance for the evidentiary obscurities notwithstanding, there is still a lack of documented private ruminations of the sort which might have accompanied a fundamental change of belief. While reason suggests the likelihood that doubts were widespread, the evidence suggests that if they existed they frequently remained unarticulated and unaddressed. The relative rarity of unambiguous conversion can be explained in part by the tendency of erstwhile opponents to adopt evasion strategies rather than concede that they had been in error, accept their wasted investment in supporting positions now being abandoned, and open themselves to accusations of inconsistency and even betrayal. The later Fabian socialist Edward Pease recalled even 'a botanical Fellow of the Royal Society who, in 1875, told me that he had no opinions on Darwin's hypothesis'.[61] Charles Bree regretted that Lyell, in partially accepting the Darwinian position, 'could, in the maturity of his age and fame, have forsaken the "principles" of his youth, of his manhood, and of his prime'.[62] As the cases of Browning, Matthew Arnold, and Trollope all illustrate, the desire to avoid cognitive dissonance remained a powerful limiting force, and Darwin was often easier to evade than to either reject or accept.[63] It took Jowett a decade to settle down to a serious examination of the *Origin*, and Gladstone even longer to thoroughly investigate the arguments for evolution.

It is true that acknowledgements of evolution spread steadily after 1859. But, at least initially, this was largely palaeontological—based on the growing fossil evidence of change; and not driven by Darwinian biology at all. It was frequently vague as to essentials, especially the extent—never mind the cause—of species transmutation, and was absorbed into diffuse evolutionary frames drawn from Lucretius and other Greek thinkers, or Kant, Hegel, and the eighteenth century German philosophers. While Darwinian debates certainly contributed to the dissemination of evidence of prehistoric life, a performance like the lecture on evolution given by Henry Cotterill (1812–86), Bishop of Edinburgh, to the Leeds Church Institute in February 1878, with its easy situation of evolution as just a variant of the 'law of growth' quite distinct from Darwin's unfounded speculations about natural selection, demonstrates just how often acceptance of evolution came as much despite, and indeed in explicit opposition to, Darwinian ideas as because of them, as well as reinforcing the need to tease out as far as possible the precise nature of the commitments being made.[64]

For the less-invested, rather than conversion, what we see is a drift which broadened zones of tolerance, without really involving positive adjustment of belief systems. The gambit of the increasing likelihood that one day the theory would be proved, but that this day had not yet arrived (and might indeed still be some significant time away), was a popular ambivalence. Deniers of evolution might concede the evidence of extinction and appearance of hitherto unknown species; those who initially stood by the immutability of species might accept a certain degree of change but only within specific boundaries. Concessions were made reluctantly and in as restricted a way as could be tallied with intellectual honesty and consistency. So, much clerical initial opposition to the developmental thesis was relaxed, but only as they recognised the resources available for an interpretation which retained notions of design. There was also no doubt a numerous group whose ascription was almost entirely passive, who like the historian Mandell

412   DARWINISM'S GENERATIONS

Creighton came to articulate a general acceptance of Darwinism largely because it seemed that everyone else had.[65] Many more invested in intellectual insurance, avoiding any decisive concession of the truth of evolution while busily constructing arguments which would ensure that even if Darwin were eventually proved true, this need not unduly disrupt the rest of their beliefs.[66]

Certainly, it would be hard to explain the shift in the balance of prevailing opinion in respect of Darwinism which was celebrated in the obituaries of Darwin in the early 1880s, and of Huxley and other prominent Darwinians thereafter, as driven primarily by these individual and compromised shifts of position. There was an element of more gradual shift towards acceptance over the decades after 1859. Those who accepted quickly were probably outnumbered by the others who proceeded more cautiously, and the progress of figures such as Edward Wrench or A. Scott Matheson towards acceptance contributes to the change in overall intellectual atmosphere, especially between the 1860s, when suspicion still ran high, and the following decade when Darwinism begins visibly to gain ground more widely. But generally retreat was slow and keenly fought, a falling back from one position to the next and defending that for as long as possible before further withdrawal. For many this involved only very slow and partial concession, which resulted in all sorts of tortured incoherence. As we have seen, John Gwyn Jeffreys at the British Association in 1877, Charles Wyville Thomson's response to the *Challenger* expedition, even George Rolleston's self-examinations through the 1860s, all provide telling examples of the intellectual origami that could ensue. Movement very often occurred only within relatively narrow parameters, and often by accretion rather absorption: components of newer ideas were taken on alongside previous positions left largely undisturbed, without any great effort to explore the possible conflicts between them. At times new ideas were expressed in the terms of older formulations in ways that suggested at best only partial recognition of changed implications, as in the way pre-1859 Lamarckians absorbed Darwin.

Instead, much of what is taken to show opinion change in the immediate aftermath of the *Origin* was in fact on closer scrutiny a combination of institutional capture, political positioning, and tonal adjustment. It was a function of power and presence not of persuasion. In part it reflected the ability of those born after 1829 to expand their 'bandwidth' through greater access to the reviews, the press, and the platform. In part it was a function of the ability of the evolutionists to marginalise their opponents, not just within the central institutions of British science, the Royal Society and the British Association, but also on the platform and press. It is no accident that gerontocracies like the Church of England, or rigid hierarchies of seniority like the Natural History Museum were two of the most visible sites both of resistance to Darwinian ideas, and of tensions between older leaders and younger members.

But these realignments were less significant in influencing the shifting intellectual and cultural context than the underlying age effects. Over time, the transformation in Darwinism's status was primarily a function of demographic churn. Opposition to Darwinism did not pass, it passed away and was replaced. As Philip Dear observed, 'The old, for the most part, hang on tenaciously to what they learnt in their youth; by degrees they are replaced by those who have been taught more modern conclusions by newer teachers'.[67] The 1870s alone brought a steady flow of new Darwinian voices passing from their teens to their thirties, including Lankester, Romanes, Meldola, Grant Allen, Annie Besant, Aubrey Moore, and R.L. Stevenson, just to name a few of those reaching their twenty-first birthday between 1866 and 1871. Meanwhile, by the time of his death in 1882, well over two-thirds of the opponents of Darwin identified in this study had died, including Murchison in 1871, Wilberforce and Gatty in 1873, Gray in 1875, Carlyle in 1881, followed in the next ten years by many more, including P.H. Gosse, Owen, Bree, Birks, and Cooper. Hull's no longer listening were in reality the no longer living.

Generations

So far, so straightforward. As was noted at the outset, historians of Darwinism have long acknowledged the age dynamics at play without fully exploring their implications.[68] The argument advanced here seeks to do this by describing systematically the extent to which Darwinian engagements were not just marked by strong *age* effects, but by *generational* effects. Of course, not all disagreements between individuals of different generations were generational divisions. But generational dynamics are frequently visible. In the tensions between parents and children. In the semi-ritual challenges of the intellectual authority/positions of their teachers by undergraduates. In the mixed deference and distance with which the young could regard older friends and mentors. In the debates and disagreements between different values and priorities which often marked institutional

414 DARWINISM'S GENERATIONS

histories. The discussions over evolution in the fifty years after the publication of the *Origin* reveal a range of semi-detached debates conducted simultaneously, each operating within its own terms, each constituting discursive communities talking past the others as much as engaging with them. Each of these communities offered a different (although only partially discrete) history. Even where different generations shared a trajectory, they did not share a common speed. The passage of time continually widened generational differences, while shifting the centre of cultural gravity by moving the demographic balance. As John Sibree told the Stroud Natural History Society in 1882, the shifting status of Darwinism was not so much a change in opinion as a change in the variety (and he could have added balance) of opinions encountered.[69] Not least for these reasons, it makes at best partial sense to talk of an 'eclipse' of Darwin at the *fin de siècle* in a context where for some contemporaries Darwin had never been accepted, others continued to accept, and only specific generational groupings were moving to challenge. In the history of Darwinism, as in historical writing more widely, what are often presented as general shifts were in fact generational shifts.[70]

Various intergenerational factors were at work. Family relations were one such. Although this study argues for the presence of generational patterns operating at a shorter periodicity than the nineteenth century family cycle, Darwinian engagements were indeed often mediated by family tensions. High Victorians often confronted Darwin in the context of the implacable opposition of their parents; late Victorians in the less direct disapproval of their grandparents. Edwardians shaped their increased scepticism in the shadow of the enthusiastic Darwinism of their parents. Operating across an intervening generational group, these relationships could help reinforce generational patterns by providing one component of the shared contexts and life course of a particular cohort.

Sociologically, other sources of intergenerational solidarity were more important: the formative experiences of early adulthood, the generational identity of much of Victorian Britain's associational life, social and correspondence networks, communities of intellectual controversy and confirmation. It is striking how far, despite the enormous variety of educational and cultural opportunities, young adults seem to have shared similar intellectual experiences, whether this was in the often intense peer-group friendships of university life, or in the debating clubs, the young men's associations, and the mutual improvement societies which feature so prominently in the intellectual autobiographies of Bagehot, Hardie and the Labour leaders,

CONCLUSION 415

and later of the Edwardian radicals. The Darwinian debates also demonstrate the strength of forces of generational attraction and repulsion throughout associational life. From the X-Club to the Rainbow Circle, informal and semi-formal groupings displayed not just the centripetal force of age, but of generational alignment. Sometimes, as in the case of the Lux Mundi group, or the late-Victorian Idealists, these clearly derived from personal relationships or intellectual preoccupations formed at university. Others, like the Sunday Tramps or the group from which the Halifax Scientific Society emerged, reflected the less defined ties of mutual sympathy and shared vigour. And where generational bounds were clearly transcended, as in the Metaphysical Society, this was not infrequently the result of a purposeful effort at intergenerational dialogue, or of deliberate processes of patronage and co-option.

The history of the reception of Darwin between 1859 and 1909 vividly illustrates the many forms of this generational solidarity, the shared temper, the characteristic modes of argument or engagement, the terms of reference held in common, the prioritised problematics, as well as prevalent points of view which created generational orthodoxies notwithstanding the inevitable intragenerational diversities of belief. Generational positions were constituted and confirmed by communities of mutual support and obligation, and by communities of controversy, natural adversaries whose exchanges drew on and were embedded in the histories of their debates. It is telling that Huxley was still aiming his guns at old opponents like Argyll and Gladstone into the 1890s, even though for his younger contemporaries they were an irrelevance not worth the powder and shot.[71] And it is equally telling that generations remain visible in the patterning of reference points, the way mid-Victorian interventions continue to cite the supporting arguments and address the opposing positions of other mid Victorians. The afterlives of seminal texts subsisted not just in their presence on study shelves and in library catalogues, and the rereadings these allowed, but in the recollections and citations of those who had read and carried their influence with them.

Such intra-generational conversations helped to consolidate and make visible generational distinctions. Generational divisions were at least one factor in the tensions within the metropolitan scientific associations, as illuminated by rows in the Linnean Society in the 1870s and the Entomological Society in the 1890s. As local scientific societies moved slowly from avoidance of Darwinism in the 1870s and 1880s, so they created a space in which early and mid-Victorian doubters debated with younger advocates. Although of course the everyday conversations and exchanges which constituted by

416   DARWINISM'S GENERATIONS

far the most important element in the intellectual life of all Victorians are almost entirely lost to us, it is telling how much of the imagined conversations of Victorian and Edwardian culture were not only deliberately structured around differences of age, but also reflected generational successions.[72] Take, for example, *Conversations on the Creation*, published anonymously by Lewis Wright in 1881, with its carefully specified dramatis personae, two young adults in their early twenties, a group aged around forty including their father, a local science lecturer and a clergyman, an older slightly old-fashioned minister, and then a more elderly and even more old-fashioned retired merchant. Given the date of publication, these map directly onto the generational cohorts under discussion.

Texts like this suggest that the Victorians were more aware of generational dynamics than might at first be thought. The evidence that they sensed the force of generational conflict is clear enough. Even though Victorian traditions of anonymity in press and periodical greatly complicated matters, and generational dynamics were camouflaged by the variable pace of displacements, this inability to clearly map the currents did not prevent Victorians from feeling their pull. They referred frequently to the generational aspects of the debate. They situated themselves in generational contexts, they anticipated and reflected on generational effects, and they reached naturally for generational languages to assess the significance of Darwinian thought. Much of this generational discourse played productively but not very precisely on the multiple meanings the Victorians ascribed to the term. There was an overwhelming consciousness of the genealogical divisions which divided children from their parents. There was a recognition of how age could bring detachment from intellectual developments, in the ways by the later 1870s that early Victorians were increasingly conscious of having drifted outside the flow of contemporary thought, and Bateson felt his imminent supersession in years before World War 1. And there were also, scattered and sometimes easy to miss, instances of generational discriminations based on responses to Darwin.[73]

Nevertheless, the Victorians did lack any meaningful stable generational *identities*. Even talk of 'my generation' overwhelmingly distinguished present versus past, rather than specific peer groups. Awareness of specific generational cohorts rarely becomes visible until after 1900. Where generational identities were used in a more directly sociological sense, they were almost always retrospective, imputed, and inconsistent. Sectional and synchronic identities were more common: constructed as much by position within a specific sequence as within an age group: so Idealist philosophers drew

CONCLUSION 417

identity more from their sense of distance from earlier traditions of Bentham and Mill, and from later Pragmatics than from a recognition of solidarity even with other intellectuals of their generation.

This is a useful warning against the subjectivist fallacy that lack of consciousness implies lack of force. The surprising strength of the generational patterns revealed by the Darwinian debates confirms the extent to which generation could operate almost entirely without the confirmatory overlay of self-consciousness. The radical nature of the challenge of the *Origin*, despite the considerable prehistory of evolutionary debates, the powerful emotional freight of its arguments, and the encouragement it gave to the adoption of stark positions is of course particularly suited to revealing as well as to creating generational effects, and so the responses examined here offer only a partial and tentative case for the existence of broad generational affiliations. But they do offer compelling support for the model advanced at the outset, of the existence of a series of cohorts, spanning a roughly fifteen- to seventeen-year age range, resolving by and large into those experiencing not just the publication of the *Origin*, but the early cultural shocks and debates it created, in their late forties and fifties (or older), in their thirties and early forties, in their twenties, as children and adolescents, or largely if not exclusively indirectly, without personal memory, that is, to cohorts which we have described as late Georgian, early Victorian, mid Victorian, high Victorian, late Victorian, and Edwardian.

As we have seen, the different generations were no more homogenous in their responses to Darwin and evolution than in any other regard. Generation did not impose a shared intellectual position. As the internal divisions of the late Victorians repeatedly demonstrate, what defined a generational cohort were aspects of congruence or commonality rather than uniformity. Nor were the components of generational distinctiveness entirely discrete. Generational character was a matter of partial novelty. Divergence of response coexisted with continuities across generations, and the lines of distinction moved into and out of focus over time. That said, it is possible to advance a schematisation of belief which does reflect surprisingly powerful generational bands. So, the early Victorians not only ranged themselves against the *Origin* on its first appearance, but largely continued to deny the fact of evolution (and in many cases even the overthrow of the Genesis chronology), and by implication at least refused to accept species mutability or natural selection.[74] They could not reconcile evolution with their religious beliefs, whereas the mid Victorians achieved an uneasy sense that compromise was possible, becoming progressively less anxious, not

418 DARWINISM'S GENERATIONS

least in the light of the greater confidence of the high Victorians that evolutionary ideas could not just be accommodated but absorbed. The mid Victorians generally came to accept the fact of evolution, in some version or other, but remained (and this was very much the case even for the minority who identified as 'Darwinists') committed to some form of design, which generally circumscribed the extent to which they accepted natural selection as the mechanism of evolutionary change. And they largely remained committed to the three decisive breaks, first cause, life, and the separation of humanity in respect of intellect and morality from the rest of nature. The high Victorians were uninterested in such intellectual gymnastics; even when they were minded to acknowledge the possible congruences of science and theology, their underlying presumption was of a single unitary natural evolutionary process in which divine action—for those who believed it operated—was confined to first cause. They saw Darwinian dynamics as unlocking the mysteries of the shape of the natural world, whatever further glosses or developments of his ideas might be required, and equally as fundamentally altering the metaphysical context of all serious thought. The shift to the late Victorians is a shift from acceptance to appropriation; they shared these commitments to the unitary evolution of human and animal, but as a given, indeed as a matter of discipleship, rather than a point of contention; for them the question was not so much whether as how, and they squabbled between themselves and the high Victorians over who could claim Darwin's imprimatur, and how broadly this identity might apply. For the late Victorians the vital areas of Darwinian implication were not just theological, but social and political; Darwinism provided a lens through which to interpret the world. With the Edwardians we move into a post-Darwinian phase; it was not generally that Darwin was thought wrong, just that he was no longer a vital guide to the really important questions, of heredity, for example. Darwin was certainly insufficient and increasingly unimportant, and to a growing minority he was even fundamentally misguided.

The tone of responses provides the most overt early signs of generational distinctiveness. The reactions of the late Georgians and early Victorians were visceral, driven by an overwhelming sense of shock and anger at the implications of the *Origin* and its successors for the whole of the Christian worldview. We see this feeling in the sneering of Disraeli or Margaret Gatty, the violent attacks of Morris and Birks, the talk of banning or even burning books associated with Whewell, Westwood, and others. This emotional charge weakened over time, but was reflected in a consistently adversarial

CONCLUSION 419

tone, even for the minority who survived into the 1880s and 1890s. The response of the mid Victorians was initially more temperate, although as we see in the dismay of Frances Power Cobbe and the distaste of Ruskin, their tone became more hostile after *Descent* in part because of its inclusion of humanity, and in part because of the growing assertiveness of the Darwinians. Otherwise, for the mid Victorians, antagonism was to be diffused. As the century progressed, space opened up for the acknowledgement of Darwin's virtues and eventually for admiration and even celebration. But while the mid Victorians tended to avoid outright hostility, urging restraint and temperance of language, they were still inclined, as Stokes' correspondence frequently demonstrated, to perform differences over Darwin as fundamental conflicts. As of course did the high Victorians, who increasingly identified as Darwinians with the zeal of converts, and made acceptance of evolution, and allegiance to Darwinism broadly conceived, a fundamental test. Theorising was embraced, differences relativised, opponents were subjected to often contemptuous rebuttal. While older figures such as Neaves or Howitt had comforted themselves by poking fun at evolutionary ideas, high Victorians focused their ridicule on evolution's opponents. Late Victorians remained aware of the charge Darwinism had carried for previous generations, but largely as a historical phenomenon. Their evolution was not so hot-headed, their emotional engagements more aspirational; for Grant Allen or Annie Besant Darwinism was less a leap of faith and more a creed, a matter of discipleship rather than mission. Their priority was not controversion or conversion, but codification; vitriol was husbanded for internal disputes rather than wasted on anti-evolutionists, as Lankester and Pearson's private intemperance often illustrated. In turn, the Edwardians abandoned dogmatism and the sectarianism it engendered, and in doing so often slid into indifference, and not infrequently into scepticism, railing against the spiritual aridity of Darwinism's dominant materialism.

The irony is that in many respects it was scientific faith positions which were the fundamental divide, especially for the mid and high Victorians, between whom there was a broad agreement about the actuality of evolution and even its causation, but who treated the points of uncertainty in diametrically opposed ways: where the high-Victorian Darwinians identified an intractable problem, they retreated behind faith that advancing knowledge would eventually produce a solution, while the mid Victorians saw exactly the same difficulties as evidence that Darwinian ideas were fundamentally compromised. At times the scientific positions of 'opponents'

420 DARWINISM'S GENERATIONS

were largely congruent (if not practically identical) to 'supporters'—in that the distinction between high-Victorian Lamarckians and mid-Victorian sceptics like Mivart was probably often as much one of approach and rhetoric as of substance. Nevertheless we should not underestimate the power of the emotional alignments which operated within theoretical positions. Emphasis could transform essentially equivalent positions: as it did in driving a wedge between Mivart and Huxley, or placing Kingsley in the Darwinian camp, and Max Müller outside.

Driven by theological fears that science was challenging conventional religious understandings, the instinct of the early Victorians was simply to deny the premises on which the scientists were seeking to operate. For them there is rarely evidence of a sustained attempt to engage with the details of the controversies; but plenty of evidence of unshaken anti-Darwinian faith. That Darwin contradicted religious authority was proof he was in error and his motives malign. Rather than engage with the substance of Darwin's arguments, they sought to ridicule his method and reasoning. Often using the sorts of extreme language that someone like Francis Orpen Morris made a speciality of, Darwin was attacked as too inductive or speculative, or inconsistent, or partial, his rhetorical concessions seized upon and his explanations as to why they were not ultimately decisive ignored, and his positions dismissed as beyond the realms of proof. The mid Victorians were unwilling to denigrate evolutionary science in this way, seeing such an approach as both wrong-headed and counterproductive. As Acland's efforts to diffuse the rows prompted by Owen's interventions showed, science was to be allowed its own validity. Although sometimes just as forthright in their disagreements, and also inclined to default to scepticism, mid Victorians generally avoided the *ad hominem*, and sought to engage with the specificity of Darwin's work, and to constrain the claims of his arguments (for example for natural selection) by challenging his science on its own terms. Particular emphasis was placed on the limits of the evidence or the magnitude of the task of demonstration; gaps in the record meant that the theory was not necessarily wrong, but certainly at best only tentative. For the high Victorians the converse was the case. They began with a general presumption of truth; even theologians such as Dods or Matheson were convinced of the need to reconcile theology to biology rather than the reverse. They set aside any intractable challenges, such as Thomson's circumscription of deep time, confident that scientific advance would ultimately provide solutions. They assumed that any

CONCLUSION 421

evidentiary gaps could be filled in, and settled down to do so, applying Darwin's methods to Darwin's problems. As the British Association address of Lubbock exemplified, embryology, homology, and physiology were all mobilised to flesh out evolutionary connections. If the high Victorians had felt that the task was still to vindicate evolution as fact, and to use Darwin's theories to effect this, the late Victorians, taking evolution itself as read, were far more willing to reopen the debate over the relative influence of its various mechanisms, and to develop new approaches to enable them to do so. They moved to exegesis, to definition and refinement, and if this involved a consideration of the limits, if any, of Darwin's mechanisms, this was with a view, as Romanes constantly stressed, to supplement, not to supplant. The Edwardians showed a similar dissatisfaction with the science of evolutionary change, not least what they saw as its speculative foundations, and sought a solution in experimental investigations into questions of heredity where Darwin ceased to be an effective guide.

Taking a long view of Darwinian responses across the fifty years from 1859 helps to demonstrate that none of these generational positions remained unchanged. Nevertheless, even these adjustments had a generational dimension: where shifts of opinion took place, they were most often simply across to the next generational orthodoxy, and took place in large part in parallel. Where early Victorians accepted evolution, the precise nature of their belief was often obscure or unarticulated, but it seems very largely to have involved a shift only to the mid-Victorian compromises, including some form of limited mutability and multiple creations. And as the early Victorians fell back to these positions and their increasingly forced reconciliations of the scientific and scriptural record, so mid Victorians were abandoning them in favour of more metaphoric interpretations which understood the act of creation in gradualist terms. The mid Victorians differed in that they tended to abandon their belief in immutability, and accept some sort of evolution as a historical fact, though often only within species types, or as a sequence of multiple creations/divine intervention. As gaps in the fossil record narrowed and the phylogenies of species were strengthened, and as Darwin's status as man of science waxed, so the basis of their resistance was undermined, and many came to occupy an evolutionary no-man's-land, their hostility refocused onto the proposed evolutionary origins of humanity. The high Victorians largely shared this trajectory, but they travelled faster and further; abandoning not just the outworks, but the inner defences, their reservations retained only for the more

422 DARWINISM'S GENERATIONS

uncompromising versions of Darwinian biology and the more expansive versions of its social applications which characterised the late Victorians. For the Edwardians we can see a reversal: the trajectory was now from Darwinian orthodoxy to degrees of doubt; at least before 1914 the rate of travel was often slow and the distance moved insignificant, but the shift in confidence was palpable, and provided the foundation for a much wider post-1918 retreat.

In any case, generational individuality was not primarily a matter of *belief*. More fundamentally, it was a matter of intellectual frameworks, of fields of implication and appropriation, of characteristic problematics. Attitudes shifted, but in the context of what we might call 'generational paradigms'. Linguistic and conceptual lenses persisted even as they came to be increasingly out of step with the terms of the debate. 'You can't make a colour-blind man see colours', as A.C. Ramsay commented to Geikie of the incapacity of early-Victorian opponents to move beyond their geological understandings to Darwin's new biological arguments.[75] Mid Victorians continued to recur to the problem of gaps in the fossil record long after the discoveries in America had rendered this line of argument all but untenable.[76] High and Late Victorians continued to place early twentieth-century evolutionary debates in the context of Lamarckian vs Darwinian contests long after these ceased to have any vital force in shaping investigation or framing understandings. Leonard Darwin, at the very end of the 1920s, was still wishing someone would 'write a good little book jumping on Lamarck'.[77] Throughout the period we see repeated evidence of the truth of Arthur Keith's self-analysis: 'the ideas which a man devotes his life to exploring are, for the greater part, those which come to him in the first tide of his inquiries'.[78] Huxley continued to be preoccupied with theological threats to the autonomy of science, W.H. Flower with the exploration of species phylogenies, Lankester and Poulton with the dangers of Lamarckian backsliding.[79] The shift from high Victorians to late Victorians was a shift from Haeckel to Weismann. Readers placed books in generationally specific contexts; they read through a prism of a particular intellectual history, but one they frequently shared with others of a similar age; the context of each reading was not just or primarily an intertextual synchronicity, but equally an 'autobiographical' historicity. Hence Thomas Hughes' response to Kidd's *Social Evolution*, that he much preferred the old Christian Socialist faith and the ideals expressed in the *Life* of F.D. Maurice: 'there is never a week in which I don't refer to it'.[80] Generational exchange may not have been a dialogue of the deaf, but it was certainly a conversation between the hard of hearing.

CONCLUSION 423

Anomalies

For the sake of schematic clarity, generations have been treated here as a sequence of largely discrete cohorts distinguished by a shared birthdate range with defined boundaries at specific year ends. Of course this is an oversimplification. Of course no historian would suggest that processes of historical change are experienced identically because they are experienced concurrently. Of course the sifts of time do not produce absolute differentiations. Generations map onto a complex field of forces—class, race, gender, place, amongst a host of others. At the risk of mixing metaphors, it might make better sense to think of generations as fields of force whose centripetal pull weakened the further from the centre (in terms of birthdate) an individual was situated. Generational dynamics are a soft structuration and they produce variability and blurred boundaries. We should perhaps envisage generational cohorts as constituting a series of steep, flat topped, and overlapping bell curves, in which figures at the margins often shared characteristics of both, and could easily incline more towards the adjacent generation than their own. There was a tendency for early-Victorian and mid-Victorian hostilities to leech into each other; and the distinctions between late Victorians and Edwardians were less marked, and yet 1830 as a line of demarcation in particular is remarkably predictive.

Perhaps somewhere between a quarter and a third of the figures included in the prosopography for this study cannot be said to fit the generational characteristics of the majority. There is nothing particularly robust about this figure because the cohort is not any sense a random sample, and 'fit' is at best an impressionistic judgement, and one which shifts across time. At times the anomaly was largely a matter of the pace of adoption. (So Herbert D. Geldart (1831–1902), President of the Norfolk and Norwich Naturalists' Society, seems eventually to have moved to a Darwinian position, but he did so unusually cautiously, by 1875 accepting the instability of species, but at that stage still distancing himself from Darwin, although he seems to have moved further subsequently.)[81] It is possible to find Darwinists and anti-Darwinists of any age across the Victorian period. There are early-Victorian scientists who displayed at least a more sympathetic view than their peers, and perhaps even a general willingness to concede the basic truth of evolution, including the clergyman botanist Miles Joseph Berkeley (1803–89).[82] There are early Victorians who rapidly accepted Darwinism, like the banker and archaeologist Henry Christie

424 DARWINISM'S GENERATIONS

(1810–65).[83] There are also mid Victorians who refused to make any real concessions to evolutionary science, and the tone of whose hostility has more in common with the previous generation than their own, including as we have seen Godfrey Thring and James Reddie, but also the Congregationalist preacher Brewin Grant (1821–92), who dismissed evolution and the survival of the fittest as 'mystic phases of juggling positivist philosophy'.[84] There are plenty of post-1830 anti-Darwinians, many visible in the debates of the Anthropological Society or the Victoria Institute, including James Hunt, or Henry George Percy, 7th Duke of Northumberland, (1846–1918).[85] The archaeologist and poet Nina F. Layard (1853–1935) was, as she herself confessed, 'very unorthodox from the Evolutionist standpoint', presenting a set of beliefs more aligned to mid-Victorian positions than anything later, maintaining strict limits on variation, and committing to the special creation of the higher organisms.[86] In the same way, such evidence as we have suggests that many Edwardians continued to espouse fairly conventional Darwinian positions, with little of the sorts of post-Darwinian angst that has been explored as a feature of the Edwardian condition.[87] Once we go beyond evolutionary science per se, and into the various intellectual implications, for example social Darwinian ideas more generally, then the picture becomes even less defined.

Many of the apparent anomalies are really just examples of the weakened pull of generational characteristics at their margins: those born in the last few years of each generational window whose seminal educational experiences were delayed might align with the succeeding generation, just as the very oldest of a generational cohort might retain the characteristics of the previous generation, especially where strong countervailing pressures of family, religion, or interest operated. So individuals who attended university later than their peer group, like the philosopher James Ward, or who, like Lubbock, were encouraged to absorb new ideas in early adolescence, could show this sort of displacement. Inevitably, the Darwinian controversies bring into particular relief the tensions and countervailing forces which made the position of those born around 1830 especially conflicted. The dilemmas of George Rolleston have been discussed extensively above. The Edwardian D'Arcy Thompson, born in May 1860, was another who illustrates how marginal figures could also display an idiosyncratic hybridity of two generational identities. In 1917 Thompson reflected on his liminal status, not really aligned to any of the late-Victorian groups, but still distanced from Mendelism: having spent too many years being overawed by 'the conventional zoologists, E.R.L[ankester]. and the rest...that is the old shell Darwinians', he felt doomed to 'live the rest of my life in very imperfect sympathy with a new school' which looked to Bergson.[88]

A more fundamental challenge is offered by examples dislocated by more than a few years from their predicted generational position. These too are not hard to find, but in relative terms they were a small proportion of the whole. They reflect the complex historical contingencies which helped shape individual character and behaviour, and in most cases were likely to have been the outcome of a complicated interaction of numerous factors. Most obvious were the small number of original thinkers, figures like Darwin and Wallace, who in developing radical new interpretations were by definition out of step with their generational cohort. Darwin and Wallace (and indeed Huxley) could draw on intense experiences of overseas expeditions not shared by their contemporaries, and in Wallace's case, as F.M. Turner, points out, this was compounded by his lack of the sort of formal education common to the mid-Victorian middle class.[89] Much the same is true for those we might describe as mavericks or iconoclasts, part of whose personality or intellectual orientation was driven by a deliberate refusal to abide by convention or orthodoxy. Figures like Samuel Butler. It is possible to see Butler as a militant Lamarckian out of time; consistent in many ways with late-Victorian neo-Lamarckism. But Butler was always something more, a figure whose whole character encouraged an oppositional stance to the received wisdom of his day, as well providing a fertile soil for the personal animus towards the Darwins that he treated as part of his patrimony.[90]

Political views, expertise in a field less amenable to evolutionary interpretations, and strong family ties could all play their part. Liberals and radicals were certainly more inclined to Darwinian ideas for much of the period. One recollection of the Northampton Natural History Society suggested that after one debate on evolution (presumably in the late 1880s or 1890s), all the Liberals voted one way and all the Conservatives the other; but there is little evidence that such stark divisions were widespread and every indication that by and large political allegiance had at best a weak and temporary correlation with fundamental attitudes.[91] One exception would be those individuals shaped by unorthodox subcultures such as the Lamarckian-infused radical cultures of the 1830s and 1840s, such as the long-time Lamarckians Robert Grant and Thomas Laycock,[92] or Samuel Laing, friend and disciple of Huxley, whose writings displayed a thoroughgoing evolutionism.[93] As we have seen, entomology was one field which

426   DARWINISM'S GENERATIONS

seemed to offer poor soil for evolutionary enthusiasm. Palaeontology (or 'paley-ontology' as one Edwardian wit termed it) was another, despite the enormous strides made in fleshing out the evolutionary record.[94] The enduring scepticism in respect of Darwinism of the marine biologist W.C. Mcintosh (1838–1931) was said to have been facilitated by his focus on the taxonomy of annelids (marine worms), which did not lend themselves to evolutionary approaches.[95] The archetypical case of filial loyalty and intellectual obeisance is Frank Buckland, whose devotion to his father William produced an uncompromising hostility to evolution despite his mid-Victorian identity. Reportedly, Frank's response whenever the theory of evolution was raised was that 'I was brought up in the principles of Church and State; and I will never admit it – I will never admit it.'[96]

In this case religion clearly also played a part, and it was by far the most powerful countervailing force. This was not exclusively negative. Those from Unitarian and freethinking traditions, for example the nondenominational preacher Charles Voysey, or the geologist and Unitarian minister Henry William Crosskey (1826–93), were more likely to take up evolutionary ideals in advance of their generational peers, just as evangelicals were under especially strong influence to reject Darwinian ideas, as was the case for high Victorians like Charles Spurgeon, Sidney E.B. Bouverie-Pusey (1839–1911), or Sir Henry Morton Stanley (1841–1904), who continued to dispute even the basic facts of organic evolution throughout their lives.[97] In the case of the Catholic Church, although for much of the period there was sufficient room for manoeuvre not just for lay scientists like Mivart, but even Catholic clergy who offered reconciliations of faith and Darwinism, the general suspicion of evolutionary thought persisted, and indeed hardened into hostility as the *fin de siècle* approached.[98] This was almost equally true of Presbyterianism, which produced individuals like Robert Wallace Murray (1836?–1904), tobacco manufacturer and shipping magnate, a Liberal with an evangelical upbringing, who, although it was said he admired Darwin immensely, presented himself as staunchly opposed to the Darwinian theory.[99]

There is of course need for a note of caution here; in the face of organised religion's often fierce hostility for much of the period, it remained difficult for ministers of religion to commit in print to positions that might lead to censure or expulsion. But this did not necessarily determine private beliefs. As Wilfrid Ward noted, in spite of official opprobrium, plenty of Catholics 'in the retirement of their studies' worked out their own (and often more complex) responses.[100] The Anglican Henry Ullyett (1838–98),

Master of St Mary's School, Folkestone, can be found at several points if not directly repudiating Darwin, then at least distancing himself from any acceptance, and yet the testimony of his son looking back from the 1920s was that his father believed not just in evolution but in the evolution of humanity, yet dared not express this belief publicly for fear of losing his position.[101] As we have suggested, similar concerns, of social ostracism or of professional excommunication, were potentially just as significant in private correspondence and even in scientific publication, albeit increasingly working in the contrary direction.

Often, of course, more than one of these factors is in play. Hence the mannered (and very un-high-Victorian) irreverence of Andrew Lang, who commented to one friend in 1885 that he hadn't a copy of one of the lives of Darwin, and didn't want one, asking 'What's the use of littering a house with new books?',[102] and who told Grant Allen around the same time, that 'Personally I don't find it easier to believe in Darwinism than in Buddhism, and I don't care a penny what conclusions physical science may come to'.[103] Lang was no doubt influenced by the fact that his Oxford undergraduate days did not start until 1864, so that as an undergraduate he was in effect mixing with the oldest of the late Victorians, but here again it was also a feature of his own carefully cultivated individuality, so that it is never entirely clear what was offered from conviction and what was uttered for effect.

There are also those figures who for whatever reason were situated outside the normal generational structuring forces. Victoria, Lady Welby is one striking example. Maid of honour to Queen Victoria, Welby later abandoned court life and threw herself into various philosophical enquiries, prosecuted via encyclopaedic reading and an extraordinary network of correspondents and visitors to her houses in Harrow and Grantham. As a high Victorian Welby is by no means entirely anomalous: despite the sceptical tone of her paper on ghosts to the 1890 British Association, she came to share an underlying commitment to evolution and natural selection; but in many ways, not least her broad commitment to social evolutionary approaches, she more effectively aligns with late Victorian positions.

The cases of Welby and Nina Layard amongst others, do pose the question of whether generational patternings operated with the same force for women as for men, perhaps for the obvious reason that they were initially excluded and then much less likely to engage in the institutionalised educational experiences which were so powerful in creating peer identities for men. By and large, the conclusion from the individuals explored here would

428 DARWINISM'S GENERATIONS

be that this is perhaps so, but that the difference is much less than might have been expected. It would seem, just as for males who did not attend university, that other channels largely make up in intellectual terms for the lack of an undergraduate experience, and that peer groups were still established, even if in part vicariously through sibling and then spousal networks. The cases of Annie Kenney, just as much as of Ethel Chamberlain and Alice Balfour, demonstrate the ways in which this could work.

Perhaps more disruptive of generational patterns for women were the dimensions of Darwinian thought which cut across the agendas of the Victorian women's movement. Hence Darwinism was a resource for Mathilde Blind or Mona Caird (1854–1932), but also a site of struggle with the arguments of Darwinists like Grant Allen that placed women in subordinate or inferior positions. Blind, for example, opposed those who mobilised Darwinian arguments in favour of the primacy of marriage and female dependence.[104] The ambivalences which could result are illustrated by Anna Kingsford (1846–1888) and Frances Swiney (1847–1922). Swiney in particular is an interesting case; steeped in the scientific literature, liberally sprinkling her writings with references to Romanes, Weismann, Haeckel; drawn both to ideas about sexual selection and mining Darwin's works for examples of female agency, described by Barbara Gates as 'a propagandist for human natural selection', but also maintaining an underlying distaste for Darwinism.[105] (Of course, this was a debate which played out more widely within Victorian progressivism. In March 1909 Herbert Burrows (1845–1922) was taking issue with what he saw as the contempt of women at the heart of the socialism of Ernest Belfort Bax and his supporters. 'The Baxites had better celebrate this Darwin centenary year', he suggested, 'by burning all of Darwin's works and going back to St Paul with his reactionary views on the position of women.')[106]

On careful examination most of these anomalies were only partial, and in many respects the individuals continued to reflect the generational position from which they diverge. This is the case with A.R. Wallace. Wallace's refusal to accept the fully natural evolutionary position of humanity, and his insistence that humanity's moral and intellectual powers required some additional divine intervention were treated as an unaccountable aberration by the Darwinians, especially given his interventions in the mid-1860s, which approached the evolution of humanity very much from the evolutionary perspective. And yet this partial exclusion is absolutely consistent with the response of the majority of his mid-Victorian contemporaries. In the same way we might say that Huxley's Darwinism was strongly inflected

CONCLUSION 429

by mid-Victorian conceptions, both in respect of his refusal to accept the absolute gradualism of Darwinian evolution, and his sustained ambivalence over natural selection.

The history of Darwinism in Britain reveals the complexity and variance of the factors at work. Just as Darwin demonstrated that the discreteness of species was an illusion ('artificial combinations made for convenience'), so is any suggested discreteness of generations. Yet, just as species continue to offer a necessary organising principle of biology, so generations can still offer an illuminating ordering of historical positioning, and the significance of generational patterning is not disproved by anomalies per se, but only if the anomalies become such a significant proportion of the whole as to undermine any claim to an overall pattern. The fact that so many 'eminent' Victorians don't neatly fit generational patterns (indeed, perhaps, that they are eminent precisely because they *don't* fit) reinforces the importance of prosopography and not simply case studies. And notwithstanding the considerable numbers of anomalies, and the difficulties of ascertaining private opinion rather than public self-positioning, the evidence of the Darwinian debates of the fifty years from 1859 suggests that generational position has a much stronger level of predictiveness than really it has any right to show.

Implications

If anything, scholarship has become less attuned to age effects in recent decades than it was at the time when protocol still expected historical figures to be provided with their life dates. And although, at least at a theoretical level, scholars have remained interested in the problematics of time, their focus has tended to gravitate towards considerations of the subjectivity of time and the disconnection of absolute geometric views of time from day-to-day experience. Where there has been attention to what we might think of as historical time, concern has been with its 'timely' qualities— pace, acceleration, tempo, direction, and seriality. A consideration of generations, though, raises questions about a different set of temporal pluralities, which don't so much problematise the envelope of physical, linear time as point instead to the debilitating attenuation of historical understanding produced by a 'thin' chronology, and a single narrative sequence. It reminds us that just as the German 'geschichte' (history) has etymological roots in 'schichte', that is layer or strata, so the social, cultural, and intellectual history of past societies can better be understood as a layered rather than a

430 DARWINISM'S GENERATIONS

unitary phenomenon, and that effective historical analysis depends not just on time's temporalities, but also, (taking up Thomas Hardy's distinction between the 'length, breadth, thickness, colour, smell, voice' of today, as opposed to the 'thin layer among many layers, without substance, colour, or articulate sound' that it tends to become as yesterday), on the *dimensions* of time, its breath, depth, structure, and composition.[107]

In this regard, the work of Reinhardt Koselleck, and particularly his notion of *Zeitschichten* (Layers of Time) is particularly suggestive.[108] Koselleck suggests that although there is an absolute temporal process—a chronology into which all things can be tied, the medium in which history takes place—there is not one *historical* time 'but...many that overlie one another'.[109] Koselleck's translators Franzel and Hoffman speak of his work as encompassing multiple historical times present at the same moment, 'layer upon layer pressed together, some still volatile, others already hardened'.[110]

The congruences of these insights with the sort of generational analysis proposed here are obvious, and Koselleck explicitly invokes generation as a way experience is structured (and accumulated variably), so that 'every historical insight emerges from generational cohorts who unavoidably have their own particular experiences and thus only grasp things in the light of them'.[111] But his conception of generations remains fairly tightly embedded in the conventional frames. Despite the strong association of his work with the idea of the simultaneity of the non-simultaneous (*Gleichzeitigkeit der Ungleichzeitigen*), his exploration of temporal multiplicity engages much more with questions of pace than of breadth, with the velocity of sedimentation rather than the coexistences which ensue.

These limitations are reinforced by the way Koselleck deploys conventional geological metaphors as a way of conceiving of the layering of historical time. There is an obvious correspondence between processes of sedimentation and the formation of generational age cohorts by the demographic cycle of births and deaths, and stratigraphical representations can help visualise the thickness of historical time, the depth created by the coexistence of overlying generations. But ultimately, the stratigraphical column remains a representation of sequence. To the extent that it models interactions between the layers it encourages a sort of tectonic understanding confined to friction along the boundary between adjacent levels, and the instabilities caused when one layer shifts across another. What it does not encourage is an exploration of the multivalent dynamics of intergenerational exchange or interpenetration, or an engagement with shifts in generational simultaneity.

CONCLUSION   431

We need a more productive metaphor. Fluid mechanics might be one, allowing for the flow of history and the coexistence and interaction of layers; but here too the science pushes primarily towards questions of progression and boundary effects. Tim Ingold has recently offered the image of the twisted rope, which emphasises the contiguity of generational strands, but not the dynamism of their interaction.[112] Music is a surprisingly neglected alternative, with its balance of diachronic and synchronic elements, and its even more promising conceptual vocabulary.[113] History, like a musical composition, is a matter of tempo, volume, and melody; but it is also a matter of harmonics, of vocal range, of cadence, and counterpoint. From this perspective, we might think of the challenge of a generational approach as requiring a move from what might be termed the 'percussive-melodic' mode of historical composition, to 'vertical-harmonic' approaches. The layering of the multipart score gives graphical representation to the separate components, with their own rhythms, melodies, and instrumentalities, but also captures both the shared progression through time and the constant interplay of the different elements. All these are visible in the history of Darwin's reception in Britain: the construction of its polyphonic character out of multiple generational voices, the prominence of which wax and wane as the responsibility for the primary melody moves between parts, shifting patterns of consonance and dissonance in the contrapuntal relations between the generations, even we might say the differences of instrumentation involved in the emergence of new periodicals or forms of knowledge.

Applying these insights more broadly requires the abandonment of reductive notions of the 'zeitgeist', of a unitary set of values or beliefs prevalent at any particular time. This would not, as some have suggested, mean the end of broad historical periodisation, although it would reinforce the already powerful arguments against conceptions of periods as a series of discrete boxes with hard/impermeable boundaries;[114] but it also encourages us to develop an understanding of periods as particular assemblages, and the points of transition between them as produced by shifts in the balances of the constituent components. From this perspective historical change becomes not a concatenated sequence, but rather an ongoing succession, modulations from movement to movement by successive partial displacements. And the importance of this partialism indicts the whiggish tendency to hurry actors and their beliefs off the historical stage, to imply if not to explicitly impute anachronism to challenges which are not deemed central to some overarching phylogeny of knowledge.

432   DARWINISM'S GENERATIONS

Summation

This still leaves the question of how far the generational dynamics revealed in the Darwinian debate were of more general validity. The fact that the generational schema being tested here was initially formulated out of explorations with little or no direct relationship with the Darwinian controversy offers some indication that the alignments are not merely a function of the specifics of the Darwinian debate. But it is clear that the timing of the publication of the *Origin*, and to a lesser extent of *Descent* and then of Darwin's death do themselves encourage a sense of generational breaks in around 1830, the later-1840s, and the mid-1860s, or between those who confronted the *Origin* largely as young adults in their twenties and early thirties, as adolescents, or whose direct experience was more shaped by the response to *Descent*. A great deal more research is needed, across a broad spread of domains and addressing the widest possible range of experiences and behaviours, before it will be possible to speak with assurance of the existence, identity, and significance of generations in Victorian Britain. The hope is that this study will help encourage this sort of work.

For the moment, close examination of the responses to Darwin between 1859 and 1909 gives powerful support to both the model of generational formation outlined at the start of this study (including the suggestion of age band widths of approximately fifteen years), and the proposed generational schema, with its indication that the birthyears around 1813, 1830, 1845, 1860, and the mid-1870s marked the points of generational transition. The initial effectiveness of this schema suggests that we need to recognise that just as modernity accelerates the pace of historical change, so it narrows generational intervals, and in doing so engineers a greater *depth* in social experience. Indications of generational alignments operating at a broad cultural level might also help historians to abandon the vagueness with which they have used 'generation', and look to more precisely indicate the constitution of the generations they invoke.

Confirmation that generational dynamics were operating in the second half of the nineteenth century also raises more acutely the question why, given the Victorians' usage of generational vocabularies, there was so little contemporary conception of particular generational identities of the sort which become a staple of the discourse of the younger generations, not merely as a result of World War 1, but in the years leading up to it as well. Part of the answer no doubt rests in the consequences of cultures of anonymity in the press and periodicals; part no doubt in the powerful forces of

CONCLUSION   433

class and religious segregation which militated against shared senses of age-based solidarity (and which were certainly losing their force by the Edwardian years).

If the Victorians were slow to develop generational consciousness, they were not slow to recognise the powerful dynamics of age, and the history of the reception of Darwin in Britain reinforces the need for greater attention to age, as a process, as a factor in attitudes and behaviour at any one moment (as 'the form time takes to dwell in us', as Marías puts it), and as a field of force shaping friendship groups, networks, and in many cases institutional composition, in a way which goes far beyond the recognition hitherto given to the importance of undergraduate experiences for the narrow university-educated elite. This encompasses both 'age effects', in the terminology deployed in generation studies (that is the typical impacts of being younger or older), but also 'cohort effects', the consequences of sharing a particular position in history with a group of similar age. As we have seen, initial responses to Darwin were heavily inflected by the age of first encounter; but the response of high Victorians was also moulded by their relative lack of direct experience of the highly charged response of older generations to the *Vestiges of the Natural History of Creation* in the mid-1840s, and their recognition from an early age of the accumulating evidences of deep time, just as late-Victorian Nonconformist ministers were the product of a very different experience of ministerial training to that of earlier generations.

The significance of age and generation challenges existing protocols, of the individual case study and the representative individual, and emphasises the desirability of addressing broad questions of historical change at the level of the population. It also directs us to take seriously the 'illusion of contemporaneity'. The lives of Gladstone and Churchill overlapped over nearly all of the final quarter of the nineteenth century, but of course they were not contemporaries, and neither can stand effectively as representative of that period. But precisely because the Victorians of the 1890s include both of them, but also John Ruskin, Leslie Stephen, and Oscar Wilde, we cannot comprehend the period through any single narrow tranche of personalities. Of course, we generally know and can place the Gladstones and Churchills of the historical world, but often we depose individuals and deploy their contributions without any thought for birthdate or generational location. The importance of age, individually, institutionally, and demographically, makes this blindness problematic. Of course, this raises practical and methodological challenges which may be even more acute for periods before the Victorian; it is not always possible to position actors,

434 DARWINISM'S GENERATIONS

authors, or artefacts in age or generational terms. There may be instances where questions of age are relatively unimportant, but even in those instances, proceeding without age identification creates the danger of overlooking important age inflections, and of mistaking age effects for other forms of social, intellectual, or cultural influence.

The generational dynamics of Darwin's reception suggests a number of fruitful lines of inquiry. In the first place the sociology of reception, which as has been demonstrated, did not just revolve around class, religion, or even geography. There seems little doubt that further exploration of the ways in which readership, reader-responses, intellectual repertoires, and even opinions and ideas were inflected by age and generation will help draw out the intertextualities of age, and the ways in which generational position and solidarities helped to shape individual attitudes and collective opinion. Each act of reading, each act of thinking and responding, occurs at the intersection of a particular historical moment and a particular autobiographical trajectory. Each is a combination of age, generation, and period.

Second, the sociology of institutions and the wider cultural landscape. In respect of periodicals, we need to attend not merely to new titles but the shifting generational make-up of owners, editors, contributors, and readers. In respect of institutions, we need to give due consideration to the ways in which their histories were shaped by the generational character of their founders, members, and leadership. Frequently, this will be the force of a single generational constituency, although as the history the Metaphysical and Synthetic Societies show, it could just as significantly operate in intergenerational ways. Although always likely to be more powerful for informal coteries or small groups of the like-minded, like the Lux Mundi group, or the Fabians, even cursory examination of provincial science associations suggests that such dynamics operated broadly.

There are also insights to be drawn in respect to periods and periodising claims. At one level, thick generational time challenges conventional periodisation by further reinforcing the powerful forces of continuity across (nearly) all historical moments, and further undermining simplistic notions defining periods by particular states of consciousness or opinions. All periods are generational assemblages in which different perspectives, experiences, and cultural investments jostle. Across timeframes of ten or twenty years, all periods are marked by a significant degree of generational continuity with periods before and after, but are also differentiated both by events and by the age effects produced as generations move from youth to prime to older age. For the Victorians, the traditional sub period of

CONCLUSION   435

mid-Victorian Britain, produced in no small measure by the economic impact of the fiscal reforms of the mid-1840s, and the political consequences of the defeat of Chartism, can also be seen to have been constituted by the greater prominence and influence of the generation born in the 1810s and 1820s, just as the particular character of Britain in the later 1880s and 1890s drew in part on the generational configuration which marginalised the mid Victorians, and saw the high Victorians slowly losing authority in the face of the maturing of that group born in the second half of the 1850s and the 1860s. There is of course a potential circularity to such arguments, and further investigation will hopefully tease out the relative importances of the event and generational effects intertwined in these shifts.

For the moment, we can at least conclude that although there was a Darwinian debate in Victorian Britain, and one—perhaps two—generations, whose general ascription to Darwinism loosely defined brought it a brief period of hegemony, there was no Darwinian revolution. Indeed, the argument presented here is that not only is this proposition indefensible, but that it is a category error. Darwin had expected the *Origin* to produce a transformation and his champions fell easily into the language of 'momentous revolution'.[115] But the changes in orthodoxy were gradual and incremental rather than saltational. Generations lead us back to an impeccably Darwinian rejection of the language of catastrophism and cultural caesura: in general *cultura non facit saltum*', not least because of the complicating effects of generational overlap, and the inertial effects of age. The transition from revolution to post-revolution was delivered by demographic churn. In that respect, an effective history of the reception of Darwinism in Victorian and Edwardian Britain, indeed any effective history, requires a double historicism, of both chronological and biographical time, one which penetrates the surface of contemporaneity and explores the generational multiplicities of each historical moment and the generational dynamics of even relatively rapid historical change.

Notes

1. Reference the debates in the *Nottingham Journal*, August 1909, and the *Sheffield Daily Telegraph*, April 1909.
2. Charles Vyse (1882–1971), *Daily Mail*, 22 July 1909.
3. Clodd's high-Victorian readers continued into the 1900s to read him through the lens of anticlericalism; see Sir Edwin Pears (1835–1919) to Clodd, 28 April, 5 December 1906, In-letters, Box 6, Edward Clodd Papers, BC MS 19c Clodd, Brotherton Library.
4. Ford Madox Ford, *Return to Yesterday* (1931), 392; the journalist Cecil Roberts (1892–1976), who was also growing up in Nottingham at this time, expressed scepticism, *The Sphere*,

436 DARWINISM'S GENERATIONS

7 November 1931; but see contemporary evidence that Lawrence at least was reading Darwin, along with Spencer, Renan, and others, Lawrence to Rev. Robert Reid, 15 October 1907, La C 35, D.H. Lawrence Collection, University of Nottingham.

5. R. Blathwayt, 'England's Taste in Literature', *FR* 91 (January 1912), 170–1.

6. Arnold Freeman, *Boy Life and Labour* (1914), 156.

7. 'Humour and Earnestness', *Morning Post*, 7 February 1908.

8. See materials in Box 6, Folder 8, Edwin Grant Conklin Papers, CO332, PUL.

9. See James to James McKeen Cattell, 25 November 1908, F.J.D. Scott, ed., *William James: Selected Unpublished Correspondence* (1986), 494.

10. See Meldola to WTTD, 21 July 1909, f.10, W.T. Thiselton-Dyer Papers, In-letters III, RBGK (where Meldola confesses that Wallace had felt his review in *Nature* was 'too mild'); J.E.B. Mayor (1825–1910) to William Sanday, 28 October 1909, ff.418–19, MS. Eng. misc d.124 (ii), Bodleian (using G.M. Humphry as his example). Compare with the claim of George Thear (1829–1918), former Vice-Chancellor of Cambridge, whose observation that 'From the first appearance of "Natural Selection" I have liked to regard myself as a humble and reverent disciple of the Great Interpreter' lost something from his failure to accurately name the *Origin*, see Thear to A.C. Seward, 6 March 1909, f.95, MS/675, Royal Society.

11. *Cheltenham Chronicle*, 27 February 1909; for Callaway see *Nature*, 21 October 1915; obituary, *Quarterly Journal of the Geological Society* 72 (1917), lvii; D.J. Cunningham, presidential address to the Royal Anthropological Institute in 1908 (see *Journal of the Royal Anthropological Institute* 38 (1908), 34).

12. *Daily Telegraph*, 26 June 1909.

13. *Daily Telegraph*, 25 November 1909.

14. Patrick Geddes, 'The Sociologist on the Streets: 1. From Nature to Human Life', *The New Age*, 11 November 1909, 32–3.

15. *Falkirk Herald*, 10 March 1909, quoting comments of Robert Durward Clarkson (1866/7–1951).

16. *The Graphic*, 28 August 1909, quoted Macgregor, 'Exhibiting Evolutionism', 90; 'Admirers of Darwin', *Daily News*, 18 August 1909.

17. E.B. Poulton to WTTD, 18 July 1908, f.134, W.T. Thiselton-Dyer Papers, In-letters III, RBGK; Poulton to Benjamin Kidd, 9 August 1908, MS Add. 8069/P55, Kidd Papers, CUL. Interestingly, responding directly to any suggestion that his position might be outmoded ('I am fully aware of the intellectual rigidity that is so prone to develop with the passing years. I have steadily endeavoured to keep my own mind elastic and flexible' (ix)), Poulton was quick to move on to the attack, lambasting the 'confusion worse confounded', of the Mendelians. See also his article in *Quarterly Review* 211 (July–October 1909), 1–38. Late Victorians like James Alexander Lindsay rallied round: 'That any of the fundamental features of the Darwinian doctrine have been subverted by 50 years of inquiry and controversy is not to my mind a tenable proposition', 'Darwinism and Medicine'.

18. W.T. Thiselton Dyer to Meldola, 22 July 1909, #977, Meldola Papers, ICL.

19. See Meldola to Wallace [sequenced in the BL collection for 1904, but from the content very likely 1909], 25 June 1909?, f.146, Wallace Papers, Add. MS 46437, BL; review of Poulton's *Darwinism and Modern Science* (1909), in *Nature*, 24 June 1909. Other late Victorians responded with delight. 'You seem to have felt the grave modest simple and always truth-clear eyes of the master upon you,' enthused Morgan, C. Lloyd Morgan to Meldola, nd (postmarked 28 June 1909), #966, Meldola Papers, ICL. For another example see Sidney Low (1857–1932), 'Darwinism and Politics', *FR* 86 (September 1909), 519–22, with its own characteristic late-Victorian attempt to distinguish between moderate Darwinism and the excesses of disciples of Spencer; Low clearly distinguishes Darwin's heritage (as manifest in Weismann, Morgan, and Thomson) and Lamarckian ideas. For the context and a broader discussion of the way some of these tensions were playing out, see Richmond, 'The 1909 Darwin Celebration', 447–84.

20. Poulton to Hooker, 8 July 1908, f.38, JDH/2/1/17, Joseph Dalton Hooker Papers, RBGK.

21. See Gillham, *Francis Galton*, 338; Bateson notes on 1908, Bateson Correspondence, Copy Collection, Vol 9/1175–1326, JIC; see report of Lankester in *Darwin–Wallace Celebration* (1908), 27–33.

CONCLUSION   437

22. E.R. Lankester to Wallace, 6 July [1909], ff.66–7, Wallace Papers, Add. MS 46438, BL.
23. Adam Sedgwick to WTTD, 8 March 1908, f.1, W.T. Thiselton-Dyer Papers, In-letters III, RBGK.
24. See *Jane Ellen Harrison: A Portrait from Letters*, ed. Jessie Stewart (1959), 111.
25. F.J. Weldon to Pearson, 8 March 1909, Pearson/11/1/22/39, UCL. Bateson rehearsed his wide-ranging antipathy in a series of letters to Seward; his issue with biometricians is 'not one of view, or theory, but of fact', etc.; A.C. Seward, 'Correspondence re Darwin and Modern Science 1909', MS Add. 7733, CUL.
26. Wallace to Frank Darwin, 28 June 1909, f.112, A.C. Seward, 'Correspondence re Darwin and Modern Science 1909', MS Add. 7733, CUL; WTTD to Seward, 16 July 1909, 'Papers Relating to the Darwin Commemoration', MS/675, Royal Society.
27. A.J. Mason (1851–1928) to A.C. Seward, 11 December 1908, 'Papers Relating to the Darwin Commemoration', MS/675/82, Royal Society (Mason recognised that Bateson's criticism of aspects of Darwinian theory might disqualify him, but suggested, naively, that 'in the pursuit of truth, no one would object on that ground').
28. As reported in Wallace to Meldola, 27 June 1909, Meldola Papers, OMNH (quoted in Raby, *Wallace*, 282).
29. See letter of M.D. in *The Times*, 2 September 1909.
30. He told Meldola, 'There seemed to me a want of sincerity about the whole thing. I doubt if there is a single Cambridge resident who holds the true Darwinian faith', WTTD to Meldola, 22 July 1909, #977, Meldola Papers, ICL.
31. See collection of ephemera, including manuscript account, collected by Woodward, listed by Gilleasbuig Ferguson Rare Books, Portree, https://www.ebay.co.uk/itm/1909-DARWIN-CELEBRATION-CENTENARY-A-UNIQUE-EPHEMERAL-COLLECTION-MANUSCRIPT-RARE/202972385862?hash=item2f4218d246:g:f24AAOSwb59dzDHn [accessed 5 January 2021].
32. Bateson to Hurst, 25 June 1909, MS Add. 8634/H17, Bateson Papers, CUL; likewise Hurst to Bateson, 29 June 1909, MS Add. 8634/H19, Bateson Papers, CUL.
33. DWT to A. Smith Woodward, tps, 15 February 1909, ms18600, D'Arcy Wentworth Thompson Papers, St Andrews; Woodward (four years younger) agreed, Smith Woodward to DWT, 23 February 1909, ms18604, D'Arcy Wentworth Thompson Papers, St Andrews.
34. W. Bateson to WTTD, 26 May 1911, f.56, W.T. Thiselton-Dyer Papers, In-letters I, RBGK.
35. Mary M. Bartley, 'Conflicts in Human Progress: Sexual Selection and the Fisherian "Runaway"', *BJHS* 27 (1994), 180; Richmond, '1909 Darwin Centenary', 479–80. See also V.B. Smocovitis, *Unifying Biology* (1996). Fisher is interesting in that his work was very much engaged with the legacy of Darwin. When offered a choice of books to mark his academic achievements at Harrow he chose the complete works of Darwin, the surviving battered version of the *Origin* suggesting it was especially closely studied; see A. Aylward, 'R.A. Fisher and the Scientific Past', in Hesketh, ed., *Imagining the Darwinian Revolution*, 188–203.
36. Quoted in Michael J.K. Walsh, *Hanging a Rebel: The Life of C.R.W. Nevinson* (2008), 232. Of course, as Henry-James Meiring has vividly demonstrated, Freud himself drew explicitly on Darwin and he and his readers considered him as developing Darwinian insights within psychology, Henry-James Meiring, 'Darwin of the Mind: Freud's Darwinian Image', in Hesketh, *Imagining the Darwinian Revolution*, 171–87.
37. For example, in the portrayal of the butter merchant Tyler Harbottle, 'president of the Literary Society, the Field Club, the Linnean Society, the Darwin Club, the Old Fogies and Ancient Codgers', in *Old Mole* (1917).
38. A. Duff Cooper, *Old Men Forget: Autobiography* (1953), 32.
39. See, for example, Grant Uden, 'Books I Have Never Read', *The Bookman* (October 1932), 21–4; Mrs Patrick Macgill (1887–1966), *Painted Butterflies* (1931).
40. *Sheffield Daily Telegraph*, 19 February 1930.
41. Huxley continued to the last to fixate on the latest discoveries filling the gaps in developmental sequences. This was a feature of the discussion in his 'Coming of Age' essay in 1880, and in his Rede Lecture of 1883; see *Nature*, 21 June 1883.

438   DARWINISM'S GENERATIONS

42. Chambers to Alexander Ireland, 24 April 1860, Dep.341/112/180–81, Letters of Robert Chambers to Alexander Ireland, NLS.

43. Hull, *Reception*, 75.

44. Huxley to George Harris, 2 November 1875, #10, George Harris Papers, UCLA.

45. Cobbe to Blanche Atkinson in *Life of Frances Power Cobbe* (2nd ed., 1904), xi–xii. Huxley, in issuing his collected essays in 1893, assured his readers that there was nothing of substance that needed altering, conceding this was either evidence of the 'soundness of his opinions' or 'of my having made no progress in wisdom for the last quarter of a century', Huxley, *Method and Results, Essays* (1893), vi. There is plenty of evidence of long-term stability of belief, of the sort which allowed W.C. McIntosh to publish in 1909, with only a few interpolations, the anti-Darwinian lecture he had given in Perth forty years earlier. W.C. McIntosh, 'The Darwinian Theory in 1867 and Now', *The Zoologist* 4th ser., 13 (1909), 81–105.

46. Notwithstanding the view of Hull on Sedgwick in *Reception* (1973), which suggests that Sedgwick abandoned belief in special creation. As well as the material above, p. 134, contrast with Philips' comment that Sedgwick continued to oppose the Darwinian theory not just with 'a pen of steel but great use of his heavy hammer'. *Nature*, 6 February 1873. Compare with accounts of his mid-1860s lectures 'drawing tears from his audience' at the wonders of creation.

47. See Sedgwick to Frederick McCoy, 16 September 1872, images 656–60, McCoy Papers, A675, SLNSW.

48. Hinchliff, *God and History*, 110.

49. *Nature*, 26 September 1895.

50. John Tyndall letter in *The Times*, 4 May 1881 (though his wording was careful and not unambiguous, and accepted that Carlyle's opinions had 'taken their final set' before the *Origin* appeared).

51. *Samoa Times and South Seas Gazette*, 11 June 1881. Carlyle, Froude observed in an interview after his death, 'hated Darwinism', Raymond Blathwayt, *Interviews* (1893), 255. Carlyle was to reported to have described Darwinism as a 'philosophy fit for dogs' (Dawson, *Darwinism, Literature and Victorian Respectability*, 23, ascribes this to Tyndall's Belfast Address, but see David R. Sorensen, '"Symbolic Mutation": Thomas Carlyle and the Legacy of Charles Darwin in England', *Carlyle Studies Annual* 25 (2009), 61–81); see also W.A. Knight, *Retrospects* (1906), 4–5, 11. There was a little cottage industry in retelling Carlylean anti-Darwinian vehemence, or in printing snippets in the press; as in, for example, Frank Harris in *English Review* (February 1911), that the survival of the fittest was 'a cowardly, sneaky evasion', or the quotation printed by *Ardrossan and Saltcoats Herald* of Carlyle to a friend: 'A good sort of man is Darwin, and well-meaning, but with very little intellect. Ah, it's a sad, a terrible thing to see nigh a whole generation of men and women, professing to be cultivated, looking around in a pur blind [sic] fashion, and finding no God in this universe', reprinted in *Pembrokeshire Herald*, 19 January 1877.

52. Letter, 26 November 1892, printed Clodd, *Grant Allen*, 143–4.

53. Letter of Freeman to W.R.W. Stephens, in Glick, *What about Darwin*, 130. As Burgon told Frederick Le Gros Clark (22 November 1883), 'The Creation of man and of women is a matter of *express* Revelation...it is mere chaff and draff [sic] for the Scientist to approach such a matter with a weak theory, unproved and unprovable. He prates of what he knows nothing. Let him keep to his *well ascertained facts...*', E.M. Goulburn, *John William Burgon* (1892), II, 230, 247–9, and his letter to Richard Owen, 12 June 1887, cited in J.W. Gruber, and J.C. Thackray, *Richard Owen Commemoration: Three Studies* (1992). See Babington to F.J. Hanbury, Esq, 13 September 1887, and Babington to Rev. W. Hunt Painter, 18 April 1891, Anna Maria Babington, *Memorials, Journal and Botanical Correspondence of Charles Cardale Babington* (1897), 414, 438.

54. H. Pearson, *The Life of Oscar Wilde* (1946), 33; see also the account of an 1884 sermon in Freeman to W.R.W. Stephens, Trinity Sunday 1884, printed in Stephens, *Life and Letters of E.A. Freeman* (1895), II, 321–2; Poulton, 'The Influence of Darwinism on Entomology', *Entomologist's Record and Journal of Variation* 13 (1901), 75. Westwood reiterated his

CONCLUSION   439

warnings in his address as Honorary Life President of the Entomological Society in 1883; see *The Zoologist* 3rd ser., 7 (1883), 350. For Westwood see Clark, *Bugs*, 111–12, 121–3.

55. Robert Patterson, *The Errors of Evolution* (1885), Rawlinson, *The Testimony of the Truth of Scripture* (1898), 10.

56. Also J.F.W. Herschel, one of whose last acts was 'to busy himself about a manuscript collection of all the passages in his own writings where he had referred to the tokens of an intelligent Will in Nature', Pritchard, *Occasional Thoughts*, 131.

57. H.C. Watson to A.R. Wallace, 24 January 1881, ff.130–1, Wallace Papers, Add. Ms 46436, BL. Watson is prominent in Desmond, *Politics of Evolution*'s account of 1830s evolutionary thought, and claimed as a Darwinist in some biographies, but this is forcefully denied in the obituary notice in *Transactions of the Edinburgh Botanical Society*, 14 (1883), 300–2, including citation of *Topographical Botany* (1873–4); see also his letter to J.D. Hooker, 1 January [1868], DCP-LETT-5077F, where he notes that his 'conviction is nearly complete, either that there is something fallacious to be eliminated from it, or else a something important to be added to it', noting that Darwin's ideas 'refuse to harmonize' with current technical classifications in zoology and botany. Although he wrote positively to Darwin, 21 November 1859, his follow up was that accepting the key premisses of the *Origin* he was left with an outcome 'so inconceivable as a reality, that it must seem in itself an improbability amounting to an absurdity almost', Watson to Darwin, [3? January 1860], DCP-LETT-2636, Darwin Correspondence. Bill Jenkins, in *Evolution before Darwin: Theories of the Transmutation of Species in Edinburgh, 1804–34* (2019), 179–82, notes there was no real evidence that Watson developed transmutationist views.

58. Lester, *Lankester* cites a review of T.H. Morgan's *Critique of the Theory of Evolution* (1917), 90–1.

59. Owen, 'Our Origin as a Species', *Longman's Magazine* 1 (1883), 64–8.

60. Rupke, *Richard Owen*, 255–6, citing Owen to Spencer H. Walpole, 5 November 1882; see also Owen to George Burrows, 15 October 1881, Temple University Misc Collection, PC5 (1040), also cited by Rupke. For Owen, 'fiercely contest[ing] Darwin's theory of natural selection', in conversation as well as in print, see the recollections in John Willis Clark, *Old Friends at Cambridge and Elsewhere* (1900), 393.

61. Edward R. Pease, *A History of the Fabian Society* (1925), 17.

62. Charles Bree, *An Exposition of the Fallacies in the Hypothesis of Mr Darwin* (1872), 290–1.

63. This explains the extraordinary indifference with which the scientific journalist James Samuelson seems to have treated the state of evolutionary thought in his later years. Despite his career as one of the most active scientific popularisers of the 1860s and 1870s, he commented in his autobiography that Darwin in his life 'had failed to trace the formation of a new species, and, although I have not of late given much attention to the subject, I have not noticed that any of his disciples has been more successful', J. Samuelson, *James Samuelson's Recollections* (1907), 149.

64. *Leeds Mercury*, 8 February 1878.

65. In 1871, having not read Darwin, he had 'gathered from scientific men that his view cannot ever claim to be more than a hypothesis', but he was prepared by 1889 to agree that 'evolution...is quite established by quiet acceptance', Creighton to W.S. Lilly, 5 August 1889, printed in Louise Creighton, *Life and Letters of Mandell Creighton* (1904), 410.

66. Liddon, *St Thomas*, 28. As he put it in 1870, evolution if proven would not be inconsistent either with an initial act of creation or plan and purpose, '"Evolution" from a Theistic point of view, is merely our way of describing what we can observe of God's continuous action upon the physical world', Liddon, *Some Elements of Religion* (1870), 56. At the same time, Liddon was particularly anxious not to abandon entirely the emotional consolations of natural theology; he praised Müller's *Origin and Growth of Religion* for its ability to remind of 'the religious functions of nature,—the services which nature has actually rendered to man in fostering and guiding the sense of an Invisible World, and in suggesting, however dimly, its awful author and ruler', Liddon to Müller, 14 January 1879, ff.78–80, Ms Eng. c.2806/1, Max Müller Papers, Bodleian.

440 DARWINISM'S GENERATIONS

67. Dear, *Are These Things So?*, 20. We should not treat this as an attitude imputed by the young of the old; it was equally an admission very commonly made by those who felt themselves slipping from the leading edge of advances in knowledge; as Hewett C. Watson told George Henslow, 'The enfeebled brain of 74 may retain knowledge of the past, and apply it more or less clearly; but will be pretty sure to fail if attempting still any onward progress', Watson to Mr [G.H.] Henslow, 27 September 1878, Asa Gray Correspondence, Wa–Wh, Harvard.

68. Judd, *Coming of Evolution*, 140.

69. *Stroud Journal*, 13 May 1882.

70. See Mark A. Largent, 'The So-Called Eclipse of Darwinism', *Transactions of the American Philosophical Society* 99.1 (2009), 3–21, which explores another way in which this might be associated with generational dynamics.

71. Desmond, *Evolution's High Priest*, 248. For example, Huxley made little attempt to engage with the more recent work of Weismann etc., and where he did, his notes suggest that Weismann was approached via Owen's *Parthenogenesis*, Desmond, *Evolution's High Priest*, 314. An equally striking case is provided by Lionel Beale, whose writings in the 1890s remain fixated on Huxley and the debates of the years around 1870 over materialism.

72. See also Harold Begbie, 'A Vision of Thought', *Daily Mail*, 28 March 1904.

73. For example, the unidentified lady who, according to Mary St. Leger Kingsley in 1884, 'belongs to an older and more dignified generation, a generation which knew not Darwin, and regarded us not as human animals, but as very wonderful creatures indeed, for whom the whole universe was made, this material world to supply us with a temporary, and heaven with an eternal, resting-place', Clifford Harrison, *Stray Records, or Personal and Professional Notes* (1892), I, 248–9.

74. For Thomas Mozley in 1891, evolution was still a 'baseless and barren philosophy', *The Son* (1891), 27. Murphy was keen to make clear he was not a Darwinian in 1893, *JTVI* 26 (1893), 75.

75. Ramsay to A. Geikie, 30 July 1864, in Geikie, *Memoir of Sir Andrew Crombie Ramsay* (1895), 282.

76. For example, Edward White to Stokes, 30 September 1894, MS Add. 7656/1W/531, Stokes Papers, CUL.

77. Leonard Darwin to EBP, 18 December 1929, Box 3, Poulton Papers, OMNH. For others see H.E. Armstrong's reading of Shaw's *Back to Methuselah* in terms of a 'reversion to Lamarck everywhere', Armstrong to Clodd, 14 August 1921, Transcripts, Box 1, Edward Clodd Papers, BC MS 19c Clodd, Brotherton Library.

78. Keith, *Autobiography*, 121.

79. Quoted in W.B. Provine, 'England', in E. Mayr and W.B. Provine, eds., *The Evolutionary Synthesis* (1980), 330. Poulton's 'Insect Adaptation as Evidence of Evolution by Natural Selection', published in 1938, ignored not just Mendelism, but the later work of R.A. Fisher, and sought to present observations on insect life which 'appeared to demand a Darwinian and exclude a Lamarckian interpretation', while admitting that his conclusions were essentially those of H.W. Bates in 1861.

80. Hughes to George Macmillan, 30 March 1894, MS Add. 8069/H86, Kidd Papers, CUL. Hence in the 1880s while Patrick Geddes was attempting to wrestle with the explanatory limits of natural selection, with a view to developing evolutionary theory, A.R. Wallace was more concerned that Geddes should present his work explicitly as supplemental, for fear of the response of Wallace's mid-Victorian contemporaries; anything that looked like an attempt to usurp its place, he warned, would merely lead outright opponents like the Duke of Argyll to parade Geddes 'as finally smashing Darwinism, while naturalists will be apt to ignore what is sound in your views', Wallace to Geddes, 2 February 1889, T-GED/9/35, Geddes Papers, Strathclyde University.

81. Geldart, 'Presidential Address', 1–11.

82. Berkeley was claimed as 'Darwinian' by Barton, though his limited endorsement is visible in his 'Address as President of the Biological Section of the British Association, Norwich 1868', *Quarterly Journal of the Microscopical Society* ns 8 (1868), 233–9, 238, and in his

CONCLUSION   441

sympathy for the anti-Darwinian ideas of Milnes-Edwards; see comments of MJB (identified as Berkeley in *Gardeners' Chronicle*, 28 December 1889), *Gardeners' Chronicle*, 21 May 1870.

83. See Kuper, *Reinvention of Primitive Society*, 69.

84. See *Birmingham Daily Post*, 23 August 1869, *Athenaeum*, 4 September 1869; for Grant see Dixon, *Altruism*, 135.

85. Northumberland, 'Religion and Physical Science', *TNC* 52 (1902), 951–6.

86. See Layard to [Henry] Sutton, 16 January 1896, 1 February 1896, Brigg Collection, Bg 122, University of Nottingham Archives; also a forthrightly anti-Darwinian paper on 'Reversions' in the Anthropology section of the BA in 1890; see *Glasgow Herald*, 6 September 1890. Layard told Welby in 1908 that she was 'still full' of evolution, though wanting to rename the 'process of differentiation', from evolution to 'the realising of improved potentialities'; see Layard to Welby, 4 February 1908, 1970-010/009(07), Lady Victoria Welby Papers, York University, Canada.

87. Take the zoologist E.S. Goodrich (1868–1946), described by his obituarist as 'a staunch upholder of Darwinian views on organic evolution', *Biographical Memoirs of Fellows of the Royal Society* 15 (1947), 477–90. Here close personal relations and opportunities for professional advancement may well have come into play; as for Lankester's favourite pupil and then Weldon's demonstrator, E.B. Ford, 'Some Recollections', 336, Radick, *Disputed Inheritance*, 381–19.

88. DWT to James Johnstone, 2 July 1917, ms19951, D'Arcy Wentworth Thompson Papers, St Andrews.

89. Turner, *Science and Religion*, 80–1.

90. Mavor, *Windows*, 255.

91. See recollections of Beeby Thompson (1848–1931), *Northampton Mercury*, 22 March 1918.

92. Laycock positioned himself as a pre-Darwinian evolutionist, brought to evolutionary ideas by teachings of Robert E. Grant and the writings of Lamarck (see his 'Reflex, Automatic and Unconscious Cerebration', *Journal of Mental Science* 21 (1875), 477–98, especially 484–5); significantly, he was not convinced of the common inheritance of man and apes, unless apes were a degeneration of one species of humanity.

93. Although he largely sidestepped the critical questions of the mechanisms of evolution, Laing is interesting because his works find their ways onto early twentieth-century rationalist reading lists, for example see J.B. Perry, ed., *A Hubert Harrison Reader* (2001), 127; Lightman, 'Spencer's British Disciples' presents Laing as much more a Spencerian than a Darwinian, and as breaking with Huxley as the latter's ideas diverged from Spencer's.

94. See above, pp. 176–77; Howard Vincent Knox (1868–1960) to Schiller, 19 November 1904, Folder H.V. Knox #1, Box 1, F.C.S. Schiller Papers, UCLA.

95. McIntosh, 'The Darwinian Theory'; Charles Pritchard to Stokes, 24 January 1884, MS Add. 7656/1P/726, Stokes Papers, CUL.

96. Quoted in G.H.O. Burgess, *The Curious World of Frank Buckland* (1967), 193; in the preface to his posthumously published *The Natural History of British Fishes*, completed only days before his death, Buckland affirmed his commitment to a single complete act of Creation.

97. S.E.B. Bouverie-Pusey, *Permanence and Evolution* (1882). A figure like Spurgeon who was an established preacher by twenty was more likely to have his opinion fixed earlier in life. Stanley lectured in Melbourne in 1892; reports suggest he was clear in his refusal to accept Darwinism: for every argument that the Darwinian system produces, he affirmed he could provide a countervailing one. Another was Ambrose Fleming (1849–1945), forty-one years Professor of Electrical Technology at UCL and first president of the Evolution Protest Movement, whose Idealism seems to have taken on a particularly strong religious flavour; see Livingstone, *Adam's Ancestors*, 202–8.

98. See Jones, 'Darwinism in Ireland'; Barry Brundell, 'Catholic Church Politics and Evolution Theory', *BJHS* 34 (2001), 81–95. Fr John Gerard (1840–1912), whose *Essays in Un-Natural History* (1900) presented a deep scepticism over evolutionary theory, are just two examples amongst many. For Gerard, see also *Evolutionary Philosophy and Common*

Sense (1902), which adopts mid-Victorian arguments seasoned with Edwardian rhetorics of the abandonment of 'Darwinism'. Gerard was willing to accept (or at least not to dispute) the fact of evolution, but challenged the exclusive significance of natural selection and continued to lament the status afforded Darwin; see his letter, *Gardeners' Chronicle*, 25 July 1908. Another is Gerald Molloy (1834–1906), Professor of Theology at St Patrick's College, Maynooth, who largely ignores Darwinism and evolution in his printed writings, but whose *Geology and Revelation* (1872) argued that the geological record gave detailed support for evolution, and was impatient with those who tried to argue away the fossil record; see T. Duddy, *A History of Irish Thought* (2002), 257–9.

99. *Northern Whig*, 26 December 1904.
100. Ward, *Men and Matters*, 308.
101. *Folkestone Express*, 11 March 1871, 31 March 1888, *Folkestone Herald*, 20 October 1928.
102. Lang to Gosse, 13 October [1885], Letters to Gosse, LAN, BC c19 Gosse, Brotherton Library.
103. Andrew Lang to Grant Allen, 'Aug 5' [1885], SUNY Library, https://sites.google.com/site/petermortonswebsite/home/grant-allen-homepage/primary-sources [accessed 24 November 2019]. Even so, Lang asked Allen to tone down his anti-Darwinian 'orgies of the biological Thermidor' in the interests of not offending readers who still hoped to combine a Darwinian interest with their religion.
104. See discussion in Diedrick, *Mathilde Blind*, 214–17. For Caird see her 'The Survival of the Fittest', *South Place Magazine* 4 (1899), 97–101, 113–17, and Angelique Richardson, '"People Talk a Lot of Nonsense about Heredity": Mona Caird and Anti-Eugenic Feminism', in Richardson and Chris Willis, eds., *The New Woman in Fiction and Fact: Fin de Siecle Feminisms* (2002), 183–211.
105. Gates, *Kindred Nature*, 156. In fact he appeared in her lists of 'great geniuses' who 'had the feminine soul very strongly developed in them' (*Awakening of Women* (rev. ed., 1908), 60). For Swiney, see *The Cosmic Procession* (1906), 57.
106. Burrows, 'The Future of Women', *South Place Magazine* 14 (March 1909), 86.
107. Diary entry, 27 January 1898, in Florence Hardy, *Thomas Hardy: Later Years*, quoted in Beer, *Darwin's Plots*, 244.
108. R. Koselleck, *Zeitschichten: Studien zur Historik (mit einem Beitrag von Hans Georg Gadamer)* (2000). Koselleck's ideas are primarily available to English-speaking scholars in two collections, *The Practice of Contemporary History* (2002) and *Sediments of Time* (2018).
109. Koselleck, 'Time and History', *The Practice of Contemporary History*, 100–14, 110; Koselleck 'Wozu noch Historie', cited and translated in Niklas Olsen, *History in the Plural: An Introduction to the Work of Reinhart Koselleck* (2012), 218.
110. S. Franzel and S.-L. Hoffman, 'Introduction', *Sediments of Time*, xiii.
111. Koselleck, '*Historik* and Hermeneutics', *Sediments of Time*, 41–59, 45; Koselleck, 'Goethe's Untimely History', ibid., 60–76, 68.
112. T. Ingold, *The Rise and Fall of Generation Now* (2023).
113. Although I acknowledge the stimulus of Nicholas Dames' hints towards the potential distinction between the 'rhythmic-melodic' and the 'vertical-harmonic' novel, in *The Physiology of the Novel* (2007), 10–11.
114. Rita Felski, 'Context Stinks', *New Literary History* 42.2 (2011), 474–91; see also the discussion in V. Jackson, *On Periodization: Selected Essays from the English Institute* (2010).
115. Archibald Geikie, 'Introduction', *Life of Alfred Newton*, ix.

The Reception of Darwinian Evolution in Britain, 1859–1909: Darwinism's Generations. Martin Hewitt, Oxford University Press. © Martin Hewitt 2024. DOI: 10.1093/9780191982941.003.0007

APPENDIX

Prosopography

Revolutionary

Mary Fairfax Somerville (1780–1872), George William Featherstonhaugh (1780–1866), John Bird Sumner (1780–1862), David Brewster (1781–1868).

Late Georgians

Sir Benjamin Brodie (1783–1862), John Crawfurd (1783–1868), Sir William Lawrence (1783–1867), Adam Sedgwick (1785–1873), George Wilkins (1785–1865), John G. Marshall (1786–1880), Marion Bell (née Shaw) (1787–1876), Richard Whately (1787–1863), Sir Henry Holland (1788–73), Francis Palgrave (1788–1861), William Clark (1788–1869), Edward Sabine (1788–1883), Charles Forster (1788–1871), Jonathan Couch (1789–1870), Saxe Bannister (1790–1877), Henry Hart Milman (1791–1868), William Sharp Macleay (1792–1865), Roderick Murchison (1792–1871), John F.W. Herschel (1792–1871), John Kelly (1791–1869), Thomas Bell (1792–1880), William Howitt (1792–1879), William Hopkins (1793–1866), Robert E. Grant (1793–1874), Walter Cooper Dendy (1794–1871), George Grote (1794–1871), William Whewell (1794–1866), Charles Daubeny (1795–1867), Thomas Carlyle (1795–1881), George Twemlow (1795–1877).

Early Victorians

Baden Powell (1796–1860), John S. Henslow (1796–1861), Robert Wight (1796–1872), Charles Lyell (1797–1875), Francis Close (1797–1882), James Scott Bowerbank (1797–1877), Robert Christison (1797–1882), Charles Girdlestone (1797–1881), Marianne Thornton (1797–1887), Thomas Griffith (*c.*1797–1883), John Benn Walsh, first Baron Ormathwaite, (1798–1881), Robert Mackenzie Beverley (1798–1868), Searles Valentine Wood (1798–1880), Charles Pinhorn Farrar (1798–1877), Alexander Thomson (1798–1868), Henry S. Boase (1799–1883), Charles Austin (1799–1874), John Lindley (1799–1865), Robert Dunn (1799–1877), George Walker Arnott (1799–1868), William Wilson (1799–1871), Edward B. Pusey (1800–82), George Bentham (1800–84), Richard Bethell, 1st Baron Westbury (1800–73), John Phillips (geologist) (1800–74), Leonard Jenyns (1800–93), Richard Greswell (1800–81), Charles Neaves (1800–76), John G. Macvicar (1800–84), Edwin Lees (1800–87), Thomas Story Spedding (1800–70), Sir Richard Rawlinson Vyvyan (1800–79), Sir William Jardine (1800–74), Charles W. Peach (1800–86), Grantley Fitzhardinge

444 APPENDIX

Berkeley (1800–81), John Edward Gray (180075), Derwent Coleridge (1800–83), John Evelyn Denison (1800–73), William Wood, Lord Hatherley (1801–81), John Henry Newman (1801–90), Henry Arthur Woodgate (1801–74), Richard Hill Sandys (1801–92), Joseph Barnard Davis (1801–81), Edward Newman (1801–76), Jane Welsh Carlyle (1801–66), William Gresley (1801–76), Robert Chambers (1802–71), Harriet Martineau (1802–76), Mungo Ponton (1802–80), Hugh Miller (1802–56), William Sharpey (1802–80), James Buckley (1801/2–83), Hensleigh Wedgwood (1803–91), James Iverach (1803–82), Archibald Boyd (1803–83), Miles Joseph Berkeley (1803–89), James Brooke (1803–68), John Longmuir (1803–83), James Challis (1803–82), Hewett C. Watson (1804–81), Richard Owen (1804–92), George Augustus Rowell (1804–92), Thomas Gough (1804–80), John Gould (1804–81), William Bennett (1804–73), Sir William Thomas Denison (1804–71), Charles Brooke (1804–79), John Scouler (1804–71), William Carus (1804–91), John O. Westwood (1805–93), George Anthony Denison (1805–96), Thomas Cooper (1805–92), Samuel Wilberforce (1805–73), John Stuart Mill (1806–73), James Martineau (1805–1900), Francis William Newman (1805–97), E.D. Girdlestone (1805–84), John Tollemache (1805–90), John Epps (1805–69), Charles Hippuff Bingham (1805/6–75), William Chapman Hewitson (1806–78), Thomas Mozley (1806–93), James Bryce (1806–77) (geologist), Neil McMichael (1806–74), Thomas Hutton (1806–75), William Bernard Ullathorne (1806–89), Sir Philip Egerton (1806–81), Alexander Anderson (1806/7–84), John Eliot Howard (1807–83), Charles Perry (1807–91), George Busk (1807–86), William MacIlwaine (1807–85), Bonamy Price (1807–88), John Cumming (1807–81), John Bullar (1807–67), Benjamin Waterhouse Hawkins (1807–94), Joshua Hughes (1807–89), Charles C. Babington (1808–95), John Hutton Balfour (1808–84), Hugh Falconer (1808–65), James Spedding (1808–81), Robert Main (1808–78), Cardinal Henry Manning (1808–92), Edward Vivian (1808–93), William Honyman Gillespie (1808–75), George Osborn (1808–91), Robert Garner (1808–90), Thomas Wharton Jones (1808–91), John Murray III (1808–92), James Michell Winn (1808–1900), John Sutherland (1808–91), Robert Stodart Wyld (1808–93), Charles James Griffith (1808–63), John Stuart Blackie (1809–95), Thomas Campbell Eyton (1809–80), Margaret Gatty (1809–73), Daniel Moore (1809–99), John Gwyn Jeffreys (1809–85), Charles Pritchard (1808–93); Willliam Rathbone Greg (1809–81), William Ewart Gladstone (1809–98), Allen Thomson (1809–84), Edward FitzGerald (1809–83), James David Forbes (1809–68), Samuel Rowles Pattison (1809–1901), William King (1809–86), Thomas Wright (1809–84), John Henry Pratt (1809–71), Charles J.F. Bunbury (1809–86), Joseph Butterworth Owen (1809–72), Philip Henry Gosse (1810–88), Edward Duke (1814–95), John Hensley Godwin (1809–89), William Keddie (1809–77), Thomas Gordon Hake (1809–95), Thomas Ragg (1809/10–81), Francis Orpen Morris (1810–93), Thomas Rawson Birks (1810–83), Thornton Hunt (1810–73), George Robert Waterhouse (1810–1888), Henry Christie (1810–65), Sir William Armstrong (1810–1900), John Brown (1810–82), Henry Wilkinson Cookson (1810–76), Nicholas Whitley (1810–91), Ebenezer Cobham Brewer (1810–97), Charles Robert Bree (1811–86), Martin Tupper (1810–89), James McCosh (1811–94), Robert Dick (1811–66), William Brown Galloway (1811–1903), James Moncreiff, 1st baron Moncreiff (1811–15), William Henry Harvey (1811–66), James Bateman (1811–97), Joseph Beete Jukes (1811–69), Andrew Dickson Murray

APPENDIX 445

(1812–78), Robert Browning (1812–89), Joseph Prestwich (1812–96), Samuel Laing (1812–97), Roundell Palmer (1812–95), William Penman Lyon (1812–77), Thomas Laycock (1812–76), Henry Cotterill (1812–86), William Josiah Irons (1812–83), George Rawlinson (1812–1902), Sara Hennell (1812–89), George Dickie (1812–82), William Bence Jones (1812–82), George James Allman (1812–98), R.G. Latham (1812–88), Edward Hoare (1812–94), Giuseppe Gagliardi (1812–81), Henry Barne (1812/13–86), John Crosse Brooks (1812/3–97), J.W. Burgon (1813–88), William B. Carpenter (1813–85), James Bowling Mozley (1813–78), George Gilfillan (1813–78), William Thomas Thornton (1813–80), Francis Polkinghorne Pascoe (1813–93), Emily Sarah Tennyson (1813–96), Henry Reeve (1813–95), William Willmer Pocock (1813–99), Osborne Gordon (1813–83), Alfred Gatty (1813–1903), Thomas Gribble (1813–c.81), William Francis Wilkinson (c.1813–c.90), John Kennedy (1813–1900), Septimus Redhead (1814–1900).

Mid Victorians

David Page (1814–79), David T. Ansted (1814–80), Aubrey Thomas De Vere (1814–1902), Henry Hugh Higgins (1814–93), William Fishburn Donkin (1814–69), Andrew Crombie Ramsay (sometimes spelled Ramsey) (1814–91), Sir James Paget (1814–99), James Buckman (1814–84), Edwin Lankester (1814–74), James Hamilton (1814–67), Alexander Goss (1814–72), Francis Roubillac Conder (1815–89), Richard Church (1815–90), Hyde Clarke (1815–95), John R. Leifchild (1815–89), Andrew Common (1815–96), Henry Wentworth Acland (1815–1900), Peter Bellinger Brodie (1815–97), William Crawford Williamson (1816–95), Charles Adolphus Row (1816–94), Edmund Beckett (1816–1905), Daniel Wilson (1816–92), Charles P. Reichel (1816–94), John Charles Ryle (1816–1900), Francis Henry Laing (1816–89), Samuel Garratt (1816–1906), Robert Milman (1816–76), Walter Mitchell (1816/17–74), Charles Morrison (1816/17–1909), Joseph Dalton Hooker (1817–1911), Thomas Smith (1817–1906), Thomas Davidson (1817–85), Brownlow Maitland (1817?–1902), George Henry Lewes (1817–78), Benjamin Jowett (1817–93), Rowland Williams (1817–70), Sir Richard Strachey (1817–1908), John James Drysdale (1817–92), Osmond Fisher (1817–1914), Andrew Whyte Barclay (1817–84), Bourchier Wrey Savile (1817–88), John Clavell Mansel-Pleydell (1817–1902), Richard Spruce (1817–93), Frederick McCoy (1817–99), Thomas Thomson (1817–78), William Allen Miller (1817–70), Michael Connal (1817–93), Alexander Bain (1818–1903), J.A. Froude (1818–94), John Ball (1818–89), George Prothero (1818–94), Harvey Goodwin (1818–91), William Guest (1818–91), Edward Meyrick Goulburn (1818–97), William Samuel Symonds (1818–87), Henry James Slack (1818–96), James Prescott Joule (1818–89), George Johnson (1818–96), R. Payne Smith (1818–95), Alfred Smee (1818–77), John Cairns (1818–92), John Griffith (1818/19–85), Edwin Brown (1818/19–76), James Fraser (1818–85), Charles Kingsley (1819–75), George Gabriel Stokes (1819–1903), George Eliot (1819–80), Alexander Macmillan (1818–96), Philip Freeman (1818–75), Charles John Ellicott (1819–1905), William Thomson (Archbishop of York) (1819–90), John Ruskin (1819–1900), Henry Cooper Key (1819–79), George Salmon (1819–1904), Henry Jeffs (1819–88), James

446 APPENDIX

Reddie (1819–71), Ella Haggard (1819–89), John Plant (1819/20–94), John Tyndall (1820–93), Herbert Spencer (1820–1903), John W. Dawson (1820–99), John Laws Milton (1820–98), Eustace R. Conder (1820–92), James Hutchison Stirling (1820–1909), Florence Nightingale (1820–1910), George Murray Humphry (1820–96), Charles Swainson (1820–87), John Duns (1820–1906), Benjamin Gregory (1820–1900), Isaac Todhunter (1820–84), Henry Thompson (1820–1904), Thomas Hood Cockburn-Hood (1820–89), Friedrich Engels (1820–94), J.W. Salter (1820–69), Willoughby Jones (1820–84), Edwin Paxton Hood (1820–85), Samuel William King (1821–68), Henry Thomas Buckle (1821–62), Felicia Mary Frances Skene (1821–99), James Harrison Rigg (1821–1909), William Connor Magee (1821–91), Gilbert Rorison (1821–69), Samuel Haughton (1821–97), Frederick Temple (1821–1902), James Croll (1821–90), Richard Howse (1821–1901), Samuel Pickworth [or Peckworth] Woodward (1821–65), Edward Thring (1821–87), Edward Hayes Plumptre (1821–91), William Baker (1821–1908), Arthur Cayley (1821–95), Henry J.S. Maine (1822–88), Matthew Arnold (1822–88), Frances Power Cobbe (1822–1904), Thomas Vernon Wollaston (1822–78), Henry Baker Tristram (1822–1906), James Hinton (1822–75), Hugh Mitchell (1822–94), Eliza Lynn Linton (1822–98), John Lynn (1822–89), Thomas Karr Callard (1822–89), Samuel Joseph Mackie (1823–1902), John Jenner Weir (1822–94), Francis Galton (1822–1911), Charles Bland Radcliffe (1822–89), John Gibbs (1822–92?), George Cupples (1822–91), Thomas Hughes (1822–96), William Purchas (1822–1903), George Dawson Rowley (1822–78), Alfred Russel Wallace (1823–1913), Max Müller (1823–1900), William K. Parker (1823–90), John Struthers (1823–99), Emelia Russell Gurney (1823–96), Goldwin Smith (1823–1910), George Douglas Campbell, 8th Duke of Argyll (1823–90), John Tulloch (1823–86), Joseph Crompton (1823/24–78), Thomas Spencer Baynes (1823–87), John Braxton Hicks (1823–97), Robert Arthington (1823–1900), William Bragge (1823–84), Godfrey Thring (1823–1903), John Lucas Tupper (1823?–79), William Bonner Hopkins (1823–90), James Spilling (1823–97), William Sweetland Dallas (1824–90), William Thomson (Lord Kelvin) (1824–1907), Frederic Bateman (1824–1904), Charles Elam (1824–89), George Herbert Curteis (1824–94), William T. Gairdner (1824–1907), William Allingham (1824–89), William Alexander (1824–1911), J.W. Ogle (1824–1905), Eustace Murray (1824–81), Samuel Wainwright (1824–99), Thomas Henry Huxley (1825–95), Alexander Robertson (1825–93), Henry W. Bates (1825–92), Samuel Osborne Habershon (1825–89), Mordecai Cubbitt Cooke (1825–1914), Henry Reynolds (1825–96), Peter Price (1825–92), George Porter S.J. (1825–89), Brooke Foss Westcott (1825–91), Anthony Wilson Thorold (1825–95), Frederick Thompson Mott (1825–1908), Thomas Francis L'Anson (1825–98), Thomas Alexander Goldie Balfour (1825–95), Edward Frankland (1825–99), James Backhouse (1825–90), Edward Joseph Lowe (1825–1900), Edward Henry Bickersteth (1825–1906), J.E.B. Mayor (1825–1910), Frank Buckland (1826–81), Robert Rainy (1826–1906), Franklin George Evans (1826–1904), Thomas Chenery (1826–84), J. Llewellyn Davies (1826–1916), Alexander Grant (1826–84), Samuel A. Steinthal (1826–1910), Richard Holt Hutton (1826–97), Adam Milroy (1826–99), Cuthbert Collingwood (1826–1908), Abraham Lichtenberg (1826–?), Samuel W. North (1826–94), Henry William Crosskey (1826–93), John Vertue (1826–1900), Samuel Kinns (1826–1903), St George Jackson Mivart (1827–1900), Joseph John Murphy (1827–94), John Ferguson McLennan (1827–81), William Binns

APPENDIX 447

(1827–1901), Lydia Ernestine Becker (1827–90), Emily Jane Pfeiffer (1827–90), Alexander Balloch Grosart (1827–99), Barbara Bodichon (1827–91), Richard St John Tyrwhitt (1827–95), William Holman Hunt (1827–1910), John Hall Gladstone (1827–1902), Daniel Hack Tuke (1827–95), John Bradford Whiting (1827–1914), Henry Fletcher Hance (1827–86), Frederic Kitton (1827–96), Lionel Smith Beale (1828–1906), Charles Voysey (1828–1912), Fenton J.A. Hort (1828–92), Charles Meynell (1828–82), Caroline Haddon (1827–99), Charles William Boase (1828–95), Harrison Branthwaite (1828–90), John Burdon Sanderson (1828–1905), Sir Edward Fry (1827–1918), Augustus Henry Lane-Fox Pitt Rivers (1827–1900), Horace Benge Dobell (1828–1917), Edward J.J. Browell (1827/8–1914), Henry Nuttall (1828–97), Balfour Stewart (1828–87), Eleanor Ormerod (1828–1901), Gavin Carlyle (1828–1919), Octavius Pickard-Cambridge (1828–1917), Gerald Massey (1828–1907), Anne Gilchrist (1828–85), Jonathan Hutchinson (1828–1913), Joseph Lightfoot (1828–89), John Stores Smith (1828–93), James Lamont (1828–1913), Matthew Forster Heddle (1828–97), William Kennedy Moore (c.1828–1905), Margaret Bell Alder (1828/9–1902), Edmund Botelier Chalmer (1828–83), George Rolleston (1829–81), James Tait (1829–99), Alfred Newton (1829–1907), Philip Lutley Sclater (1829–1913), Edward Hull (1829–1917), Henry P. Liddon (1829–90), James Fitzjames Stephen (1829–94), Edward King (1829–1910), James Samuelson (1829–1918), Thomas Goadby (1829–89), George Thear (1829–1918), James MacGregor (1829–94), John Mugford Quicke (1829–1914), Charles Pierrepont Cleaver (later Peach) (1829–86), Francis Day (1829–89), William Robinson Clark (1829–1912), Thomas Wilkinson Norwood (1829–1908), George Phear (1829–1918), Joseph Timbrell Fisher (1829–83).

High Victorians

William Carruthers (1830–1922), Charles Wyville Thomson (1830–82), Charles Clement Coe (1830–1921), David W. Simon (1830–1909), Thomas Archer Hirst (1830–92), Robert Henry Codrington (1830–1922), Marianne North (1830–90), Eliza Brightwen (1830–1906), Peter Bayne (1830–96), William Walter Roberts (1830–1911), Sidney Biddle (1830–1911), Brooke Herford (1830–1903), John Whitaker Hulke (1830–95), John Brown Paton (1830–1911), Frank H. Hill (1830–1910), Peter Guthrie Tait (1831–1901), Charles Barnes Upton (1831–1920), William Henry Flower (1831–99), James Knowles (1831–1908), Thomas Hodgkin (1831–1913), Frederic Harrison (1831–1923), George Charles Brodrick (1831–1903), Henry Bradshaw (1831–86), Samuel Hawksley Burbury (1831–1911), Herbert D. Geldart (1831–1902), Edmund Yates (1831–94), Henry George Day (1831–1900), Frederick Merrifield (1831–1924), Neville Goodman (1831–90), William Wynne Peyton (1831–1924), Henry Gyles Turner (1831–1920), Isa Knox (1831–1903), T. Wickham Tozer (1832–1908), Sir William Turner (1832–1916), Leslie Stephen (1832–1904), J. Allanson Picton (1832–1910), Thomas Belt (1832–78), Jane Hume Clapperton (1832–1914), Henry Seebohm (1832–95), John Laidlaw (1832–1906), E.B. Tylor (1832–1917), Herbert Alfred Vaughan (1832–1903), George Stewardson Brady (1832–1921), Lorimer Fison (1832–1907), Anne Walbank Buckland (1832–99), James A. Cotter Morrison (1832–88), Henry Sturt (1832–1922), James Hurd Keeling

448 APPENDIX

(1832–1909), Edwin Arnold (1832–92), George Stewardson Brady (1832–1921), William Salmond (1832/3–?), Frederick Arnold (1833–91), Thomas G. Bonney (1833–1924), Elizabeth Wolstenholme-Elmy (1833–1918), Maxwell T. Masters (1833–1907), Walter Onions Purton (1833–92), Fleeming Jenkin (1833–85), Hugh Macmillan (1833–1903), James Hunt (1833–69), Frederic Seebohm (1833–1916), Henry Fawcett (1833–84), Edward Wrench (1833–1912), Edmund Knowles Muspratt (1833–1923), Joseph Goddard (1833–1911), Augustus Henry Keane (1833–1912), Horace Waller (1833–96), Alfred William Bennett (1833–1902), George Howell (1833–1910), Falconer Larkworthy (1833–1924), Isaac Whitwell Wilson (1833–81), Charles Egerton Fitzgerald (1833–98), John Lubbock (1834–1913), Sabine Baring-Gould (1834–1924), Gerald Molloy (1834–1906), Charles Haddon Spurgeon (1834–92), John James Lias (1834–1923), John Porritt (1834–1904), Francis Peek (1834–99), William G. Wheatcroft (1834–93), James Bell Pettigrew (1834–1908), George Macloskie (1834–1920), Sydney Thelwall (1834–1922), William Morris (1834–96), Compton Reade (1834–1909), Alexander Mair (1834–1911), William Durham (1834–93), George du Maurier (1834–96), John Watts (1834–66), Samuel Edward Peal (1834–97), Samuel Butler (1835–1902), Frederic William Harmer (1835–1923), Edward Caird (1835–1908), George Henslow (1835–1925), Henry Maudsley (1835–1918), George Birkbeck Hill (1835–1903), James Franklin Fuller (1835–1924), Mary Elizabeth Braddon (1835–1915), Percy Wyndham (1835–1911), William Fiddian Moulton (1835–98), James Maurice Wilson (1836–1931), Michael Foster (1836–1907), Archibald Geikie (1835–1924), John Clelland (1835–1924), Robert Munro (1835–1920), William Stanley Jevons (1835–82), Albert Venn Dicey (1835–1922), Thomas Welbank Fowle (1835–1903), John Hughlings Jackson (1835–1911), Alfred Comyn Lyall (1835–1911), George St Clair (1836–1908), Frederick Wollaston Hutton (1836–1905), Joseph Norman Lockyer (1836–1920), John Magens Mello (1836–1903), William Page Roberts (1836–1928), Ellice Hopkins (1836–1904), Joseph Reay Greene (1836–1903), Robert Swinhoe (1836–77), John Clifford (1836–1923), Henry Mason Bompas (1836–1909), Henry Wace (1836–1924), Walter Besant (1836–1901), Samuel Smith (1836–1906), John Urquhart (c.1836/39–1914), Robert Baker Girdlestone (1836–1923), John Snaith (1836–1923), Percy Greg (1836–89), Edward Henry Winfield (1836–1922), James Gurnhill (1836–1928), William Knight (1836–1916), George Newman (1836–1911), Arthur Philip Morres (c.1836–1900), Drummond Grant (c.1836–1909), William Boyd Dawkins (1837–1929), Edmund Symes-Thompson (1837–1906), H. Charlton Bastian (1837–1915), John Cuthbert Hedley (1837–1915), George Deane (1837–91?), Coutts Trotter (1837–87), Alexander Hay Japp (1837–1905), James Ross (1837–92), John Richard Green (1837–83), William Powell James (1837–85), John Ellor Taylor (1837–95), James Blaikie Keith (1837–1915), Victoria, Lady Welby (1837–1912), Benjamin Harrison (1837–1921), Charles Codrington Pressick Hobkirk (1837–1902), Richard Nicholls Worth (1837–96), Mary Elizabeth Lawrence (1838/9–?), Charles Swinhoe (1838–1923), James Philip Mansel Weale (1838–1911), John Morley (1838–1923), Henry Sidgwick (1838–1900), A.M. Fairbairn (1838–1912), Edwin Abbott Abbott (1838–1926), Charles Callaway (1838–1916), Lewis Wright (1838–1905), Hugh Reginal Haweis (1838–1901), Lionel Arthur Tollemache (1838–1919), Henry Ullyett (1838–98), William Carmichael McIntosh (1838–1931), James Bryce (1838–1922), W.E.H. Lecky (1838–1903), Winwood Reade (1838–75), William Lonsdale

APPENDIX   449

Watkinson (1838–1925), Robert Flint (1838–1910), Henry Colley March (1838–1916), Theodore Hook (1838–?), Murray A. Matthew (1838–1908), H.G. (Harry) Seeley (1839–1909), William Graham (1839–1911), Benjamin Thompson Lowne (1839–93), James Geikie (1839–1915), Phillip Henry Pye-Smith (1839–1914), Laura Forster (1839–1924), William Gooderidge (*c.*1839–?), James Iverach (1839–1922), George Jennings Hinde (1839–1918), Henry Heathcote Statham (1839–1924), John Neale Dalton (1839–1931), Katherine Euphemia Wedgwood (1839–1931), Edward Clodd (1840–1930), Robert Stawell Ball (1840–1913), David Sharp (1840–1922), Arabella Buckley (1840–1929), Alice Bodington (1840–97), John W. Judd (1840–1912), Thomas Hardy (1840–1928), John Rhys (1840–1915), Wilfrid Scawen Blunt (1840–1922), John Addington Symonds (1840–93), John Gerard (1840–1912), James Crichton-Browne (1840–1938), Charles Booth (1840–1916), Charles Carter Blake (1840–97), Lemuel Howard (1839/40–71), Frances Thompson (1840–1926), Thomas S. Clouston (1840–1915), Marmaduke Alexander Lawson (1840–96), M.B. Moorhouse (1840–1925), Mathilde Blind (1841–96), John Milner Fothergill (1841–88), Henry Lawson (1841–77), William Henry Hudson (1841–1922), Herbert McLeod (1841–1923), Emily Spender (1841–1922), Alfred Ewen Fletcher (1841–1915), John Gray McKendrick (1841–1926), George Francis Millin (*c.*1841–*c.*1910), Henry Elford Luxmoore (1841–1926), Thomas Whiteside Hime (1841/2–1920), Alfred Marshall (1842–1924), James Sully (1842–1923), Louis Compton Miall (1842–1921), Henry H. Howorth (1842–1923), George Croom Robertson (1842–92), Robert Brown (1842–95), Matthew Horace Hayes (1842–1904), Alexander William Bickerton (1842–1929), John Ross (1842–1915), James Macdonell (1841–79), Robert Edmonstone (*c.*1842–1914), Charles Robert Lloyd Engström (1842?–1922), Arthur Roope Hunt (1843–1914), William Turner Thiselton-Dyer (1843–1928), Edward Dowden (1843–1913), Edward Lee Hicks (1843–1919), Charles Thomas Druery (1843–1917), Alexander Henry Craufurd (1843–1917), Alfred William Benn (1843–1915), Frederic William Henry Myers (1843–1901), Mandell Creighton (1843–1901), Richard Acland Armstrong (1843–1905), James Ward (1843–1925), John Wordsworth (1843–1911), John Sidney Turner (1843–1920), James Oliver Bevan (1843–1930), William Woods Smyth (1843–1928), Arthur James Dadson (1843–1908), Thomas Lauder Brunton (1844–1916), Alexander Macalister (1844–1919), William Forsell Kirby (1844–1912), William Platt Ball (1844–1917), Henry Nottidge Moseley (1844–91), H. Alleyne Nicholson (1844–89), Andrew Lang (1844–1912), Philip Henry Wicksteed (1844–1927), James Orr (1844–1913), William Wallace (1844–97), Caroline Ann Martineau (1844–1902), William Chester Tait (1844–1928), Francis Reginald Statham (1844–1908), Edward Stuart Talbot (1844–1934), William Cole (1844–1922), Gerard Manley Hopkins (1844–89), Robert Bevan Hull (1844–1900), William Stewart Ross (1844–1906), Archibald Henry Sayce (1845–1933), William Kingdon Clifford (1845–79), Frederick Pollock (1845–1937), Lawson Tait (1845–89), Sybil Anne Tait (1845–1909), George Ferris Whidbourne (1845–1910), Walter Crane (1845–1915), Louisa Sarah Bevington (1845–95), Francis Ysidro Edgeworth (1845–1926), William Lucas Distant (1845–1922), Horatio Bryan Donkin (1845–1927), William Lowe Walker (1845–1930), Herbert Burrows (1845–1922), Robert Eyton (1845–1908), Henry Worsley Seymour Worsley-Bennison (1845–1918), Patrick William Stuart-Menteath (1845–1925).

450   APPENDIX

Late Victorians

William Robertson Smith (1846–94), Francis Herbert Bradley (1846–1924), Samuel Rolles Driver (1846–1914), Percy Gardner (1846–1937), Frederick Rogers (1846–1915), Albert Fleming (1846–1923), John Woodhouse Sanders (1846–1917), Thomas Gunn Selby (1846–1910), William Spiers (1846–1930), Harry Croft Hiller (1846–1934), Josiah Nicholson Shearman (1846–1915), Alfred Henry Garrod (1846–79), Wiliam Carr Crofts (1846–94), James Collier (1847–1925), Edmund Gurney (1847–88), E. Ray Lankester (1847–1929), Henry Scott Holland (1847–1914), Frances Swiney (1847–1922), Archibald Liversidge (1847–1927), James Allen Harker (1847–94), Joseph Compton-Rickett (1847–1919), William Holbrook Gaskell (1847–1914), Richard Jefferies (1847–87), Annie Besant (1847–1933), Franklin Thomas Richards (1847–1905), Alexander Graham Bell (1847–1922), Hugh Price Hughes (1847–1902), John Buchan (1847–1911), Frederick William Thomas Burbidge (1847–1905), John Andrewes Reeve (1847–1911), George J. Romanes (1848–94), Aubrey Lackington Moore (1848–90), Grant Allen (1848–99), Arthur Balfour (1848–1930), Mary Emily Dowson (1848–1941), Constance E. Plumptre (1848–1929), James Rodway (1848–1926), John Richardson Illingworth (1848–1915), Charles Lewis Tupper (1848–1910), Carveth Read (1848–1931), John Edmondson Manning (1848–1910), Alfred Williams Momerie (1848–1900), Ambrose Fleming (1849–1945), George Skelton Stephenson (1848/49–1929), Raphael Meldola (1849–1915), William H. Mallock (1849–1923), Sydney Howard Vines (1849–1934), Edward Aveling (1849–98), William Johnson Sollas (1849–1936), Percy Watkins Fenton Myles (1849–91), Constance Louisa Maynard (1849–1935), William Francis Barry (1849–1930), Edmund Gosse (1849–1928), John Cook Wilson (1849–1915), Mary Katharine Stanley (1849–1929), Helen Gladstone (1849–1925), Henry Bellyse Baildon (1849–1907), W.T. Stead (1849–1912), Walter Percy Sladen (1849–1900), G.B. Longstaffe (1849–1921), William Neish Walker (1849–1927), Richard Lyddeker (1849–1915), William Cecil Marshall (1849–1921), T. Jeffrey Parker (1850–97), Robert Louis Stevenson (1850–94), Richard Burdon Haldane (1850–1928), James Dennis Hird (1850–1920), George Claridge Druce (1850–1932), George Buckston Browne (1850–1945), William Leonard Courtney (1850–1928), Thomas Whitelegge (1850–1927), Alice Balfour (1850–1936), Theodore Cooke Taylor (1850–1952), Joseph Shield Nicholson (1850–1927), George Thomas Bettany (1850–91), Herbert Junius Hardwicke (1850–c.1930?), Sir John Alexander Cockburn (1850–1929), William Henry Muncaster (c.1850–1921), Eliza Margaret Humphreys (1850–1938), Frederick York Powell (1850–1904), Oliver Lodge (1851–1940), Alfred John Jukes Browne (1851–1914), Johnson Symington (1851–1924), Robert Blatchford (1851–1943), Francis Maitland Balfour (1851–82), William Robertson Nicholl (1851–1923), Walter Leaf (1852–1927), James Cossar Ewart (1851–1933), Henry Drummond (1851–97), Frederick Courteney Selous (1851–1917), Marcus Manuel Hartog (1851–1924), Mary Augusta Ward (1851–1921), Hector C. Macpherson (1851–1924), Sylvanus Philips Thompson (1851–1916), Charles Arthur Mercier (1851–1919), Robert Radclyffe Dolling (1851–1902), Frank Ballard (1851–1931), John Ferguson Nesbit (1851–99), Alexander Shewan (1851–1941), Robert Ramsay Wright (1852–1933), Arthur Milnes Marshall (1852–93), Andrew Wilson (1852–1912), Herbert Asquith (1852–1928), Thomas Okey (1852–1935),

APPENDIX   451

Ernest Huntington Dykes (1852–1924), Henry Jones (1852–1923), Everard im Thurn (1852–1932), Emma Marie Caillard (1852–1927), James Tochatti (1852–1928), James Trengrove Nance (1852–1942), Mary St. Leger (Kingsley) Harrison (1852–1931), Frank Young (c.1852–1941), Nina F. Layard (1853–1935), David G. Ritchie (1853–1903), Charles Gore (1853–1932), Cecil Rhodes (1853–1902), Thomas George Bond Howes (1853–1905), Bernard B. Woodward (1853–1930), Ernest Bruce Iwan-Müller (1853–1910), Isaac Bayley Balfour (1853–1922), Walter Aubrey Kidd (1853–1929), Alfred Cecil Cruttwell (1853–1901), George Lindsay Johnson (1853–1943), John Greenwood Tasker (1853–1936), Frederick Walker Mott (1853–1926), Charles Forrington (1853/54–1937), Conwy Lloyd Morgan (1854–1936), Patrick Geddes (1854–1932), Oscar Wilde (1854–1900), James Mavor (1854–1925), Dukinfield Henry Scott (1854–1934), Mary Higgs (1854–1937), Henry R. Knipe (1854–1918), Joseph Bamfylde Fuller (1854–1935), Adam Sedgwick (1854–1913), Alfred Milner (1854–1925), Frederick Natusch Maude (1854–1933), Alfred Cort Haddon (1854–1940), Harry Marshall Ward (1854–1906), Joseph Jacobs (1854–1916), Reginald John Ryle (1854–1922), Albert Wilson (1854–1928), Hubert Handley (1854–1943), Philip John Dear (1854–1934), John Scott Lidgett (1854–1953), William Roger Williams (1854–1948), John Wynn (1854–1909), Edward Mayrick (1854–1938), Frederick Orpen Bower (1855–1948), William Ritchie Sorley (1855–1935), Walter Heape (1855–1929), John Bland-Sutton (1855–1936), Frederick Augustus Dixey (1855–1935), Mary Dendy (1855–1933), James Oliphant (1855–1921), John Trevor (1855–1930), Francis Jeffrey Bell (1855–1924), Florence Caroline Dixie (1855–1905), Hans Gadow (1855–1928), John Henry Muirhead (1855–1940), Edward Step (1855–1931), Robert Forman Horton (1855–1934), Hubert Bland (1855–1914), Robert Allen Rolfe (1855–1921), Edward Bagnall Poulton (1856–1943), Edward Kay Robinson (1855–1928), Frederick Webb Headley (1856–1919), Keir Hardie (1856–1915), John Mackinnon Robertson (1856–1923), Wilfrid Philip Ward (1856–1916), Vernon Lee (Violet Paget) (1856–1935), Henry W. Nevinson (1856–1941), James Denney (1856–1917), Tom Mann (1856–1941), David W. Forrest (1856–1918), Andrew Seth Pringle-Pattison (1856–1931), Charles Herbert Hurst (1856–98), Rider Haggard (1856–1925), George Beesley Austin (1856–1936), Percy Fry Kendall (1856–1936), Henry Neville Hutchinson (1856–1927), Bernard Henry Holland (1856–1926), Edward Langridge Lunn (1856–1922), Lewis R. Farnell (1856–1934), James Alexander Lindsay (1856–1931), Charles John Bond (1857–1939), Karl Pearson (1857–1936), Alfred Lyttelton (1857–1913), Louis Robinson (1857–1928), Ernest Albert Parkyn (1857–1941), Edmund Selous (1857–1934), Archibald Edward Garrod (1857–1936), Sidney Low (1857–1932), Thomas Duckworth Benson (1857–1926), George Gissing (1857–1903), Frederick Howard Collins (1857–1910), Clement King Shorter (1857–1926), Edward R. Pease (1857–1955), Joseph Conrad (1857–1924), Charles Scott Sherrington (1857–1952), Arthur B. Moss (1855–1937), Morley Roberts (1857–1942), Herbert Wildon Carr (1857–1931), Richard Warwick Bond (1857–1943), Agnes Mary Frances Robinson (1857–1944), William Hardy Harwood (1857?–1924), Charles Dixon (1858–1926), James W. Tutt (1858–1911), Constance Naden (1858–89), Benjamin Kidd (1858–1916), Frank Evers Beddard (1858–1925), Robert Mackintosh (1858–1933), Graham Wallas (1858–1932), John A. Hobson (1858–1940), Archibald Bryan Macallum (1858–1934), William Abbott Herdman (1858–1924), Joseph Armitage Robinson (1858–1933),

452 APPENDIX

Winfrid Burrows (1858–1929), Bertram Coghill Alan Windle (1858–1929), Coulson Kernahan (1858–1943), Frank Byron Jevons (1858–1936), Henry (Harry) Quelch (1858–1913), John Alexander Clapperton (1858–1942), George Murray (1858–1911), William Watson (1858–1935), Hastings Rashdall (1858–1924), Frank Clement Offley Beaman (1858–1928), Jonas Bradley (1858–1943), Samuel Alexander (1859–1938), Havelock Ellis (1859–1939), Charles Frederick D'Arcy (1859–1928), Beatrice Webb (1859–1947), John Bruce Glasier (1859–1920), John W. Graham (1859–1932), Joseph Thomas Cunningham (1859–1935), Arthur Conan Doyle (1859–1930), Arabella Kenealy (1859–1938), Sydney John Hickson (1859–1940), Mary Cholmondeley (1859–1925), Walter Gardiner (1859–1941), Harold Fielding-Hall (1859–1917), Frederick Rowbotham (1859–1925), William A. Manning (1859–1924), Arthur Chandler (1859?–1939).

Edwardians

Sydney Savory Buckman (1860–1929), William Blaxland Benham (1860–1950), D'Arcy Wentworth Thompson (1860–1948), George Archdall Reid (1860–1929), Walter Frank Raphael Weldon (1860–1906), Walter Baldwin Spencer (1860–1929), William Ralph Inge (1860–1954), Harry Campbell (1860–1938), Thomas Bailey Saunders (1860–1928), (Thomas) Rhondda Williams (1860–1915), Ebenezer Griffith Jones (1860–1942), William Paterson Paterson (1860–1939), Clara E. Collet (1860–1948), Margaret McMillan (1860–1931), Alfred Leslie Lilley (1860–1948), William J. Ashley (1860–1927), John Stuart Mackenzie (1860–1935), Lawrence Pearsall Jacks (1860–1955), John Wood Oman (1860–1939), George F. Stout (1860–1944), Ben Tillett (1860–1943), John George Kelly (1860–1924), Henry William Massingham (1860–1924), Alfred Denny (1860/61–1947), J. Arthur Thomson (1861–1933), Karl Jordan (1861–1959), William Bateson (1861–1926), Arthur Shipley (1861–1927), Gilbert Charles Bourne (1861–1933), John Bagnall Bury (1861–1927), May Kendall (Born Emma Goldworth Kendall) (1861–1943), Henry Head (1861–1940), Oliver Elton (1861–1945), William Hall Calvert (1861–1917), Mary Elizabeth Coleridge (1861–1907), Charles Stewart Addis (1861–1945), James G. Adderley (1861–1942), Thomas Luther Davies (1861–1936), William John Reynolds (1861–1922), George Galloway (1861–1933), Alfred Ernest Knight (1861–1934), Huntly Carter (1861/2–1942), Mary Kingsley (1862–1900), James Wilson (1861–1941), John George Adami (1862–1926), John Alfred Spender (1862–1942), Philip Napier Waggett (1862–1939), Goldsworthy Lowes Dickinson (1862–1932), Frederick William Bussell (1862–1944), Edith Helen Sichel (1862–1914), Beatrice Chamberlain (1862–1918), George Francis Scott-Elliot (1862–1934), Robert Charles Fillingham (1862–1908), William Gordon Burn Murdoch (1862–1939), David Smith Cairns (1862–1946), Sidney Frederic Harmer (1862–1950), Eden Phillpotts (1862–1960), George Dawes Hicks (1862–1941), Ethel Brilliana Tweedie (1862–1940), Robert Locke Bremner (1862–1918), William Boothby Selbie (1862–1944), George A.B. Dewar (1862–1934), Francis Arthur Bather (1863–1934), Henry Balfour (1863–1939), Albert Charles Seward (1863–1941), Agnes Isabella Mary Elliot (1863–1946), Henry Brereton Marriott Watson (1863–1921), Anthony Hope Hawkins (Antony Hope) (1863–1933), Alan Campbell Swinton

APPENDIX   453

(1863–1930), Allen Upward (1863–1926), Victor Branford (1863–1930), Arthur Quiller-Couch (1863–1944), Francis Edward Younghusband (1863–1942), Bellerby (Harry) Lowerison (1863–1935), Herbert Hensley Henson (1863–1947), George Wyndham (1863–1913), Francis William Bain (1863–1940), Mary Amelia St Clair (1863–1946), John Saxon Mills (1863–1929), Henry Cecil Sturt (1863–1946), Timothy M. Donovan (c.1863–1950), George Martius Macdermott (1863/4–1939), Leonard Trelawny Hobhouse (1864–1929), John Walter Gregory (1864–1932), Richard A. Gregory (1864–1952), Gilbert Slater (1864–1938), Gustav Spiller (1864–1940), Cosmo Gordon Lang (1864–1945), William Bate Hardy (1864–1934), William H.R. Rivers (1864–1922), Edward Aurelian Ridsdale (1864–1923), Arthur Smith Woodward (1864–1944), William Henry Winch (1864–?), T. Dixon Rutherford (1864–1943), Francis Crawford Burkitt (1864–1935), Ferdinand Canning Scott Schiller (1864–1937), John Henry Jowett (1864–1923), Oswald Hawkins Latter (1864–1948), Alfred Milne Gossage (1864–1948), Arthur Dendy (1865–1925), William Butler Yeats (1865–1939), John Bretland Farmer (1865–1944), George Herbert Carpenter (1865–1939), Mrs Patrick Campbell (1865–1940), Joseph Hannay Leckie (1865–1935), Andrew John Herbertson (1865–1915), James Leatham (1865–1945), Edith Rebecca Saunders (1865–1945), Margaret Benson (1865–1916), Frederick Ernest Weiss (1865–1953), Henry Snell (1865–1944), Arthur Samuel Peake (1865–1929), Frederick Robert Tennant (1866–1957), Ernest William MacBride (1866–1940), Robert Cyril Layton Perkins (1866–1955), Robert Ranulph Marett (1866–1943), David Joseph Scourfield (1866–1949), Walter William Skeat (1866–1953), Ernest Starling (1866–1927), Herbert George Wells (1866–1946), Lionel Earle (1866–1948), Arthur Keith (1866–1955), Robert Broom (1866–1951), J. Ramsay MacDonald (1866–1937), George Herbert Perris (1866–1920), George Gilbert Aimé Murray (1866–1957), Richard Le Gallienne (1866–1947), James O'Grady (1866–1934), Willie Herbert Utley (1866–1918), Robert Durward Clarkson (1866/67–1951), Ronald Campbell Macfie (1867–1931), Reginald John Campbell (1867–1956), Joseph McCabe (1867–1955), Arnold Bennett (1867–1931), Percival (Percy) Dearmer (1867–1936), Arthur Willey (1867–1942), James Douglas (1867–1940), Benchara Branford (1867–1944), George William Russell (1867–1935), Kathryn Glasier (1867–1950), George Tyrrell (1861–1909), Harold Anson (1867–1954), Gertrude Clarke Nuttall (1868–1929), James Louis Garvin (1868–1947), Gerald Rowley Leighton (1868–1953), Edwin Stephen Goodrich (1868–1946), John Christopher Willis (1868–1958), Walter Garstang (1868–1949), Joseph Morgan Lloyd Thomas (1868–1955), Edward Garnett (1868–1937), William Beach Thomas (1868–1957), Walter Rothschild (1868–1937), Sydney Herbert Mellone (1869–1956), Edgar Chamberlain (1868–1931), Frederic Carrel (1869–1928), Alfred Ernest Crawley (1869–1924), Mary Patricia Willcocks (1869–1952), Marion Isabel Newbigin (1869–1934), Arthur Compton-Rickett (1869–1927), Ellis Thomas Powell (1869–1922), John Graham Kerr (1869–1957), Vernon Faithfull Storr (1869–1940), Helena Maude Caulfield (née Martyn) (1869–1951), Conrad Noel (1869–1942), Florence Margaret Durham (1869–1948), Neville Chamberlain (1869–1940), Laurence Binyon (1869–1943), Hilaire Belloc (1870–1953), Hebert Samuel (1870–1963), Charles J. Patten (1870–1948), George Percival Mudge (1870–1939), Charles Chamberlain Hurst (1870–1947), Harold Christopherson Morton (1870–1936), Wynfried Lawrence Henry Duckworth (1870–1956), Herbert Samuel (1870–1963), Alfred Frank Tredgold (1870–1952), Herbert Gresford Jones

454   APPENDIX

(1870–1958), Arthur George Tansley (1871–1955), John Millington Synge (1871–1909), William MacDougall (1871–1938), Grafton Elliot Smith (1871–1937), J.W. Jenkinson (1871–1915), George Udney Yule (1871–1951), Edward Harold Begbie (1871–1929), Thomas John Gerrard (1871–1916), W.J. Cunningham Pike (1871–1925), Bertrand Russell (1872–1970), Norman Angell (1872–1967), Gertrude Elles (1872–1960), Wilfred Trotter (1872–1939), William Lawrence Schroeder (1872–1950), John Lawrence Le Breton Hammond (1872–1949), Laurie Magnus (1872–1933), Charles Samuel Myers (1873–1946), Frederick James Chittenden (1873–1950), G.K. Chesterton (1873–1936), Alfred Richard Orage (1873–1934), Henry Noel Brailsford (1873–1958), Alfred Salter (1873–1945), Ford Madox Ford (1873–1939), Lawrence J. Tremayne (*c.*1873–1959), John Lionel Tayler (1874–1930), Ernest William Barnes (1874–1953), Hewlett Johnson (1874–1966), Rowland Harry Biffen (1874–1949), Solomon Herbert (1874–?), Cecil Henry Desch (1874–1958), Redcliffe Nathan Salaman (1874–1955), William Somerset Maugham (1874–1965), James Gilchrist Lawson (1874–1946), Sidney Dark (1874–1947), Ernest Shackleton (1874–1922), Margery Fry (1874–1958), Frederick E.M. Docker (1874–1936), Holbrook Jackson (1874–1948), Mary Hart-Davis (1875–1934), Reginald Crundall Punnett (1875–1967), Percy Redfern (1875–1958), Ernest Benn (1875–1954), Cyril Arthur Edward Ranger Gull ('Guy Thorne') (1875–1923), Robert John Strutt (1875–1947), Arthur Joseph Penty (1875–1937), Edmund Clerihew Bentley (1875–1956).

Georgians

Hilda Blanche Killby (1877–1962), Caleb Williams Saleeby (1878–1940), Edward Thomas (1878–1917), John Masefield (1878–1967), Robert Heath Lock (1879–1915), Annie Kenney (1879–1953), Neville Stuart Talbot (1879–1943), E.M. Forster (1879–1970), Reginald Philip Gregory (1879–1918), Alfred Noyes (1880–1958), Harold Herbert Williams (1880–1964), Ethel Georgina Romanes (1880–1918), Harold Herbert Williams (1880–1964), Maurice B. Talbot-Crosbie (1881–?), Eric Gill (1882–1940), Charles Vyse (1882–1971), John Grant McKenzie (1882–1963), John Maynard Keynes (1883–1946), Thomas Ernest Hulme (1883–1917), George Chatterton-Hill (1883–1947), Nora Barlow (1885–1988), David Herbert Lawrence (1885–1930), Thomas Quelch (1886–1954), Christopher R.W. Nevinson (1889–1946), Sewall Wright (1889–1988), Hermann Joseph Müller (1890–1967), Ronald A. Fisher (1890–1962), Ellen Wilkinson (1891–1947), John Burdon Sanderson Haldane (1892–1964).

Select Bibliography

Main Archival Collections

H.W. Acland Papers, Bodleian
Samuel Alexander Papers, MJRL
Argyll Papers, Inveraray Castle
William Bateson Papers, CUL
William Bateson Collection, JIC
George Bentham Papers, RBGK
J.S. Blackie Papers, NLS
J.S. Bowerbank Papers, Wellcome Library, London
Samuel Butler Papers, BL
Samuel Butler Papers, St John's College, Cambridge
William Carruthers Papers, NHM
Chamberlain Papers, Birmingham
Edward Clodd Papers, Brotherton
Coleridge Papers, BL
Evolution Collection, University of California, Santa Barbara
James Cossar Ewart Papers, Edinburgh
A. Campbell Fraser Papers, NLS
Francis Galton Papers, UCL
Patrick Geddes Papers, NLS
Patrick Geddes Papers, Strathclyde University
W.E. Gladstone Papers, BL
Edmund Gosse Papers, Brotherton
Asa Gray Papers, Harvard
Gunther Papers, NHM
Haldane Papers, NLS
Frederic Harrison Papers, BLPES
Joseph D. Hooker Papers, RBGK
C.C. Hurst Papers, CUL
T.H. Huxley Papers, APS
T.H. Huxley Papers, ICL
William James Papers, Harvard
Benjamin Kidd Papers, CUL
Charles Kingsley Correspondence (M.L. Parrish Collection), PUL
John Lubbock, Lord Avebury Papers, BL
Charles Lyell Papers, Edinburgh
Raphael Meldola Papers, ICL
Raphael Meldola Papers, OMNH
Aubrey L. Moore Papers, APS
Conwy Lloyd Morgan Papers, University of Bristol
Max Müller Papers, Bodleian
Francis W. Newman Papers, Senate House, UL
Alfred Newton Papers, CUL

456 SELECT BIBLIOGRAPHY

Richard Owen Papers, NHM
Karl Pearson Papers, UCL
Edward Bagnall Poulton Papers, OMNH
E.B. Pusey Papers, Pusey House, Oxford
H.N. Ridley Papers, RBGK
J. Burdon Sanderson Papers, NLS
F.C.S. Schiller Papers, UCLA
Adam Sedgwick Papers, CUL
Herbert Spencer Papers, Senate House, UL
G.G. Stokes Papers, CUL
William Thiselton-Dyer Papers, RBGK
D'Arcy Wentworth Thompson Papers, St Andrews
A.R. Wallace Papers, BL
Graham Wallas Papers, BLPES
Lester F. Ward Papers, Brown University, Providence, RI
Wilfrid Ward Papers, University of St Andrews
Victoria, Lady Welby Papers, York University, Toronto

Note on transcription. Deleted matter is denoted by strikethrough; inserted matter thus:/ insertion\. No attempt has been made to retain the distinction between '&' and 'and'.

Online Resources

Biodiversity Heritage Library (including *Gardeners' Chronicle*)
British Newspaper Archive
Charles Darwin Correspondence Project
Darwin Online
Modernist Journal Project, Brown University
National Library of Wales, Welsh Newspapers
Olive Schreiner Correspondence Online
Times Online
Trove (Australian Newspapers online)
Wallace Letters Online
Yellow Nineties Online

Note on referencing from online collections. Where citation is to transcriptions of texts accessed online, this is made clear in referencing, but without the standard date of access information, which I have deemed superfluous for these stable texts. No attempt has been made to identify in the notes instances where facsimile copies of original manuscripts or printed texts have been accessed online via collections like the Internet Archive.

Periodicals

Academy
British Medical Journal
British Quarterly Review
Church Quarterly Review
Contemporary Review
Economic Review
The Expositor
Fortnightly Review
Gardeners' Chronicle
Geological Magazine

SELECT BIBLIOGRAPHY   457

Good Words
Hardwicke's Science Gossip
Ibis
Intellectual Observer
International Journal of Ethics
Journal of Anatomy and Physiology
Journal of the Linnean Society
Journal of Mental Science
Journal of Microscopy
Journals of the Transactions of the Victoria Institute
Land and Water
Man
Medical Times and Gazette
Mind
The Monist
The National Review
Natural Science
Nature
The Nineteenth Century
Pall Mall Magazine
The Philosophical Review
Popular Science Review
Proceedings of the Aristotelian Society
Proceedings of the Royal Colonial Institute
Proceedings of the Royal Institution
Proceedings of the Royal Philosophical Society of Glasgow
Proceedings of the Zoological Society
Quarterly Review
Sociological Review
Transactions of the Ethnological Society
Transactions of the Norfolk and Norwich Naturalists' Society
Westminster Review
The Zoologist

Primary Printed

Duke of Argyll, *The Reign of Law* (1867).
Duke of Argyll, *Primeval Man* (1869).
Duke of Argyll, *Unity of Nature* (1884).
Duke of Argyll, 'A Great Lesson', *TNC* 22 (Jul–Dec 1887), 293–309.
Duke of Argyll, 'Life of Darwin', *Murray's Magazine* 3 (1888), 145–59.
Duke of Argyll, *Organic Evolution Cross-Examined* (1898).
Barbour, G.F., *Life of Alexander Whyte* (1925).
Bateson, W., Presidential Address Section D. Zoology, *Report of the British Association for the Advancement of Science 1904* (1905), 574–89.
Bateson, W., 'The Progress of Genetic Research', in *Report of the Third International Conference 1906 on Genetics* (1907), 90–7.
Bateson, W., *Biological Fact and the Structure of Society* (1912).
Beverley, R.M., *The Darwinian Theory of the Transmutation of Species* (1867).
Birks, T.R., *Scripture Doctrine of Creation* (1872).
Birks, T.R., *Modern Fatalism* (1876).

458   SELECT BIBLIOGRAPHY

Bonney, T.G., *Memories of a Long Life* (1921).
Bosanquet, B. *Aspects of the Social Problem* (1895).
Bree, C.R., *Species not Transmutable* (1860).
Bree, C.R., *Popular Illustrations of the Lower Forms of Life* (1868).
Bree, C.R., *On Darwinism and its Effects upon Religious Thought* (1873).
Brightwen, E., *Life and Thoughts of a Naturalist* (1909).
Bryce, J., 'Personal Reminiscences of Charles Darwin and of the Reception of the "Origin of Species"', *Proceedings of the American Philosophical Society* 48 (193) (September 1909), iii–xiv.
Buchanan, R., 'Lucretius and Modern Materialism', *New Quarterly Magazine* 6 (1876), 1–30.
Butler, S., *Life and Habit* (1878).
Butler, S., *Evolution Old and New* (1879).
Caird, E., 'St Paul and the Idea of Evolution', *The Hibbert Journal* 2.1 (October 1903), 1–19.
Cairns, D.S., *An Autobiography* (1950).
Campbell, I. Duchess of Argyll, *George Douglas, Eighth Duke of Argyll, K.G., K.T. (1823–1900): Autobiography and Memoirs* (2 vols, 1906).
Carpenter, E., *Civilisation: Its Cause and Cure and Other Essays* (1889).
Carpenter, E., *Intermediate Types Among Primitive Folk: A Study In Social Evolution* (1914).
Carpenter, W.B., 'Charles Darwin', *Modern Review* 3 (1882), 500–24.
Carpenter, W.B., 'Darwinism in England', in *Nature and Man. Essays Scientific and Philosophical. With an Introductory Memoir by J. Estlin Carpenter* (1889).
Cline, C.L., ed., *The Letters of George Meredith* (3 vols, 1970).
Clodd, E., *The Story of Creation* (1888).
Clodd, E., *Pioneers of Evolution* (1897).
Clodd, E., *Grant Allen* (1900).
Clodd, E., *Memories* (1916).
Cobbe, F.P., *Life of Frances Cobbe, Written by Herself. Posthumous Version* (1904).
Compton-Rickett, A., *I Look Back: Memories of Fifty Years* (1933).
Courtney, J.E., *Recollected in Tranquility* (1926).
Cunningham, J.T., *Charles Darwin*, in Baildon, H.B., *The Round Table Series* (1887).
Cunningham, J.T., *The Evolution of Man; an Address Delivered to the Mendel Society February 1908* (1908).
Darwin, C., *The Voyage of the Beagle* (1839 etc.).
Darwin, C., *On the Origin of Species* (1859 etc.).
Darwin, C., *The Variation of Animals and Plants Under Domestication* (1868).
Darwin, C., *The Descent of Man* (1871).
Darwin, C., *The Expression of the Emotions in Man and Animals* (1872).
Darwin, F., *The Life and Letters of Charles Darwin* (1887).
Dewar, D. and Finn, F., *The Making of Species* (1909).
Drummond, H., *Natural Law in the Spiritual World* (1889).
Drummond, H., *The Ascent of Man* (1894).
Elam, C., *Winds of Doctrine, Being an Examination of the Modern Theories of Automatism and Evolution* (1876).
Elam, C., 'The Gospel of Evolution', *Contemporary Review* 37 (1880), 713–40.
Elam, C., *Facts and Fancies in Modern Science* (1882).
Farnell, L.R., *An Oxonian Looks Back* (1934).

SELECT BIBLIOGRAPHY 459

Farrar, F.W., *Chapters on Language* (1865).

Fry, A., *Memoir of Rt Hon Sir Edward Fry* (1921).

Geddes, P., 'A Naturalists' Society and its Work', *Scottish Geographical Magazine* 19 (1903), 89.

Geikie, A., 'Life and Letters of Charles Darwin, *Contemporary Review* (1887), in his *Landscape and History. And Other Essays* (1905).

Gould, F., *Chats with Pioneers of Modern Thought* (1898).

Gould, F.J., *The Life Story of a Humanist* (1923).

Gunther, A., *A Century of Zoology at the British Museum* (1975).

Hardwick, C.S., ed., *Semiotics and Significs: Correspondence Between Charles S. Peirce and Victoria, Lady Welby* (1977).

Harrison, J.E., *A Portrait from Letters*, ed. Jessie Stewart (1959).

Henslow, G., 'Present Day Rationalism with an Examination of Darwinism', in W.W. Seton, ed., *Christian Apologetics* (1903), 1–24.

Henslow, G., *Present Day Rationalism Critically Examined* (1904).

Henslow, G., 'The Mutation Theory: A Criticism and an Appreciation', *Journal of the Royal Horticultural Society* 37 (1911–2), 144–8.

Horton, R.F., *Autobiography* (1918).

Howard, J.E., *An Examination of the Belfast Address of the British Association 1874 from a Scientific Point of View* (1874).

Howard, J.E., *Scientific Facts and Christian Evidence* (1876).

Howard, J.E., *Creation and Providence, with Especial Reference to the Evolutional Theory* (1878).

Hudson, W.H., *Men, Books and Birds* (1923).

Hutchison, H.G., *The Life of John Lubbock, Baron Avebury* (2 vols, 1914).

Hutton, F.W., *Darwinism* (1887).

Hutton, F.W., *The Lesson of Evolution* (1902).

Huxley, T.H., *Man's Place in Nature* (1863).

Huxley, T.H., *Lay Sermons* (1870).

Huxley, T.H., *Critiques and Addresses* (1873).

Huxley, T.H., *Darwiniana* (1894).

Inge, W.R., *The Idea of Progress* (1920).

Inge, W.R., 'Evolution and the Idea of God', in H.A. Wilson, ed., *The Anglican Communion* (1929), 251–60.

Japp, A.H., *Darwin Considered Mainly as Ethical Thinker, Human Reformer and Pessimist* (1901).

Kent, W., *The Testament of a Victorian Youth: An Autobiography* (1938).

Kidd, B., *Social Evolution* (1894).

Kidd, B., *Principles of Western Civilization* (1902).

Kingsley, F.E., *Charles Kingsley. His Letters and Memories of his Life* (1877).

Kropotkin, P., 'Recent Science', *TNC* 50 (Jul–Dec 1901), 417–38.

Kropotkin, P., 'The Ethical Need of the Present Day', *TNC* 56 (Jul–Dec 1904), 207–26.

Kropotkin, P., 'The Morality of Nature', *TNC* 57 (Jan–Jun 1905), 407–26.

Lankester, E.R., *Degeneration. A Chapter in Darwinism* (1880).

Laurence, D.H., ed., *George Bernard Shaw. Collected Letters* (1965–88).

Liddon, H.P., *The Recovery of St. Thomas. A Sermon [St. John xx. 27.] Preached in St. Paul's Cathedral on the Second Sunday after Easter April 23, 1882. With a Prefatory Note on the Late Mr. Darwin* (1882).

Liddon, H.P., *Life of Edward Bouverie Pusey* (4 vols, 1897).

460 SELECT BIBLIOGRAPHY

Lidgett, J.S., *My Guided Life* (1936).

Lindsay, J.A., 'Darwinism and Medicine, Bradshaw Lecture, 1909', *Lancet*, 6 November 1909, 1327–33.

Lodge, O., *The Substance of Faith Allied with Science* (1907).

Lodge, O., *Advancing Science: Personal Reminiscences of the British Association in the Nineteenth Century* (1931).

Lodge, O., *Past Years. An Autobiography* (1931).

Lubbock, J., *Addresses, Political and Educational* (1879).

Lubbock, J., *The Pleasures of Life* (1887–89).

Lyell, K.M., ed., *Memoir of Leonard Horner* (2 vols, 1890).

Macfie, R.C., *The Faiths and Heresies of a Poet and Scientist* (1932).

Marchant, J., *Raphael Meldola. Reminiscences of his Worth and Work...* (1916).

Marett, R.R., *A Jerseyman at Oxford* (1941).

Marett, R.R. and Penniman J.K., eds. *Spencer's Scientific Correspondence with Sir J.G. Frazer and Others* (1932).

Marrot, H.V., *The Life and Letters of John Galsworthy* (1930).

Martineau, H., *Letters*, ed. Valerie Sanders (1990).

Marvin, F.S., 'Biology and Evolution', in his *The Century of Hope: a Sketch of Western Progress from 1815 to the Great War* (1919), 135–60.

Matheson, G., *Can the Old Faith Live with the New? Or, the Problem of Evolution and Revelation* (1885).

Mitchell, P.C., 'A Word with Mr Huxley', *National Review* 21 (1893), 713–5.

Mitchell, P.C., 'T.H. Huxley', *New Review* 13 (1895), 147–55.

Mivart, St G., *On the Genesis of Species* (1871).

Mivart, St G., *Man and Apes: An Exposition of Structural Resemblances and Differences Bearing upon Questions of Affinity and Origin* (1873).

Mivart, St G., *Lessons from Nature* (1876).

Mivart, St G., *Essays and Criticisms* (2 vols, 1892).

Mivart, St G., 'Evolution in Professor Huxley', *TNC* 34 (1893), 198–211.

Moore, A.L., 'The Relations Between Science and Religion', *Quarterly Review* 159 (1885), 360–86.

Moorhouse, J., *Some Modern Difficulties Respecting the Facts of Nature and Revelation, Considered in Four Sermons* (1861).

Morgan, C.L., 'The Philosophy of Evolution', *The Monist* 8 (1898), 481–501.

Morgan, C.L., 'Vitalism', *The Monist* 9 (1899), 179–96.

Morris, F.O., *All the Articles of the Darwinian Faith* (1875).

Muirhead, J.H., ed, *Bernard Bosanquet and his Friends* (1935).

Muirhead, J.H., *Reflections of a Journeyman Philosopher* (1942).

Müller, F.M., *Introduction to the Science of Religion* (1873).

Müller, F.M., 'My Reply to Mr Darwin', *Contemporary Review* 25 (Dec 1874–May 1875), 305–26.

Müller, F.M., 'Forgotten Bibles', *TNC* 15 (1884), 1004–22.

Müller, F.M., 'Why I Am Not Agnostic', *TNC* 36 (1894), 890–5.

Müller, F.M., *Autobiography: A Fragment* (1901).

Murphy, J.J., *Habit and Intelligence* (1869, 1879).

Murphy, J.J., *Scientific Bases of Faith* (1873).

Murphy, J.J., *Natural Selection and Spiritual Freedom* (1893).

Myers, F.W.H., *Fragments of Inner Life: An Autobiographical Sketch* (1961).

Newbigin, M., *Colour in Nature* (1896).

Newbolt, H., *My World as in My Time* (1932).

SELECT BIBLIOGRAPHY   461

Newton, A., 'The Early Days of Darwinism', *Macmillan's Magazine* 57 (February 1888), 241–9.

Okey, T., *A Basketful of Memories* (1930).

Owen, R., *Instances of the Power of God as Manifested in His Animal Creation* (1864).

Owen, R., *Anatomy of Vertebrates* (1869).

Paget, S., *Memoirs and Letters of Sir James Paget* (1902).

Palmer, W.S., *An Agnostic's Progress* (1908).

Pascoe, F.P., *The Darwinian Theory of the Origin of Species* (1890).

Pearson, K., *The Chances of Death and Other Studies of Evolution* (1897).

Pearson, K., *National Life from the Standpoint of Science* (1901).

Pearson, K., *Life, Letters and Labour of Francis Galton* (1926).

Phillips, J.A., *Life on the Earth. Its Origin and Succession* (1860).

Picton, J.A., *Man and the Bible* (1910).

Pollock, F., 'Evolution and Ethics', *Mind* 1.3 (1876), 334–45.

Ponton, M., *The Beginning: Its When and Its How* (1871).

Poulton, E.B., 'A Naturalist's Contribution to the Discussion Upon the Age of the Earth', in *Annual Report of the British Association for the Advancement of Science* 66 (1896), 808–28.

Poulton, E.B., *Charles Darwin and the Theory of Natural Selection* (1896).

Poulton, E.B., *Essays on Evolution 1889–1907* (1908).

Poulton, E.B., *Charles Darwin and the Origin of Species* (1909).

Poulton, E.B., 'Darwin and his Modern Critics', *Quarterly Review* 211 (1909), 1–38.

Poulton, E.B., 'Huxley and Natural Selection', in O. Lodge, ed., *Huxley Memorial Lectures to the University of Birmingham* (1914), 45–51.

Pringle-Pattison, A.S., 'Professor Huxley on Nature and Man', *Blackwoods* 154 (1893), 823–34.

Pringle-Pattison, A.S., *The Philosophical Radicals and Other Essays* (1907).

Pringle-Pattison, A.S., et al., 'Symposium: is there evidence of design in Nature', *Proceedings of the Aristotelian Society* 1 (1889–91), 49–76.

Proctor, D., ed., *The Autobiography of G. Lowes Dickinson* (1973).

Pusey, E.P., *Unscience, Not Science, Adverse to Faith* (1878).

Quinn, E.V., and Prest, J., eds., *Dear Miss Nightingale; Letters of Benjamin Jowett* (1987).

Reid, G.A., 'Human Evolution and Alcohol', *British Journal of Inebriety* 1 (1903–4), 186–201.

Reid, G.A., 'The Biological Foundations of Sociobiology', *Sociological Review* 3 (1906), 3–52.

Richards, G., *Memories of A Misspent Youth* (1932).

Richardson, E.W., *A Veteran Naturalist: Being the Life and Work of W.B. Tegetmeier* (1916).

Ritchie, D.G., *Darwinism and Politics* (1889, 1891).

Ritchie, D.G., 'Natural Selection and the Spiritual World', *Westminster Review* 133 (May 1890), 459–69.

Ritchie, D.G., 'Social Evolution', *International Journal of Ethics* 6 (1895–6), 165–81.

Ritchie, D.G., *Studies in Political and Social Ethics* (1902).

Romanes, G.J., 'A Reply to the "Fallacies of Evolution"', *Fortnightly Review* 26 (1879), 492–504.

Romanes, G.J., *The Scientific Evidences of Organic Evolution* (1882).

Romanes, G.J., 'Recent Critics of Darwinism', *Contemporary Review* 53 (1888), 836–54.

Romanes, G.J., 'Mr Wallace on Darwinism', *Contemporary Review* 56 (1889), 244–58.

Romanes, G.J., 'Primitive Natural History', *TNC* (1890), 297–308.

462 SELECT BIBLIOGRAPHY

Romanes, G.J., *Animal Intelligence* (1892).

Romanes, G.J., *The Darwinian Theory* (1892).

Romanes, G.J., *An Examination of Weismannism* (1893).

Romanes, G.J., *Mental Evolution in Animals* (1893).

Romanes, G.J., 'The Darwinism of Darwin, and of the Post-Darwinian Schools', *The Monist* 6.1 (October 1895), 1–27.

Romanes, G.J., *Post-Darwinian Questions: Heredity and Utility* (1895).

Romanes, G.J., *Post-Darwinian Questions: Isolation and Physiological Selection* (1897).

Ross, J., *Graft Theory of Disease: An Application of Mr Darwin's Theory of Pangenesis to the Explanation of Zymotic Diseases* (1874).

Ross, J., *On Protoplasm: Being an Examination of Dr J.H. Stirling's Criticism of Professor Huxley's Views* (1874).

Row, C.A., *Principles of Modern Pantheistic and Atheistic Philosophy* (1874).

Russell, B., 'My Religious Reminiscences', in Egner, R.E. and Denonn, L.E., eds., *The Basic Writings of Bertrand Russell* (1961), 31–6.

Samuelson, J., *James Samuelson's Recollections: Being Some Experiences and Reflections Mainly on Subjects of the Day, By an Old Author and Traveller* (1907).

Schiller, F.C.S., *Humanism. Philosophical Essays* (1903, 1912).

Seth, J., *English Philosophers and Schools of Philosophy* (1912).

Seward, A.C., ed., *Darwin and Modern Science* (1909).

Skrupskelis, I.K., et al., eds., *Correspondence of William James* (1993–2004).

Snell, H., *Movements, Men and Myself* (1936).

Sollas, W.J., 'On Evolution in Geology', *Geological Magazine* 4 (1877), 1–7.

Sollas, W.J., 'The Influence of Oxford on the History of Geology', *The Age of the Earth and Other Geological Studies* (1905), 219–56.

Spencer, H., 'Progress: Its Law and Cause', *Westminster Review* 67 (April 1857), 445–85.

Spruce, R., *Notes of a Botanist on the Amazon and Andes* (1908).

Stebbing, T.R.R., *Darwinism. A Lecture* (1871).

Stephen, L., *Essays in Freethinking* (1879).

Stephen, L., *Essays to Ethical Societies* (1896).

Stephen, L., 'The Growth of Toleration', in *Encyclopedia Britannica Volume 28. Elections to Glamorgan* (1902).

Stephenson, G., *Edward Stuart Talbot, 1844–1934* (1936).

Stirling, A.H., *James Hutchison Stirling, His Life and Work. With Pref. by the Right Hon. Viscount Haldane of Cloan* (1912).

Sully, J., *My Life and Friends. A Psychologist's Memories* (1918).

Talbot, E.S., *Memories of Early Life* (1924).

Tayler, J.L., *Aspects of Social Evolution* (1904).

Tennant, F.R., 'The Influence of Darwinism Upon Theology', *Quarterly Review* 211 (1909), 418–40.

Thiselton-Dyer, W.T., 'The Rise and Influence of Darwinism', *Edinburgh Review* 196 (1902), 366–407.

Thomson, J.A., *Herbert Spencer* (1906).

Thomson, J.A., *Heredity* (1908).

Thomson, J.A., *Darwinism and Human Life* (1909).

Thomson, J., 'The Influence of Darwinism on Thought and Life', in Marvin, F.S., *Science and Civilization* (1923), 203–22.

Tillett, B., *Memories and Reflections* (1931).

Tollemache, L.A., *Old and Odd Memories* (1908).

SELECT BIBLIOGRAPHY 463

Waggett, P.N., *Religion and Science. Some Suggestion for the Study of the Relations Between Them* (1904).

Wallace, A.R., *The Malay Archipelago* (1869).

Wallace, A.R., *Contributions to the Theory of Natural Selection* (1870).

Wallace, A.R., *Darwinism* (1890).

Wallace, A.R., 'Darwinism Versus Wallaceism', *Contemporary Review* 94 (1908), 716–7.

Wallace, A.R., *My Life* (1905).

Ward, J., *Naturalism and Agnosticism: The Gifford Lectures Delivered Before the University of Aberdeen in the Years 1896–1989* (1899).

Wedgwood, J., 'Ethics and Science', *Contemporary Review* 72 (August 1897), 220–33.

Weiss, F.E., *Bearings of the Darwinian Theory of Evolution on Moral and Religious Progress* (1909).

Welby, V., *Echoes of Larger Life. A Selection From the Early Correspondence of Victoria Lady Welby. Edited by … Cust, Mrs. Henry* (1929).

Welby, V., *Other Dimensions. A Selection From the Later Correspondence of Victoria Lady Welby. Edited by … Cust, Mrs. Henry* (1931).

Wilson, J.M., *Essays and Addresses* (1887).

Wilson, J.M., *Some Contributions to the Religious Thought of our Times* (1888).

Wilson, J. M., *Evolution and the Holy Scriptures* (1903).

Secondary

Ackerman, R.A., *J.G. Frazer. His Life and Work* (1987).

Ackerman, R.A., *The Myth and Ritual School: J.G. Frazer and the Cambridge Ritualists* (1991).

Allen, David E., 'Arcana ex Multitudine: Prosopography as a Research Tool', *Archives of Natural History* 17.3 (1990), 349–59.

Allen, G.E., 'Mechanism, Vitalism and Organicism in Late Nineteenth and Twentieth Century Biology', *Studies in the History and Philosophy of Biological and Biomedical Science* 36 (2005), 261–83.

Altholz, J., 'A Tale of Two Controversies: Darwinism in the Debate over "Essays and Reviews"', *Church History* 63 (1994), 50–59.

Amigoni, D., *Colonies, Cults and Evolution: Literature, Science and Culture in Nineteenth Century Writing* (2007).

Amundson, R., *The Changing Role of the Embryo in Evolutionary Thought* (2005).

Annan, N., *Sir Leslie Stephen: His Thought and Character in Relation to His Time* (1951).

Asprem, E., *The Problem of Disenchantment. Scientific Naturalism and Esoteric Discourse, 1900–1939* (2014).

Baldwin, M., *Making* Nature: *The History of a Scientific Journal* (2015).

Barmann, L.F., *Baron Friedrich Von Hügel and the Modernist Crisis in England* (1972).

Bartholomew, M., 'Lyell and Evolution: an Account of Lyell's Response to the Prospect of an Evolutionary Ancestry for Man', *British Journal for the History of Science* 6.3 (1973), 261–303.

Barton, R., 'John Tyndall, Pantheist: A Rereading of the Belfast Address', *Osiris* 3 (1987), 111–34.

Barton, R., 'Just Before Nature: The Purposes of Science and the Purposes of Popularization in Some English Popular Science Journals of the 1860s', *Annals of Science* 55 (1998), 1–33.

Barton, R., *The X Club: Power and Authority in Victorian Science* (2018).

Barton, S.G. and Wilkinson, D., eds., *Reading Genesis After Darwin* (2009).

464 SELECT BIBLIOGRAPHY

Beasley, E., *The Victorian Reinvention of Race: New Racisms and the Problem of Grouping in the Human Sciences* (2010).

Bebbington, D.W., 'Henry Drummond, Evangelism and Science', in Corts, Thomas E., ed., *Henry Drummond: A Perpetual Benediction* (1999), 19–38.

Bebbington, D.W., 'Science and Evangelical Theology in Britain from Wesley to Orr', in Livingstone, et al., eds., *Evangelicals and Science in Historical Perspective*, 120–41.

Beer, G., *Darwin's Plots. Evolutionary Narrative in Darwin, George Eliot and Nineteenth Century Fiction* (1983).

Bevir, M., ed., *Historicism and the Human Sciences in Victorian Britain* (2017).

Boddice, R., *The Science of Sympathy: Morality, Evolution, and Victorian Civilization* (2016).

Boulter, M., *Bloomsbury Scientists: Science and Art in the Wake of Darwin* (2017).

Bowler, P.J., *The Eclipse of Darwin. Anti-Darwinian Evolution Theories in the Decades Around 1900* (1983).

Bowler, P.J., *Theories of Human Evolution* (1986).

Bowler, P.J., *The Non-Darwinian Revolution. Reinterpreting a Historical Myth* (1992).

Bowler, P.J., *Reconciling Science and Religion. The Debate in Early-Twentieth Century Britain* (1996).

Bowler, P.J., 'Evolution and the Eucharist: Bishop E.W. Barnes on Science and Religion in the 1920s and 1930s', *British Journal of the History of Science* 31 (1998), 453–67.

Bowler, P.J., *Science for All. The Popularization of Science in Early Twentieth Century Britain* (2009).

Brooke, J.H., 'The Natural Theology of the Geologists: Some Theological Strata', in Jordanova, L.J. and Porter, R., eds., *Essays in the History of the Environmental Sciences* (1979), 39–64.

Brooke, J.H. and Cantor, G., *Reconstructing Nature: The Engagement of Science and Religion* (1998).

Browne, J., 'Darwin in Caricature: A Study of Popularization and Dissemination in Evolutionary Theory', in Larson, B. and Brauer, F., eds., *The Art of Evolution: Darwin, Darwinisms, and Visual Culture* (2009), 18–39.

Bulmer, M., *Francis Galton: Pioneer of Heredity and Eugenics* (2003).

Burckhardt, F., 'England and Scotland: The Learned Societies', in Glick, *Comparative Reception*, 32–74.

Burrow, J.W., *Evolution and Society* (1966).

Cantor, G., *Quakers, Jews, and Science: Religious Responses to Modernity and the Sciences in Britain, 1650–1900* (2005).

Carroll, J., *Literary Darwinism: Evolution, Human Nature, and Literature* (2004).

Caruana, L., ed., *Darwin and Catholicism: The Past and Present Dynamics of a Cultural Encounter* (2009).

Cashdollar, C.D., *The Transformation of Theology 1830–1890* (2014).

Churchill, F.B., *August Weismann. Development, Heredity, Evolution* (2015).

Claeys, G., 'The "Survival of the Fittest" and the Origins of Social Darwinism', *Journal of the History of Ideas* 61.2 (2000), 223–40.

Clark, J.F.M., *Bugs and the Victorians* (2009).

Clarke, P.F., *Liberals and Social Democrats* (1978).

Clifford, D., ed., *Repositioning Victorian Sciences: Shifting Centres in Nineteenth Century Scientific Thinking* (2006).

Cobham, E.M., *Mary Everest Boole. A Memoir With Some Letters* (1951).

Cock, A.G., 'Bernard's Symposium—The Species Concept in 1900', *Biological Journal of the Linnean Society* 9.1 (March 1977), 1–30.

SELECT BIBLIOGRAPHY 465

Cock, A.G. and Forsdyke, D.R., *Treasure your Expectations: The Science and Life of William Bateson* (2008).

Cohen, B., *The Newtonian Revolution* (1980).

Cohen, I.B. 'Three Notes on the Reception of Darwin's Ideas on Natural Selection (Henry Baker Tristram, Alfred Newton, Samuel Wilberforce)', in Kohn, *The Darwinian Heritage*, 589–607.

Collard, D.A., *Generations of Economists* (2011).

Collini, S., *Liberalism and Sociology: L.T. Hobhouse and Political Argument in England, 1880–1914* (1979).

Collini, S., *Public Moralists. Political Thought and Intellectual Life in Britain, 1850–1930* (1991).

Collins, K.K., 'G.H. Lewes Revised: George Eliot and the Moral Sense', *Victorian Studies* 21.4 (Summer, 1978), 463–92.

Conlin, J., *Evolution and the Victorians: Science, Culture and Politics in Darwin's Britain* (2014).

Costa, J.T., *Radical By Nature. The Revolutionary Life of Alfred Russel Wallace* (2003).

Craik, A.D.D., *Mr Hopkins' Men. Cambridge Reform and British Mathematics in the Nineteenth Century* (2008).

Crawforth, A., *The Butterfly Hunter: Henry Walter Bates FRS, 1825–1892* (2009).

Crook, D., *Benjamin Kidd. Portrait of a Social Darwinist* (1984).

Crook, D., *Darwinism, War and History: The Debate Over the Biology of War From the 'Origin of Species' to the First World War* (1994).

Crook, D., *Darwin's Coat-tails: Essays on Social Darwinism* (2007).

Daston, L., 'British Responses to Psycho-Physiology, 1860–1900', *Isis* 69 (1978), 192–208.

Daston, L., 'The Theory of Will Versus the Science of Mind', in Woodward, W.R., ed., *The Problematic Science: Psychology in Nineteenth century Thought* (1982), 88–115.

Davis, M., *George Eliot and Nineteenth Century Psychology* (2006).

Dawson, G., 'Aestheticism, Immorality and the Reception of Darwinism in Victorian Britain', in Zwierlein, A.-J., *Unmapped Countries. Biological Visions in Nineteenth Century Literature and Culture* (2005), 43–54.

Dawson, G., *Darwinism, Literature and Victorian Respectability* (2007).

Day, M., 'Godless Savages and Superstitious Dogs: Charles Darwin, Imperial Ethnography and the Problem of Human Uniqueness', *Journal of the History of Ideas* 69 (2008), 49–70.

Delisle, R.G. and Tierney, J,, *Rereading Darwin's Origin of Species. The Hesitations of an Evolutionist* (2023).

Desmond, A., *Archetypes and Ancestors* (1984).

Desmond, A., *The Politics of Evolution. Morphology, Medicine and Reform in Radical London* (1989).

Desmond, A., *Huxley. I. The Devil's Disciple* (1994).

Desmond, A., *Huxley. II. Evolution's High Priest* (1997).

Desmond, A. and Moore, J., *Darwin's Sacred Cause* (2014).

Diedrick, J., *Mathilde Blind: Late-Victorian Culture and the Woman of Letters* (2018).

Dixon, T., *The Invention of Altruism. Making Moral Meanings in Victorian Britain* (2008).

Donald, D. and Munro, J., eds., *Endless Forms: Charles Darwin, Natural Science and the Visual Arts* (2009).

Duncan, I., *Human Forms. The Novel in the Age of Evolution* (2018).

Durant, J., ed., *Darwinism and Divinity* (1985).

Egerton, F.N., *Hewett Cottrell Watson: Victorian Plant Ecologist and Evolutionist* (2017).

466 SELECT BIBLIOGRAPHY

Eisenman, S. and Granof, C., *Design in the Age of Darwin. From William Morris to Frank Lloyd Wright* (2008).

Elder, G.P., *Chronic Vigour: Darwin, Anglicans, Catholics and the Development of a Doctrine of Providential Evolution* (1996).

Ellegard, A., *Darwin and the General Reader. The Reception of Darwin's Theory of Evolution in the British Periodical Press, 1859–72* (1990, or. 1958).

Ellison, R.H., *A New History of the Sermon: the Nineteenth Century* (2010).

Endersby, J., *Imperial Nature: Joseph Hooker and the Practices of Victorian Science* (2010).

Engels, E.-M. and Glick, T.F., eds., *The Reception of Charles Darwin in Europe* (2008).

England, R., 'Natural Selection, Teleology and the Logos', *Osiris* 16 (2001), 270–87.

England, R., 'Interpreting Scripture, Assimilating Science: Four British and American Christian Evolutionists on the Relationship Between Science, the Bible, and Doctrine', *Nature and Scripture in the Abrahamic Religions, 1700–Present* 37 (2009), 183–224.

English, M., *Mordecai Cubitt Cooke: Victorian Naturalist, Mycologist, Teacher & Eccentric* (1987).

Epperson, G., *The Mind of Edmund Gurney* (1997).

Fara, P., *A Lab of One's Own: Science and Suffrage in the First World War* (2017).

Farrall, L., *The Origins and Growth of the English Eugenics Movement, 1865–1915* (1985).

Feuer, L.S., *Einstein and the Generations of Science* (1972).

Feuer, L.S., 'The Generational Basis for Ideological Waves', in *idem, Ideology and Ideologists* (1975), 69–95.

Fichman, M., *Evolutionary Theory and Victorian Culture* (2002).

Forrester, J. and Cameron, L., *Freud in Cambridge* (2017).

Forsdyke, D.R., *The Origin of Species Revisited: A Victorian Who Anticipated Modern Developments in Darwin's Theory* (2001).

Francis, K.A., 'A Quintessential Clergyman-scientist? George Henslow, Darwin's Theory of Natural Selection, and Nineteenth-century Science', *Anglican and Episcopal History* 79.1 (2010), 1–33.

Francis, M., et al., *Herbert Spencer: Legacies* (2014).

Freeden, M., 'Biological and Evolutionary Roots of the New Liberalism in England', *Political Theory* 4 (1976), 471–90.

Freeden, M., 'Eugenics and Progressive Thought', *Historical Journal* 22 (1979), 959–62.

French, R.D., *Anti-vivisection and Medical Society in Victorian Britain* (1975).

Gahan, P., *Bernard Shaw and Beatrice Webb on Poverty and Equality in the Modern World, 1905–1914* (2017).

Gange, D., *Dialogues with the Dead: Egyptology in British Culture and Religion, 1822–1922* (2013).

Gates, B., *Kindred Nature. Victorian and Edwardian Women Embrace the Living World* (1998).

Gauld, A., *The Founders of Psychical Research* (1968).

Gay, H., 'No 'Heathen's Corner' Here: The Failed Campaign to Memorialize Herbert Spencer in Westminster Abbey', *British Journal for the History of Science* 31.1 (1998), 41–54.

Gay, H., '"The Declaration of Students of the Natural and Physical Sciences", revisited: Youth, Science and Religion in Mid-Victorian Britain', in Feist, R. and Sweet, W., eds., *Religion and the Challenges of Science* (2007), 19–37.

Gay, H., 'Chemist, Entomologist, Darwinian, and Man of Affairs: Raphael Meldola and the Making of a Scientific Career', *Annals of Science* 67 (2010), 79–119.

Gayon, J., *Darwinism's Struggle for Survival* (1998).

SELECT BIBLIOGRAPHY 467

Geison, G.L., 'The Protoplasmic Theory of Life and the Vitalist/Mechanist Debate', *Isis* 60 (1969), 273–92.

Geison, G.L., *Michael Foster and the Cambridge School of Physiology. The Scientific Enterprise in Late Victorian Society* (1978).

Ghiselin, M.T., *The Triumph of the Darwinian Method* (1984).

Gillham, N.W., *A Life of Sir Francis Galton: From African Exploration to the Birth of Eugenics* (2001).

Gillham, N.W., 'Evolution by Jumps: Francis Galton and William Bateson and the Mechanism of Evolutionary Change', *Genetics* 159.4 (2001), 1383–92.

Gillot, D., *Samuel Butler Against the Professionals: Rethinking Lamarckism, 1860–1900* (2017).

Glendening, J., *The Evolutionary Imagination in Late Victorian Novels. An Entangled Bank* (2016).

Glick, T.F., ed., *The Comparative Reception of Darwinism* (1972).

Glick, T.F., *What About Darwin. All Species of Opinion from Scientists, Sages, Friends, and Enemies Who Met, Read, and Discussed the Naturalist Who Changed the World* (2010).

Glick, T.F. and Shaffer, E., eds., *The Literary and Cultural Reception of Charles Darwin in Europe* (2014).

Goodall, J., *Performance and Evolution in the Age of Darwin: Out of the Natural Order* (2002).

Gooday, G., '"Nature" in the Laboratory: Domestication and Discipline with the Microscope in Victorian Life Sciences', *BJHS* 24 (1991), 307–41.

Gouldstone, T., *The Rise and Decline of Anglican Idealism in the Nineteenth Century* (2005).

Greene, J.C., 'The Kuhnian Paradigm and the Darwinian Revolution in Natural History', in idem, *Science, Ideology and World View: Essays in the History of Evolutionary Ideas* (1981), 30–59.

Greenslade, W. and Rodgers, T., eds., *Grant Allen. Literature and Cultural Politics at the Fin de Siecle* (2005).

Gregory, F., 'The Impact of Darwinian Evolution on Protestant Theology in the Nineteenth Century', in Lindberg, D. and Numbers, R.L., eds., *God and Nature* (1986), 369–90.

Griffiths, D., *The Age of Analogy. Science and Literature Between the Darwins* (2016).

Grosskurth, P., *Havelock Ellis: A Biography* (1981).

Gruber, J.W., *A Conscience in Conflict. The Life of St George Jackson Mivart* (1960).

Gruber, J.W. and Thackray, J.C., *Richard Owen Commemoration: Three Studies* (1992).

Gunther, A., *A Century of Zoology* (1975).

Hale, P.J., *Political Descent. Malthus, Mutualism and the Politics of Evolution in Victorian England* (2014).

Hale, P.J., 'Rejecting the Myth of the Non-Darwinian Revolution', *Victorian Review* 41.2 (2015), 13–18.

Hall, L., 'Hauling Down the Double Standard: Feminism, Social Purity and Sexual Science in Late Nineteenth-Century Britain', *Gender and History* 16.1 (2004), 36–56.

Hamilton, T., *Immortal Longings: F.W.H. Myers and the Victorian Search for Life After Death* (2009).

Harris, J., 'Between Civic Virtue and Social Darwinism: The Concept of the Residuum', in O'Day, R. and Englander, D., *Received Riches. Social Investigation in Britain, 1840–1914* (1995), 67–87.

Harwood, J., *Styles of Scientific Thought: The German Genetics Community, 1900–1933* (1993).

Hawkins, M., *Social Darwinism in European and American Thought, 1860–1945* (1997).

468 SELECT BIBLIOGRAPHY

Hearnshaw, L.S., *A Short History of British Psychology* (1964).

Heath, K., *Ageing By the Book: The Emergence of Midlife in Victorian Britain* (2009).

Herbert, C., *Culture and Anomie. Ethnographic Imagination in the Nineteenth Century* (1991).

Hesketh, I., *Of Apes and Ancestors. Evolution, Christianity and the Oxford Debate* (2009).

Hesketh, I., *The Science of History in Victorian Britain: Making the Past Speak* (2016).

Hesketh, I., *Victorian Jesus: J.R. Seeley, Religion, and the Cultural Significance of Anonymity* (2017).

Hesketh, I., 'John Robert Seeley, *Natural Religion*, and the Victorian Conflict Between Science and Religion', *Journal of the History of Ideas* 79 (2018), 309–29.

Hesketh, I., ed., *Imagining the Darwinian Revolution. Historical Narratives of Evolution from the Nineteenth Century to the Present* (2022).

Hinchliff, P., 'Darwinism and Faith', in idem, *Benjamin Jowett and the Christian Religion* (1987), 182–208.

Hinchliff, P., *God and History. Aspects of British Theology, 1875–1914* (1992).

Hinchliff, P., *Frederick Temple. A Life* (1998).

Hodge, M.J.S., 'Against "Revolution" and "Evolution"', *Journal of the History of Biology* 38 (2005), 101–24.

Hodge, M.J.S., and Radick, G., eds., *The Cambridge Companion to Darwin* (2003).

Holmes, J., *Darwin's Bards: British and American Poetry in the Age of Evolution* (2009).

Holmes, J.*The PreRaphaelites and Science* (2018).

Howsam, L., 'An Experiment with Science for the Nineteenth-century Book Trade: The International Scientific Series', *British Journal of the History of Science* 33 (2000), 187–207.

Hughes, R., *The Red Dean: The Life and Riddle of Dr. Hewlett Johnson* (1987).

Hull, D., *The Reception of Darwin's Theory of Evolution by the Scientific Community* (1973).

Ingold, T., *The Rise of Generation Now* (2023).

Irvine, W., *Apes, Angels and Victorians* (1963).

Jacyna, L.S., 'Science and Social Order in the Thought of A. G. Balfour', *Isis* 71 (1980), 11–34.

Jacyna, L.S., 'The Physiology of Mind, the Unity of Nature, and the Moral Order in Victorian Thought', *British Journal for the History of Science* 14 (1981), 109–32.

Jacyna, L.S., *Lost Words: Narratives of Language and the Brain, 1825–1926* (2000).

James, S. and Saul, N., eds., *The Evolution of Literature* (2011).

Jann, R., 'Darwin and the Anthropologists: Sexual Selection and its Discontents', *Victorian Studies* 37.2 (1992), 287–306.

Jones, G., 'The Social History of Darwin's *Descent of Man*', *Economy and Society* 7 (1978), 1–23.

Jones, G., *Social Darwinism and English Thought: The Interaction Between Biological and Social Theory* (1980).

Jones, G., 'Contested Territories. Alfred Cort Haddon, Progressive Evolutionism and Ireland', *History of European Ideas* 24.3 (1998), 195–211.

Jones, G., 'Darwinism in Ireland', in Attis, D. and Mollan, C., eds., *Science and Irish Culture* 1 (2004), 115–37.

Jones, J.E., et al., eds., *Darwin in Atlantic Cultures. Evolutionary Visions of Race, Gender and Sexuality* (2010).

Jones, P.d'A., *The Christian Socialist Revival, 1877–1914. Religion, Class and Social Conscience in Late-Victorian England* (1968).

SELECT BIBLIOGRAPHY 469

Kelly, A., *The Descent of Darwin. Popularization of Darwinism in Germany* (1981).

Kent, J., *From Darwin to Blatchford. The Role of Darwinism in Christian Apologetic 1875–1910* (1966).

Kohn, D., ed., *The Darwinian Heritage* (1985).

Knoll, E., 'The Science of Language and the Evolution of Mind: Max Müller's Quarrel with Darwinism', *Journal of the History of the Behavioural Sciences* 22 (1986), 3–22.

Kuhn, T.S., *The Structure of Scientific Revolutions. Fiftieth Anniversary Edition* (2012).

Kuklick, H., *The Savage Within: The Social History of British Anthropology, 1885–1945* (1991).

Langham, I., *The Building of British Social Anthropology: W.H.R. Rivers and his Cambridge Disciples in the Development of Kinship Studies* (1981).

Larsen, T., *Contested Christianity: The Political and Social Contexts of Victorian Theology* (2004).

Larsen, T., *The Slain God: Anthropologists and the Christian Faith* (2014).

Larson, B., *The Art of Evolution: Darwin, Darwinisms and Visual Culture* (2009).

Laurent, J., 'Science, Society and Politics in Late Nineteenth Century England', *Social Studies of Science* 14 (1984), 595–608.

Laurent, J. and Nightingale, J., eds., *Darwinism and Evolutionary Economics* (2001).

Lenoir, T., 'Generational Factors in the Origin of *Romantische Naturphilosophie*', *Journal of the History of Biology* 11.1 (1978), 57–100.

Lenoir, T., *The Strategy of Life. Teleology and Mechanics in Nineteenth Century German Biology* (1987).

Leopold, J., *Culture in Comparative and Evolutionary Perspective: E.B. Tylor and the Making of Primitive Culture* (1980).

Lester, J., *E. Ray Lankester and the Making of Modern British Biology* (1995).

Levine, G., *Darwin and the Novelists* (1988).

Levine, G., *Darwin the Writer* (2011).

Levy, P., *Moore: G.E. Moore and the Cambridge Apostles* (1981).

Lightman, B., ed., *Victorian Science in Context* (1987).

Lightman, B., *Victorian Popularizers of Science. Designing Nature for New Audiences* (2007).

Lightman, B., ed., *Evolutionary Naturalism in Victorian Britain: The 'Darwinians' and Their Critics* (2009).

Lightman, B., 'Darwin and the Popularization of Evolution', *Notes and Records of the Royal Society* 64.1 (2010), 5–24.

Lightman, B., 'The International Scientific Series and the Communication of Darwinism', *Journal of Cambridge Studies* 5 (December 2010), 27–38.

Lightman, B., 'The Many Lives of Charles Darwin: Biographies and the Definitive Evolutionist', *Notes and Records of the Royal Society* 64.4 (2010), 339–58.

Lightman, B. and Dawson, G., eds., *Victorian Scientific Naturalism: Community, Identity, Continuity* (2014).

Lightman, B. and Reidy, M.S., eds., *The Age of Scientific Naturalism: Tyndall and His Contemporaries* (2016).

Lightman, B. and Zon, B., eds., *Evolution and Victorian Culture* (2014).

Livingstone, D.N., *Darwin's Forgotten Defenders. The Encounter Between Evangelical Theology and Evolutionary Thought* (1984).

Livingstone, D.N., 'Darwin in Belfast: The Evolution Debate', in Wilson, J., ed., *Nature in Ireland. A Scientific and Cultural History* (1997), 387–408.

Livingstone, D.N., 'Situating Evangelical Responses to Evolution', in Livingstone, D.N., ed., *Evangelicals and Science in Historical Perspective* (1999), 193–219.

470 SELECT BIBLIOGRAPHY

Livingstone, D.N., 'Public Spectacle and Scientific Theory: William Robertson Smith and the Reading of Evolution in Victorian Scotland', *Part C: Studies in History and Philosophy of Biological and Biomedical Sciences* 35.1 (2004), 1–29.

Livingstone, D.N., *Adam's Ancestors: Race, Religion, and the Politics of Human Origins* (2008).

Livingstone, D.N., *Dealing with Darwin. Place, Politics and Rhetoric in Religious Engagements with Evolution* (2014).

Lubenow, W.C., 'Intimacy, Imagination, and the Inner Dialectics of Knowledge Communities: The Synthetic Society, 1896–1908', in Daunton, M., ed., *The Organization of Knowledge in Victorian Britain* (2005), 357–70.

Lyons, S.L., *Species, Serpents, Spirits, and Skulls: Science at the Margins in the Victorian Age* (2009).

McCartney, M., Whitaker, A., and Wood, A., eds., *George Gabriel Stokes: Life, Science and Faith* (2019).

McIvor, T., *Anti-Evolution; A Reader's Guide to Writings Before and After Darwin* (1988).

MacKillop, I.D., *The British Ethical Societies* (1986).

MacLeod, R.M., 'Evolutionism and Richard Owen', *Isis* 56 (1965), 259–80.

MacLeod, R.M., 'Macmillan and the Young Guard', *Nature* 224 (1969), 435–61.

MacLeod, R.M., 'Evolutionism, Internationalism and Commercial Enterprise in Science: The International Scientific Series 1871–1910', in Meadows, A.J., ed., *The Development of Science Publishing in Europe* (1980), 63–93.

MacLeod, R.M., *Darwin's Laboratory: Evolutionary Theory and Natural History in the Pacific* (1994).

McLaren, A., *Birth Control in Nineteenth-century England* (1978).

Mairet, P., *A.R. Orage: A Memoir* (1966).

Mannheim, K., 'The Problem of Generations' (1928), in Kecskemeti, P., ed., *Karl Mannheim: Essays* (1972, or. 1952), 276–322.

Mathieson, S., 'Faith and Thought: Stokes as a Religious Man of Science', in McCartney, et al., eds., *George Gabriel Stokes*, 179–96.

Mathieson, S., *Evangelicals and the Philosophy of Science. The Victoria Institute, 1865–1939* (2021).

Mayr, E., *One Long Argument. Charles Darwin and the Genesis of Modern Evolutionary Thought* (1991).

Meadows, A., *Science and Controversy. A Biography of Sir Norman Lockyer* (2008).

Meller, H., *Patrick Geddes. Social Evolutionist and City Planner* (1990).

Mill, H.R., *Life Interests of a Geographer, 1861–1944: An Experiment in Autobiography* (1945).

Miskell, L., *Meeting Places: Scientific Congresses and Urban Identity in Victorian Britain* (2013).

Moore, J.R., *The Post-Darwinian Controversies: A Study of the Protestant Struggle to Come to Terms with Darwin in Great Britain and America, 1870–1900* (1979).

Moore, J.R., '1859 and All That: Remaking the Story of Evolution and Religion', in Chapman, R.G. and Duval, C.T., eds., *Charles Darwin, 1809–1882: A Centennial Commemorative* (1982), 167–94.

Moore, J.R., 'Charles Darwin Lies in Westminster Abbey', *Journal of the Linnean Society* 17 (1982), 97–113.

Moore, J.R., 'Evangelicals and Evolution: Henry Drummond, Herbert Spencer and the Naturalisation of the Spiritual World', *Scottish Journal of Theology* 38 (1985), 383–417.

Moore, J.R., 'The Erotics of Evolution: Constance Naden and Hylo-Idealism', in Levine, G., ed., *One Culture: Essays in Science and Literature* (1987), 225–57.

SELECT BIBLIOGRAPHY 471

Moore, J.R., 'Deconstructing Darwinism: The Politics of Evolution in the 1860s', *Journal of History of Biology* 24 (1991), 353–408.

Morgan, B., *The Outward. Mind: Materialist Aesthetics in Victorian Science and Literature* (2017).

Morton, P., *The Vital Science. Biology and the Literary Imagination* (1984).

Nias, J.C.S., *Flame From an Oxford Cloister. The Life and Writings of Philip Napier Waggett* (1961).

Nicolls, A., *Chesterton and the Modernist Crisis* (1989).

Nixon, J.V., *Gerard Manley Hopkins and His Contemporaries: Liddon, Newman, Darwin and Pater* (1994).

Northcott, M., 'The Biometric Defence of Darwinism', *Journal of the History of Biology* 6 (1973), 283–316.

Northcott, M. and Berry, R.J., eds., *Theology After Darwin* (2009).

Numbers, R.L. and Stenhouse, J., *Disseminating Darwinism: The Role of Place, Race, Religion and Gender* (1999).

Nyhart, L.K., *Biology Takes Form. Animal Morphology and the German Universities, 1800–1900* (1995).

O'Hanlon, R., *Joseph Conrad and Darwin: The Influence of Scientific Thought on Conrad's Fiction* (1984).

Orel, H., ed., *Charles Darwin. Interviews and Recollections* (2000).

Owen, J., *L.T. Hobhouse, Sociologist* (1974).

Owen, J., 'A Significant Friendship: Evans, Lubbock and a Darwinian World Order', in MacGregor, A., ed., *Sir John Evans 1823–1908: Antiquity, Commerce and Natural Science in the Age of Darwin* (2003), 206–20.

Palladino, P., 'Between Craft and Science: Plant Breeding, Mendelian Genetics, and British Universities, 1900–1920', *Technology and Culture* 34.2 (1993), 300–23.

Paradis, J.G., *Samuel Butler. Victorian Against the Grain* (2007).

Paradis, J.G. and Williams, George C., *Evolution and Ethics: T. H. Huxley's Evolution and Ethics (1893), With New Essays on Its Victorian and Sociobiological Context* (1989).

Patton, M., *Science, Politics and Business in the Work of Sir John Lubbock* (2007).

Paxton, N., *George Eliot and Herbert Spencer: Feminism, Evolutionism and the Construction of Gender* (1991).

Petrilli, S., *Signifying and Understanding: Reading the Works of Victoria Welby and the Signific Movement* (2009).

Phipps, P.A., *Constance Maynard's Passions: Religion, Sexuality, and an English Educational Pioneer, 1849–1935* (2015).

Pick, D., *Faces of Degeneration: A European Disorder, c.1848–c.1918* (1989).

Platt, J., *Subscribing to Faith? The Anglican Parish Magazine 1859–1929* (2015).

Pleins, J.D., *The Evolving God: Charles Darwin on the Naturalness of Religion* (2013).

Pleins, J.D., *In Praise of Darwin: George Romanes and the Evolution of a Darwinian Believer* (2014).

Porrovecchio, M.J., *F.C.S. Schiller and the Dawn of Pragmatism: The Rhetoric of a Philosophical Rebel* (2011).

Porter, T.M., *Karl Pearson: The Scientific Life in a Statistical Age* (2004).

Porter, T.M., *Genetics in the Madhouse: The Unknown History of Human Heredity* (2018).

Proctor, D., ed., *The Autobiography of G. Lowes Dickinson and other Unpublished Writings* (1973).

Prodger, P., *Darwin's Camera: Art and Photography in the Theory of Evolution* (2009).

Provine, W.B., *The Origins of Theoretical Population Genetics* (1971).

Punnett, R.C., *Scientific Papers of William Bateson* (1928).

472 SELECT BIBLIOGRAPHY

Purton, V., ed., *Darwin, Tennyson and Their Readers: Explorations in Victorian Literature and Science* (2014).

Quiggin, A.H., *The Head Hunter: A Short Sketch of the Life of A.C. Haddon* (1942).

Qureshi, S., *Peoples on Parade: Exhibitions, Empire, and Anthropology in Nineteenth-Century Britain* (2011).

Raby, P., *Alfred Russel Wallace. A Life* (2001).

Radick, G., *The Simian Tongue: The Long Debate about Animal Language* (2007).

Radick, G., *Disputed Inheritance. The Battle Over Mendel and the Future of Biology* (2023).

Reid, J., *Robert Louis Stevenson and the Evolutionary Sciences* (2002).

Reid, J., *Robert Louis Stevenson, Science and the Fin de Siecle* (2006).

Renwick, C., *British Sociology's Lost Biological Roots. A History of Futures Past* (2012).

Richards, E., 'A Question of Property Rights: Richard Owen's Evolutionism Re-Assessed', *British Journal of the History of Science* 20 (1987), 129–71.

Richards, E., *Darwin and the Making of Sexual Selection* (2017).

Richards, R.J., 'Lloyd Morgan's Theory of Instinct: From Darwinism to neo-Darwinism', *Journal of the History of the Behavioural Sciences* 13.1 (1977), 12–32.

Richards, R.J., *The Tragic Sense of Life: Ernst Haeckel and the Struggle over Evolutionary Thought* (2008).

Richards, R.J., *Darwin and the Emergence of Evolutionary Theories of the Mind* (2014).

Richardson, A., *Love and Eugenics in the Late Nineteenth Century: Rational Reproduction and the New Woman* (2003).

Richardson, A., *After Darwin. Animals, Emotions and the Mind* (2013).

Richmond, M.L., 'Women in the Early History of Genetics: William Bateson and the Newnham College Mendelians', *Isis* 92 (2001), 422–55.

Richmond, M.L., 'The 1909 Darwin Celebration: Re-examining Evolution in the Light of Mendel, Mutation, and Meiosis', *Isis* 97 (2006), 447–84.

Richmond, M.L., 'The "Domestication" of Heredity: The Familial Organisation of Geneticists at Cambridge University, 1895–1910', *Journal of the History of Biology* 39.3 (2006), 565–605.

Richter, V., *Literature After Darwin: Human Beasts in Western Fiction, 1859–1939* (2011).

Rupke, N.A., *Richard Owen. Biology without Darwin* (2nd ed., 2009).

Ruse, M., *From Monad to Man* (1996).

Ruse, M., *The Darwinian Revolution. Science Red in Tooth and Claw* (2nd ed., 1999).

Ruse, M., *Mystery of Mysteries. Is Evolution a Social Construction?* (1999).

Ruse, Michael, 'The Darwinian Revolution: Rethinking Its Meaning and Significance', *Proceedings of the National Academy of Sciences* 106 (2009), 10040–7.

Ruse, M., *Darwinism as Religion: What Literature Tells Us About Evolution* (2017).

Ryan, R.M., *Charles Darwin and the Church of Wordsworth* (2016).

Rylance, R., *Victorian Psychology and British Culture 1850–1880* (2000).

Schabas, M., *The Natural Origins of Economics* (2006).

Schlossberg, H., *Conflict and Crisis in the Religious Life of Late Victorian England* (2009).

Schultz, B., ed., *Essays on Henry Sidgwick* (1994).

Schultz, B., *Henry Sidgwick—Eye of the Universe: An Intellectual Biography* (2004).

Schuman, H. and Scott, J., 'Generations and Collective Memories', *American Sociological Review* 54 (1989), 359–81.

Scott, J. 'The Geddes Circle in Sociology: Ideas, Influence, and Decline', *Journal of Scottish Thought* 5 (2012), 121–34

Scott, J. and Bromley, R., *Envisioning Sociology: Victor Branford, Patrick Geddes, and the Quest for Social Reconstruction* (2013).

SELECT BIBLIOGRAPHY 473

Scott, J. and Husbands, C.T., 'Victor Branford and the Building of British Sociology', *Sociological Review* 55.3 (2007), 460–85.

Scotti, P., *Out of Due Time: Wilfrid Ward and the Dublin Review* (2006).

Searle, G.R., *Eugenics and Politics in Modern Britain, 1900–1914* (1976).

Seaward M.R.D. and FitzGerald, S.M.D., eds., *Richard Spruce (1817–1893), Botanist and Explorer* (1996).

Secord, J.A., *Victorian Sensation. The Extraordinary Publication, Reception, and Secret Authorship of* Vestiges of the Natural History of Creation (2000).

Secord, J.A., 'How Scientific Conversation Became Shop Talk', in Fyfe, A. and Lightman, B., eds., Science in the Marketplace (2007), 23–59.

Secord, J.A., ed., *Charles Darwin: Evolutionary Writings* (2010).

Secord, J.A., *Visions of Science. Books and Readers at the Dawn of the Victorian Age* (2015).

Sell, A.P.F., *Defending and Declaring the Faith* (1987).

Sell, A.P.F., *Philosophy, Dissent and Nonconformity, 1689–1920* (2009).

Shanahan, T., *The Evolution of Darwinism* (2009).

Shepherd-Barr, K.E., *Theatre and Evolution from Ibsen to Beckett* (2015).

Sera-Shriar, E. ed., *Historicizing Humans: Deep Time, Evolution, and Race in Nineteenth-Century British Sciences* (2018).

Shaw, C., 'Eliminating the Yahoo: Eugenics, Socialism and Five Fabians', *History of Political Thought* 8 (1987), 521–44.

Sheets-Pyenson, S., *John William Dawson: Faith, Hope and Science* (1996).

Sheffield, S.L., *Revealing New Worlds: Three Victorian Women Naturalists* (2001).

Shepherd, J.A., 'Lawson Tait – Disciple of Charles Darwin', *BMJ* 284 (1982), 1386–7.

Shermer, M., *In Darwin's Shadow: The Life and Science of Alfred Russel Wallace: A Biographical Study on the Psychology of History* (2002).

Shuttleworth, S., *George Eliot and Science* (1984).

Slotten, R.A., *A Heretic in Darwin's Court: The Life of Alfred Russel Wallace* (2004).

Small, H., *The Long Life* (2007).

Smith, C.H., Beccaloni, G., and Dickson, A., eds., *Natural Selection and Beyond: The Intellectual Legacy of Alfred Russel Wallace* (2008).

Smith, C., *The Science of Energy. A Cultural History of Energy Physics in Victorian Britain* (1998).

Smith, C.U.M., 'Evolution and the Problem of Mind: Part I: Herbert Spencer', *Journal of the History of Biology* 15 (1982), 55–88.

Smith, C.U.M., 'Evolution and the Problem of Mind: Part II: John Hughlings Jackson', *Journal of the History of Biology* 15 (1982), 241–62.

Smith, D., *The Correspondence of H.G. Wells* (4 vols, 1998).

Smith, J., 'Grant Allen, Physiological Aesthetics, and the Dissemination of Darwin's Botany', in Cantor, G. and Shuttleworth, S., eds., *Science Serialized: Representations of the Sciences in Nineteenth Century Periodicals* (2004), 285–305.

Smith, J., *Charles Darwin and Victorian Visual Culture* (2006).

Smith, R., 'The Human Significance of Biology: Carpenter, Darwin and the *Vera Causa*', in Knoepflmacher, U.C. and Tennyson, G.B., eds., *Nature and the Victorian Imagination* (1977), 216–30.

Smocovitis, V.B., 'Singing His Praises: Darwin and His Theory in Song and Musical Production', *Isis* 100 (2009), 590–614.

Sponsel, A., *Darwin's Evolving Identity. Adventure, Ambition and the Sin of Speculation* (2018).

Stack, D.A., *The First Darwinian Left: Socialism and Darwinism, 1859–1914* (2003).

474 SELECT BIBLIOGRAPHY

Stanley, M., *Huxley's Church and Maxwell's Demon: From Theistic Science to Naturalistic Science* (2015).

Stenhouse, J., 'Darwin's Captain: F.W. Hutton and the Nineteenth Century Darwinian Debates', *Journal of the History of Biology* 23.3 (1990), 411–22.

Stiles, A., *Popular Fiction and Brain Science in the Late Nineteenth Century* (2012).

Stocking, G.W. Jnr, *Victorian Anthropology* (1987).

Stocking, G.W. Jnr, *After Tylor. British Social Anthropology, 1888–1951* (1995).

Stone, A., *Women Philosophers in Nineteenth Century Britain* (2023).

Stone, D., *Breeding Superman. Nietzsche, Race and Eugenics in Edwardian and Interwar Britain* (2002).

Straley, J., *Evolution and Imagination in Victorian Children's Literature* (2016).

Strick, J., *Sparks of Life: Darwinism and the Victorian Debates over Spontaneous Generation* (2000).

Sulloway, F.J., 'The Darwinian Revolution's Legacy to Psychology and Psychoanalysis', in his *Freud, Biologist of the Mind: Beyond the Psychoanalytic* (1982), 238–276.

Sulloway, F.J., *Born to Rebel: Birth Order, Family Dynamics and Creative Lives* (1996).

Thomson, M., *The Problem of Mental Deficiency: Eugenics, Democracy and Social Policy in Britain, c.1870–1959* (1998).

Titterington, E.J.G., 'Early History of The Victoria Institute', *Journal of the Transactions of the Victoria Institute* 82 (1950), 53–69.

Tjoa, H. G., *George Henry Lewes. A Victorian Mind* (1977).

Tomaiuolo, S., *In Lady Audley's Shadow: Mary Elizabeth Braddon and Victorian Literary Genres* (2010).

Topham, J.R., *Reading the Book of Nature. How Eight Best Sellers Reconnected Christianity and the Sciences on the Eve of the Victorian Age* (2022).

Torrens, H.S., 'James Buckman (1814–1884): The Scientific Career of an English Darwinian', in Kölbl-Ebert, M., ed., *Geology and Religion. A History of Harmony and Hostility* (2009), 245–58.

Travis, A.S., 'Raphael Meldola and the Nineteenth-century Neo-Darwinians', *Journal for General Philosophy of Science* 41 (2010), 89–118.

Turbil, C., 'In Between Mental Evolution and Unconscious Memory: Lamarckism, Darwinism, and Professionalism in Late Victorian Britain', *Journal of the History of the Behavioural Sciences* 53.4 (2017), 347–63.

Turner, F.M., *Between Science and Religion. The Reaction to Scientific Naturalism in Late Victorian England* (1974).

Turner, F.M., *Contesting Cultural Authority. Essays in Victorian Intellectual Life* (1993).

Voigts, E., Schaff, B., and Pietrzak-Franger, M., eds., *Reflecting on Darwin* (2014).

von Arx, J.P., *Progress and Pessimism: Religion, Politics and History in Late Nineteenth Century Britain* (1985).

Voss, J., *Darwin's Pictures: Views of Evolutionary Theory, 1837–1874* (2010).

Waller, J.C., 'Becoming a Darwinian: The Micro-politics of Sir Francis Galton's Scientific Career 1859–65', *Annals of Science* 61.2 (2004), 141–63.

Wheeler-Barclay, M., *The Science of Religion in Britain, 1860–1915* (2010).

Wiener, M.J., *Between Two Worlds: The Political Thought of Graham Wallas* (1971).

Wilson, D.B., *Kelvin and Stokes: A Comparative Study in Victorian Physics* (1987).

Wilson, L.G., ed., *Lyell. The Man and His Times* (1998).

Wilson, M., *Moralising Space: The Utopian Urbanism of the British Positivists, 1855–1920* (2018).

Wolfe, W., *From Radicalism to Socialism. Men and Ideas in the Formation of Fabian Socialist Doctrines, 1881–1889* (1975).

Wollaston, A.F.R., *Life of Alfred Newton* (1921).

Yanni, C., *Nature's Museums. Victorian Science and the Architecture of Display* (1999).

Yorty, T., 'The English Methodist Response to Darwinism Reconsidered', *Methodist History* 32.2 (January 1994), 116–25.

Young, R.M., *Darwin's Metaphor: Nature's Place in Victorian Culture* (1985).

Young, R.M., *Mind, Brain and Adaptation in the Nineteenth Century* (1990).

Zon, B., *Music and Metaphor in Nineteenth Century British Musicology* (2000).

Zon, B., '"Spiritual Selection": Joseph Goddard and the Music Theology of Evolution', in Clarke, M.V., ed., *Music and Theology in Nineteenth Century Britain* (2012), 215–35.

Index

Since the index has been created to work across multiple formats, indexed terms for which a page range is given (e.g., 52–53, 66–70, etc.) may occasionally appear only on some, but not all of the pages within the range.

Aberdeen 92–3, 101–2, 114, 292–3
abiogenesis 209–11
Academy 261–2, 275–6
Acland, Henry W. 63–4, 130–1, 135, 250–1, 420–1
agnosticism 203–5, 254, 258, 262–3, 288–9, 292–3, 296, 332–3, 370–1
Alexander, Samuel 287–8, 364–5
Alexander, William 253–4
Allbutt, Clifford 202–3, 210, 384 n.182
Allen, Grant 14–15, 188, 241–3, 260–2, 275, 279–81, 418–19, 428
'Professor Milliter's Dilemma' 195–6
Evolution of the Idea of God (1897) 326
Allingham, William 119, 177–8
Allman, George J. 101, 158–9 n.298
Alloa 113, 192
Alpine Club 125
America 165, 275, 329, 398, 422
American Association for the Advancement of Science 398
anatomy 40–1, 47–8, 53–4, 214, 265, 340
Angell, Norman 362–3
Anglicanism, *see* Church of England
Annals of Botany 110–11
Anthropological Institute 109
Anthropological Society 109, 423–4
anthropology 96, 109, 165–6, 202, 204, 259–60, 278–9, 333, 350–1, 366, 373
anti-vivisection movement 336
archaeology 49, 89, 360
Argentina 191
Argyll, Duke of 212–13, 222 n.112, 250
response to the *Origin* 46, 67–8
developing evolutionary beliefs 123–5, 352, 409–10
and natural selection 127–30

responses to 99–100, 103–4, 154 n.239, 190, 248–9, 329–30, 415
and Benjamin Kidd 283–4
'Darwinism as a Philosophy' 248–9
Primeval Man; an examination of some recent speculations (1869) 120
The Reign of Law (1867) 120, 130, 146 n.115, 170
Organic Evolution Cross-Examined (1898) 327–9
Aristotelian Society 364–5
Armstrong, William 71, 124–5
Arnold, Matthew 46–7, 118–19, 146 n.114, 410–11
artificial selection 40–1, 57–8, 67–8
Athenaeum 53–4, 104–5, 112, 133
Aveling, Edward 225 n.155, 241, 278–9

Bagehot, Walter 8, 11, 89–91, 414–15
Physics and Politics (1873) 93–4
Bain, Alexander 46, 97, 186
Baldwin, Melanie 258, 264–5
Balfour, Alice 347–8, 427–8
Balfour, Arthur 15–16, 238, 254, 342–3, 401–2
A Defence of Philosophic Doubt (1879) 370
Balfour, F.H. 263–4
Balfour, Frank Maitland 255–6, 265
Balfour, Isaac Bayley 88, 346–7
Balfour, John Hutton 31–2, 57–8, 88
Ball, Robert S. 69–70
Bampton Lectures 126–7, 129–30, 135, 210–11, 289–90
Baptists 97, 200–1, 207
Barton, Ruth 45–6, 239–40
Bastian, H.C. 184–5, 210–11
Bateman, Frederic 181–2

478 INDEX

Bates, H.W. 62, 106, 119, 270
   The Naturalist on the River Amazons (1863) 89
Bateson, William 343–9, 353, 357–8, 378 n.80, 399–400, 402, 416
   early career 256, 265, 338–9, 344–5
   discontinuous variation 339, 349–50
   progressive variation 352–3
   Evolution Committee 340–3, 348
   and Mendelism 343
   and the 1909 commemorations 400–2
   Materials for the Study of Variation (1894) 295, 339
   Mendel's Theory of Heredity; a defence (1902) 348
Bath 107–8, 215
Bayne, Peter 191, 213–14
Beale, Lionel 153 n.221, 176, 184–5, 210, 323–4 n.436, 440 n.71
Beddard, F.E. 171, 237, 255, 269–70, 339
Bedrock 329–31
Beer, Gillian 21, 25–6, 40–1, 121–2, 199–200
Belfast 113–14, 173–4, 248–9, 253–4
Belfast Address, *see* John Tyndall
Bell, Alexander Graham 285, 308 n.174
Bell, Thomas 55–6, 70
Belloc, Hilaire 331
Bennett, Arnold 8, 354, 394 n.350
Benson, A.C. 247–8
Benson, E.W. 192, 240, 250–1
Bentham, George 62–3, 100–1, 105–6, 110–11, 185
Bergson, Henri 290, 353–6, 363, 368–9
Besant, Annie 14–15, 188, 238, 266, 275–6, 278–9, 413, 418–19
Besant, Walter 188–9, 208, 211–12
Bettany, G.T. 241, 257–8
Beveridge, William 335–6, 365–6
Bible 1–2, 67, 192–3, 205–6, 290, 358–9
   Genesis 23, 46–7, 49, 58, 62, 116–17, 170, 192–3, 250–2, 356, 417–18
   biblical chronology 134–5, 408
   biblical criticism 58, 97, 116, 250, 291, 356–7, 409–10
   biblical inspiration 191
   See also Mosaic cosmogeny
biometrics 277, 343–7
   Biometrika 343–4
Birks, Thomas R. 60–1, 115, 122–3, 170, 172–3
Birmingham 194, 239–40, 397–8

Blackie, John S. 42, 167, 169–70, 209–10, 212–13, 248–9, 355–6
Bland-Sutton, John 162 n.364, 242–3, 257–8
Blatchford, Robert 238, 282, 369–72
Blind, Mathilde 194, 197, 211–12, 428
Bloomsbury group 8–9, 335–6
Blunt, Wilfrid Scawen 69–70, 284
Boer War 284, 349, 362–3
Bonar, James 285–6
Bonney, T.G. 69–70, 129, 185, 194, 208, 246–7, 250, 340
Boole, Mary Everest 16, 198, 204–5
Booth, Charles 214–15, 282–3
Bosanquet, Bernard 278–81, 287
botany 59, 176–7, 237–8, 337–8, 340–2
   See also George Bentham, Linnean Society
Bower, F.O. 336–7, 346–7
Bowerbank, James S. 105–6, 110, 112–13, 134–5
Boyle Lectures 340
Braddon, M.E. 89–91, 211–12, 237–8, 243–4, 288–9, 332
Bradley, F.H. 265–6, 287–8, 364–5
Brain 186
Braithwaite, James 214
Bree, C.R. 60–1, 112–13, 115, 168–70, 410–11, 413
Bridgewater Treatises 47–8
Bristol 278–9
British Academy 398–9
British Association 26–7, 40, 49, 71, 92, 100–1, 107–8, 123–5, 131–2, 172, 185, 197–8, 212–14, 238–41, 253–4, 267–8, 271, 275–6, 340, 342–4, 346–7, 373, 406, 412, 427
   Oxford (1860), *see* Huxley-Wilberforce debate
   Norwich (1868) 88–9, 131–2, 162 n.355
   Edinburgh (1871) 107–8, 137–9
   Salisbury Address (1894) 327
British Medical Association 288–9
   British Medical Journal 167
British Museum 61, 71, 106, 259
Brodie, Benjamin 51–2, 55–6
Brooke, Stopford 132–3, 368
Browning, Robert 95, 363–4, 410–11
Bryce, James 304 n.120
Buchanan, Robert W. 191, 197, 211–12, 243–4, 407

INDEX 479

Buckland, Frank 112–13, 134–5, 167, 171, 425–6
Buckland, William 49, 62
Buckley, Arabella 103, 197, 199, 241–2, 409–10
Bunbury, Sir Charles 43–4, 57–8, 67, 99–100, 133, 149 n.162, 168–9
Burnley 376 n.42
Bury 317 n.337
Bury, J.B. 351, 401–2
Butler, Samuel 70, 158 n.296, 198–9, 211–12, 268–9
  Evolution Old and New (1879) 200
  The Way of All Flesh (1903) 8, 189–90, 396–7
  as a Lamarckian 199–200, 425
  influence of 272, 275, 335–6, 370–1

Caird, Edward 212–13, 285–6
Caird, Mona 428
Calderwood, Henry 181
Cambridge 55
  University of 46, 56–7, 69–70, 73, 88–9, 106–7, 169, 185, 255–6, 332–3, 338–9, 342–5, 399–400
  Cambridge Philosophical Society 44, 56
  Cambridge Commemorations of 1909 400–2
  See also British Association
Campbell, George Douglas. *See* Duke of Argyll
Campbell, R.J. 357–9, 376 n.54
Cardiff 137–9
Carlisle 166–7
Carlyle, A.J. 357
Carlyle, Jane 46
Carlyle, Thomas 13–14, 55, 182–3, 408–10
Carpenter, Edward 274, 282
Carpenter, J.E. 206–7
Carpenter, W.B. 45–6, 66–7, 95–6, 102–5, 112–13, 128–9, 180–1, 184–5, 244
Carroll, Lewis 13–14, 211–12
Carruthers, William 99–100, 110–11, 176–7, 190, 249–50, 259, 406
Cassell 326
Catholicism 140 n.25, 333–4, 359, 426–7
  See also G.K. Chesterton, St George Jackson Mivart
censorship 56–7, 298 n.24
Challenger expedition 412

Chamberlain, Ethel 16, 427–8
Chamberlain, Neville 16–17
Chambers Brothers 112
Chambers, Robert 42–3, 96–7, 405
  See also *Vestiges of the Natural History of Creation*
Chapman, John 48, 97–8, 272–3, 405
Cheshunt College 147 n.136
Chester 73
Chesterton, G.K. 13–14, 333–4, 349, 358–9, 371–2
Church of England 56–7, 102, 107–8, 117, 192, 197, 205–7, 214, 289–90, 409–10, 412–13, 426–7
  Church Congress 119–20, 123, 149 n.159, 177–8, 231–2 n.267, 234 n.309, 322–3 n.423
  pan-Anglican congress 357–8
  See also *Lux Mundi*
Clapperton, Jane 202–3
Clapperton, John Alexander 1–2
Clarion 336, 369–72
Clifford, John 206–7, 239
Clifford, W.K. 69–70, 105–6, 183–4, 188, 195, 199, 272–3
Clodd, Edward 43–4, 139 n.14, 183–4, 188, 251–2, 275, 282, 311 n.230, 329, 405
Cobbe, Frances Power 102–3, 178–9, 182, 199, 407–10
Cobbold, T. Spencer 82 n.156, 101–2
Collier, James 219 n.61, 274
Collins, Mortimer 150 n.166, 235 n.321
Collins, Wilkie 89–92
Comte, Auguste, *see* Comtism
Comtism 10–11, 48, 97, 103–4, 188–9, 202, 243–4, 266, 281, 288–9
Congregationalism 192, 255, 289–90, 356–7, 376 n.42
Contemporary Review 167–8, 174, 181, 186, 275–6
Contentio Veritatis (1902) 357
Conway, Moncure Daniel 188
Cooke, Mordecai C. 112, 126–7
Cooper, Thomas 53, 170, 213–14, 247–8, 413
Corelli, Marie 263–4
Cotteswold Naturalists' Field Club 52–3, 113–14, 215
Couch, Jonathan 57
Crane, Agnes 265
Crane, Walter 201
Crawfurd, John 86 n.225, 107–8

480   INDEX

Creighton, Mandell 411–12
Crichton-Browne, James 94, 314 n.293
Croll, James 129, 158 n.295, 232 n.288
Crompton, Joseph 76 n.71, 161 n.351
Cunningham, J.T. 241, 275–6, 341–2, 346–7
cytology 276–7, 336–7

Dallas, W.S. 95, 167–8
Dallinger, W.H. 196, 200–1, 215, 289, 292–3, 327–8
Darwin, Charles 66–7, 117–18
  and age 3
  evolutionary writings 20–4, 89, 257–8, 326, 331–2
  and the Darwinians 19, 83 n.174
  responses to the evolutionary debate 45–6, 129
  personal reputation 107, 113, 119–20, 182–3, 237–48, 256–7, 396–8
  as a personal presence in evolutionary debates 94–6, 99–100, 104–5, 194, 333
  and Wallace 49
  and Romanes 269–70
  honours 88–9, 106–7, 212–13, 401–2
  death and burial of 237, 239–40
  loss of musical appreciation 368
  statues of 240, 282, 329
  1909 commemorations 398–402
  Formation of Vegetable Mould Through the Action of Worms (1881) 20, 244, 247–8
  Journal of Researches into the Natural History and Geology of the countries visited during the voyage of H.M.S. Beagle round the world (1845 etc) 16, 141 n.30, 248–9, 257–8, 331–2
  Life and Letters of Charles Darwin (1887) 20, 206, 244–7
  The Expression of the Emotions in Man and Animals (1872) 164–7, 169
  The Fertilisation of Orchids (1862) 92, 113
  The Variation of Animals and Plants Under Domestication (1868) 19, 60–1, 89, 92, 133
  See also *On the Origin of Species*, *Descent of Man*, Down House
Darwin, Erasmus 46–7
Darwin, Francis (Frank) 373, 398
Darwin, Leonard 284–5, 349–50, 422
Darwinian Revolution 24–30, 435
Darwinism
  definitional issues 17–24

  as a philosophy 19, 59, 115–16, 182–3, 205, 219–20 n.76, 248–9, 266, 288–9, 356, 358–9, 371, 379–80 n.106, 392 n.311, 423–4, 440 n.74, See also scientific naturalism
  institutionalisation of 105–15
  See also British Association, Royal Society, and *passim*
Daubeny, Charles 44, 51–2
Davidson, Thomas 64, 68, 176
Dawkins, William Boyd 31–2, 94, 133, 185, 405
Day, H.G. 301 n.77, 406
de Vries, Hugo 343, 346–7, 349–51, 356–7
'Declaration of Students of the Natural and Physical Sciences' 135
degeneration 238–9, 269–70, 284–5, 349
Derry 52–3
Descent of Man
  content 164–5
  circulation 20, 166–7, 213–14, 241–2, 247, 326
  readings 92–3, 166–9, 181–2, 202–3, 214, 304 n.122, 308 n.179
  impacts 103–4, 135–9, 165, 171–3, 178–80, 211–12, 230–1 n.245, 253–4, 291–2, 418–19
  status 51, 213–14, 243–4, 278–9, 330
design, theories of 18–19, 23–4, 92, 117–18, 249, 292–3, 327
  as a reason for resisting Darwinism 59, 71, 117–18, 122–3, 126–9, 137–9, 170, 207
  reconciliations with Darwinism 62, 66–7, 102–3, 128–30, 134–5, 200–1, 294–5, 411–12, 417–18
  rejections of 64–5
  See also Darwin; *The Fertilisation of Orchids* (1862)
Dicey, Albert Venn 97, 123–4, 246–7
Dick, Robert 43, 58
Dickie, George 31–2, 117–18
Dickins, F.V. 192–3, 270–1, 370
Dickinson, G.L. 378 n.87
Disraeli, Benjamin 115–16, 188–9, 418–19
Distant, William L. 226 n.182, 303 n.104
distribution, *see* geographical distribution
Dixey, F.A. 252–3, 267–8, 291–2, 294, 375 n.27
Dods, Marcus 283–4, 420–1
Down House 186–7, 240–3, 342
Dowson, Mary Emily 262–3
Doyle, Arthur Conan 251–2, 263
drama 213–14, 263–4, 353–4

Driver, Samuel R. 291
Drummond, Henry 15–16, 173, 278–9, 291–3
Du Chaillu, Paul 15–16, 53–4
Du Maurier, George 211–12, 215, 298 n.32
Dublin 306 n.155, 373
  Trinity College, Dublin 46–7, 69–70, 73, 127–8
  Catholic University 140 n.25
Duffy, Bobby 3
Duke, Edward 123, 219 n.74
Duncan, Peter Martin 111, 128–9
Dundee 249–50, 256, 306–7 n.157
Dunfermline 278–9

early Victorians 13–14, 417–22
  responses to the *Origin* 57–63, 101, 106–7, 114–18, 122–4, 127–8, 131, 134–5
  responses to *Descent* 168–72, 209–10
  late-century interventions 250–2, 374 n.14
  eclipse of 98, 112, 137, 368–9, 416
earth, age of 23, 54–5, 76 n.61, 107–8, 129, 196, 420–1
Edinburgh 53
  Edinburgh Botanic Society 101
  Edinburgh Botanical Garden 93–4
  Edinburgh Philosophical Institution 96–7, 113
  Edinburgh Philosophical Society 167, 181
  Royal Society of Edinburgh 44, 52–3
  Edinburgh University 1–2, 255, 274, 276–7, 285–6
  Edinburgh University Darwinian Society 304 n.128
  Edinburgh University Philosophical Society 212–13
  See also Gifford Lectures, British Association
Edinburgh Evening News 274–5
Edinburgh Review 44, 133, 344–5, 399–400
Edward Arnold (publishers) 257
Edwardians 13–14, 417–22
  emergence of 254–5, 260, 295, 329–30
  and High Victorian ideas 283–4
  evolutionary beliefs 296, 329–36, 349–55
  evolutionary questions 336–43
  evolutionary doubts 336, 367–72
  applications of evolutionary beliefs 355–67
  and the 1909 commemorations 401–2
  See also Mendelism
Elam, Charles 174

Elgin 74 n.19
Eliot, George 42–3, 48, 272–3, 279
Ellicott, Charles James 119–20, 173–4
Ellis, Havelock 239, 257, 262–3, 274, 277–9, 308 n.177
Elwes, H.J. 252–3
Elwin, Whitwell 63–4, 81–2 n.147
embryology 40–1, 101, 103, 165, 197–8, 255–6, 265, 420–1
Endersby, Jim 27–8, 126
Engels, Friedrich 124–5
entomology 109–10, 176, 198–9, 252–3, 346–7, 425–6
  Entomological Society 109–10, 252–3, 406, 415–16
  Entomologist's Monthly Magazine 109–10
  Entomologists' Weekly Intelligencer 149 n.156
essay competitions 185, 247–8, 374 n.11, 379–80 n.106
Essays and Reviews 68, 93, 98, 106–7, 188–9, 243–4, 332–3, 409–10
Ethnological Society 109
eugenics 280, 284–5, 345–6, 349, 371
  Eugenics Education Society 349
  Eugenics Review 349
Evans, John 95, 107–8
Ewart, James Cossar 255, 276–7, 340, 344–5, 347–8
Exeter 88–9
Exeter Hall 54–5, 250

Fabian Society 272, 280–2, 370–1
Fairbairn, A.M. 205–6, 208
Falconer, Hugh 42, 72–3, 107
Farrar, F.W. 51–2, 121–3, 188–9, 210, 248–9
Fawcett, Henry 69
Fernley Lecture 215, 289
fiction 89–91, 191, 211–14, 216 n.11, 237–8, 243–4, 262–4, 284, 295, 333–6, 354–5, 363–4, 367–8
Field 112–13, 348
field clubs 52–3, 113–14, 256–7
Figgis, J.N. 324 n.445, 354
Fisher, R.A. 402
Flower, W.H. 54–5, 123–4, 136–7, 190–1, 259, 268–9, 336–7, 405, 422
Ford, Ford Madox 335–6, 396–7
Forster, E.M. 115–16, 189–90
Fortnightly Review 89–91, 93–4, 186, 262–3
fossil record, *see* palaeontology

482 INDEX

Foster, Michael 107–8, 131, 185–6, 202, 249, 268–9, 303 n.102, 341–2
Frankland, Edward 95, 374 n.13
Fraser, Alexander Campbell 97, 181, 244
Frazer, James G. 275, 282
Free Church 206, 289, 291–2
　Free Church College 52–3
　Free Church Congress 247–8
　See also Edward White, Henry Drummond
Freud, Sigmund 366, 402
Froude, J.A. 96, 106–7, 178–9
Fry, Edward 120, 311 n.228, 377 n.64

Gadow, Hans F. 255–6, 314 n.285
Galsworthy, John 326–7, 367–8
Galton, Francis 98–9, 125–6, 276, 280, 339, 341–3, 384 n.183, 400, 407
　Hereditary Genius (1869) 133, 284–5
　Natural Inheritance (1889) 277
Gardeners' Chronicle 112–13, 216 n.17, 347–8, 399–400
Gardiner, Walter 314 n.285, 351
Garrod, Alfred 265
Garstang, Walter 309 n.197, 377 n.73, 386 n.215
Gascoyne-Cecil, Robert, see Lord Salisbury
Gatty, Alfred 216 n.23
Gatty, Margaret 57–8, 60–1, 418–19
Geddes, Patrick 11, 14–15, 31–2, 238–9, 249–50, 278–9, 285, 341–2, 353
　and Lamarckism 267, 275
　and sociology 280–1, 365–6, 398–9
　influence of 352–3, 360, 440 n.80
　Patrick Geddes and Colleagues 257
Geikie, Archibald 69–70, 401–2, 422
Geikie, James 256
generations
　identities 416–17
　theories of 5–7, 9–12
　languages of 4–5
　summary of differences over evolution 417–22
　Victorian generational schema 12–14
　Victorians and generations 8–9
　women and 427–8
　See also early Victorians, mid Victorians, high Victorians, late Victorians, Edwardians
geographical distribution 40–1, 52–3, 70, 99–100, 106, 124–5

Geological Magazine 111
Geological Society 111, 179, 406
Geologist 111
geology 49, 57, 62, 116, 130–3, 175–6, 188, 249–50, 291–2, 356
　Darwin's geology 401–2
　primacy of geological understandings 422
　uniformitarians 21–2
　See also Charles Lyell, William Thomson, palaeontology
George Tyrrell 359–60
Germany 13–14, 32–3, 58, 244–6, 255, 352, 411
　biblical criticism 93, 291, 319 n.367
　militarism 391 n.310
　See also Ernst Haeckel, August Weismann
Gifford Lectures 158 n.294, 213–14, 220 n.86, 249, 287, 336, 355–6
Gilfillan, George 121–2
Girdlestone, Robert Baker 207
Gissing, George 262
Gladstone, Helen 300 n.65
Gladstone, John Hall 67–8
Gladstone, W.E. 171–2, 192, 295–6, 433–4
　and T.H.Huxley 251–2, 333, 415
Glasgow 69–70, 288–9
　Free Church College 52–3, 291–2
　Science Lectures Association 128–9
　University of Glasgow 214, 285–6, 346–7, 356–7
Glasgow Herald 263–4
Glasier, J. Bruce 165, 262–3, 407
Glasier, Kathryn 375 n.38
Gloucester 53
Godman, F. DuCane 70, 111
Good Words 173, 248–9
Gore, Charles 254–5, 293–4, 323 n.434, 357
gorillas 15–16, 53–4, 164, 188, 198, 317 n.333
Gosse, Edmund 8–9, 62–3, 363–4, 396
　Father and Son (1907) 8, 189–90, 396–7
Gosse, Philip H. 48–9, 413
Gould, Frederick J. 257–8
Gould, John 99–100, 122–3, 203–4
Graham, William 268–9
Grant, Alexander 167–8, 181
Graphic 184, 399–400
Gray, Asa 41–2, 59, 253–4, 277–8, 324 n.437
Gray, John E. 61, 224 n.145
Green, T.H. 255–6, 285–6, 309 n.204

INDEX 483

Greening, Linnaeus 256–7
Greg, W.R. 97–8
Gresham College 327
Grove, William R. 107–8, 318 n.347
Guardian 107–8, 131–2, 224 n.145, 357–8
Guest, William 216 n.24, 300 n.55
Gunther, Albert 110–11, 147 n.127, 156 n.267, 259

Haddon, Alfred Cort 16, 174–5, 259–62, 278–9, 291–2, 366, 393 n.328
Haddon, Caroline 174–5
Haeckel, Ernst 17–18, 100–1, 195, 207, 244–6, 249, 327–8, 331–2, 422
  Evolution of Man (1883) 241–2
  The History of Creation (1876) 209, 268
  The Riddle of the Universe (1900) 282, 354–5
  responses to 135–6, 140 n.25, 237, 290–1, 324 n.437, 371, 428
Haggard, H. Rider 284
Haldane, John S. 255, 298 n.35
Haldane, Robert Burdon 255, 309 n.192, 319 n.367
Hale, Piers 24, 27–9, 47–8, 270
Halifax 186–8, 414–15
Hardie, Keir 278–9, 282, 350–1, 414–15
Hardy, Thomas 89–91, 183–4, 197, 199–200, 429–30
  'Drinking Song' 198
  Return of the Native (1878) 211–12
Harris, Frank 262–3, 438 n.51
Harrison, Frederic 186, 189–90, 204–5, 231 n.251, 367–8
Harrison, Mary St Leger (Kingsley) 299 n.49, 440 n.73
Hartog, Marcus 235 n.320, 314 n.285
Haughton, Samuel 127–8, 213–14
Hawkins, Benjamin Waterhouse 53, 71–2, 213–14
Hegel, G.W.F. 21–2, 285–6
  Hegelianism 97, 127–8, 202, 206, 290, 294, 355–7, 363, 411
Henley, W.E. 279–80
Henslow, George 154 n.236, 199–200, 352–3, 357–8
Henslow, John S. 44, 55–6, 102, 208
heredity 133, 268, 275–8, 286–7, 336–41, 349–50, 400–1, 409–10
  See also Mendelism

Herschel, J.F.W. 55–6, 59–60, 122–3, 156 n.264, 210, 409–10
Hesketh, Ian 18–19, 25, 29
high Victorians 13–14, 21–4
  emergence of 183–8, 256–7
  responses to the *Origin* 29–69, 93–4, 114, 137
  evolutionary beliefs 167–8, 188–200, 214–15, 273–4
  applications of evolution 201–8, 211–12, 358–9
  and abiogenesis 210–11
  dismissal of pre-evolutionary ideas 100–1, 114–15
  resistance to later Victorian developments 264–9, 283–4, 288, 292–3, 296, 329, 336–7, 344–5, 368, 434–5
  and the 1909 commemorations 398–9, 401–2
Hirst, Thomas Archer 73, 95, 107–8
Hobhouse, L.T. 316 n.321, 361, 365–7, 395 n.375, 400–1
Hobson, J.A. 274–5, 280–1, 354–5
Hodgkin, Thomas 248–9
Hodgskin, Thomas 93
Holland, Henry 153 n.216
Holland, Henry Scott 232 n.277
Holmes, John 3, 24–7, 29–30, 171, 203–4
homological arguments 60–1, 127–8, 198, 214, 420–1
Hooker, Joseph D. 24, 48–9, 69–70, 95, 110–11, 184–5, 244–6
  response to the *Origin* 41–2, 45–6, 95–6, 101–2
  British Association (1868) 88–9, 107–8
  and natural selection 126
  and pangenesis 133
  and the Linnean Society 110–11
  and Weismannism 311 n.230
  and the 1909 Commemorations 400–2
Hopkins, Gerard Manley 192–3
Hopkins, William 48, 55–6
Hort, Fenton 72–3, 135
Horton, R.F. 247–8
Howard, John Eliot 107–8, 170
Howard, Lemuel 166–7
Howell, George 237–9
Howitt, William 169–70, 418–19
Huddersfield 114, 306–7 n.157

484   INDEX

Hudson, William Henry 191, 237, 407
Huggins, William 102, 249–50, 407
Hughes, Hugh Price 289–90, 292–3
Hughes, Thomas 283–4, 422
Hull 52–3
Hull, David L. 27–8, 406, 413
Hulsean Lectures 154 n.230
human evolution 21–4, 40–1, 46, 53–5,
    97–100, 114–15, 119–20, 164–8, 180
  early Victorians 40, 48–9, 114–15, 134–5,
    169–70, 410
  mid Victorians 121–3, 135–6, 172–4,
    178–81, 374 nn.11, 14, 428–9
  high Victorians 197–200, 204–5, 211–12
  late Victorians 265–6, 272, 278–9, 282
  Edwardians 363, 369
  See also language, morality, natural
    selection
Humanitarian League 288
Humphry, G.M. 112–13, 136–7
Hunt, William Holman 179
Hunt, James 109, 153 n.222, 423–4
Hunt, Thornton 61
Hurst, Charles Chamberlain 348
Hurst, Charles Herbert 276–7
Hutchinson, H.N. 306 n.145
Hutton, Frederick W. 69, 202–3
Hutton, R.H. 73, 82 n.159, 119–20, 133–4
Huxley, Julian 268–9, 402
Huxley, Leonard 350
Huxley, Thomas Henry 2–3, 24, 46–8, 71–2,
    186–7, 291, 407–8, 425
  Darwinian campaigner 19, 43–4, 46–7,
    53, 64–6, 88–9, 105–6, 113, 120–1, 131–2,
    161 n.336, 165, 173–4, 180–1, 224 n.145,
    239–40, 251, 327, 408–9
  response to the Origin 41–2, 45–6
  debate with Richard Owen 53–5, 63–4
  anti-theological animus 64–7, 105–6, 116,
    128–9, 131, 135, 207, 223 n.139, 422
  and natural selection 103–4,
    157 n.280, 404
  as teacher 106–7, 275–9, 340
  as mid Victorian 126, 128–9, 280–1, 296,
    415, 428–9
  and Lamarckians 271–2, 341–2
  and Spencer 273–4
  and materialism 209–10
  and Gladstone 251–2
  saltationist beliefs 64–5, 349–50

organisational influence 71, 88–9, 107,
    109–11, 123, 184–5
intellectual influence 92–3, 98, 102–4,
    118–19, 133–7, 176–7, 189–90, 192, 195–6,
    202–3, 241–4, 247, 261–2, 290, 332–4,
    370–1
responses to 179, 214, 229 n.231, 235 n.318,
    251–2, 329–30, 335–6
Lectures to Working Men (1863) 70–1
Man's Place in Nature (1863) 43–4, 89,
    164, 325–6
'The Physical Basis of Life' (1868) 182–3
Darwiniana (1893) 251–2
See also X-Club, Royal School of Mines,
    Romanes Lectures
Huxley-Wilberforce debate 50–2, 73, 88–9,
    98–9, 101, 188–9, 239, 300 n.60

Ibis 112
Idealism 255–7, 274, 285–8, 295–6, 355–6,
    361, 364–5, 416–17
Illingworth, J.R. 293–4, 357, 359
Illustrated London News 138, 243
immanentism 294–6, 357–9
imperialism 284, 360, 362–3
Independent Labour Party 279–80, 336
India 64, 93–4
Inge, W.R. 333, 349, 357
Iverach, James 206–7, 292–3

Jackson, John Hughlings 109, 202, 272–3
Jacobs, Joseph 235 n.321, 307 n.166
James, William 355–6, 364–5, 398
Jeffreys, J.G. 49–50, 172, 412
Jenkin, Fleeming 123–4
Jenyns, Leonard 44, 61–2
Jevons, W. Stanley 201–3
John Murray (publishers) 20, 40, 253
John O'Groat's Debating Society 327–8
Johns Hopkins University 398
Jones, Henry Arthur 241–2, 263–4
Jones, Henry Bence 105–6
Jones, Henry Festing 335–6
Jones, Thomas Wharton 117–18
Journal of Anatomy and Physiology 112–13
Journal of Travel and Natural History
    112–13
Jowett, Benjamin 92–3, 96–7, 160 n.316,
    166–7, 178–9, 295–6, 410–11
Jowett, John H. 356–7

INDEX   485

Judd, J.W. 150 n.173, 157 n.280, 194, 264–5, 299 n.51, 354
Justice 282

Kant, Emmanuel 21–2, 285–6, 356, 411
Keddie, William 52–3, 122–3
Keith, Arthur 332, 377 n.60, 422
Kelly, John 56–7
Kendal Literary and Scientific Institution 132–3
Kendall, May 251–2, 360
Kenealy, Arabella 263–4, 310 n.210, 367–8
Kenney, Annie 327–8, 427–8
Kidd, Benjamin 278–9, 282–4, 287, 296, 334–5, 366, 422
Kidd, Walter A. 316 n.314
King's College London 92–3, 261–2
Kingsley, Charles 43–4, 46–7, 54–5, 76 n.61, 92, 95–6, 130, 135–6, 152 n.199, 169, 208, 327–8, 409–10, 419–20
   accounts of the spread of Darwinism 70–1, 73, 88–9
   doubts about Darwinism 67, 101–4, 131, 135, 158 n.296
   The Water Babies (1862) 53–4, 211–12
Kingsley, Mary 331
Kinns, Samuel 102, 250–1, 300 n.55
Knowles, James 186, 250
Koselleck, Reinhardt 430
Kropotkin, Peter 188, 282

Labour Church 288, 336
Laing, Samuel 178, 325, 425–6
Lamarck, Jean-Baptiste 21
Lamarckism
   and pre-Darwinian evolution 21–2, 46–8, 425–6
   readings of the *Origin* 52–3, 56–7, 61, 73 n.4, 86 n.231, 94–7, 122–3, 412
   influences 105–6, 126, 141 n.35, 154 n.236, 172, 195–6, 206, 219–20 n.76, 282, 287, 306 n.145, 337–8, 395 n.384
   neo-Lamarckism 237, 260–1, 267–76, 344–8, 353–4
   Darwinian opposition to 329–31, 344–5, 368–9, 398–9, 422
   See also Herbert Spencer, Samuel Butler
Land and Water 112–13, 167, 406
Lang, Andrew 25–6, 106–7, 186–7, 202, 285, 397–8, 427

Lang, Cosmo Gordon 356–7, 390 n.287
language, capacity for 181–2, 198–9, 202, 265–6, 327
Lankester, Edwin 71–2
Lankester, E. Ray 6–7, 110–11, 249, 251–2, 277–8, 294, 340–1, 345, 353–4, 383 n.169, 400–1, 418–19
   and the British Association 163 n.370, 346–7
   academic career 255, 304–5 n.129, 338
   and the Natural History Museum 259
   and neo-Darwinism 267–8, 275–6, 329, 340, 398–402, 422
   and Romanes 270–1
   and Mendelism 346, 383 n.159
   and Bergson 388 n.245
   Degeneration (1880) 238–9
   See also Evolution Committee
late Georgians 55–7, 113–14, 137
late Victorians 13–14, 417–22
   emergence of 110–11, 204–5, 238–9, 254–60
   evolutionary beliefs 260–6
   applications of Darwinism 278–95, 362–3, 365–6
   sectarian debates 266–78, 329, 398–9
   and Mendelism 339, 346–7, 351
   and laboratory science 340
   and abiogenesis 211
   and vitalism 353–4
Layard, Nina F. 423–4, 427–8
Laycock, Thomas 96, 425–6
Le Gallienne, Richard 326–7, 354–5
League of Progressive Thought and Social Service 358–9
Lear, Edward 14–15, 179
Lecky, W.E.H. 46–7, 94, 169
lectures and lecturing 2, 44, 174–5, 237–8, 327
   academic lectures 14–15, 92–3, 255–6, 292–3
   provincial lecture platform 53–5, 71–2, 113–14, 132–3, 137, 168, 170, 187–8, 213–15, 256–7, 290, 411
   British Association 184–5, 238–9
   museum lectures 260
   Sunday Lectures 188
   University Extension 256–7
   See also Bampton Lectures, Fernley Lecture, Gifford Lectures, Rede Lectures, Romanes Lectures, Royal Institution, Thomas Henry Huxley

486 INDEX

Lee, Henry 134–5, 150 n.166
Lee, Vernon 186–7, 279
Leeds 54–5, 135–6, 214, 411
Leighton, Gerald R. 348, 384 n.181
Levine, George 24–6, 203–4
Lewes, George Henry 48, 89–91, 95–7, 130–1, 180–1, 186–7, 272–3
libraries
  personal libraries 91–2, 241–2, 369–70
  public libraries 15–17, 241–2, 278–9, 292–3, 331, 376 n.51, 393–4 n.349, 397–8
  school libraries 16, 213–14, 332
  subscription libraries 49, 246–7, 282–3
Liddon, Henry 126–7, 135, 166–7, 169, 180, 212–13, 222 n.120, 249, 293–4
Lightman, Bernard 24, 112, 116, 130
Lindsay, James A. 436 n.17
Lindsay, William Lauder 198–9
Linnean Society
  Darwin-Wallace papers 1858 41–2, 49–50, 98–101
  conflicts within 100–1, 110–11, 185, 415–16
  1908 celebration 396, 400
Liverpool 52–5, 119–20
  See also British Association
Lockyer, Norman 112, 186–7, 202, 244–6, 271
Lodge, Oliver 92–3, 288–9
London 47–8, 89–91, 105–6, 112, 166–7, 289, 326, 376 n.51
London Ethical Society 320 n.381
London Institution 310 n.215
London Medical School 348–9
London School of Sociology 365–6
London Working Men's College 338
Longman, William 14–15
Longman (publishers) 257
Lowne, B.T. 195, 197
Lubbock, John 16–17, 38 n.73, 45–6, 65–6, 105–6, 109–10, 179, 186–7, 241–2, 424–5
  and the British Association 107–8, 163 n.370, 197–8, 213–14, 420–1
  and entomology 198–9
  X-Clubber 64–5, 70, 95, 184–5, 239–40
  Prehistoric Times (1865) 89
Lucretius 21, 91–2, 118–19, 213–14, 411
Lux Mundi 291–4, 357, 414–15, 434
Lyall, A.C. 254–5, 226 n.173
Lyell, Charles 43–4, 49, 55, 59
  evolutionary doubts 96–7, 99–100, 102–3, 133–4, 150 n.173

  conversion 99–100
  Antiquity of Man (1863) 86 n.225, 89, 99–100, 121–2
  Principles of Geology (1830–33) 46–7, 99–100
  influence of 36 n.43, 57–8, 122–3, 227 n.189, 410–11
Lyttelton, Edward 262, 378 n.82

M'Kinney, S.B.G. 288–9
Macalister, Alexander 233 n.295, 309 n.196
Maccoll, D. Stewart 369–70
MacColl, Norman 186
MacDonald, Ramsay 350–1, 360, 376 n.51
MacIlwaine, William 113–14, 248–9
Mackenzie, John S. 283–4
Mackinder, Halford 360
Mackintosh, Daniel 53, 71–2
Macmillan (publishers) 120–1, 247, 257
Macmillan, Alexander 63, 74 n.11
Macmillan's Magazine 69, 98–9, 125, 129
Macpherson, Hector 274–5, 288
Macvicar, John G. 43, 56–7, 147 n.127
Madeira 68, 158 n.289
Magee, William Connor 216 n.13, 303 n.111
Mallet, Lucas. *See* Mary St. Leger (Kingsley) Harrison
Mallock, W.H. 153–4 n.226, 265, 278–9, 327–8
Malthusianism 25–6, 40–1, 196, 279–80, 316 n.319, 349
Manchester 41–2, 117, 290, 317 n.333
  Manchester Grammar School 213–14
  Manchester School 97–8, 280, 360–1, 363
Mannheim, Karl 5–9
Marett, R.R. 333, 366, 391 n.304
marginalia 55–6, 59–60, 126, 166–9, 214, 221–2 n.105, 285, 317 n.333, 326
Marías, Julian 5–6, 9, 433
marine biology 95–6, 176, 206, 425–6
  Marine Biological Association 180, 333–4
Marshall, Alfred 202–3, 272–3
Martineau, Harriet 42
Martineau, James 59, 61, 93, 98, 122–3, 127–8, 156 n.266, 254, 292–3
Marx, Karl 74 n.11, 363, 372
Massingham, H.W. 163 n.371, 328–9
Masters, Maxwell T. 112–13, 385 n.192
Matheson, A. Scott 192, 412
Matheson, George 69–70, 206–8, 420–1
Maugham, W. Somerset 331

INDEX 487

Maurice, F.D. 98, 101–2, 396–7
Maurier, George du, *see* Du Maurier, George
McCosh, James 101, 123–6
McMillan, Margaret 336
Meldola, Raphael 112–13, 149 n.156, 341–3, 394 n.359
    as neo-Darwinian 267–8, 270, 353, 399–400
    and the Entomological Society 252–3
Men and Women's Club 280–1
Mendelism 28–9, 330–1, 342–50, 356–7, 361, 399–402, 409–10, 424–5
    Mendel Journal 348–9
    Mendelian Society 348–9
Meredith, George 25–6, 91–2, 143 n.71
Metaphysical Society 133–4, 254, 414–15
Methodism 1–2, 205, 213–15, 232 n.271, 257–8, 289–90, 305 n.135, 375 n.20
Miall, Louis C. 197–8
mid Victorians 63–8, 88–9, 417–22
    responses to the *Origin* 97, 101–5, 112, 114–15, 118–31, 135–7
    responses to *Descent* 167–9, 172–83, 187–8, 208
    and abiogenesis 209–10
    responses to Darwin's death 239, 247–8, 251, 253–4, 266–7
    and High-Victorian Darwinism 283–4, 292–3
    persistence and decline 238, 328–9, 368, 434–5
Midland Institute 137
Midland Union of Natural History Societies 194
Mill, John Stuart 10–11, 35 n.24, 48, 61, 132–3, 212–13
    influence of 186, 225 n.158, 230 n.238, 266, 308 n.173, 376 nn.49, 56, 416–17
Millais, John Everett 203–4
Miller, Hugh 48–9, 58, 113
Miller, William Allen 107, 123
mimicry, *see* protective mimicry
Mind 186
Mitchell, Hugh 147 n.136
Mitchell, P.C. 329–32, 343–4, 352–3
Mivart, St George Jackson 67–8, 120–1, 167–8, 183–4, 198, 295–6, 359, 426
    On the Genesis of Species (1871) 120, 127–8, 196

Lessons from Nature as manifested in Mind and Matter (1876) 181
    limits of transmutation 123, 125–6
    and natural selection 124–5, 196
    teleology 127–8
    materialism 183
    relations with the Darwinists 131, 173–4, 184–5, 328–9
    influence of 158 n.296, 168, 171–2, 181–2, 200, 217 n.31, 222 n.119, 236 n.342, 419–20
Moore, Aubrey L. 103–4, 289, 293–4, 413
Moore, F.F. 213–14
Moore, G.E. 317 n.326, 364–5
Moore, George 263–4
Moore, James 18–19, 24, 120, 408–9
morality
    capacity for 61–2, 165, 169, 180–1
    evolution of 197, 199, 286–8, 296, 357
Morgan, Conwy Lloyd 238–9, 255–7, 265–6, 278–9, 288, 340, 353, 367–8, 410
    Spencerianism 274–5
    Gifford Lectures 351
    and Mendelism 436 n.19
Morley, Henry 127–8
Morley, John 17, 85 n.214, 167–9, 186, 202–3, 360
Morning Post 397–8
Morris, Francis Orpen 60–1, 107–8, 112–13, 115–16, 170, 192, 213–14, 418–21
Morris, William 203–4, 326–7, 368
Mosaic cosmogeny 46–7, 56–8, 123, 136–7
Moseley, H.N. 194–5, 206, 360
Mudge, G.P. 348–9
Muirhead, John H. 255, 257, 279–80, 285–8
Müller, Fritz 192–3
Müller, Max 18–19, 48–9, 71, 120, 135–6, 249, 302 n.94, 355–6, 409–10
    and the origins of language 53–4, 121–2, 168, 181–2
    influence of 198–9, 202, 265–6, 320 n.384, 328–9, 333, 439 n.66
    Chips from a German Workshop (1895) 327–8
    Introduction to the Science of Religion (1873) 173–4
    Gifford Lectures 249
Murchison, Roderick 55–6, 71, 111
Murphy, Joseph J. 120, 123, 125–6, 160 n.316, 173–4, 236 n.342

488   INDEX

Murray, Andrew 57–8, 112–13
Murray, George 150 n.167, 249–50, 406
Murtle Lectures 233 n.295
museums 258–60
  See also Natural History Museum
Museums Association 260
music 279, 290–1, 431
  See Darwin, loss of musical appreciation
mutation theory 330–1, 343–5

Naden, Constance 14–15, 242–3, 266
National Reformer 69, 275–6
National Review 73
Natural History Museum 110–11, 171, 240,
  249–50, 259, 282, 329, 399–400, 406,
  412–13
Natural History Review 112
Natural Science 267–8
natural selection
  origins of 46–7, 49
  place within Darwinian thought 22–4,
    40–1, 96, 126
  initial responses to 42–3, 88–9, 107–8
  early-Victorian challenges 57–60, 62,
    67–8, 123–4, 170, 411
  mid-Victorian scepticism 54–5, 59, 64–8,
    93–4, 99–101, 103–4, 119–21, 124–30,
    173–4, 178–9, 249, 417–18, 428–9
  high Victorians 85 n.210, 93–4, 133, 192–3,
    196, 200–3, 206, 214–15
  late-Victorian debates 267–72, 274–5,
    277–8, 284–8, 291–2, 344–7, 428
  and the Edwardians 296, 333–4, 337–9,
    350, 352–3, 361–2
  and humanity 97–8, 180–1, 365–6
  applications of 181–2, 202–3, 259–60,
    278–80
  See also age of the earth
natural theology 47–8, 53, 59, 71–2, 92, 101–2,
  106–7, 109–10, 112–14, 123–4, 130,
  230 n.236, 291–2, 409–10, 439 n.66
  See also Gifford Lectures
Nature 112–13, 186, 242–6, 258, 260–1, 268,
  270–1, 275–6, 327, 339, 406
Nevinson, C.R.W. 402
Nevinson, H.W. 402
New Age 350–1, 363, 398–9
New Century Magazine 242–3
New Liberalism 280–1, 285, 335–6
New Zealand 48–9, 69–70, 203–4, 260
Newcastle 68, 136–7, 162 n.356, 282–3, 314 n.293

Newman, Edward 109–10, 112–13
Newman, F.W. 96
Newman, John Henry 95–6, 359
newspaper correspondence 69, 187–8, 193–4,
  247, 327–8, 355–6
Newton, Alfred 49–50, 54–5, 89, 98–9, 120–1,
  180–1, 255–6, 340–1, 407
  and Cambridge 106–7, 255–6
  responses to the *Origin* 66–7
  review of *Descent* 246–7
Newton Hall 241–2
Nietzsche, Friedrich 333, 363, 396–7
Nightingale, Florence 96–7, 166–7
Nineteenth Century 186, 333
Nonconformity 132–3, 203–4, 331
Norfolk and Norwich Naturalists'
  Society 187–8
Normal School of Science 333
North British Physicists 107, 173
North British Review 93–4
North, Marianne 194
Northampton 256, 425–6
Norwich 54–5, 76 n.71, 163 n.371
  See also British Association, Norfolk and
    Norwich Naturalists' Society
Nottingham 40, 69, 187–8, 396–7
  See also British Association

Observer group 279–80
Olivier, Sydney 275, 281, 316 n.316, 378 n.85
On the Origin of Species
  historiography 25–9, 178–9, 244–6, 435
  contents 22–3, 40, 167
  initial responses 15–16, 41–6, 50–3, 55–73,
    97, 110, 117, 171–2
  circulation of 16, 40, 71, 92–3, 257–8,
    278–9, 326, 331–2
  influence in the 1860s 95–103, 124–5,
    130–1
  subsequent editions 125, 196, 269, 398
  later readings 191, 205–6, 246–7, 266, 331,
    336, 367–8
  later responses 215, 237, 260, 343, 363–4,
    369–71
  status 93–4, 192, 237, 262, 281, 325–6, 330,
    335–6, 372, 398–9, 402, 409–10
  emotional charge 131–2, 261–4, 333,
    418–19
ornithology 111, 126–7, 164–5
  British Ornithological Union
    111, 406

INDEX   489

Orr, James 200–1, 208
Ortega y Gasset, José 5–6
orthogenesis, *see* variation
Owen, Richard 16, 31–2, 104–5, 209–10,
   399–400
   evolutionary ideas 47–50, 62–3, 171
   opposition to Darwin 51–5, 61, 63–4, 117,
      133, 160 n.317, 439 n.60
   influence of 71, 99–100, 103–4, 162 n.361,
      219 n.66, 248–9, 251, 332
   See also Natural History Museum
Oxford 259–60, 268, 273–4, 357
   University of 43–4, 69–70, 88–9, 106–7,
      135–6, 185, 212–13, 249, 256, 309 n.197,
      329, 346–7, 360, 398, 405, 409–10
   Hope Museum 240–1
   Oxford Junior Scientific Club 255–6
   Ruskin Hall 332
   See also British Association

Page, David 53, 66–7, 153 n.222
Paget, Violet, *see* Vernon Lee
Paisley 52–3, 327–8
palaeobotany 14–15, 126, 220 n.80
palaeontology 48–9, 55, 125–6
   and 'deep time' 46–8, 68, 78 n.100, 173
   reconciliations with the Bible 23, 48–9,
      55–6, 62, 170–2, 188–9, 251, *See also*
      Samuel Kinns
   imperfections of the fossil record 41–2,
      46, 59, 61–2, 67–8, 99–102, 121–3, 125–6,
      135–6, 149–50 n.165, 176–7, 195–6, 198,
      214, 233 n.289, 320 n.384, 422
   contribution to evolutionary belief 64–7,
      71–2, 145 n.103, 165, 175–7, 190, 195–6,
      265, 411, 421–2
Palgrave, Francis Turner 77 n.94, 154 n.230
Pall Mall Gazette 133, 167–8, 186, 325–6
pangenesis 133, 145 n.99, 170, 199–200,
   236 n.338, 276
panmixia 269–70, 384 n.185
Parker, Thomas J. 145–6 n.111, 260
Parker, William K. 103, 409–10
Pater, Walter 69–70, 201–2
Patmore, Coventry 98
Patrick Geddes and Colleagues 257
Pattison, Mark 10, 97–8
Pearson, Karl 188, 256, 266, 270–1, 277–9,
   285, 327–8, 342–3, 391 n.298
   and Bateson 343–4, 400–1
   and Mendelism 346–9

'National Life from the Standpoint of
   Science' (1900) 279–80
   statistical approaches 277
   See also biometrics
Pease, Edward R. 244, 263, 410–11
Pengelly, William 101
periodisation 431, 434–5
Petrie, Flinders 363
Pfeiffer, Emily 177–8
Phillips, John 44, 123–4
philology 48–9, 121–2, 181–2, 198–9, 202
Philosophical Transactions. See Royal Society
philosophy 21, 48, 186, 191, 202, 212–13, 254,
   337–8, 364–5
   See also Herbert Spencer, Idealism,
      Pragmatism, Henri Bergson
phrenology 45–6, 62
phylogeny 157 n.280, 174, 197–8, 260, 265,
   338–9, 421–2
physiology 112–13, 180, 185–6, 202
Physiological Society 186
Picton, J.A. 41–2, 92–3
Pitt Rivers, Augustus Henry Lane-Fox 96,
   259–60
Plant, John 156 n.255, 258–9
Plymouth 340–1
poetry 46–7, 91–2, 98, 177–8, 211–13, 251–2,
   261–2, 360–1, 363–4, 368, 372
Popular Science Review 112
Porcupine expedition 133–4
Positivism 48, 97, 283–4
   See also Comtism
Poulton, Edward Bagnall 238–9, 263,
   315 n.304, 323 n.435, 400
   Essays on Evolution (1908) 346–7
   influence of 192, 331–2
   and Bateson 339, 341–3, 395 n.384
   and Darwin's reputation 240–1, 247, 354,
      368, 398
   and neo-Darwinism 267–8, 329, 422
   and entomology 252–3
   and eugenics 285
   and Huxley 324 n.443
   and Mendelism 330–1, 344–7, 382 n.149,
      399–400
   and G.J. Romanes 310 n.218
Powell, Baden 42–3, 48, 118–19
Powell, Frederick York 279, 307 n.166
Pragmatism 287–8, 364–5, 416–17
preaching 92–3, 159 n.311, 170–1, 205, 207,
   239–40, 289–90, 356–7

490 INDEX

Pre-Raphaelites 12, 179
Presbyterianism 93, 114, 181, 191–2, 288–9,
    327–8, 370, 426
Prestwich, Joseph 117–18, 250, 302 n.98
Pringle-Pattison, A.S. 255, 316 n.324,
    319 n.365, 320 n.380
Pritchard, Charles 116–17, 126–7, 140 n.20, 250
protective mimicry 109–10, 113, 259, 339
protoplasm 182–3, 209, 347–8
providence 99–100, 103–4, 120, 122–3,
    129–30, 156 n.265, 157–8 n.284, 165–6,
    208, 290, 360–1
psychology 186, 202–3, 272–4, 319 n.356, 361,
    364–7
public libraries, *see* libraries
publishing 112, 257
Punch 51, 53–4, 90, 180, 215, 242–3, 245, 397
Pusey, Edward B. 58, 106–7, 115, 168–71, 199,
    233 n.294, 238, 249, 408–9
Pye-Smith, P.H. 217 n.34, 235 n.318

Quarterly Journal of Science 112
Quarterly Review 44, 63–4, 244
Quiller-Couch, Arthur 331, 376–7 n.56,
    393–4 n.349

race 202–3, 221–2 n.105, 279–80, 284–7, 361–2
Radick, Greg 265–6
Rainbow Circle 280–1, 362–3, 414–15
Ramsay, A.C. 42–3, 45–6, 184–5, 422
Rashdall, Hastings 273–4, 319 n.367
Rationalist Press Association 326, 374 n.6
Reade, Winwood 196, 211–12, 318 n.348
Reader 112, 135–6
reading practices 16–17, 41–2, 64, 92–3, 113,
    191–2, 243–4, 262–3, 279, 298 n.24,
    331–2, 336, 362–3, 367–8, 422, 434
    See also marginalia
recapitulation 165, 275–6, 339, 360
Reddie, James 114–15, 423–4
Rede Lectures 44, 49–50, 118–19, 401–2
Reid, George Archdall 285, 345–6, 365–6
Research Defence Society 261–2
Rhodes, Cecil 279–80, 284
Richards, Grant 16, 36 n.42
Ridgeway, William 373, 390–1 n.294
Ritchie, D.G. 266, 280–1, 287–8
    and Spencer 274, 278–9, 286–7
    and vitalism 387 n.235
    Darwinism and Politics (1889) 286–7
    responses to 360, 364–5

Rivers, W.H.R. 350–1, 366
Robertson, J.M. 194–5, 274–5, 280–1
Robinson, Edward Kay 235 n.320, 266–7,
    308 n.179
Rolleston, George 31–2, 107–8, 249,
    304–5 n.129
    as Darwinian 41–2, 53–5, 66–7
    Darwinian doubts 107–8, 133–6,
    301 n.75, 412
Roman Catholicism, *see* Catholicism
Romanes Lectures 295–6, 345
Romanes, George J. 145 n.102, 190, 238–9,
    244–6
    An Examination of Weismannism
    (1893) 276
    and abiogenesis 211
    and Bateson 380 n.114
    and biometrics 345
    and the development of Darwinism
    265–6, 269–70, 420–1
    responses to 213–14, 270–1, 428
Row, C.A. 128–9, 151–2 n.196
Royal Academy 242–3
Royal Botanic Gardens, Kew 110–11
Royal College of Art 396–7
Royal Commission on Scientific Instruction
    and Advancement of Science 213–14
Royal Geographical Society 106, 111
Royal Horticultural Society 343
Royal Institution 43–4, 83 n.174, 100–1,
    106–8, 165, 175–6, 212–13, 234 n.314,
    269–70, 321 n.397
Royal School of Mines 53–4, 70–1, 106–7,
    185, 255
Royal Society 88–9, 107, 110, 184–5, 346,
    404–6, 410–11
    Darwin Fund 244–6
    Evolution Committee 341–3
    Philosophical Transactions 107
rudimentary organs 124–5, 166–7, 197–8
Ruskin, John 119, 179, 304 n.115, 409–10,
    418–19
    Unto This Last (1860) 165
    influence of 187–8, 242–3, 363, 372
Ruskin Hall 332
Russell, Bertrand 332–3, 350, 354–5, 362–5,
    375 n.38, 395 n.374
Russell, G.W. 367–8
Russell, Lord John 9
Ryle, R.J. 273–4, 277–8, 313 n.270,
    385 n.194

INDEX

Sabine, Edward 88–9, 107
Salford 258–9
Salisbury, Lord 13–14, 157–8 n.284, 327
saltation 23, 120–1, 125–6, 176, 287–8,
    349–51, 357
Saltcoats Literary Society 241
Salter, J.W. 71–2, 150 n.166
Samuelson, James 112
Sanderson, John Burdon 176, 249
Saturday Review 25, 167
Savile Club 346
Sayce, A.H. 188–9, 202
Schiller, F.C.S. 337–8, 349, 352–3, 357–8,
    360–1, 389 n.262, 391 n.305
    and Pragmatism 364–5
School of Mines, *see* Royal School of Mines
Schreiner, Olive 275
Science Gossip 187–8
Science Progress 373
scientific naturalism 19, 23–4, 47–8, 66–7,
    124–5, 155 n.252, 182–3, 190, 214–15,
    355–6, 403
Sclater, Philip Lutley 95, 112
Scotland 59, 249–50, 261–2
Scott, D.H. 263, 269–70
Scott, Robert Falcon 15–16, 331–2
secularism 69, 94–5, 105–6, 255
Sedgwick, Adam (1785–1873) 46, 51–2, 56–7,
    62, 80–1 n.136, 134–5, 408
Sedgwick, Adam (1854–1913) 277–8, 338–9,
    381 n.133, 400–1
Seeley, J.R. 93, 288
Seth, Andrew. *See* A.S. Pringle-Pattison
sexual selection 89, 164–5, 167–70, 211–12,
    267, 428
Sharp, David 265, 303 n.105
Shaw, George Bernard 1, 188, 262, 272, 275,
    280, 353–4, 361, 369–72
Sheffield 52–3, 238–9, 242–3
Shorter, Clement 137–9
Shrewsbury 240–1, 398
Sibree, John 236 n.342, 414
Sidgwick, Arthur 164, 252–3
Sidgwick, Eleanor 401–2
Sidgwick, Henry 10–11, 169, 188–9,
    201–3, 288
Smiles, Samuel 298–9 n.37
Smith, Grafton Elliot 336–7, 350–1, 360
Snaith, John 300 n.55, 388 n.253
social Darwinism 200–1, 279–80, 282, 284,
    286–7, 335–6, 360–1, 424–5

socialism 165, 262–3, 272, 274, 278–9, 282,
    287, 292–3, 336, 350–1, 428
    See also Fabian Society
Society for Psychical Research 204–5
Society of Amateur Botanists 126–7
Sociological Review 365–6
Sociological Society 280
sociology 365–6
Sollas, W.J. 275–6, 307–8 n.168
Somerville, Mary 77 n.94
Sorley, W.R. 273–4, 398–9
South America 99–100
South Place Chapel 188
Southsea 345–6
Speaker group 374 n.12
special creation 42–3, 46–7, 81 n.142, 95–6,
    104–5, 124–5, 127–8, 145–6 n.111, 170,
    182, 199–200, 226 n.173, 250, 263, 294,
    344–5, 423–4
species distribution, *see* geographical
    distribution
Spencer, Herbert 48, 106–7, 288–9, 327–8,
    340–1, 372, 409–10
    responses to 18–19, 96, 103–4, 106, 165,
        208, 405
    and Darwin 45–6, 133–4
    and Huxley 295–6
    and the mid Victorians 97, 126, 129–30
    and the high Victorians 195–6, 201,
        204–6, 268
    and the late Victorians 238–9, 241–4,
        255–6, 260–4, 266, 278–81, 285–7,
        290–3
    and the Edwardians 330–3, 338, 363, 366,
        370–1
    Autobiography (1904) 273–4
    First Principles (1860–2) 89, 279
    Man versus the State (1884) 280
    Principles of Biology (1863–4) 89, 338
    Principles of Psychology (1855) 273–4
    Study of Sociology (1873) 165
    Principles of Sociology (1898) 273–4
    See also X-Club
Spencerian evolution 21–3
Spencerians 271–5, 436 n.19, 441 n.93
    See also Lamarckism
Spencer–Weismann debate 239–40, 276
Spencer Lectures 360
spontaneous generation, *see* abiogenesis
Spottiswoode, William 65–6, 95,
    107, 184–5

492 INDEX

Spruce, Richard 110, 160 n.329
Spurgeon, Charles 207, 426
St Andrews 379–80 n.106
St Paul's Magazine 132–3
Stanley, Arthur P. 135
Stanley, Henry Morton 426
Stanley, Henrietta Maria 247–8
Stebbing, Thomas R.R. 191, 195, 210
Stephen, James Fitzjames 174–5
Stephen, Leslie 69–70, 98, 174–5, 194, 199, 201, 204
 Sunday tramps 186–7
 and Weismann 268–9
 and Henry Drummond 292–3
Stevenson, Robert Louis 186–7, 274–5, 413
Stewart, Balfour 51–2, 173
Stirling, James Hutchison 127–8, 167, 183, 210, 355–6
 citation of 153 n.221, 300 n.55, 388 n.253
Stokes, George G. 103, 107, 177–8, 184–5, 234 n.309, 409–10
Stopes, Marie 14–15
Strachey, Lytton 367–8
Stroud 215, 298 n.24, 413–14
Struthers, John 62–3, 101–2
Sturt, Henry 227 n.185, 233 n.289
Sturt, Henry Cecil 360–1
Sunday Evenings for the People 188
Sunday Lecture Societies 188
Sunday Magazine 112
Sunday Tramps 414–15
Sunderland 241–2
Swinburne, Algernon C. 89–92, 211–12
Swindon 372
Swiney Lectures 145 n.103
Synge, J.M. 331
Synthetic Society 254–5, 434

Tait, Lawson 137, 194
Tait, P.G. 173, 182–3
Talbot, Edward S. 192, 207, 254–5, 331–2
Talbot, Neville 330–1, 372–3
Tegetmeier, William B. 104–5, 242–3
telegony 276–7
teleology 127–9, 156 n.268, 160 n.315, 175–6, 200–1, 288, 294–5, 323 n.435, 353
Temple, Frederick 120–8, 189, 192, 250
 Bampton Lectures 129–30, 210–11, 289–90
Tennant, F.R. 324 n.445, 357–8

Tennyson, Alfred 25–6, 91–2, 261–2, 298–9 n.37, 363–4
 In Memoriam 146 n.120
 'Locksley Hall Sixty Years After' (1886) 91–2
 'Lucretius' (1868) 91–2
theology
 theological resistances to Darwin 58, 72–3, 82 n.159, 93, 97, 102, 114–17, 119–20, 193–4, 207, 239, 244, 368–9
 anti-theological animus 64–7, 105–6, 131, 288–9, 329–30, 372
 evolutionary accommodations with 128–9, 165
 impact of evolution on 200, 204–8, 254, 288–95, 327–8, 355–9
 theological training 106–7, 326–7
 See also natural theology
Times 25, 43–4, 46, 167, 224 n.145, 239, 246–7
Thiselton-Dyer, William T. 110–11, 192, 234 n.313, 296, 314 n.285, 327–8, 408–9
 and neo-Darwinism 268
 and the Evolution Committee 340–3
 and G.J. Romanes 270–1
 and Mendelism 344–7, 399–400
 and W. Bateson 339
 and the 1909 commemorations 400–2
Thomas, William Beach 335, 376 n.47
Thompson, D'Arcy W. 324 n.447, 334–5, 401–2, 424–5
Thomson, Allen 101, 184–5, 234 n.312
Thomson, Charles Wyville 124–5, 412
Thomson, J.A. 257, 302 n.97, 337–8, 340–1, 349–50, 352–3, 382 n.142
Thomson, William (1819–90) 154 n.236, 158 n.294
Thomson, William (1824–1907) 16, 120, 126–7, 137–9, 173, 195, 209–10, 327–8
 and abiogenesis 209–10
 and deep time 54–5, 68, 107–8, 129, 420–1
 and protoplasm 234 n.308
Thring, Edward 118–19, 423–4
Thring, Godfrey 49
Tierra del Fuego 247
Torres Straits expedition 366
transmutation of species 21–3, 40–1
 refusal to accept 49–50, 62, 67–8, 122–3, 134–5, 153 n.220, 171–2, 209–10
 doubts over 93–4, 99–100, 410–11

INDEX 493

limited acceptance of 48–9, 118–19, 123–5, 129, 134–5
acceptance of 178–9, 264–5, 333–4
Trevor, John 282, 288
Trinity College Dublin, *see* Dublin
Tristram, H.B. 72–3, 107, 135, 311 n.230
Trollope, Anthony 132–3, 410–11
Trotter, Wilfred 350–1, 366–7, 390–1 n.294
Turner, Frank M. 183–4, 425
Turner, William 112–13, 185, 198
Tylor, E.B. 164, 202, 333
Tyndall, John 5–6, 105–8, 133–4, 372
and Darwinism 64–5, 126, 144 n.93
Belfast Address 109–10, 164, 209, 376 n.51
on germs 210–11
influence of 123, 133–7, 186–8, 192, 241–3, 255, 261–2, 268–9, 370–1, 393–4 n.349
and Carlyle 408–9
See also X-Club
Tyneside Naturalists' Field Club 86 n.231

Unitarianism 394 n.366, 426
University College London 255
Uppingham School 118–19

variation 48–50, 276–7
doubts 62, 123–5, 269–70
limits of 59, 68, 71–2, 123, 176, 250, 303 n.106, 423–4
acceptance of 95–6, 277–8
variation as saltation 125–6
purposefulness of 126–7, 129–30, 196, 200–1, 352–3
experimental explorations of 336–9, 342–3
orthogenesis 352–3
See also rudimentary organs, Mendelism, William Bateson
Vestiges of the Natural History of Creation 27–8, 41–2, 46–8, 52–3, 63, 94–7, 188–9, 433
Victoria Institute 114–15, 119–20, 170, 177–8, 209–10, 255, 327–8
Vienna 92–3
Vines, Sydney H. 223 n.136, 255
vitalism 272, 336, 352–5
See also Henri Bergson
Vyvyan, Richard Rawlinson 71, 96–7

W. Swan Sonnenschein & Co 257
Waggett, P.N. 347–8, 378 n.89, 401–2

Wagner, Richard 327, 396–7
Wales 137–9, 257–8, 327–8
Walker, W.L. 208, 232 n.276
Wallace, A.R. 45–6, 48–9, 62, 89, 109–10, 131–2, 157 n.278, 197, 242–3, 284–5, 309 n.202, 333, 339–41, 400, 425, 440 n.80
Linnean Society presentation 41–2, 55–6
and natural selection 49, 126, 270–1
professional marginalisation 106
influence 93–4, 97–8, 103–4, 269–70, 290–1, 409–10
as a mid Victorian 426, 428–9
and the Edwardians 339–40, 346, 398–401
and the evolution of humanity 133–4, 180–1, 284–5
'On the Law which has regulated the Introduction of New Species' (1855) 49
Contributions to the Theory of Natural Selection (1870) 59
Darwinism (1889) 16
Malay Archipelago (1869) 89
Wallas, Graham 256, 280–1, 309 n.204, 313–14 n.282, 350
Walter Scott Company 257
Ward, James 199, 287, 355–6, 424–5
Ward, Lock, and Co. 257–8
Ward, Mrs Humphry 262–3, 278–9, 290
Ward, Wilfrid 254–5, 290–1, 313 n.271, 359, 426–7
Warrington 113, 192, 256–7
Watkinson, W.L. 321 n.393
Watson, Hewett C. 45–6, 409–10, 440 n.67
Webb, Sidney 272, 280–1, 285
Wedgwood, Hensleigh 59–60, 182
Wedgwood, Julia 16, 69, 201–2, 239
Weekly Review 191
Weismann, August 237, 269–70, 276–7, 286–7, 327, 329, 332, 337–8, 344–5, 422, 428
and neo-Darwinism 268–9, 373
and the Spencerians 275–7, 410
and eugenics 284–5
Weismannism 268–9, 360
Welby, Victoria, Lady 16, 192, 204–5, 365–6, 398
Weldon, Raphael 270, 277–8, 342–5, 348
Wells, H.G. 14–15, 295, 333, 336–7, 349, 354–5, 400–1
Welsh National Eisteddfod 327
Westminster Abbey 155 n.254, 239–40, 244–6, 273–4, 313 n.266

494 INDEX

Westminster Review 48, 64–5, 167–8, 374 n.14

Westwood, John O. 57–9, 106, 109–10, 267–8, 409–10, 418–19

Whewell, William 56–7, 71, 418–19
　On the Philosophy of Discovery (1860) 59–60
　The Philosophy of the Inductive Sciences (1840) 59–60

White, Edward 36 n.29, 127–8, 131, 173–4, 302 n.98

Wight, Robert 41–2, 71–2

Wilberforce, Samuel 44, 50–1, 61, 413
　See also Huxley-Wilberforce debate

Wilde, Oscar 256, 332–3, 433–4

William Scott Palmer, *see* Mary Emily Dowson

Williamson, W.C. 126, 176–7, 242–3, 408–9

Wilson, Andrew 256, 314 n.285

Wilson, James M. 205–7

Wilson, William 59

Winkworth, Catherine 16, 36 n.40

Wollaston, Thomas V. 49–50, 64, 68, 86 n.231, 123, 136–7, 176

Woodward, Henry 94, 111, 301 n.86

Woolf, Virginia 8

Woolhope Field Naturalists' Club 113–14

Wordsworth, Christopher 249–50, 406

World War 1 350–1, 366–7, 396–8

Wrench, Edward 40–2, 137, 244, 412

Wright, Lewis 193–4, 415–16

Wright, Thomas 150 n.172, 213–14

X-Club 64–7, 95, 103–8, 112, 132–4, 179, 184–6, 213–14, 239–40, 258, 405

Yeats, William Butler 8–9, 375 n.34

York 16, 52–3

Young, Frank 301 n.80, 305 n.136

Younghusband, Francis 333, 360

zoology 261–2, 310 n.218, 339, 341–4
　Zoologist 109–10, 406
　Zoological Gardens 166–7
　Zoological Society 51–2, 226 n.174